Advances in Aquatic Invertebrate Stem Cell Research

Advances in Aquatic Invertebrate Stem Cell Research

From Basic Research to Innovative Applications

Editors

Loriano Ballarin
Baruch Rinkevich
Bert Hobmayer

Basel • Beijing • Wuhan • Barcelona • Belgrade • Novi Sad • Cluj • Manchester

Editors

Loriano Ballarin
University of Padova
Italy

Baruch Rinkevich
Israel Oceanographic and
Limnological Research
Israel

Bert Hobmayer
University of Innsbruck
Austria

Editorial Office
MDPI
St. Alban-Anlage 66
4052 Basel, Switzerland

For citation purposes, cite each chapter independently as indicated below:

Author 1, and Author 2. 2022. Chapter Title. In *Advances in Aquatic Invertebrate Stem Cell Research: From Basic Research to Innovative Applications*. Edited by Loriano Ballarin, Baruch Rinkevich, and Bert Hobmayer. Basel: MDPI, Page Range.

ISBN 978-3-0365-1636-3 (Hbk)
ISBN 978-3-0365-1635-6 (PDF)
doi.org/10.3390/books978-3-0365-1635-6

Cover Design by Ms. Oshrat Ben-Hamo, 2021, used with permission.
This research and publication were funded by European Cooperation in Science & Technology (COST).

Contents

About the Editors

Loriano Ballarin (ORCID: 0000-0002-3287-8550) is a Professor of Zoology at the Department of Biology, University of Padova (Italy). He is a member of the editorial board of the *Invertebrate Survival Journal* and the *European Journal of Zoology*, and is a founding member of the Italian Association of Developmental and Comparative Immunobiology (IADCI). His main research interests are the evolution of innate immunity and the study of the cellular and molecular basis of immune responses in marine invertebrates, with particular reference to the role of hemocytes/coelomocytes in immune defense and, more generally, in the stress response. Most of his studies were carried out using the colonial ascidian *Botryllus schlosseri* as a model organism. His interest in stem cells is directly related to their role in hematopoiesis, as they assure the continuous renewal of the circulating immunocytes of marine invertebrates. He is the chair of the COST Action 16203 MARISTEM "Stem cells of marine/aquatic invertebrates: from basic research to innovative applications" that supported the publication of this book. He is the author or co-author of more than 130 peer-reviewed publications in scientific journals, co-editor of scientific books, including "Lessons in immunity: from single-cell organisms to mammals" (Elsevier, 2016), and guest editor of Special Issues, including: "Ancient immunity. phylogenetic emergence of recognition-defence mechanisms" (Biology (Basel), 2020–2021).

Baruch (Buki) Rinkevich has been a professor and senior scientist at the National Institute of Oceanography, Haifa, since 1988, the University of Haifa, Israel, since 2008, and an adjunct professor at the Biology Department, Ben-Gurion University in Beer-Sheva, Israel, since 2006. He received his PhD at Tel Aviv University, Israel, on coral physiology, and spent two postdoctoral periods in the United States: at Scripps Oceanographic Institute, La Jolla, California, studying silicification in marine invertebrates, and at Stanford University Medical School, California, studying ascidian immunity. A marine biologist, he works on the immunity, allorecognition, and developmental biology of marine invertebrates (primarily corals and colonial tunicates); on cell cultures/stem cell biology of marine invertebrates; on pattern formation of colonial organisms; on the unit of selection; and on coral reef restoration. Other aspects studied by Rinkevich in the laboratory include marine genotoxicity, DNA barcoding, the eDNA of marine organisms, and studies on thraustochytrids. He was the head of the MINERVA Center on "Invertebrate immunology and developmental biology" (1995–2002) and the head of the Department of Marine Biology and Biotechnology, National Institute of Oceanography (2002–2008). He has served on the editorial boards of 18 scientific, peer-reviewed journals, has been the mentor of > 60 graduate students (already graduated and in process), and is the author of > 360 publications and three edited books.

Bert Hobmayer is a Professor of Zoology at the University of Innsbruck, Austria (since 2009). He received his doctoral degree (Dr. rer. nat.) at the University of Munich, Germany, working on mechanisms of regeneration and neurogenesis in the freshwater polyp *Hydra*. After spending two post-doctorate years at the National Institute of Genetics in Mishima, Japan (characterization of cell-cell adhesion proteins in *Hydra*), he continued as a zoologist and developmental biologist to study cellular and molecular mechanisms underlying regeneration, axial patterning, and adult stem cell behavior in simple animal models at the Universities of Frankfurt, Darmstadt, and Innsbruck. Focus areas of his research include the roles of Wnt signaling pathways in providing positional information and in orchestrating tissue morphogenesis, and the impact of Myc transcription factors on adult stem cell decision making. At the University of Innsbruck, he served as head of the Department of Zoology (2006–2013) and head of the Special Research Cluster Center for Molecular Biosciences Innsbruck (CMBI) (2016–2021). He also served as a member of the Austrian Ministry of Science Panel "Future Life Sciences" in 2016 and 2017. He mentored more than 20 graduate students and is the author of more than 50 publications, many of which were published in high-impact journals.

Acknowledgements

This publication is based upon work from COST Action '16203 MARISTEM Stem cells of marine/aquatic invertebrates: from basic research to innovative applications', supported by COST (European Cooperation in Science and Technology).

COST (European Cooperation in Science and Technology) is a funding agency for research and innovation networks. Our Actions help connect research initiatives across Europe and enable scientists to grow their ideas by sharing them with their peers. This boosts their research, career and innovation.

www.cost.eu

Contributors

ALESSANDRA SALVETTI
Assoc. Prof., Department Clinical and
Experimental Medicine, University of Pisa,
Pisa, Italy.

ALEXANDER ERESKOVSKY
[1] Research Director, Institut Méditerranéen
de Biodiversité et d'Ecologie marine
et continentale (IMBE), Aix Marseille
University, CNRS, IRD, Avignon
University, Station Marine d'Endoume,
Marseille, France.
[2] Prof. Dr., Department of Embryology,
Faculty of Biology, Saint-Petersburg State
University, Saint-Petersburg, Russia.
[3] Koltzov Institute of Developmental
Biology of Russian Academy of Sciences,
Moscow, Russia.

AMALIA ROSNER
Dr., Israel Oceanographic & Limnological
Research, National Institute of
Oceanography, Haifa, Israel.

ANDY QARRI
PhD student, Helmholtz Zentrum
München, Regenerative Biology and
Medicine Institute, Munich, Germany.

ANNA PERONATO
PhD student, Department of Biology,
University of Padova, Padova, Italy.

ANTONIETTA SPAGNUOLO
Dr., Department of Biology and Evolution
of Marine Organisms, Stazione Zoologica
A. Dohrn, Naples, Italy.

AYELET VOSKOBOYNIK
Dr., Hopkins Marine Station, Stanford
University, Pacific Grove, CA, USA.

BENYAMIN ROSENTAL
Dr., The Shraga Segal Department of
Microbiology, Immunology, and Genetics,
Faculty of Health Sciences, Regenerative
Medicine and Stem Cell Research Center,
Ben-Gurion University of the Negev, Beer
Sheva, Israel.

CHIARA ANSELMI
Dr., Hopkins Marine Station, Stanford
University, Pacific Grove, CA, USA.

DAN BOBKOV
Dr., Collection of Vertebrate Cell Cultures,
Institute of Cytology, Russian Academy of
Sciences, St. Petersburg, Russia.

FABIO GASPARINI
Dr., Department of Biology, University of
Padova, Padova, Italy.

FAN ZENG
Dr., Institute of Zoology, University
Innsbruck, Innsbruck, Austria.

GAETANA GAMBINO
Dr., Department Clinical and Experimental
Medicine, University of Pisa, Pisa, Italy.

HELIKE LÕHELAID
[1] Researcher, Department of Chemistry
and Biotechnology, Tallinn University of
Technology, Tallinn, Estonia.
[2] Postdoc., Neuroscience Center/HiLIFE,
University of Helsinki, Helsinki, Finland.

ILDIKO M. L. SOMORJAI
Dr., School of Biology, University of St
Andrews, St Andrews, Fife, Scotland, UK.

ISABELLE DOMART-COULON
Assoc. Prof., Microorganism
Communication and Adaptation Molecules
(MCAM) Laboratory UMR 7245CNRS-
MNHN, Museum National d'Histoire
Naturelle, Paris, France.

KONSTANTIN V. KISELEV
Dr., Laboratory of Biotechnology, Federal
Scientific Center of the East Asia Terrestrial
Biodiversity, FEB RAS, Vladivostok, Russia.

LAURA DRAGO
PhD student, Department of Biology,
University of Padova, Padova, Italy.

LEONARDO ROSSI
Prof., Department Clinical and
Experimental Medicine, University of Pisa,
Pisa, Italy.

LUCIA MANNI
Prof. Department of Biology, University of
Padova, Padova, Italy.

MARION LECHABLE
PhD student, Department of Zoology,
University of Innsbruck, Innsbruck,
Austria.

MAREN KRUUS
M.Sc., Department of Zoology, University
of Innsbruck, Innsbruck, Austria.

MATTHIAS ACHRAINER
B.Sc., Department of Zoology, University of
Innsbruck, Innsbruck, Austria.

NATALIA SHARLAIMOVA
Dr., Collection of Vertebrate Cell Cultures,
Institute of Cytology, Russian Academy of
Sciences, St. Petersburg, Russia.

NATALYA V. AGEENKO
Dr., Laboratory of Cytotechnology,
National Scientific Center of Marine
Biology, The Far Eastern Branch of the
Russian Academy of Sciences, Vladivostok,
Russia.

NELLY A. ODINTSOVA
Prof., Laboratory of Cytotechnology,
National Scientific Center of Marine
Biology, The Far Eastern Branch of the
Russian Academy of Sciences, Vladivostok,
Russia.

OLGA PETUKHOVA
Dr., Collection of Vertebrate Cell Cultures,
Institute of Cytology, Russian Academy of
Sciences, St. Petersburg, Russia.

SERGEY SHABELNIKOV
Dr., Cell Technologies Centre, Institute of
Cytology, Russian Academy of Sciences, St.
Petersburg, Russia.

SIMON BLANCHOUD
Dr., Department of Biology, University of
Fribourg, Fribourg, Switzerland.

TAL GORDON
Dr., Department of Biochemistry and
Molecular Biology, George S. Wise Faculty
of Life Sciences, Tel-Aviv University, Tel-
Aviv, Israel.

TARVI TEDER
Postdoc., Department of Medical
Biochemistry and Biophysics, Karolinska
Institute, Stockholm, Sweden.

UTE ROTHBÄCHER
Dr., Institute of Zoology, University
Innsbruck, Innsbruck, Austria.

VIRGINIA VANNI
PhD Student, Department of Biology,
University of Padova, Padova, Italy.

WILLI SALVENMOSER
BMA, Department of Zoology, University
of Innsbruck, Innsbruck, Austria.

YUVAL RINKEVICH
Dr., Helmholtz Zentrum München,
Regenerative Biology and Medicine
Institute, Munich, Germany.

Preface

Why publish a new book on invertebrate stem cells—and particularly one with a focus on aquatic invertebrates? The answer lies in a rapidly evolving stem cell discipline, driven by ever-advancing molecular tools and imaging techniques, today being one of the most dynamic areas in biology and biomedicine. This inevitably influences the research on invertebrates, with a noteworthy reference to aquatic organisms. Indeed, aquatic invertebrates represent the greatest majority of animal biodiversity. They exhibit various biological features that are of vast interest to stem cell researchers and biologists in general. These include the high regenerative power displayed by a broad range of taxa, the lack of early germ cell sequestering, and a widespread presence of asexual reproduction, dormancy, postponed aging and rejuvenation. All of these phenomena are associated with the action of pools of adult stem cells throughout the animals' life cycles (Ballarin et al. 2018).

Current research on aquatic invertebrate stem cells takes advantage from new experimental approaches and an increasing number of sequenced animal genomes and transcriptomes available in databases. The accumulated results reveal that aquatic invertebrate stem cells have unique features not recorded in the vertebrates and model terrestrial invertebrates, such as their high abundance (up to 40% of the entire body cells in some taxa) and their often indeterminate capacity for growth. They further express typical stemness genes, previously considered an attribute of germ cells, including piwi, vasa, nanos, and more (Rinkevich et al. 2022), and they are directly involved in the control and modulation of innate immune responses (Ballarin et al. 2021). Finally, aquatic invertebrate stem cells challenge the concept of the stem cell niche as defined in vertebrates and ecdysozoans (Martinez et al. 2022). The following chapters elaborate on several of these aspects.

This book stems from the activities within the COST Action 16203 MARISTEM Stem cells of marine/aquatic invertebrates: from basic research to innovative applications, which will end on 1 April 2022. It represents one of the final deliverables of four years of interaction and collects the contributions of a relevant part of its members. It holds 11 chapters dealing with sponges, cnidarians, flatworms, echinoderms and tunicates, highlighting the best-studied adult stem cell lineages among aquatic invertebrates. Four chapters review stem cell dynamics in regeneration, development, tissue homeostasis, and symbiosis. Another four chapters discuss profiles of stem cell-specific gene expression and the action of glycoproteins and fatty acids. Three chapters describe efforts to approach the long-term goal of establishing invertebrate stem cell cultures.

We thank all the contributors to this volume and Oliva Andereggen and Jelena Milojevic for their friendly support and for cautiously handling all editorial issues. Our hope is that this book can stimulate researchers to pay closer attention to organisms from aquatic environments, as those—due to their simple Bauplan and to the high potentialities of their stem cells—will advance our knowledge in basic biological processes.

References

Ballarin, Loriano, Baruch Rinkevich, Kerstin Bartscherer, Artur Burzynski, Sebastien Cambier, Matteo Cammarata, Isabelle Domart-Coulon, Damjana Drobne, Juanma Encinas, Uri Frank, and et al. 2018. Maristem—Stem cells of marine/aquatic invertebrates: from basic research to innovative applications. *Sustainability* 10: 526. https://doi.org/10.3390/su10020526.
Ballarin, Loriano, Arzu Karahan, Alessandra Salvetti, Leonardo Rossi, Lucia Manni, Baruch Rinkevich, Amalia Rosner, Ayelet Voskoboynik, Benyamin Rosental, Laura Canesi, and et al. 2021. Stem cells and innate immunity in aquatic invertebrates: Bridging these seemingly disparate disciplines for new discoveries in biology. *Frontiers in Immunology* 12: 688106. https://doi.org/10.3389/fimmu.2021.688106.
Martinez, Pedro, Loriano Ballarin, Alexander Ereskovsky, Eve Gazave, Bert Hobmayer, Lucia Manni, Eric Rottinger, Simon G. Sprecher, Stefano Tiozzo, Ana Varela-Coelho, et al. 2022. Articulating the "stem cell niche" paradigm through the lens of non-model aquatic invertebrates. *BMC Biology* 20: 23 https://doi.org/10.1186/s12915-022-01230-5.
Rinkevich, Baruch, Loriano Ballarin, Pedro Martinez, Ildiko Somorjai, Oshrat Ben-Hamo, Ilya Borisenko, Eugene Berezikov, Alexander Ereskovsky, Eve Gazave, Denis Khnykin, and et al. 2022. A pan-metazoan concept for adult stem cells: The wobbling Penrose landscape. *Biological Reviews* 97: 299–325. https://doi.org/10.1111/BRV.12801.

From Primary Cell and Tissue Cultures to Aquatic Invertebrate Cell Lines: An Updated Overview

Isabelle Domart-Coulon and Simon Blanchoud

Abstract: The stem cells discipline represents one of the most dynamic areas in biology and biomedicine. The vast majority of research on stem cells is being conducted in vertebrate models. Currently, over 98% of all cell lines are of mammalian origin, which represent only 0.4% of the extant identified metazoan evolution. In particular, aquatic invertebrates as a whole show the largest biodiversity and the widest phylogenetic radiation on Earth but have not yet significantly contributed to cell lines. Yet, with over 500 publications since the 1960s, the current lack of cell lines does not result from a lack of attempts at cultivating these cells but rather from fragmented research efforts in highly taxonomically diverse model species, a paucity in reports of negative results and persistent knowledge gaps in their in vitro metabolic requirements. To promote the establishment of aquatic invertebrate cell lines, there is thus a need for comprehensive knowledge mapping across taxa to identify adequate, possibly cell type-specific, protocols. Here, we review strategies for preparing an optimal inoculum, for optimizing culture conditions and for cell lineage authentication to monitor the quality of cell cultures. Finally, we conclude with our view on promising research perspectives towards establishing aquatic invertebrate cell lines.

1. Introduction

Currently, the origins of in vitro cell lines are highly biased towards humans. Around 75% of the total number of established cell lines are from Hominidae origin (96,862/128,799) and over 97% are of mammalian origin (126,033/128,799) (Bairoch 2018) (Figure 1). However, mammals represent only 0.4% (1.3% when excluding the Insecta taxon) of the extant identified metazoan evolution (Zhang 2013; Wilson and Reeder 2011; Chapman 2009) (Figure 1). In addition to the scientific interest relative to their sheer diversity, non-mammalian cells have multiple potential applications, including as a source for bio-active molecules or as assays for eco-toxicological tests (e.g., Ribeiro et al. 2018; Rosner et al. 2021). Yet, with over 500 publications on aquatic invertebrate cell culture alone (Figure 1), the current limited number of invertebrate cell lines does not result from a lack of attempts at cultivating these cells but most likely from inappropriate techniques to cultivate these cells (reviewed in

Rinkevich 2005; Yoshino et al. 2013; Cai and Zhang 2014). As exemplified in insects, a breakthrough in culturing conditions (Grace 1962) initiated the emergence of a huge variety of cell lines (Bairoch 2018) (895 cell lines from 104 genera in around 50 years). There is thus a need for a sustained research effort in non-insect invertebrate cell culture to identify adequate culturing conditions and promote the establishment of cell lines. In particular, aquatic invertebrates as a whole show the largest biodiversity and the widest phylogenetic radiation on Earth but have currently contributed to only six cell lines (Figure 1).

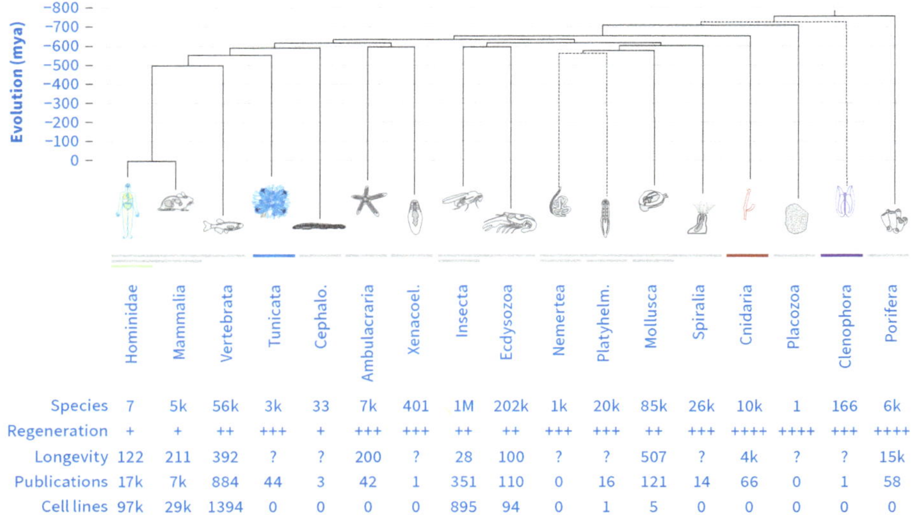

	Hominidae	Mammalia	Vertebrata	Tunicata	Cephalo.	Ambulacraria	Xenacoel.	Insecta	Ecdysozoa	Nemertea	Platyhelm.	Mollusca	Spiralia	Cnidaria	Placozoa	Clenophora	Porifera
Species	7	5k	56k	3k	33	7k	401	1M	202k	1k	20k	85k	26k	10k	1	166	6k
Regeneration	+	+	++	+++	+	+++	+++	++	++	+++	+++	++	+++	++++	++++	+++	++++
Longevity	122	211	392	?	?	200	?	28	100	?	?	507	?	4k	?	?	15k
Publications	17k	7k	884	44	3	42	1	351	110	0	16	121	14	66	0	1	58
Cell lines	97k	29k	1394	0	0	0	0	895	94	0	1	5	0	0	0	0	0

Figure 1. Comparison of the diversity between metazoan taxa. Depicted is the phylogenetic relation between metazoan taxa and their characteristics with respect to evolutionary radiation, regenerative capacity, lifespan and in vitro cell culture (see Appendix A Table A1 for exact values). Dashed lines represent branchings for which speciation timings have not yet been determined. Colored taxa highlight those whose publication timelines are detailed in Figure 2. Regeneration capacity of the taxa is depicted as follows: + tissue regeneration, ++ appendage regeneration, +++ whole-body regeneration, ++++ from cell aggregates. Longevity is given in years from the maximum reported characteristic of the taxon in the AnAge database (Magalhães et al. 2007). Phylogenetic tree based on Halanych (2004), species numbers on Zhang (2013), regeneration potential on Bely and Nyberg (2010), publications on manually curated online searches (Appendix B Table A2) and cell line numbers on Bairoch (2018). Source: Graphic by authors.

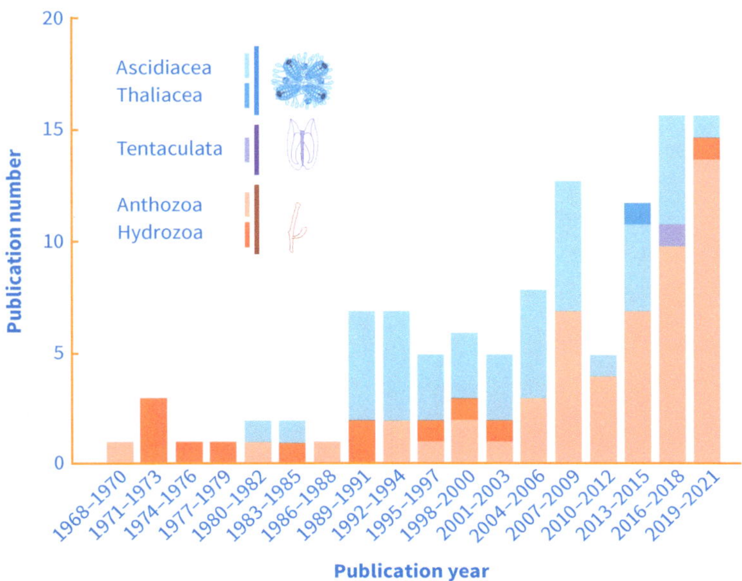

Figure 2. Five decades of research on isolation and primary culture of cells from aquatic invertebrates. The figure shows the number of publications for the phyla cnidaria (66 in total), ctenophora (1 in total), and tunicata (44 in total). Publications are grouped by classes of the species used for cell isolation, color-coded as indicated. Publications were manually curated from online searches, and detailed references are available in the Appendix B Table A2. Source: Graphic by authors.

Cell lines have been established through two main strategies (Cai and Zhang 2014; Rinkevich 2011): either by the isolation of proliferating and self-renewing cells, typically from an embryonic (Hansen 1979) or cancerous origin (Scherer 1953), or by immortalizing proliferating cells, typically through mutagenesis (Earle et al. 1943) or transfection (Russell et al. 1977). Both strategies thus require, at least transiently, a proliferating primary cell culture. The long-term culture (up to 22 months) of cells from various aquatic invertebrate phyla has been achieved by using a variety of culturing environments (Rinkevich and Rabinowitz 1993; Daugavet and Blinova 2015; Chen and Wang 1999; Kingsley et al. 1987). However, most of these in vitro primary cultures show an apparently ubiquitous cellular quiescence within three days that leads to an absence of proliferation within 1–4 weeks of primary culture (Rinkevich 2011; Cai and Zhang 2014). Yet, transient proliferation events, limited to a subset of acclimated cells, are persistently recorded across most marine invertebrate taxa ~2–4 weeks after the establishment of primary cultures at high seeding density from larval or regenerating adult tissue. For instance, DNA synthesis and mitosis have been observed both in primary cultures of explanted ectodermal tissue monolayers of regenerating *Nematostella vectensis* (Rabinowitz et al. 2016), as well as in dissociated cell culture from regenerating tentacles of *Anemonia viridis* (Ventura et al. 2018), and dividing cells have been reported in primary culture of regenerating tissues of *Apostichopus japonicus* (Odintsova et al. 2005). The only established mollusc cell line, *Bge*, was initiated from the long-term culture of embryonic tissue of the freshwater snail *Biomphalaria glabrata* (Hansen 1979). Taken together, these results suggest that a key to setting efficient primary cultures are to use tissue with high proliferation capacity, potentially due to the presence of stem-like cells. Conveniently, aquatic invertebrates display a variety of asexual reproduction, aging and regeneration phenomena (Figure 1) that indicate high cellular plasticity, cellular proliferation and a likely involvement of stem-like cells (Bely and Nyberg 2010; Slack 2017; Bodnar 2009; Tomczyk et al. 2015; Rinkevich et al. 2022). However, established guidelines for the isolation and identification of stem-like cells are currently only available for very few species (Hayashi et al. 2006; Sun et al. 2007; Hemmrich et al. 2012; Kassmer et al. 2020). The recent improvements in next-generation sequencing techniques, and in single cell transcriptomics in particular, are enabling researchers to characterize stem-like cells in an increasing number of taxa (Hayashi et al. 2010; Siebert et al. 2019; Rinkevich et al. 2022), a first important step for their isolation and in vitro culture.

There is an ample body of work that provides numerous quantitative assessments of culturing conditions (e.g., Toullec 1999; Khalesi 2008; Dessai 2012; Maselli et al. 2018), without highlighting one ideal consensus. Given that aquatic invertebrates are phylogenetically very distant, the development of a ubiquitous culturing environment appears rather unlikely. Nevertheless, each phylum could benefit from the advances in primary cell culture made in other phyla. However, a significant fraction of the relevant research data remains unpublished in conventional peer-reviewed journals, being only accessible as chapters in master's or doctoral dissertations, conference proceedings and specialized books. Consequently, in the last five decades, the publication of research efforts has been uneven across phyla, and temporally fragmented, as illustrated for the cnidaria and tunicata phyla (Figure 2).

Here, we review three major drawbacks and limitations of this field of research and their most promising work-around (Rinkevich 2005; Cai and Zhang 2014; Rinkevich 2011; Yoshino et al. 2013): (1) seeding the cell culture with a population enriched in proliferating and potentially stem-like cells; (2) devising marine invertebrate specific in vitro culturing environment, including management of oxidative stress and cell adhesion requirements; (3) preventing culture contamination with other cell types and microbes. This review is intended to be accessible both to the non-experts and newcomers to the field of primary cell culture, while providing an updated and curated list of references on the primary cell culture of aquatic invertebrates compiled for the experienced reader.

Given the huge scope of this review (>360,000 species, >60 years of research, >510 publications), we set out to illustrate previous work on aquatic invertebrate cell culture with three summarizing tables (Tables 1–3), filled with a selection of representative publications in each taxon and focusing on stem cell cultures whenever these have been described. This review is, by nature, not exhaustive and omits, by necessity, many publications, which thus limits generalizations. We conclude this review by providing perspectives on how to solve this limitation, mainly through dramatically extending the present effort in the data mining and metacoding of published work to build an exhaustive knowledge database on aquatic invertebrate cell culture. We also highlight abiotic factors that should be further investigated. We hope that the provided perspectives will help researchers to develop robust and reproducible approaches for culturing dividing aquatic invertebrate cells, a first step towards the possible establishment of cell lines.

Table 1. Established cell isolation techniques across taxa. Indicative examples were selected among the references listed in Appendix B Table A2.

Phylum	Class	Species	Inoculum Type	Target Cells	Isolation Technique	Isolation Medium	Enzyme Digestion	Cell-Type Enrichment Strategy	Reference
Chordata (Tunicata)	Ascidiacea	*Botryllus schlosseri*	adult	circulating blood cells	mechanical, teasing apart colonial zooids	buffered washing solution with HEPES	none	none	Rinkevich and Rabinowitz (1993)
	Ascidiacea	*Botryllus schlosseri*	embryo	all cell types	mechanical, chemical	buffered washing solution with HEPES	none	none	Rinkevich and Rabinowitz (1994)
	Ascidiacea	*Botryllus schlosseri*	zooids and buds	epithelial	mechanical, enzymatic dissociations	incubation medium	collagenase	none	Rinkevich and Rabinowitz (1997)
	Ascidiacea	*Botryllus schlosseri*	zooids and buds	cup cell disease cells	mechanical, cell strainer	Fisher's medium	none	none	Moiseeva et al. (2004)
	Ascidiacea	*Botryllus schlosseri*	zooids and buds	epithelial	mechanical	filter sea water and antibiotics	none	none	Rabinowitz and Rinkevich (2004)
	Ascidiacea	*Botryllus schlosseri*	zooids and buds	epithelial	mechanical, cell strainer	buffered washing solution with HEPES	none	none	Rabinowitz et al. (2009)
	Ascidiacea	*Botryllus schlosseri*	paleal buds	epithelial	mechanical, cell strainer	buffered washing solution with HEPES	none	none	Rabinowitz and Rinkevich (2011)
	Ascidiacea	*Botrylloides leachii*	adult, hibernating colonies blood cells		mechanical	buffered washing solution with HEPES	none	none	Hyams et al. (2017)

Table 1. *Cont.*

Phylum	Class	Species	Inoculum Type	Target Cells	Isolation Technique	Isolation Medium	Enzyme Digestion	Cell-Type Enrichment Strategy	Reference
Chordata (Cephalochordata)	Leptocardii	*Branchiostoma belcheri Japanese*	Adult epidermis, gill, gut, spermary, ovary	all cell types	mechanical and enzymatic for ovary and spermary	FSW	100 U/mL type I collagenase	none	Cai et al. (2013)
	Leptocardii	*Branchiostoma belcheri tsingtauense*	adult, buccal cirri, tail, gill, gut and metapleural fold	all cell types	mechanical dissociation	none	none	none	Wang et al. (2009)
Echinodermata (Ambulacraria)	Echinoidea	*Anthocidaris crassispina, Pseudocentrotus depressus, Hemicentrotus pulcherrimus*	eggs	egg cortex cells	mechanical dissociation	hypotonic $MaCl_2$ (0.1 M) solution	none	sedimentation by centrifuging at 1000 RPM	Sakai (1960)
	Echinoidea	*Anthocidaris crassispina*	embryo (blastula stage)	all cell types	spontaneous dissociation	Ca^{2+} and Mg^{2+} free SW (CMF-SW) (1 h)	none	removal of suspended cells	Fujita et al. (1972)
	Echinoidea	*Strongylocentrotus purpuratus*	embryo (blastula stage)	all cell types	mechanical dissociation	Ca^{2+} and Mg^{2+} free SW (CMF-SW)	trypsin 0.1 mg/mL	two successive centrifugation	Oppenheimer and Meyer (1982)
	Asteroidea	*Asterias amurensis*	embryo (mesenchyme migration stage)	all cell types	mechanical dissociation	1.2 M glycine supplemented with 1% (*v/v*) nystatin-filtered sea water, 6% newborn bovine serum	none	centrifugations	Kaneko et al. (1995)

Table 1. *Cont.*

Phylum	Class	Species	Inoculum Type	Target Cells	Isolation Technique	Isolation Medium	Enzyme Digestion	Cell-Type Enrichment Strategy	Reference
	Echinoidea	*Strongylocentrotus intermedius*	embryo	all cell types	spontaneous dissociation	Ca^{2+} and Mg^{2+}-free SW + antibiotics (100 IU/mL penicillin and 100 mg/mL streptomycin)	0.25% collagenase	none	Odintsova et al. (2015)
	Holothuroidea	*Apostichopus japonicus*	adult	guts	Mechanical/ enzymatic	sea water	collagenase	none	Odintsova et al. (2005)
Xenacoelomorpha	Acoela	*Isodiametra pulchra*	adults	all cell types	enzymatic dissociation	nutrient-enriched f/2 ASW	CMF/1% trypsin	none	De Mulder et al. (2009)
Arthropoda (Ecdysozoa)	Malacostraca	*Pacifasticus leniusculus, Homarus americanus*	adults	hematopoietic and testicular tissue	mechanical dissociation	Medium 19	type 2 collagenase	none	Brody and Chang (1989)
	Malacostraca	*M. ensis*	adults	lymphoid organ	mechanical	sea water	none	sedimentation	Han et al. (2013)
	Malacostraca	*Penaeus vannamei*	juvenile	all cell types	Mechanical/ enzymatic	ASW	collagenase	none	Toullec et al. (1996)
Nematoda (Ecdysozoa)	Chromadorea	*C. elegans*	eggs	larval cells	Mechanical/ enzymatic	L-15 medium	pronase	none	Zhang et al. (2011)
Platyhelminthes	Turbellaria	*Girardia ystati, Schmidtea mediterranea*	adult regenerating prepharyngeal zone	neoblasts	chemical dissociation	citric acid 0.1 M with 0.5% Tween 20, 10 min RT	none	FACS-sorting (DNA-stained)	Ermakov et al. (2012)

Table 1. *Cont.*

Phylum	Class	Species	Inoculum Type	Target Cells	Isolation Technique	Isolation Medium	Enzyme Digestion	Cell-Type Enrichment Strategy	Reference
	Turbellaria	*Dugesia japonica*	adult (sexual and asexual strains)	all cell nuclei	mechanical dissociation	1/3 PBS hypotonic solution with 1% Triton X100	none	FACS-sorting (DNA content)	Hoshino et al. (1991)
	Turbellaria	*Dugesia japonica*	adult head	neuronal cells	chemical dissociation	5/8 Holtfreter's solution containing 30 µg/mL trypsin inhibitor	trypsin 0.25%	FACS-sorting (DNA-stained)	Asami et al. (2002)
	Turbellaria	*Dugesia japonica*	adult	neoblasts	chemical dissociation	5/8 Holtfreter's solution containing 30 µg/mL trypsin inhibitor	trypsin 0.25%	FACS-sorting (X-ray sensitive cells)	Hayashi et al. (2006)
	Turbellaria	*Schmidtea polychroa*	adult	neoblasts	mechanical dissociation	1% Digest-Eur in hypotonic solution	none	centrifugation (Percoll-density gradient)	Schürmann and Roland (2001)
	Turbellaria	*Schmidtea mediterranea*	adult tail	neoblasts and other cell types	mechanical dissociation	Ca-Mg-free buffer, with 1% BSA	none	FACS sorting (vital dye SiR-DNA; size and complexity gating)	Lei et al. (2019)

9

Table 1. *Cont.*

Phylum	Class	Species	Inoculum Type	Target Cells	Isolation Technique	Isolation Medium	Enzyme Digestion	Cell-Type Enrichment Strategy	Reference
Mollusca	Gastropoda	*A. californica*	adult pleural ganglia	neurons	mechanical	artificial seawater	none	none	Bedi et al. (1998)
	Gastropoda	*A. californica*	adult and juvenile	ganglia	enzymatic dissociation	L-15 medium	pronase	none	Schacher and Proshansky (1983)
	Gastropoda	*A. californica*	adult	ganglia	enzymatic dissociation	L-15 medium	collagenase	none	Montgomery et al. (2002)
	Gastropoda	*Helix aspersa aspersa, Helix aspersa maxima*	adult	gonadal cells	enzymatic dissociation	PBS	collagenase-dispase	none	Monnier and Bride (1995)
Annelida (Spiralia)	Clitellata	*Hirudo medicinalis*	adult, Matrigel implants	myoendothelial cells	mechanical	DMEM	none	none	Grimaldi et al. (2009)
	Clitellata	*Hirudo medicinalis*	adult, ganglia	retzius cells	enzymatic dissociation	L-15 medium	collagenase-dispase	none	Bookman and Liu (1990)
	Clitellata	*Hirudo medicinalis*	adult	microglial cells	enzymatic dissociation	L-15 medium	collagenase-dispase	none	Masuda-Nakagawa et al. (1994)
Cnidaria	Hydrozoa	*Polyorchis penicillatus*	adult nerve rings	neurons	enzymatic dissociation	Ca^{2+} and Mg^{2+} free ASW	collagenase 100 U/mL (3–5 h)	removal of unattached cells	Przysiezniak and Spencer (1989)
	Hydrozoa	*Podocoryne carnea*	adult umbrella	smooth muscle cells	mechanical dissociation	Ca^{2+} and Mg^{2+} free ASW (15 min)	collagenase 150 U/mL (6–8 h)	microdissection of striated muscle tissue	Schmid (1992)
	Anthozoa	*Leptogorgia virgulata*	adult	scleroblasts	mechanical dissociation	FSW with antibiotics	none	removal of suspended cells	Kingsley et al. (1987)

Table 1. *Cont.*

Phylum	Class	Species	Inoculum Type	Target Cells	Isolation Technique	Isolation Medium	Enzyme Digestion	Cell-Type Enrichment Strategy	Reference
	Anthozoa	*Nematostella vectensis*	adult oral fragment	ectodermis	mucolytic agent (2% NAC for 10 s)	phosphate-buffer-saline	none	separation of epithelial layers by a reducing agent	Rabinowitz et al. (2016)
	Anthozoa	*Anemonia sulcata*	adult tentacle	epithelial and interstitial cells		Ca^{2+} and Mg^{2+} free ASW	papain 10 U/mL (1.5 h)	Percoll continuous gradient centrifugation	Apte et al. (1996)
	Anthozoa	*Anemonia viridis*	adult regenerating tentacle	adherent small round cell types	enzymatic dissociation	Ca^{2+} and Mg^{2+} free ASW then ASW	0.05% collagenase I (1 h)	fragments from 3d post-cutting tentacle	Barnay-Verdier et al. (2013)
	Anthozoa	*Anemonia viridis*	adult regenerating tentacle	gastrodermal differentiated cells	enzymatic dissociation	Ca^{2+} and Mg^{2+} free ASW with 1 mM glycine, then ASW	0.15% collagenase type I (30 min)	microdissected epithelia from 3d post-cutting tentacle	Ventura et al. (2018)
	Anthozoa	*Pocillopora damicornis*	adult	gastrodermal cell containing zooxanthellae	mechanical disruption	Ca^{2+} free ASW (3–4 h)	none	none	Gates and Muscatine (1992)

Table 1. *Cont.*

Phylum	Class	Species	Inoculum Type	Target Cells	Isolation Technique	Isolation Medium	Enzyme Digestion	Cell-Type Enrichment Strategy	Reference
	Anthozoa	*Stylophora pistillata, Porites lutea, Dendronephthya hemprichi, Paraerythropodium fulvum fulvum, Heteroxenia fuscescence, Clathraria rubrinoides, Plexaura* sp., *Millepora dichotoma*	adult and larva	all cell types	spontaneous or mechanical fragmentation	Ca^{2+} and Mg^{2+} free ASW	collagenase 0.05% (4 h)/pronase 0.1% (1 h)/EDTA 0.02% (2 h)	Percoll continuous gradient centrifugation	Frank et al. (1994)
	Anthozoa	*Acropora microphtalma, Pocillopora damicornis, Montipora digitata*	adult	all cell types	spontaneous dissociation	Ca^{2+} free ASW (2–4 h)	none	none	Kopecky and Ostrander (1999)
	Anthozoa	*Pocillopora damicornis*	adult	all cell types	spontaneous dissociation	Ca^{2+} free (and Mg^{2+} free) ASW (2–3 h)	none	Percoll step gradient centrifugation	Domart-Coulon et al. (2001)
	Anthozoa	*Pocillopora damicornis*	adult	all cell types	spontaneous dissociation	Ca^{2+} free ASW (3 h)	none	none	Domart-Coulon et al. (2004)
	Anthozoa	*Montipora digitata*	adult	all cell types	enzymatic dissociation	Ca^{2+} free ASW (2.5 h)	collagenase 0.15%	none	Helman et al. (2008)
	Anthozoa	*Xenia elongata*	adult	all cell types	enzymatic dissociation	Ca^{2+} free ASW (2.5 h)	collagenase 0.15%	none	Helman et al. (2008)

Table 1. *Cont.*

Phylum	Class	Species	Inoculum Type	Target Cells	Isolation Technique	Isolation Medium	Enzyme Digestion	Cell-Type Enrichment Strategy	Reference
	Anthozoa	*Sinularia flexibilis*	adult	all cell types	mechanical/ enzymatic dissociation	ASW with 1% gentamycin– streptomycin	trypsine-EDTA 0.05%	none	Khalesi (2008)
	Anthozoa	*ystatin, Pavona divaricata*	adult	all cell types	mechanical dissociation	FSW	none	spontaneous cell aggregation into spheroids	Nesa and Hidaka (2009)
	Anthozoa	*Pocillopora damicornis*	adult	coral cell types without zooxanthellae	enzymatic dissociation	ASW	dispase 2 U/mL, lysozyme 3 U/mL, a amylase 2 U/mL, a glucosidase 0.5 U/mL, b galactosidase 0.5 U/mL, endoglycosidase H 0.25 U/mL	two successive Percoll step gradient centrifugations	Downs et al. (2010)
	Anthozoa	*Acropora millepora*	larva (planula)	"interstitial" cells	spontaneous dissociation	spontaneous dissociation	none	none	Reyes-Bermudez and Miller (2009)
	Anthozoa	*Pocillopora damicornis*	adult	all cell types	mechanical dissociation	3.3× PBS without Ca and Mg, 2% FCS, 20 mM Hepes, pH 7.4	none	FACS-sorting (size, autofluorescence, enzyme-activated fluorescence, organelle-specific fluo markers)	Rosental et al. (2017)

Table 1. *Cont.*

Phylum	Class	Species	Inoculum Type	Target Cells	Isolation Technique	Isolation Medium	Enzyme Digestion	Cell-Type Enrichment Strategy	Reference
	Anthozoa	*Fungia granulosa*	adult regenerating tissue (peeled off skeleton)	all cell types (coral and zooxanthellae)	mechanical dissociation	FSW with gentamicin and kanamycin (both 50 µg/mL) for 2 days only	none	spontaneous spheroid formation	Gardner et al. (2015)
	Anthozoa	*Pocillopora damicornis*	adult	all cell types	mechanical dissociation	$3.3\times$ PBS without Ca and Mg, 2% FCS, 20 mM Hepes, pH 7.4	none	FACS-sorting (size, autofluorescence, enzyme-activated fluorescence, organelle-specific fluo markers)	Rosental et al. (2017)
	Anthozoa	*Acropora tenuis*	planula	gastrodermal secretory cells, undifferentiated cells, neuronal cells, and epidermal cells	enzymatic dissociation	FSW	trypsin-EDTA + collagenase I (1–4 h)	none	Kawamura et al. (2021)
	Anthozoa	*Acropora digitifera*	adult	pluripotent cells	spontaneous dissociation (after thermal bleaching)	Ca^{2+} free ASW (2 h)	none	none	Reyes-Bermudez et al. (2021)
Ctenophora	Tentaculata	*Mnemiopsis leidyi*	adult healing lobe	all cell types	spontaneous	ctenophore mesogleal serum	0.25% trypsin/EDTA	none	Vandepas et al. (2017)

Table 1. *Cont.*

Phylum	Class	Species	Inoculum Type	Target Cells	Isolation Technique	Isolation Medium	Enzyme Digestion	Cell-Type Enrichment Strategy	Reference
Porifera	Demospongiae	*Axinella* sp.	adult	all cell types	mechanical dissociation	Calcium-/magnesium-free ASW + EDTA	none	none	Song et al. (2021)
	Demospongiae	*H. panicea, H. aquaeductus, H. dujardinii*	adult	all cell types	mechanical dissociation	FSW	none	none	Lavrov and Kosevich (2016)
	Demospongiae	*Spongosorites, Cinachyrella, Haliclona*	adult	all cell types	mechanical dissociation	CMFASW	none	sedimentation	Robinson (2015)
	Demospongiae	*H. oculata, H. xena, D. avara, A. polypoides*	adult	all cell types	mechanical dissociation	FSW	none	none	Schippers et al. (2011)
	Demospongiae	*Crambe crambe*	adult	all cell types	mechanical dissociation	sodium hypochlorite seawater, gentamicin, ystatin, penicillin	none	none	Garcia Camacho et al. (2006)
	Demospongiae	*S. domuncula*	adult	all cell types	mechanical dissociation	CMFSW-E/AB (penicillin, streptomycin)	none	none	Zhang et al. (2004)
	Demospongiae	*Ircinia muscarum, Dysidea avara, Suberites domuncula*	adult	all cell types	mechanical dissociation	CMFSW, ampicillin, gentamycin, kanamycin, tylosin, tetracyclin	none	none	De Rosa et al. (2001)

Table 1. *Cont.*

Phylum	Class	Species	Inoculum Type	Target Cells	Isolation Technique	Isolation Medium	Enzyme Digestion	Cell-Type Enrichment Strategy	Reference
	Demospongiae	*Suberites domuncula*	adult	all cell types	mechanical dissociation	CMFSW EDTA	none	none	Müller et al. (1999)
	Demospongiae	*Geodia cydonium, Suberites domuncula*	Adult, outer cortical layer	spherulous cells	mechanical dissociation	PBSA-SW, EDTA	none	density gradient centrifugation using Percoll solution	Koziol et al. (1998)
	Demospongiae	*S. domuncula*	adult	all cell types	mechanical dissociation	CMFSW-E	none	none	Custodio et al. (1998)

Table 2. Established media and culture conditions across taxa. Indicative examples were selected among the references listed in Appendix B Table A2. When different methods were compared, only the selected optimized protocol is listed in the table. Basal commercial formula was adjusted to seawater osmolarity by dilution of concentrated formula in seawater or by the addition of mineral salts (except for GIM and GMIM). % indicates dilution in filtered natural (FSW) or artificial seawater (ASW). DMEM: Dulbecco's Modified Eagle Medium; MEM: Modified Eagle Medium; M199: medium 199; GIM: Grace's insect medium; GMIM: Grace's modified insect medium; L-15: Leibovitz medium L-15; RPMI: xx; IPM: Isotonic Planarian Medium; TTP: Teshirogi & Tohya Planarian Medium.

Phylum	Class	Species	Inoculum Type	Target Cells	Adherent/ Suspended Cell Culture	Basal Formula	Supplement	Antioxidant	Trace Elements and Vitamins	Antimicrobials	Substrate	Light	Atmosphere	Temperature	Medium Change	Reference
Chordata (Tunicata)	Ascidiacea	*Botryllus schlosseri*	adult zooids	circulatory blood cells	adherent/ suspended	DME supplemented with salts	glutamine solution	none	$MgSO_4$, $MgCl_2$	penicillin, streptomycin, nystatin	plastic	dark	air	19 °C	weekly	Rinkevich and Rabinowitz (1993)
	Ascidiacea	*Botryllus schlosseri*	embryo	all cell types	adherent/ suspended	DME supplemented with salts	glutamine solution, hemolymph from Botrylloides, chick embryo extract	none	$MgSO_4$, $MgCl_2$	penicillin, streptomycin, nystatin	plastic, gelatin, poly-l-lysin, collagen	dark	air	20 °C	weekly	Rinkevich and Rabinowitz (1994)
	Ascidiacea	*Botryllus schlosseri*	colonial buds	epithelium	adherent/ suspended	RPMI-1640	fetal calf serum, HEPES	none	$MgSO_4$, $MgCl_2$	penicillin, streptomycin, nystatin	plastic	dark	air	20 °C	partly replaced every other day	Rinkevich and Rabinowitz (1997)
	Ascidiacea	*Botryllus schlosseri*	adult zooids, buds	cup cell disease	suspended	Fisher's medium	fetal calf serum		salts	gentamycin	plastic	dark	air	20 °C	partly replaced every 2w	Moiseeva et al. (2004)
	Ascidiacea	*Botryllus schlosseri*	palleal buds	epithelium	adherent/ suspended	RPMI-1640, DMEM, HAM, Fischer's, Schnider's, Grace, Iscov's, DCCM-1	fetal calf serum, HEPES, L-glutamine, EFG		salts	penicillin, streptomycin, amphotericin, gentamycin	Plastic, collagen, Matrigel, methocel, fibronectin, poly-l-lysine	dark	air	19–23 °C	partly replaced every 2w	Rabinowitz and Rinkevich (2004)
	Ascidiacea	*Botryllus schlosseri*	palleal buds	epithelium	adherent	L-15	fetal calf serum, HEPES, L-glutamine	none	salts	penicillin, streptomycin, amphotericin	plastic	dark	air	20 °C	unknown	Rabinowitz et al. (2009)

Table 2. *Cont.*

Phylum	Class	Species	Inoculum Type	Target Cells	Adherent/ Suspended Cell Culture	Basal Formula	Supplement	Antioxidant	Trace Elements and Vitamins	Antimicrobials	Substrate	Light	Atmosphere	Temperature	Medium Change	Reference
	Ascidiacea	*Botryllus schlosseri*	palleal buds	epithelium	adherent	L-15	fetal calf serum, HEPES, L-glutamine		salts	penicillin, streptomycin, amphotericin	plastic	dark	air	20 °C	unknown	Rabinowitz and Rinkevich (2011)
	Ascidiacea	*Botrylloides leachii*	adult, hibernating colonies	blood cells	adherent	DME supplemented with salts	glutamine solution	none	MgSO$_4$, MgCl$_2$	penicillin, streptomycin, nystatin	plastic	dark	air	19 °C	weekly	Hyams et al. (2017)
Chordata (Cephalo-chordata)	Leptocardii	*Branchiostoma belcheri Japanese*	Adult epidermis, gill, gut, spermary, ovary	all cell types	suspended cells	2x Leiboviz's L-15 medium	10% (FBS) + 1% BPE	none	none	streptomycin (100 U/mL), rifampicin (50 µg/mL)	unknown	unknown	unknown	25 °C	unknown	Cai et al. (2013)
	Leptocardii	*Branchiostoma belcheri tsingtauense*	adult, buccal cirri, tail, gill, gut and metapleural fold	all cell types	adherent	L-15 and F-12	20% FBS	none	none	none	plastic	unknown	air	25 °C	every 10–15 days	Wang et al. (2009)
Echinodermata (Ambulacraria)	Asteroidea	*Asterias amurensis*	embryo (mesenchyme migration stage)	all cell types	suspended cells	Millipore-filtered seawater	4% (v/v) newborn bovine albumin	none	none	penicillin G potassium (50 units/mL) /streptomycin sulfate (50 µg/mL)	plastic	ambient	air	18 °C	none	Kaneko et al. (1995)
	Echinoidea	*Strongy-locentrotus intermedius*	embryo	all cell types	adherent cells	seawater	2 or 8% FCS	none	none	none	glass	unknown	unknown	17 °C	unknown	Odintsova et al. (2015)
	Holothuroidea	*Apostichopus japonicus*	adult	guts	adherent	L-15	2% FCS	none	vitamin E, insulin	penicillin, streptomycin, gentamycin	plastic	unknown	unknown	15 °C	every 3 days	Odintsova et al. (2005)
Xenacoelo-morpha	Acoela	*Isodiametra pulchra*	adults	all cell types	suspended	ASW	none	none	none	none	none	unknown	unknown	unknown	cells used directly for experimen-tation	De Mulder et al. (2009)
Arthropoda (Ecdysozoa)	Malacostraca	*Pacifasticus leniusculus, Homarus americanus*	adults	hematopoietic and testicular tissue	suspended	Medium 199	10% fetal bovine serum	none	NaHCO$_3$	penicillin, streptomycin	plastic	unknown	air	20 °C	weekly	Brody and Chang (1989)

Table 2. *Cont.*

Phylum	Class	Species	Inoculum Type	Target Cells	Adherent/ Suspended Cell Culture	Basal Formula	Supplement	Antioxidant	Trace Elements and Vitamins	Antimicrobials	Substrate	Light	Atmosphere	Temperature	Medium Change	Reference
	Malacostraca	*M. ensis*	adults	lymphoid organ	adherent	L-15	15% FBS, glucose	none	NaCl, NaHCO$_3$	Penicillin, streptomycin	plastic	dark	5% CO$_2$	26 °C	unknown	Han et al. (2013)
	Malacostraca	*Penaeus vannamei*	juvenile	all cell types	adherent	L-15 or M199	FBS 10%	none	none	penicillin, streptomycin	plastic	unknown	unknown	unknown	unknown	Toullec et al. (1996)
Nematoda (Ecdysozoa)	Chromadorea	*C. elegans*	eggs	larval cells	adherent	L-15	15% FBS	none	none	penicillin, streptomycin	glass	dark	without CO$_2$	20 °C	unknown	Zhang et al. (2011)
Platyhe-lminthes	Turbellaria	*Dugesia japonica*	adult head	neuronal cells	adherent cells	modified TTP medium; Teshirogi and Tohya, 1998	unknown	unknown	unknown	none	fibronectin or laminin-coated glass	dark	air	20 °C	none (4 d culture)	Asami et al. (2002)
	Turbellaria	*Schmidtea polychroa*	adult	neoblasts	adherent cells	isosmotic saline (Hepes-buffered)	FCS and BSA ("BMS"), MEM amino acids (essential and non-essential), glucose, trehalose, sodium pyruvate, glutamine	none	d-biotin, MEM vitamins	neomycin sulfate 100 mg/L	collagenI/ fibronectin coated glass and plastic; planarian homologous matrix	dark	air	18 °C	every 3rd day	Schürmann and Roland (2001)
	Turbellaria	*Schmidtea mediterranea*	adult tail	neoblasts	adherent cells	IPM, KnockOut DMEM, diluted L-15, diluted Grace's medium	FBS 5%	none	unknown	none	poly-D-lysine (50 µg/mL) coated plastic	dark	5% CO$_2$–95% air (for L-15: air)	22 °C	none (6 d culture)	Lei et al. (2019)
Mollusca	Gastropoda	*A. californica*	adult pleural ganglia	neuronal cells	adherent	L-15	hemolymph supplemented	none	none	none	plastic	dark	unknown	18 °C	unknown	Bedi et al. (1998)
	Gastropoda	*A. californica*	adult and juvenile	ganglia	adherent	L-15	1% FBS	none	dextrose, glutamine	none	Plastic, poly-l-lysine	dark	air	RT	every 2 days	Schacher and Proshansky (1983)

Table 2. *Cont.*

Phylum	Class	Species	Inoculum Type	Target Cells	Adherent/ Suspended Cell Culture	Basal Formula	Supplement	Antioxidant	Trace Elements and Vitamins	Antimicrobials	Substrate	Light	Atmosphere	Temperature	Medium Change	Reference
	Gastropoda	*A. californica*	adult	ganglia	adherent	L-15	FBS	none	none	none	glass	unknown	unknown	18 °C	none	Montgomery et al. (2002)
	Gastropoda	*Helix aspersa aspersa, Helix aspersa maxima*	adult	gonadal cells	adherent	Medium 199	20% FCS	none	EGF, methyl cellulose	penicillin, streptomycin	plastic	unknown	unknown	unknown	unknown	Monnier and Bride (1995)
Annelida (Spiralia)	Clitellata	*Hirudo medicinalis*	adult, Matrigel implants	myoend-othelial cells	adherent	DMEM	FBS 10%	none	glutamine	none	plastic	unknown	unknown	20 °C	unknown	Grimaldi et al. (2009)
	Clitellata	*Hirudo medicinalis*	adult, ganglia	retzius cells	adherent	L-15	FBS 2%	none	none	gentamycin	plastic, poly-L-lysine	unknown	unknown	20 °C	unknown	Bookman and Liu (1990)
	Clitellata	*Hirudo medicinalis*	adult	microglial cells	adherent	L-15	Glutamine, glucose	none	none	gentamycin	plastic	unknown	unknown	unknown	unknown	Masuda-Nakagawa et al. (1994)
Cnidaria	Hydrozoa	*Polyorchis penicillatus*	adult nerve rings	neurons	adherent cells and clusters	ASW	none	none	none	1% gentamycin	native mesoglea	ambient	air	10–15 °C	daily	Przysiezniak and Spencer (1989)
	Hydrozoa	*Podocoryne carnea*	adult umbrella	smooth muscle cells	adherent cells	ASW	unknown	unknown	unknown	unknown	remnants of native mesoglea	ambient	air	12 °C	none	Schmid (1992)
	Anthozoa	*Leptogorgia virgulata*	adult	scleroblasts	adherent cells	TC199	30% horse serum	none	none	penicillin + streptomycin + colimycine + bacitracine (first 2 weeks)	plastic	12L: 12D	air	21 °C	daily, then every 2–3 days	Kingsley et al. (1987)
	Anthozoa	*Nematostella vectensis*	adult	ectodermis	adherent cells	20% L-15	3% FCS	none	none	penicillin, streptomycin, amphotericin b		dark	air	20 °C	weekly, half medium	Rabinowitz et al. (2016)
	Anthozoa	*Anemonia sulcata*	adult tentacle	epithelial and interstitial cells	suspended cells	ASW	none	none	none	1% streptomycine	plastic	ambient	air	16 °C	none	Apte et al. (1996)

20

Table 2. *Cont.*

Phylum	Class	Species	Inoculum Type	Target Cells	Adherent/ Suspended Cell Culture	Basal Formula	Supplement	Antioxidant	Trace Elements and Vitamins	Antimicrobials	Substrate	Light	Atmosphere	Temperature	Medium Change	Reference
Anthozoa		*Anemonia viridis*	adult regenerating tentacle	anemone small round cell types	adherent clusters of cells	80% modified DMEM	5% FBS	none	none	1% kanamycin (100 μg/mL), 1% Amphotericin B (2.5 μg/mL) (after 3–7 d MycoKill 1× initial supplementation)	plastic	dark	air	20 °C	weekly	Barnay-Verdier et al. (2013)
Anthozoa		*Anemonia viridis*	adult regenerating (whole) tentacle	anemone gastrodermal differentiated cells	suspended	20% GMIM	5% FBS, 1% L-glutamate	none	none	1% kanamycin (100 μg/mL), 1% Amphotericin B (2.5 μg/mL), 1% Antibiotics-Antimycotics (Sigma)	plastic	dark	air	20 °C	weekly	Ventura et al. (2018)
Anthozoa		*Pocillopora damicornis*	adult	gastrodermal cell containing zooxanthellae	suspended cells	filtered seawater	none	none	none	none	polylysine 0.1% coated slides	NA	air	25 °C	none	Gates and Muscatine (1992)

Table 2. *Cont.*

Phylum	Class	Species	Inoculum Type	Target Cells	Adherent/Suspended Cell Culture	Basal Formula	Supplement	Antioxidant	Trace Elements and Vitamins	Antimicrobials	Substrate	Light	Atmosphere	Temperature	Medium Change	Reference
	Anthozoa	*Stylophora pistillata, Porites lutea, Dendron-ephthya hemprichi, Paraerythro-podium fulvum fulvum, Heteroxenia fuscescence, Clathraria rubrinoides, Plexaura* sp., *Millepora dichotoma*	adult and larva	all cell types	adherent and suspended cells	10% L-15/DMEM/M199 + taurine 40 mg/L	10% FCS	none	none	1% gentamycin, penicillin, streptomycin, amphotericin, nystatin	plastic	NA	air	23–24 °C	weekly	Frank et al. (1994)
	Anthozoa	*Acropora microphtalma, Pocillopora damicornis, Montipora digitata*	adult	all cell types	suspended multicellular isolates	25–50% DMEM	10% FBS	none	Mg^{2+}, Fe^{2+}, Mn^{2+}	1% Antibiotic-Antimycotic Gibco	Primaria plastic	24 h light	5% CO_2–95% air	24 °C	none	Kopecky and Ostrander (1999)
	Anthozoa	*Pocillopora damicornis*	adult	all cell types	suspended cells and adherent multicellular isolates	50% DMEM + taurine 40 mg/L	10% FBS	none	Sr^{2+}	1% Antibiotic-Antimycotic Gibco	Primaria plastic	24 h light	5% CO_2–95% air	24 °C	once every 2 weeks	Domart-Coulon et al. (2001)
	Anthozoa	*Pocillopora damicornis*	adult	all cell types	adherent multicellular isolates	12.5% DMEM + taurine 40 mg/L	1.25% FCS	none	Sr^{2+}	1% Antibiotic-Antimycotic Gibco	RGD peptide-coated plastic/glass	12 hL:12 hD	air	27 °C	weekly	Domart-Coulon et al. (2004)
	Anthozoa	*Montipora digitata, Xenia elongata*	adult	all cell types	adherent cells and cell aggregates	ASW + MEM aminoacids + taurine 10 mM, glutamine 2mM, aspartic acid 20 μg/mL + glucose 0.1–3 mM	2% FBS	ascorbic acid 50 μg/mL	MEM vitamins	1% Antibiotic-Antimycotic Gibco	Primaria plastic	12 hL:12 hD	air	26 °C	weekly (every other day ascorbic acid)	Helman et al. (2008)

Table 2. *Cont.*

Phylum Class	Species	Inoculum Type	Target Cells	Adherent/ Suspended Cell Culture	Basal Formula	Supplement	Antioxidant	Trace Elements and Vitamins	Antimicrobials	Substrate	Light	Atmosphere	Temperature	Medium Change	Reference
Anthozoa	*Sinularia flexibilis*	adult	all cell types	suspended cells and cell clusters	GIM, GMIM without salt addition	none	none	none	1% gentamycine-1% streptomycin	plastic	12 hL:12 hD	5% CO_2–95% air	24 °C	weekly	Khalesi (2008)
Anthozoa	*Fungia* sp., *Pavona divaricata*	adult	all cell types	suspended tissue balls	seawater	none	ascorbic acid 125 μM + catalase 250 U/mL; mannitol 10 mM	none	none	plastic	24 h light	air	25 °C	daily	Nesa and Hidaka (2009)
Anthozoa	*Pocillopora damicornis*	adult	coral cell types without zooxanthellae	suspended cells	RPMI + $CaCl_2$ 1 mM, NaPyruvate 1 mM, glucose 2g/L, galactose 2g/L, Na-lactic acid 0.25 g/L, Bovine Albumin 1 g/L, cysteine 0.5 mM, methionine 0.5 mM, glutamine 0.25 g/L	none	ascorbate 0.25 mM	hydroxy-cobalamin 0.01 mM, Na-folate 0.001 mM, succinate 0.05 g/L	none	plastic or Teflon	light?	air	26 °C	every 3rd day	Downs et al. (2010)
Anthozoa	*Acropora millepora*	larva (planula)	"interstitial" cells	suspended cells	10% DMEM	5% FBS	none	2% marine enrichment F/2 (Sigma)	3% Antibiotics-Antimycotics Sigma	plastic	dark	air	23 °C	every 2 weeks	Reyes-Bermudez and Miller (2009)
Anthozoa	*Fungia granulosa*	adult regenerating tissue (peeled off skeleton)	all cell types	suspended spheroids	FSW without antibiotics	coralline algae whole fragment	none	none	none	glass	10 hL:14 hD 100 μmol photons m^{-2} s^{-1}	air	25 °C	every 3rd day	Gardner et al. (2015)

Table 2. *Cont.*

Phylum	Class	Species	Inoculum Type	Target Cells	Adherent/ Suspended Cell Culture	Basal Formula	Supplement	Antioxidant	Trace Elements and Vitamins	Antimicrobials	Substrate	Light	Atmosphere	Temperature	Medium Change	Reference
	Anthozoa	*Acropora tenuis*	planula	gastrodermal secretory cells, undifferen- tiated cells, neuronal cells, and epidermal cells	cell aggregates (brown- pigmented)	10% DMEM	1.5% FBS	none	plasmin 2 µg/mL (modular protease)	penicillin 100 U/mL, streptomycin 100 µg/mL, amphotericinB (0.25 µg/mL)	plastic	dark	air	20 °C	every 3rd day	Kawamura et al. (2021)
	Anthozoa	*Acropora digitifera*	adult	pluripotent cells	suspended cell aggregates	30% DMEM	10% FBS	none	1% Glutamax	1% penicillin/ streptomycine, 0.1% fungizone	plastic	dark	air	23 °C	when 60% confluence	Reyes-Bermudez et al. (2021)
Cteno- phora	Tentaculata	*Mnemiopsis leidyi*	adult wound- healing lobe	all cell types	round ectodermal cells/ adherent giant smooth muscle cell/ suspended digestive endodermal cells	10–50% ctenophore mesogleal homogenate (in seawater)	ctenophore mesogleal serum (CMS) homogenate	none	none	1% penicillin/ streptomycin	plastic, glass	dark	air	14–16 °C	every 3rd day	Vandepas et al. (2017)
Pori- fera	Demos- pongiae	*Axinella* sp.	adult	all cell types	suspended cells and cell clusters	NSW (natural see water)	none	none	none	400 mg/L gentamicin	plastic	unknown	air	24 °C	every 1–4 days	Song et al. (2021)
	Demos- pongiae	*H. panicea, H. aquaeductus, H. dujardinii*	adult	all cell types	suspended cells	FSW	none	none	none	none	plastic	unknown	air	8–10 °C	every 48 h	Lavrov and Kosevich (2016)
	Demos- pongiae	*Spongosorites, Cinachyrella, Haliclona*	adult	all cell types	suspended cells	CMFASW	none	none	none	none	plastic	unknown	reduced air	unknown	none	Robinson (2015)

24

Table 2. *Cont.*

Phylum	Class	Species	Inoculum Type	Target Cells	Adherent/ Suspended Cell Culture	Basal Formula	Supplement	Antioxidant	Trace Elements and Vitamins	Antimicrobials	Substrate	Light	Atmosphere	Temperature	Medium Change	Reference
	Demos-pongiae	*H. oculata, H. xena, D. avara, A. polypoides*	adult	all cell types	suspended cells	Calcium- and magnesium-free ASW EDTA	none	none	none	none	plastic	unknown	unknown	2 °C	unknown	Schippers et al. (2011)
	Demospongiae	*Crambe crambe*	adult	all cell types	suspended cells	RPMI 1640	octopus extract 5–20%	none	inorganic salts, amino acids	none	plastic	unknown	4% CO_2	17–19 °C	1–4 per week	Garcia Camacho et al. (2006)
	Demospongiae	*S. domuncula*	adult	all cell types	suspended cells	SW-AB	none	none	L-glutamine, pyruvate, ferric citrate, sodium silicate, RPMI 1640, Marine Broth	penicillin, streptomycin	plastic	unknown	unknown	16 °C	none	Zhang et al. (2004)
	Demospongiae	*Ircinia muscarum, Dysidea avara, Suberites domuncula*	adult	all cell types	suspended cells	CMFSW, DMEM after 3rd week	sterols in DMEM	none	none	ampicillin, gentamycin, kanamycin, tylosin, tetracyclin	plastic	Dark or light	unknown	18 or 22 °C	Every 3 days	De Rosa et al. (2001)
	Demospongiae	*Suberites domuncula*	adult	all cell types	suspended cells	CMFSW EDTA	none	none	none	penicillin, streptomycin	plastic	unknown	unknown	16 °C	daily	Müller et al. (1999)
	Demospongiae	*Geodia cydonium, Suberites domuncula*	Adult, outer cortical layer	spherulous cells	suspended cells	PBSA-SW	none	none	none	none	plastic	unknown	unknown	17 °C	none	Koziol et al. (1998)
	Demospongiae	*S. domuncula*	adult	all cell types	suspended cells	CMFSW-E	none	none	none	penicillin, streptomycin	plastic	unknown	unknown	16 °C	daily, once a week after the 2nd week	Custodio et al. (1998)

Table 3. Molecular tools to detect the most frequently recorded unicellular microeucaryote contaminants. Sequences from cloned PCR products amplified with these eukaryote primers are blasted against National Center for Marine Biotechnology (NCBI) nucleotide databases to check the potential microbial nature of long-term cultured candidate aquatic invertebrate stem cells.

Host Metazoan Taxon	Associated Microeukaryote Taxon	Opportunistic In Vitro Growth	Gene Marker	Amplicon Size (bp)	Primer Pairs	Oligonucleotide Sequences	Reference
Scleractinian coral	*Chromeria velia* (Chromerida: Alveolata)	mixotrophy: use of medium-provided mono/disaccharides and aminoacids (glutamate and glycine)	psbA	~2000	psbAf6	5′-GARCAC AACATHYTNA TGCAYCC-3′	Moore et al. (2003)
					psbAL1	5′-CRTGCA TWACTTC CATWCC-3′	
Scleractinian coral	Stramenopiles (Thraustochytriidae)	use of C & N from Leibovitz L-15 formula diluted in filtered seawater (and organics from mucus of coral)	18S rRNA	1500	Lab-490F	5′-TTCGG TTCCGG AGAGGG AGCCTGAGAG-3′	Siboni et al. (2010)
					Lab-2004R	5′-GCAGA ATCCGA AGATT CACCGG-3′	
all aquatic invertebrates	unicellular holozoan Opisthokonts	predation on invertebrate cells and debris, osmotrophy	specific alpha-Amylases, alpha-glucosidases	variable	18SFU	5′-ATGC TTGTC TCAAAG RYTAA GCCATGC-3′	Tikhonenkov et al. (2020b)

Table 3. *Cont.*

Host Metazoan Taxon	Associated Microeukaryote Taxon	Opportunistic In Vitro Growth	Gene Marker	Amplicon Size (bp)	Primer Pairs	Oligonucleotide Sequences	Reference
					18SRU	5'-CWGG TTCAC CWACGG AAACC TTGTTACG-3'	
all aquatic invertebrates	Fungi (including chytrids)	saprotrophy (feed on cellular debris)	18S rDNA	2000	SR1R	5'-TACCT GGTT GATQC TGCCAGT-3'	Fliegerová et al. (2008)
					NS8.1	5'-CCGC AGGTTC ACCTACG-3'	
all aquatic invertebrates	Volvocacean algae/zooxanthellae	phototrophy (use light and inorganic C & N)	ITS rDNA	700	ITSa	5'-GGGA TCCGTTTCCG TAGGTGAAC CTGC-3'	Coleman et al. (1994)
					ITSb	5'-GGGA TCCAT ATGCT TAAG TTCA GCGGG-3'	
all aquatic invertebrates	Eukaryotes (including chytrids)	undetermined organic C & N	18S rDNA	1800	#328	5'-ACCTG GTTGAT CCTGC CAG-3'	Kupper et al. (2006)
					#329	5'-TGAT CCTTC YGCA GGTTCAC-3'	

2. Isolating Stem-like Cell Types Suitable for In Vitro Culture

Inoculum type is highly diverse, depending on targeted aquatic invertebrate taxon, tissue and life cycle stage. At the same time, selecting the proper inoculum for a given in vitro culture is certainly the most important decision towards establishing a suitable cell culture.

2.1. Selecting Suitable Sources of Cells

Based on the number of publications on primary cultures and the number of reported cell lines across aquatic invertebrate taxa (Figure 1), most research has focused on cell culture establishment from mollusks (mostly bivalves, and comparatively fewer gastropods and cephalopods), followed by porifera (mostly demosponges), cnidaria (historically hydrozoans and currently anthozoans, Figure 2), crustacea (pennaeid shrimps, crabs and crayfish), echinoderms and tunicates. Episodic attempts have also been made for one or two species representatives of ctenophores (*Mnemiopsis*), annelids (*Lumbricidae; Nereidae*), nematodes (*Caenorhabditis*), chelicerates (*Limulus*) and cephalochordates (*Amphioxus*). These differences in research efforts illustrate differences in the attractivity of specific taxa for cell culture, which stem from three complementary considerations, detailed below, that every researcher has to take into account when selecting the origin of the cells to be cultured in vitro.

The first pertinent consideration is whether to work on tissue isolated from established experimental models maintained in controlled aquarium or laboratory facilities. These animals, in contrast to wild animal sampling, provide both access to early life stages, as well as increased reproducibility for cell culturing experiments. Their use also meets the biodiversity protection regulations and traceability requirements of the Nagoya protocol. More and more clonal lineages of genotyped animals are becoming available across taxa, and establishing cell cultures from this traceable material is an additional source of reproducibility that reduces complexity and facilitates comparisons between intervention protocols. Ultimately, these biological models will help the optimization of the culturing conditions, for guidelines specific to a few model species. In this context, two attractive taxa for fundamental research are cnidaria (e.g., *Hydra vulgaris, Hydractinia echinata, Exaiptasia pallida*) and platyhelminthes (e.g., *Schmidtea mediterranea*) with well-established strains of animals. Mollusks or crustaceans of commercial importance, and with a complete life cycle obtained in captivity and traceable across generations, also represent important taxa to develop stem cell cultures.

A second important consideration when selecting a model species for cell culturing experiments is the wealth of genomic, transcriptomic and metabolomic information available for that species. In addition to allowing the identification of cell-type-specific markers, post-genomic information gained on metabolic pathways and cellular adhesion systems can be used to formulate working hypotheses on taxon-specific in vitro cellular requirements of media components and substrates. Similarly, knowledge acquired on in vivo tissue homeostasis, dynamics of cell proliferation, and somatic stem cell niches (Martinez et al. 2022) could help in selecting a seed tissue of high proliferative potential. In this respect, the continuous decrease in sequencing costs, as well as in other omics techniques, is allowing more and more research groups working with marine aquatic invertebrates to characterize their favorite species, suggesting that omics information will fast become available for almost all taxa.

The third point to consider is the desired approach for obtaining immortalized cells. Similar to mammalian cell culture, potential sources for immortalized cell lines are artificially reprogrammed cells and spontaneously tumor-like tissue. However, immortalization methodologies are currently limited in aquatic invertebrate cells by low yields and poor stability, as observed in sponges (Pomponi et al. 2013; Revilla-I-Domingo et al. 2018), bivalve mollusks (Hetrick et al. 1981; Boulo et al. 1996), and crustaceans (Claydon and Owens 2008; Xu et al. 2018). As suggested by Odintsova et al. 2011, natural tumor-like tissue, characterized by increased (hyperplasia) or altered (neoplasia) cell proliferation patterns, is thus a promising inoculum to initiate primary cultures. However, tumor-like lesions in wild or captive aquatic invertebrate taxa have low registered frequencies (Peters 2006; Tascedda and Ottaviani 2014), and there has been repeated unexpected failures at maintaining the hyperproliferation of successfully isolated cancerous cells in vitro. For instance, in transmissible soft-shell clam (*Mya arenaria*) leukemia, cancerous hemocytes rapidly undergo in vitro apoptosis, triggered by the release of mortalin-based cytoplasmic sequestration of p53 (Walker et al. 2006). Other attempts at primary culture initiation from artificially induced tumors of carcinogen-exposed bivalves (*Crassostrea virginica*) also failed to maintain persistent in vitro cell division (Hetrick et al. 1981).

Consequently, the use of stem-like cells for seeding in vitro cultures appears key to setting dividing primary cultures. Marine invertebrates display a wide variety of intriguing cellular phenomenon, such as asexual reproduction, striking regenerative capacity, reduced aging and dormant stages, which upon arousal restore fully functional individuals (Figure 1). These mechanisms indicate high cellular plasticity, proliferation and a likely involvement of stem-like cells. Although the potency of these cells remains largely uncharacterized in most species, and the orthology between these stem-like cells remains to be assessed (Rinkevich et al. 2022), they represent a promising source of proliferating and self-renewable cell types. However, the identification, isolation and characterization of aquatic invertebrate stem cells remains a major, typically species-specific, technical challenge. With few species having established protocols for the isolation of identified stem-like cells (Hayashi et al. 2006; Sun et al. 2007; Hemmrich et al. 2012; Kassmer et al. 2020;

Reyes-Bermudez et al. 2021), their generalization and transfer by taking advantage of the vast diversity of specific approaches explored in other taxa (Table 1) appear particularly promising.

2.2. Selecting a Suitable Type of Inoculum

Aquatic invertebrates have indirect development cycles, with widespread asexual propagation strategies, including colonial budding and the generation of dormant stages, as well as high regenerative abilities, including whole-body regeneration. These developmental properties are suggestive of the presence of proliferative cells, including potential stem-like cells, which are of particular interest to establish proliferating cell cultures. Hence, they provide the following range of theoretically ideal inoculum material: embryonic/larval tissue, regenerating tissue, asexually propagating tissue and dormant stages.

Dissociated tissue from whole embryo/larva consistently yields primary cell cultures dividing over 2–3 weeks, allowing a few rounds of successive subcultures. For example, when applied to cnidarian models, whole dissociated *Acropora* planula larvae yielded subsets of dividing coral cells that could undergo several successive subcultures (see Reyes-Bermudez and Miller (2009) for *A. millepora*, and Kawamura et al. (2021) for *A. tenuis*).

Dissociated somatic adult tissue sampled from regenerating tissue has also been observed to yield dividing cell cultures that could be subcultured for several weeks. Among other examples, cultures based on regenerating tentacle tips of the sea anemone *Anemonia viridis* (Barnay-Verdier et al. 2013) could be subcultured for 2–4 weeks, and primary cultures from regenerating intestinal tissue of the holothurian *Apostichopus japonicus* displayed limited but active in vitro proliferation at ~2 weeks after evisceration (Odintsova et al. 2005).

The dissociation of asexually growing tissue similarly gave rise to cell cultures with observable proliferative activity for a few weeks. For instance, using fast-growing branch tip fragments of the *Acropora millepora* coral (Reyes-Bermudez et al. 2021), cells could be subcultured for 2–4 weeks, and delayed senescence was reported in primary cultures from extracted buds of tunicates (Rabinowitz and Rinkevich 2004).

As implied in the "live slow, grow old" adage, cold adapted hibernating freshwater sponge species from lake Baikal yielded primary cell reaggregate (termed primmorph) cultures with record (max 8 months) longevity (Chernogor et al. 2011). Sponge gemmules also represent dormant hibernation/aestivation stages rich in multipotent stem cells (Simpson 1984) that, upon hatching, regenerate a functional adult. Activated gemmules could thus constitute a promising inoculum for primary cultures. Similarly, in the colonial tunicate *Botrylloides leachii*, arousal from a cold-induced dormancy (Burighel et al. 1976) leads to the restoration of multiple adults by proliferating *piwi*$^+$/*pl10*+ cells, two markers suggestive of stem-like properties (Hyams et al. 2017). In both cases, investigating the mechanisms regulating arousal from dormancy may yield cues to stimulate the tissues established in vitro to switch from quiescence to active cell cycling.

In conclusion, despite their initial abundance in cells with proliferative stemness-like properties, the shared in vitro fate of all four above-cited inoculum categories is terminal cell cycle arrest and the gradual accumulation of senescent, necrotic cells in primary culture and subsequent subcultures.

2.3. Selecting Suitable Cell Isolation Techniques

Inoculum type is highly diverse, depending on targeted aquatic invertebrate taxon, tissue and life cycle stage. Although no cell culture has yet been observed to sustain its proliferative activity for long, short-term functional primary cultures are routinely established from terminally differentiated cell types of aquatic invertebrates. Differentiated cells being arrested in G_0 can survive in vitro for a limited time with intact function, and hence are best used within hours to ~3 days of isolation. Nevertheless, comparing their tissue-isolation protocols offers opportunities to survey tissue sampling and dissociation methods (Table 1), as well as the cellular interactions and defense mechanisms that may support their in vitro viability, even for short periods of time. Emblematic examples of short-term invertebrate primary culture from quiescent cells include neuron-like cells and circulating hemocytes.

Giant neuronal cells from gastropod mollusks, such as the sensory and motor neurons from the sea hare *Aplysia californica*, are used to study growth cone motility and synapse plasticity (Kaczmarek et al. 1979; Lee et al. 2008; Zhao et al. 2009; Ren et al. 2019; Suter 2011). Cultured neurons from the pond snail *Lymnea stagnalis* are also routinely used for studies on synapse formation, neuronal aging and memory (Magoski et al. 1994; Prinz and Fromherz 2000; Walcourt and Winlow 2019). The in vitro establishment of nerve cells from jellyfish bell tissue (Przysiezniak and Spencer 1989; Schmid 1992) or from the solitary tunicate *Ciona intestinalis* (Zanetti et al. 2007) have also been reported. These neuronal cell types are usually micro-dissected from their ganglion, enzyme digested with protease, immobilized on positively charged polylysine-coated coverslips and then used for short-term electrophysiology assays, providing non-conventional in vitro models in neuroscience.

Circulating cells sampled from internal fluids, are another major category of cultured aquatic invertebrate cells. When seeded at high density ($>10^6$ cells/mL), cultured adherent hemocytes can form partly complete confluent monolayers, with clusters forming in suspension above the monolayer that may then be detached and transferred to new culture dishes. Such cultures have been routinely established since the late 1960s from a wide range of species, including mollusks, crustaceans, tunicates and echinoderms, typically for in vitro cell/microbe interactions and immunopathology assays. Such cultures display short-term conserved functionality, as shown by phagocytosis or immunomodulatory assays. Proliferation may be induced by stimulation with bacterial antigens, as shown for the bivalve *Mytilus galloprovincialis* (Cao et al. 2003), the tunicate *Styela* (Raftos and Cooper 1991) and the earthworm *Lombricus* (Bilej et al. 1994). These differentiated cell types are drawn directly from internal cavities, lacunae and sinuses using a syringe. To counter their spontaneous self-aggregation (clotting) behavior, hemocytes are collected in syringes half-filled with species-specific anti-clotting saline solution, such as artificial seawater without calcium or magnesium, artificial seawater with a calcium chelator, or Na-citrate based "Alsever" saline solutions. Indeed, hemocytes secrete their own set of taxon-specific lectins (e.g., Matsumoto et al. 2001) and extracellular-matrix components (ECM) (e.g., a fibronectin-like ECM in bivalve hemocytes (Dyachuk et al. 2015)) that support rapid adherence, within hours of sampling, to glass or poly-lysine-coated coverslips.

Aside from the two isolation techniques described above, the quantitative evaluation of various approaches for cell extraction in different species suggests that, for the rapid obtention of single-dissociated cells from soft tissues for RNAseq cell phenotyping, and thus to obtain cells as close as possible to their wild-type state, mechanical isolation is the most efficient method (Khalesi 2008; Dessai 2012; Daugavet and Blinova 2015; Maselli et al. 2018). For example, cnidarian larval tissue or demosponge and calcisponge adult tissue fragments are dissociated within minutes via shearing in calcium-free seawater and passage through a 40–70 µm nylon mesh. However, species with tough cuticles (e.g., *Lombricidae*), important extracellular matrices (e.g., *Styelidae*) or abundant surface mucus (e.g., *Dugesiidae*) necessitate treatments with specific enzymes to liberate the cells. For instance, in the stony coral *Pocillopora damicornis*, chemical treatment with a divalent cation chelator followed by a mix of glycosidases and collagenase was reported to help dissolve the mucus and improve the yield of released cells (Downs et al. 2010). Proteolytic treatments (trypsine, dispase and other protease mixes) are routinely used to dissociate cells from solid tissues dissected from mollusks and crustaceans. Interestingly, protease treatment may induce cellular reprogramming, as shown by the collagenase-induced transdifferentiation of in vitro explanted striated muscle of jellyfish (Alder and Schmid 1987; Schmid and Alder 1984; Schmid and Reber-Müller 1995).

2.4. Selecting Suitable Cell-Type Enrichment Strategies

Cells of interest are typically mixed with other cell types after the dissociation of the inoculum. The enrichment of specific cell types, typically proliferative or multipotent ones, relies on the prior development of taxon-specific and custom-designed cell separation methods. For instance, in *Stylophora pistillata*, stem cells were not identified in the cell atlas established from both larval and adult tissues, following either enzymatic or mechanical dissociation methods (Levy et al. 2021), which severely limits the development of stem cell-enriched primary cultures.

Sorting methods for enriching inoculum suspensions in proliferative or multipotent cell types are thus required. Initial methods were based on differential sedimentation on density gradients, including sucrose, Percoll, or mixtures of Ficoll and polyethylene glycol. To further discriminate between morphologically similar cell types, and thus target specific cell types, Fluorescence Activated Cell Sorting (FACS) methods have recently been developed and have become highly prominent. For instance, FACS has been used to separate vital-stained coral cells (Rosental et al. 2017), and to isolate cell-type subpopulations for their single-cell gene expression characterization in hydrozoan (Siebert et al. 2019), as well as in anthozoan species (Levy et al. 2021; Sebé-Pedrós et al. 2018). In these diblastic animals, which lack a circulatory system, FACS is necessary to enrich dissociated tissue suspensions in hexacorallian putative immune cells, the amoebocytes recovered from the inter-epithelial mesogleal layer typical of cnidarian, for short-term functional phagocytosis characterization (Snyder et al. 2021). In triploblastic animals, FACS has also been refined to sort tunicate cell subpopulations to study the hematopoietic system (Rosental et al. 2018). Echinoderm coelomocyte subpopulations have been further separated by FACS into distinct cell types, such as the red pigment autofluorescent spherulocytes (Hira et al. 2020).

Consequently, there is a need for stem cell markers suitable for non-invasive stemness tracing in live cells to enable their enrichment. One promising perspective comes from the few aquatic invertebrate experimental models that can be genetically manipulated for which transgenic reporters of stemness properties can be engineered (e.g., in *Hydra* (Juliano et al. 2014)). Another direction of interest is the usage of fluorescent markers conjugated with antibodies specifically labeling stem cells. However, the identification of such markers remains extremely rare for aquatic invertebrates, with the recent notable exception of the colonial tunicate *Botrylloides diegensis* for which integrin-alpha-6 was shown to specifically label pluripotent cells (Kassmer et al. 2020). Whether this specific marker can be used in other species of aquatic invertebrates to label stem cells will be important to assess.

2.5. Selecting Cleansing Techniques to Minimize Contamination

There is a wide consensus across the scientific community that the highest obstacle to continuous marine/freshwater invertebrate cell culture propagation is overgrowth by aquatic microbial contaminants (Rinkevich 2005). This problem is critical in marine invertebrate primary cell cultures for two main reasons. First, it is because the tissues sampled to initiate the primary cultures come typically from areas directly or semi-directly exposed to environmental microbes, such as the thin epithelial structures at the interface with water (e.g., in porifera and cnidaria), tissues irrigated by a semi-open circulatory system (e.g., in mollusks, echinoderms and tunicates) or digestive and other internal tissues hosting their own microbiota (e.g., gills and hepatopancreas of mollusks). Second, commercial antibiotics/antimycotics/antiparasitic drugs have been designed against microbes isolated from terrestrial animals and mostly from humans and are thus largely ineffective against the mostly underexplored diversity of environmental aquatic microbes. To control the contamination of cell cultures by these aquatic microbes, three main strategies can be attempted.

First, microbial load can be reduced before cell isolation. The inoculum can be sampled from starved animals depurated in oxygenated sterile-filtered seawater to limit environmental microbial contaminants (e.g., for abalone mantle cell culture (Suja et al. 2014)). Microdissecting internal tissues that naturally protect from seawater by epithelial envelopes (e.g., molluscan heart tissue), and thus from aquatic microbes, would also reduce the initial microbial load of the inoculum. Collecting cells that possess natural antiseptic defenses, such as innate immune hemocytes (e.g., from mollusks, crustaceans, tunicates or echinoderms), would also have a positive impact on reducing the contamination of the culture. Alternative strategies include using short-term ubiquitous surface sterilization methods on the surface-exposed tissue, such as dipping for up to 1 hour in 10–70% ethanol (e.g., for molluscan abalone mantle, see Suja et al. (2014), and for oyster tissue, see Stephens and Hetrick (1979)) or a few seconds in $KmnO_4$ (e.g., in sea anemone tissue (Doumenc personal communication)) and treating the dissected tissue for up to days in sterile-filtered seawater enriched with a mixture of concentrated large-spectra commercial antibiotics/antimycotics/antiprotist compounds (e.g., molluscan mantle, gill or hepatopancreas tissue).

Second, if specific invertebrate cell types need to be recovered from contaminated primary cultures, the cell-type enrichment strategies established for preparing a suitable inoculum (see Section 2.4) could be reused. For instance, this approach successfully retrieved accessory nidamental gland cells pelleted from native bacteria through a 2% sucrose layer (Figure 3). In addition, the selective rinsing of adherent invertebrate cell types could help to remove cellular debris, toxins and suspended microbes.

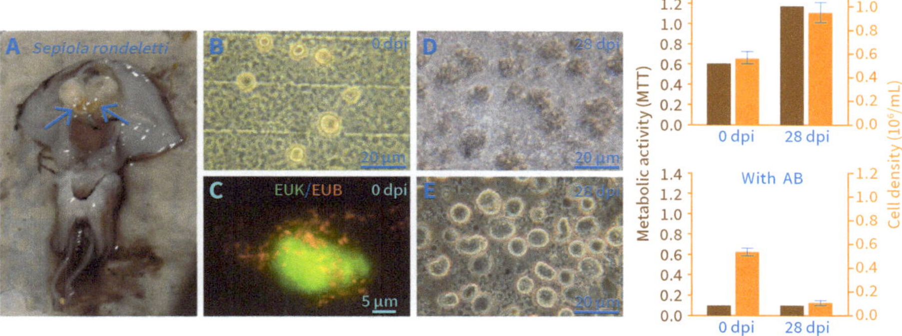

Figure 3. Primary co-culture of squid gland cells with native bacteria. (**A**) Accessory nidamental gland tissue (white arrows) from *Sepiola rondeletti* is enzymatically dissociated by trypsine (0.2% 30 min at 25 °C). (**B**) Gland cells are enriched via centrifugation through a sucrose cushion (2% in seawater), and their seeding density is controlled by Malassez hemocytometer numeration. (**C**) Glandular cell types visualized via Fluorescence In Situ Hybridization (EUK, universal eukaryote probe, fluorescein, green) are covered with surface-associated symbiotic bacteria (EUB, universal bacterial probe, Cy3, red). (**D**) Four-week-old primary culture (without antibiotics) showing high bacterial density around the cultured glandular cell types, (**E**) which can be re-enriched via sucrose cushion centrifugation. (**F**) Gland cell viability (mitochondrial enzyme activity assessed by MTT reduction assay, DO 580/630) is higher in the absence than in the presence of antibiotics (AB) and increases in primary co-culture with native bacteria, along with cell density, indicating the beneficial effect of native bacteria on the survival of cell cultures. Times are given as days post-inoculation (dpi). Source: Graphic by authors.

Third, contamination can be controlled during the primary culture itself. The main strategy for this step to reduce the unwanted mixotrophic growth of contaminants is to use a nutrient-poor basal medium formula, hence limiting the provision of carbon and nitrogen sources that typically exceeds the in vitro energy requirements of the target cells. The culture medium can also be supplemented with antibiotics/antimycotics/antiprotist drugs and changed frequently until the culture appears clean. Proliferating cultures should be closely monitored, and the primary cell cultures containing visible ciliates, bacteria, or clusters of cells with characteristic chytrid-like rhizoid morphology should be discarded.

However, these methods may rescue a subset of the targeted cell-type populations from contaminant overgrowth but carry a high cost in terms of time-consumption and cell yield reduction, for overall limited efficiency.

3. Defining Optimal Culture Conditions

While obtaining a high-quality inoculum is essential for establishing healthy cell cultures, the culturing conditions used are equally crucial. Indeed, even highly proliferative tissue will undergo terminal cell cycle arrest, typically within weeks after inoculation. Moreover, a breakthrough in the culturing conditions used for the *Bge* cell line was at the origin of a large expansion in cell lines.

3.1. Selecting Suitable Culture Media Composition

Media formulation should strive to provide adequate levels of carbon and nitrogen sources to meet the nutritional needs of each isolated aquatic invertebrate cell-type population. However, the metabolism of stem cells and their nutrient requirements are poorly documented across aquatic invertebrate taxa. Consequently, a large variety of culture media have been tested for their in vitro culture (Table 2). Based on the hypothesis of the conservation of major metabolic pathways across animal phyla, a widespread approach is to use commercial basal formulas originally designed for vertebrate cells, typically MEM, DMEM or Leibovitz L-15, supplemented with salts to adjust to the targeted osmolarity of the specimen's original environment and generally diluted to 10–50% (Maramorosch and Mitsuhashi 1997; Mothersill et al. 2000). An even simpler option is to provide a minimal medium composed of seawater with pyruvate as a carbon source, and glutamic acid as a nitrogen source. This approach has been used with sponge primmorph spontaneously aggregated from dissociated cells. These media have, however, persistently failed to sustain the in vitro division of cells of aquatic invertebrates. Another much more complex option is to entirely custom design the media's formula based on an extensive biochemical characterization of internal tissue or fluid composition from the targeted animal species (e.g., molluscan hemolymph). However, these taxon-specific media have not yet demonstrated sufficient benefits to justify their development cost.

A more integrated and personalized approach is to adapt the media formulations to meet the needs of the targeted invertebrate cell subpopulations. To check nutrient consumption in vitro, individual uptake experiments of targeted organic carbon (glucose, lipids, etc.) or nitrogen (amino-acid) substrates (see Apte et al. (1996) for amino-acid transport into sea anemone cells, and Heude-Berthelin et al. (2003) for glucose uptake and glycogen metabolism in oyster cells) may now be updated to metabolomics-based global approaches. Indeed, the search for changes in the metabolite profiles of media sampled at various timepoints in cultured mammalian CHO-CK1 cell lines has helped identify factors that sustain growth and affect in vitro behavior (Mohmad-Saberi et al. 2013). A recent breakthrough was reached using this approach to develop an amino-acid-enriched sponge cell culture medium that sustains cell division in primary cultures (Conkling et al. 2019). The team used a genetic algorithm to identify suitable amino acid components to supplement a commercial basal formula (M199) for improving the in vitro metabolic activity of *Dysidea avara* sponge cells (Munroe et al. 2019).

A striking feature of successful insect culture media that support proliferating primary cultures and cell lines is the addition of lipid-rich supplements, with a trophic role and potential protection against oxidative stress. Lipid addition has been shown to transiently increase metabolic activity (mitochondrial MTT reduction) in cultured oyster heart cells (Domart-Coulon et al. 1994). The lipid-rich "Grace" commercial formula was shown in cnidarian primary cultures to increase octocoral cell numbers (Khalesi 2008), and is used to obtain a subset of dividing cells and a few rounds of subcultures from cultured sea anemone tentacle (Barnay-Verdier et al. 2013). However, a more global picture of the impact of lipids on culture media for aquatic invertebrate cells is currently lacking.

Medium renewal strategy should be aimed at striking a balance between a conditioned medium supply of undefined trophic factors and cytokines and the removal of senescent cells, debris and toxins from the aging primary cultures. Manipulating inoculum cell densities is an efficient way to facilitate confluence and thus maintain cell-to-cell contacts necessary for the secretion of cytokines that, although currently undefined, are certainly necessary for sustaining cell survival. Old-time tissue explantation methods that rely on the slow outward migration of mixed cell types from a dissected tissue fragment adherent to a culture dish yield the successive outgrowth of distinct morphotypes characterized at minima by their in vitro shape and behavior. These cells can broadly be classified by the following three categories: fibroblast-like, epithelial-like and amoeboid-like cell types (Vago 2012), and can be selected for their ability to survive in vitro on residual native extra-cellular-matrix components. Insect cell lines have emerged from such long-term maintained explant cultures of lepidopteran imaginal discs (Echalier 1997). More recently, the explantation of ectodermal monolayers of regenerating starlet sea anemone yielded mitotically active, mixed cell types, primary cultures (Rabinowitz et al. 2016).

In addition, culture medium can be complemented with a number of factors to promote cell proliferation: C-type lectins have been shown to have cytostatic effects on the hemocytes of the tunicate *Polyandrocarpa misakiensis* (Matsumoto et al. 2001); lectins from another tunicate, *Didemnum ternatanum*, promote the adhesion of a range of marine invertebrate cells (Odintsova et al. 1999); insulin and insulin growth factor, as well as other vertebrate growth factors, were shown to have a positive impact on the transient proliferation of molluscan bivalve cells (Domart-Coulon et al. 1994; Giard et al. 1998); and retinoic acid-related molecules are known to be involved in the dedifferentiation process of multipotent cells as reported for tunicate hemocyte cultures (*Polyandrocarpa misakiensis*) (Kawamura and Fujiwara 1995).

3.2. The Oxidative Stress Problem

Very few and exclusively freshwater taxa among the large diversity of aquatic invertebrates have given rise to cell lines, including the snail *Biomphalaria* (*Gastropoda*). Salinity is thus a major difference between the primary culture systems that have given rise to cell lines and the unsuccessful attempts based on aquatic invertebrate species. One possible cause for this difference is that higher salinity correlates with lower dissolved oxygen. Consequently, dissolved oxygen levels in the cell cultures might be an important yet overlooked physico-chemical parameter of culture conditions. To date, primary cultures of aquatic invertebrate cells are indeed mostly conducted under standard atmospheric conditions (i.e., ~20% O_2), with the cells covered by a thin layer of culture medium where dissolved oxygen is equilibrated by diffusion with the surrounding air. Except for a few cases of full-strength Modified Eagle Medium (or derivatives, osmotically adjusted by salt addition), which requires a bicarbonate/5% CO_2 buffer system, the gaseous atmosphere of most cell cultures is thus composed of air (Table 2). The widely used, amino-acid rich, Leibovitz L-15-based media do not require a 5% CO_2 atmosphere. Seawater/freshwater diluted commercial or custom-made media rely on the addition of Hepes (~20 mM) for pH buffering at 7.4–7.6, depending on species (Tris-HCl is used for sponge cells grown at pH ~8.0). Hence, under typical laboratory conditions (air and 15–25 °C), in vitro aquatic invertebrate cells are exposed to ~20% O_2, which is largely more than in their natural aquatic environment, and could likely expose them to in vitro oxidative stress.

To circumvent this potential problem, the first step will be to monitor invertebrate intracellular oxidative stress, for instance, via a fluorescent general oxidative stress indicator, such as CM-H_2DCFDA, which has been used on the spheroid tissue of *Fungia* coral exposed to short-term acute thermal stress (Gardner et al. 2017). Upon the confirmation of oxidative stress, the second step could be medium supplementation with exogenous antioxidants (for example, ascorbic acid (Helman et al. 2008), catalase enzyme (Domart-Coulon et al. 1994)) or native pigments with high antioxidant properties (e.g., sea urchin spinochrome (Ageenko et al. 2014) and shrimp astaxanthin (Lee et al. 2021)). Both approaches have reproducibly led to the increased maintenance of the primary cell cultures. An alternative when establishing cultures of tissues containing photosynthetic endosymbionts (e.g., Cyanobacteria-containing sponges, and Symbiodiniaceae-containing sea anemones, corals and octocorals) is to maintain the cultures in the dark to inhibit the photosynthetic processes that generate oxygen and thus increase oxidative stress (Table 2).

3.3. Understanding Adhesion and Cell-to-Cell Contact Requirements of Aquatic Invertebrate Stem Cells

To optimize the proliferation of culture cells, transferring genomic knowledge obtained in each aquatic invertebrate taxon on cell-to-cell and cell-to-ECM adhesion systems will be particularly useful for selecting suitable ECM-coatings of culture dishes.

Shifting from classical 2D monolayer culture to 3D "spheroid" culture systems offers opportunities to facilitate the maintenance of cell-to-cell interactions and of native secretions within the cell cluster. Cells from the earliest branching aquatic metazoans, such as poriferans and cnidarians, display spontaneous aggregation properties after tissue dissociation into single cells, sometimes leading to whole-body regeneration (see Simpson (1984) for sponge, Gierer et al. (1972) for hydra and Vizel et al. (2011) for coral). This re-aggregation property is being harnessed for spheroid formation (coral "tissue balls", sponge primmorphs) and their establishment for primary culture (Figure 4A–C). Hemocytes drawn from mollusks, crustaceans, tunicates or echinoderms also self-aggregate into clusters, through sequential migrations of adherent cells on the culture substrate followed by the putative secretion of self-recognition lectins (Figure 4D–F). A recent breakthrough using 3D cultures of sponge cells in ultra-low-gel agarose hydrogel microdroplets has been reported to support cell–ECM interactions and to facilitate the survival of differentiated *Geodia neptuni* demosponge cells (Urban-Gedamke et al. 2021).

Similarly, improving the in vitro microenvironment of isolated stem cells could potentially sustain their proliferation. Adapting the cellular microenvironment to mimic stem cell niches of a target organism should be pursued in each model taxon, as such information becomes available. In addition, primary cultures that gave rise to cell lines (e.g., insect imaginal disc cells) can provide mechanistic insights into the cellular microenvironment needed to maintain stem cell self-renewal. As shown in mammalian systems, multidirectional signaling by co-culturing stromal "feeder" cells with the target cells (e.g., neurons from the gastropod *Aplysia californica* (Montgomery et al. 2002); stem-like cells on a monolayer of confluent cephalopod hemocytes (Figure 4E)) might help to generate a microenvironment suitable for stem cell maintenance, proliferation and differentiation, typically by providing cell adhesion molecules, growth factors, hormones and other secreted proteins (see Girard et al. (2021) for hematopoietic stem cell niche, and Ootani et al. (2009) for intestinal stem cell niche). Furthermore, supplementing the culture media with specific growth factors (e.g., Wnt fusion proteins for ISC (Ootani et al. 2009)) can lead to the expansion of stem cells with sustained proliferation and multilineage differentiation. As in vivo information on the regulation of aquatic invertebrate stemness becomes available, transferring such information will be particularly important to design optimized cell culture media.

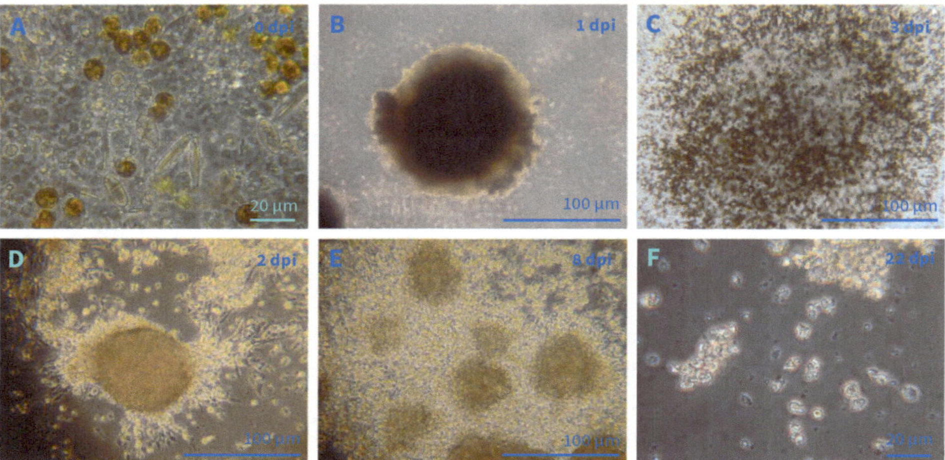

Figure 4. Aggregate vs. dissociated primary tissue and cell culture, on plastic dish-culture substrates. The figure shows micrographs of individual cells or multicellular aggregates at the indicated days post-inoculation (dpi). (**A**) Scleractinian coral cell types and their *Symbiodiniaceae endosymbionts* (within coral gastrodermal host cell, or free-living in the culture medium). (**B**) Suspended coral multicellular aggregates spontaneously formed in explant culture of colonies of *Pocillopora damicornis*. (**C**) Spontaneous dissociation into multilayered, mixed-cell-type culture, containing translucent coral cells and brown-pigmented microalgal symbionts (*Symbiodiniaceae*). (**D**) Cephalopod hemocytes from *Nautilus pompilius* aggregate in cell culture when seeded at high seeding density ($>10^6$ cells/mL, 2 dpi). (**E**) Confluent primary culture 8 dpi, showing networks of adherent hemocytes and proliferating cell clusters (300–500 µm in diameter), which can be detached and transferred (passaged) to new culture dishes. (**F**) Subcultured hemocytes (14 days post transfer from cells detached from clusters in 8 days post-inoculation-primary culture) remain quiescent and do not grow to confluence. Source: Graphic by authors.

To further mimic the in vivo microenvironment of the isolated cells, and their interactions with their environment in particular, new "physiomimetic" approaches should be developed using, for instance, versatile hydrogels to concentrate cells in a 3D microenvironment (see Otero et al. (2021) for a review on such experimental approaches for vertebrate cell systems). The ongoing development of commercial hydrogels (synthetic or derived from jellyfish, i.e., "Jellagel") provides new 3D substrates to test on aquatic invertebrate cells. To determine whether these cells behave in vitro similarly to in vivo, live-cell or live-tissue observations based on the micropropagation of tissue in microfluidic devices should be further established (Januszyk et al. 2015).

4. Controlling the Purity and Quality of Cultured Cells

Upon isolation from their initial tissue microenvironment for establishment in culture dishes, aquatic invertebrate cells change morphology and are notoriously difficult to identify by their in vitro shape and behavior (Rinkevich 2011; 2005; Cai and Zhang 2014). Moreover, cultured cells are morphologically highly plastic, changing shape, granularity and sometimes pigmentation with culture age and substratum composition (i.e., with or without surface coating with positive charges or ECM compounds). For cell lineage authentication, checking phenotype and genetic identity is imperative, not only upon culture initiation but also throughout the primary culture and derived subcultures, at least at the time of use for functional assays and before/after cryopreservation.

4.1. Proliferation

The monitoring of in vitro cell proliferation is traditionally based on monitoring cell densities (via subsampling a fraction of the culture followed by cell numeration on Malassez- or Neubauer-type hemocytometers, or time-lapse image analysis of microscopy fields of view) and attentive changes in the total protein content or DNA content extracted from cell pellets or monolayers. These methods overestimate live cell densities as they integrate dying cells to the viable cells. Another widely used method relies on the miniaturized high-throughput colorimetric quantification of mitochondrial oxidative phosphorylation (MTT or XTT reduction assays) by the cultured cells. First adapted for screening medium nutritional factors and physico-chemical parameters for molluscan cells (bivalve oyster *Crassostrea gigas* (Domart-Coulon et al. 1994)), it has also been adapted to sponge cell mitochondrial activity evaluation in primary culture (Zhang et al. 2004) and to the monitoring of coral larval cell density in primary cultures (Kawamura et al. 2021). However, this type of MTT test detects not only oxidative phosphorylations of the animal cells but also that of bacterial associates in primary culture (e.g., of cephalopod holobiont) tissue (Pichon et al. 2007). The fluorescence monitoring of cellular esterase activity is also a common method to quantify viable cells in cultures. However, their use for aquatic invertebrate taxa can be limited by the widespread co-occurrence of autofluorescent cell types with fluorescence spectra overlapping those of the enzyme substrates.

By quantifying the proportions of cells in each phase of the cell cycle, flow cytometry allows us to check the proliferative status of the collected tissue sample before culture establishment, and to monitor cell cycling in the derived primary cultures and potential sub-cultures. Applied, for example, in the early 2010s to primary cell cultures from five demosponge species, this flow-cytometry-based approach revealed rapid changes in the cell cycle distribution of a mixed-cell-type suspension over time in primary culture (over a short-term 2–10-day timescale)

(Schippers et al. 2011). The rapid accumulation of cells with low DNA content together with a drop in the proportion of quiescent (G_1/G_0) and cycling cells (S & G_2/M) could be visualized, supported by the parallel detection of activated (caspase3) apoptosis pathways. This evidence supported the hypothesis of the rapid senescence of cultured sponge cells, with the accumulation of cellular debris (demonstrated by widely scattered cell size distribution), despite stable or slowly declining cell counts, by only minus ~20% over the 10-day culture period. This observation calls to cautious interpretations of stable or slightly growing cell densities, counted from image analyses of microscopic fields or enumerated on Malassez-type slides, as round empty cell bodies cannot be unambiguously discriminated from living cells based on morphology only, even when using vital stains assays (e.g., neutral red or trypan-blue). Another point of caution when using this approach is the ploidy of the studied samples, and in particular, the presence of mixoploid cell populations that could bias their cell-cycle profile (Ermakov et al. 2012).

4.2. Phenotyping

Autofluorescent markers (e.g., Green Fluorescent Protein-rich intracellular granules of cnidarian cells,) or chromophore/pigments of specific cell types (e.g., red "echinochrome" pigments of echinoderm coelomocytes) can be used to sort cell types among a mixed cell suspension. However, care should be taken to minimize irradiance energy during fluorescence microscopy examination as it may damage the living cells by DNA photodamage or lipid peroxidation, and thus limit their subsequent in vitro survival. Enzyme activity assays (e.g., phenoloxidase of mollusk and crustacean immune cells), biochemical phenotyping and phagocytosis assays have also been used to characterize the in vitro functionality of hemocytes from molluscan hemolymph, tunicate hemolymph and echinoderm coelomic fluid.

Immunophenotyping requires the prior development and validation of polyclonal or monoclonal antibodies against epitopes of cell-type-specific proteins or membrane preparations. Although labor-intensive and time consuming, this strategy provides the advantage of the unambiguous localization of immuno-positive phenotypes in initial tissue and in primary tissue or mixed cell culture. For instance, low abundant small round coral skeletogenic (calicoblast) cell types were labeled with a polyclonal antibody raised against the biomineral organic matrix (Puverel et al. 2005) and antibodies were raised against the *Botrylloides piwi* sequence to label a specific population of hemocytes (Rinkevich et al. 2010). This antibody-based approach has been successfully used to trace self-sorting processes during cell-to-cell aggregation from mixed-cell-type dissociated tissue suspensions (Schmid et al. 1999) and for cell fusion experiments (Pomponi et al. 2013). This has an interesting yet still overlooked potential for cell-type enrichment via the antibody panning of immuno-positive cell types (Auzoux-Bordenave and Domart-Coulon 2010).

Novel phenotyping methods have recently been developed from cutting-edge single-cell RNA sequencing methods, which are applied to cultured cells. However, these techniques currently have the three following drawbacks: (1) the prior definition of cell-type-specific markers is needed, through the data-mining of single-cell RNAseq libraries obtained from dissociated tissue suspensions, which is still available for a limited number of species of established model organisms (e.g., for the starlet sea anemone (Sebé-Pedrós et al. 2018), for planarians (Hayashi et al. 2010) or for the scleractinian coral (Levy et al. 2021)); (2) molecular markers should be specific to metabolic pathways restricted to the targeted invertebrate taxon and exclude pathways that are also active in potential contaminating protists/microeucaryotes; (3) assessing the polyclonality (mixture of cell types) versus clonality (single cell line) of the culture requires the quantification of the percentage of reads obtained for each claimed phenotypic marker, relative to the total number of reads.

4.3. Genotyping

Despite the proper isolation and cleansing of cells, cultures can easily be overgrown by undesired cells. An undetermined fraction of these aquatic microbes survives the tissue aseptization treatments prior to dissociation or explantation and co-occurs along with metazoan cells in the mixed-cell-type suspensions obtained from soft tissue dissociation or hemolymph syringe-drawings. This large diversity of aquatic microbes is hard to monitor as it requires taxa-specific specialist microbiology knowledge and molecular tools for accurate identification. It is especially difficult to recognize their morphological traits in a mixed-cell-type primary culture that combines the morphological and behavioral plasticities of both the microbe and microbial life stage, and the targeted invertebrate cell types.

To address this problem, genetic markers specific to a species (e.g., *Axinella corrugata* demosponge, (Lopez et al. 2002)) or to a genus (e.g., *Acropora* scleractinian coral (Shinzato et al. 2014)) have been developed and validated for identifying cells from the targeted taxon in the initial tissue and over time in primary cultures and subcultures. Marker development is based on molecular genetics methods, such as DNA fingerprinting, amplified fragment length polymorphism (AFLP), single-locus DNA sequence analyses and microsatellites markers designed by next-generation sequencing population genetics methods.

As microbes tend to proliferate more actively than the cells of primary interest of the in vitro culture, it is crucial to check the potential microbial nature of long-term cultured candidate aquatic invertebrate stem cells.

4.4. Microbial Contaminants Authentication

Detecting genetic markers specific to the invertebrate taxa of interest does not exclude the potential co-occurrence of microbial contaminants. In fact, because molecular detections are highly sensitive, the detected invertebrate cells could even represent a very small fraction of the cell culture. Thus, it is highly recommended to also systematically use molecular probes for microbial taxa to detect potential culture contaminants (Table 3).

Culture contamination is a major obstacle to the development of aquatic invertebrate in vitro models. Indeed, it is widely acknowledged that microbes persistently take over the cultured aquatic invertebrate cell types (Rinkevich 2005), putatively as a result of antagonistic interactions (predation and competition for nutrients) or metabolic plasticity and better adaptation to the in vitro growth conditions. Culture media are commercially designed for vertebrates (e.g., DMEM, Leibovitz L-15) or insects (e.g., Grace Insect Medium) and partly diluted in seawater (or freshwater) or formula custom-prepared to mimic the microenvironment of the sampled tissue, and they are nutrient-rich. Although they may not adequately meet the largely unknown growth requirements of the cultured invertebrate cell types, they provide abundant organic carbon and nitrogen sources that facilitate the overgrowth of opportunistic resident microbial associates. Indeed, epibiotic or endobiotic microbiota (especially unicellular microeukaryotes that are hard to discriminate from animal cells) have repeatedly been shown to take advantage of the medium-derived nutritional resources to fuel their fast heterotrophic growth; see, for example, the consumption of mono and disaccharides, glycerol, glutamate and glycine by the opportunistic unicellular Alveolate *Chromera velia* (Foster et al. 2014). Predatory opisthokonts, ubiquitous in aquatic environments, have a highly plastic morphology, with in vitro growth alternating between a unicellular 'spindle-shape' stage and aggregative or clonal (partly fused) multicellular stages, and they are known to feed on metazoan tissue or derived cells (Tikhonenkov et al. 2020a).

Such eukaryotic microbes, collectively defined as protists, may feed on cellular debris from senescent or dead host/aquatic invertebrate cells, taking over the initial host cell population in long-term primary cultures or their successive subcultures.

5. Perspectives

While marine invertebrates as a whole show the largest biodiversity and the widest phylogenetic radiation on Earth, they have contributed very little to the in vitro cell lines discipline. The culture of marine invertebrate stem cells and/or their progenitors could thus create new perspectives for fundamental research as well as for biomedical applications. To reach this objective, we thus recommend two main actions.

First, a systematic map of knowledge, built in the form of a database of publications with metacoded information on taxon population, intervention strategies (e.g., cell isolation methods, culture media and physico-chemical conditions) and outcomes (e.g., cell viability, proliferation and differentiation) would be an important tool for increasing the visibility of protocols and know-how in the fragmented scientific community. Furthermore, it would help to incorporate typically unpublished results, including negative results, a scientific status that is rarely highlighted in the refereed literature (Grasela et al. 2012). Such database would help to build a comprehensive knowledge map to identify optimized culturing conditions for each aquatic invertebrate taxon and cell type, adapted to the expected timescale of utilization. The multiple usage of primary cell and tissue cultures from aquatic invertebrates ranges from short-term use (within hours to <7 days for physiology and cytotoxicity testing) to the long-term (bi-weekly to monthly) selection of subpopulations of dividing cells for serial sub-culturing attempts. Each type of inoculum implies distinct culture media and condition strategies to balance cellular yield, functional stability and proliferation potential. The curated list of 511 relevant publications compiled in Appendix B Table A2 provides a start to this database, allowing us to assess by taxon the extent of research efforts to initiate or develop cell cultures. It should be maintained and completed by the scientific community, for more exhaustive listing and optimized visibility.

Second, best practices would be to develop and adopt robust cell-type authentication protocols applicable to insect or vertebrate cell lines and primary cultures (Lynn 2001; Dominici et al. 2006), and to systematically deposit live or cryopreserved vouchers of "cell lines" in cell repositories. This could lead to the identification of more general stem cell markers for aquatic invertebrates, which would be crucial for obtaining a robust inoculum for in vitro cell cultures. Most recurrent past claims of successfully established aquatic invertebrate cell lines have turned out *in fine* to be cultures overgrown by microeukaryote contaminants, with examples in each taxon (porifera, colonial cnidarians, crustaceans and others).

Overall, these two actions taken together could help to standardize aquatic invertebrate cell culture to facilitates comparisons between intervention protocols and thus help to optimize the standardized protocols. Given that aquatic invertebrates are phylogenetically very distant, the development of a ubiquitous culturing environment appears rather unlikely. Nevertheless, each phylum could benefit from the scientific and technological advances in primary cell culture made in other phyla.

In particular, the assessment of whether the list of three identification criteria, defined for vertebrate stem cells, are conserved in aquatic invertebrate stem cells, would be of particular interest. The first criterion is whether the stem cells adhere to plastic, and more generally if a specific culture method, such as 3D Matrigel, could lead to decisive improvements (Urban-Gedamke et al. 2021). The second criterion explores the expression of specific surface markers that would allow the robust isolation and enrichment of stem cells/progenitors, as has been attempted by using a single marker in a colonial tunicate (Kassmer et al. 2020). The third criterion aims to define protocols for assessing stem cells' potency differentiation potential, typically by using predefined induction cocktails combined with markers for differentiated cell types, both of which require a precise characterization of gene expression profiles specific to each cell-type for every species of interest (Sebé-Pedrós et al. 2018).

An alternative to identifying suitable stem cells is to immortalize cells of interest in a reproducible manner. One suggested approach is to manipulate adult stem cells of aquatic invertebrates similarly to the approach implemented in mammalian induced pluripotent stem (iPS) cells (Rinkevich 2011). The second route, probably the most promising and reliable approach, is to control the process of tumorigenesis in aquatic invertebrates, as already suggested (Odintsova et al. 2011). Research on this topic is currently very scarce (Gardner 1993; Robert 2010) primarily due to the facts that tumorigenesis in aquatic invertebrates is not as commonly observed as in vertebrates (Vogt 2008; Tascedda and Ottaviani 2014), that tumor-like lesions in aquatic invertebrates possess a low mitotic index (Odintsova et al. 2011) and that the definitions of tumors and tumor cells in aquatic invertebrates are less familiar to pathologists (Tascedda and Ottaviani 2014). Yet, the tool of tumorigenesis may constitute a very important route for future research, and a potential approach is to use the trait of the vertebrates' cancer cells (Vincent 2012) as a guiding list for tumors in aquatic invertebrates. A third concept proposes the use of regeneration processes as the source of tumor development (Oviedo and Beane 2009), which is particularly interesting given the broad involvement of aquatic invertebrates' stem cells in regeneration processes, including whole-body regeneration (Rinkevich et al. 2022). For each one of these three approaches, the development of suitable tools for the controlled editing of genetic material of cells, typically through viral transfection, could enable the knockdown of suppressor genes, similar to standard approaches in mammalian cells (Yang et al. 2007). One such advance is the successful induced stem cell neoplasia in the marine hydrozoan *Hydractinia echinata* by the ectopic expression of a POU domain transcription factor (Millane et al. 2011). However, even immortalized tumor-like cells will need appropriate culturing conditions to proliferate properly. Lessons may be drawn from the failure to sustain in vitro the neoplastic hemocyte proliferation observed in vivo in spontaneously occurring clam leukemia, with research pointing to a role for the stress protein mortalin in the induction of apoptosis in cancerous hemocytes (Walker et al. 2013). The RNA-seq approach may be applied to compare gene expression patterns in cultured cells and initial tissue, with a focus on essential cell proliferation and cell cycle arrest regulator genes, in order to develop future strategies for immortalization, as recently explored for developing shrimp cell lines (Thammasorn et al. 2020).

As a supplementary approach to support the development of cell lines from aquatic invertebrate stem cells, studies on metabolomes of cultured cells, and their secretomes in particular, could be considered. Such an approach may provide important insights into the requirements in media composition that support proliferative activities. Ample information has been gained on this issue in mammalian cell cultures (Čuperlović-Culf et al. 2010; Mohmad-Saberi et al. 2013). Yet, the study of the secretome of aquatic invertebrates has seldomly been undertaken (Kocot et al. 2016), but data on the metabolome of whole organisms in the context of marine natural product discoveries are becoming quite common (Reverter et al. 2020). Furthermore, high-precision tool development specific to seawater are now available (Sogin et al. 2019).

Finally, future research should also address the still largely overlooked abiotic factors, such as testing hypoxia and pressure stimuli, on primary cultures.

Regarding hypoxia, parallel research in cultured mammalian cell models has highlighted the better survival and proliferation of stem cells in low oxygen environments (Zhu et al. 2005; Hung et al. 2012; Ramirez et al. 2011). A shift from oxidative phosphorylation to aerobic glycolysis, known as the Warburg effect, has been documented in the context of proliferating cancer cells: the glucose consumed in high amounts to fuel the growing biomass of cancer cells is fermented to lactate rather than oxidized, even when there is sufficient oxygen to convert glucose to CO_2, although the process is less efficient in terms of ATP synthesis (reviewed by DeBerardinis and Chandel 2020). Hyperactive glycolysis involving lactate supports the tumor energy metabolism of cancer stem cells in mostly hypoxic environments, and similar pathways might support the metabolism of aquatic invertebrate stem cells. A similar Warburg effect has indeed been documented in *Crassostrea gigas* oyster tissue. First discovered during the response to viral infection with ostreid herpesvirus-1 (Corporeau et al. 2014), it is thought to be a mechanism to adapt the oyster metabolism to extreme (salinity and oxygen) changes in the intertidal environment (Corporeau et al. 2019). In agreement with this finding, preliminary data obtained on oyster heart primary cell cultures showed transient increased proliferation between 2 and 4 weeks post-inoculation in a 2% O_2 atmosphere (obtained by incubation in a 95% N_2/5% air incubator), compared with 20% O_2 atmosphere (air) (Domart-Coulon, unpublished), when medium was supplemented with growth factors, lipids and antioxidants (Domart-Coulon et al. 1994).

Regarding pressure, research on the primary cultures of vertebrate (Wharton Jelly's) mesenchymal stem cells has shown the combined positive effects of pressure and hypoxia (Park et al. 2020). In response to pressure stimuli, cell proliferation was increased, and stemness was maintained. Cellular adhesion and confluency were higher in 5% O_2 hypoxia with 2.0 PSI pressure conditions relative to standard 5% CO_2–95% air conditions, and hypoxia alone yielded a mild increase in stem cell adhesion and confluency. Thus, we propose the inclusion of these abiotic parameters in future invertebrate stem cell culture optimization efforts.

Author Contributions: Conceptualization, I.D.-C. and S.B.; Resources, I.D.-C. and S.B.; Data Curation, I.D.-C. and S.B.; Writing—Original Draft Preparation, I.D.-C. and S.B.; Writing—Review and Editing, I.D.-C. and S.B.; Visualization, I.D.-C. and S.B. All authors have read and agreed to the published version of the manuscript.

Funding: I.D.-C. received funding from the French National Museum of Natural History (MNHN) and CNRS funds to MCAM laboratory (UMR 7245CNRS/MNHN). S.B. received funding from the Swiss National Science Foundation (SNF) (grant number PZ00P3_173981). Both I.D-C. and S.B. received funding from the MARISTEM Cost Action 16203 for collaborations.

Acknowledgments: We would like to thank Buki Rinkevich, Marta Wawrzyniak, Nathalie Weber and Aude Blanchoud for their help in compiling the data used for Tables 1 and 2. We also thank Buki Rinkevich and Loriano Ballarin for constructive comments on this manuscript.

Conflicts of Interest: The authors declare no conflict of interest.

Appendix A

Table A1. All the detailed values used for building Figure 1.

Taxon	Species	Regeneration	Longevity	Publications	Cell Lines
Hominidae	7	organ	122	16,851	96,862
Mammalia	5480	organ	211	7238	29,171
Vertebrata	56,508	appendage	392	884	1394
Tunicata	2760	WBR	-	44	0
Cephalochordata	33	organ	-	3	0
Ambulacraria	7111	WBR	200	42	0
Xenacoelomorpha	401	WBR	-	1	0
Insecta	1,015,897	appendage	28	351	895
Ecdysozoa	202,423	appendage	100	110	94
Nemertea	1200	WBR	-	0	0
Platyhelminthes	20,000	WBR	-	16	1
Mollusca	85,000	appendage	507	121	5
Spiralia	26,099	WBR	-	14	0
Cnidaria	9795	aggregates	4265	66	0
Placozoa	1	aggregates	-	0	0
Ctenophora	166	WBR	-	1	0
Porifera	6000	aggregates	15,000	58	0

Appendix B

Table A2. The full curated list of 511 references, sorted per taxa. (This table was not included in the print version of this book, to view the table please visit https://www.mdpi.com/books/pdfview/edition/5071).

References

Ageenko, Natalya V., Konstantin V. Kiselev, Pavel S. Dmitrenok, and Nelly A. Odintsova. 2014. Pigment cell diferentiation in sea urchin blastula-derived primary cell cultures. *Marine Drugs* 12: 3874–91. [CrossRef] [PubMed]

Alder, Hansjürg, and Volker Schmid. 1987. Cell Cycles and in Vitro Transdifferentiation and Regeneration of Isolated, Striated Muscle of Jellyfish. *Developmental Biology* 124: 358–69. [CrossRef]

Apte, Smita, Fares Khoury, Wera Roth, and Dietrich Sehlichter. 1996. Transport of Amino Acids into Freshly Isolated Cells from a Sea Anemone. *Endocytobiosis and Cell Research* 11: 129–46.

Asami, Maki, Tsuyoshi Nakatsuka, Tetsutaro Hayashi, Kenji Kou, Hiroaki Kagawa, and Kiyokazu Agata. 2002. Cultivation and Characterization of Planarian Neuronal Cells Isolated by Fluorescence Activated Cell Sorting (FACS). *Zoological Science* 19: 1257–65. [CrossRef] [PubMed]

Auzoux-Bordenave, Stéphanie, and Isabelle Domart-Coulon. 2010. Marine Invertebrate Cell Cultures as Tools for Biomineralization Studies. *Journal des Sciences Halieutiques et Aquatiques* 2: 42–47.

Bairoch, Amos. 2018. The Cellosaurus, a Cell-Line Knowledge Resource. *Journal of Biomolecular Techniques: JBT* 18: 2902–002. [CrossRef] [PubMed]

Barnay-Verdier, Stéphanie, Diane Dall'Osso, Nathalie Joli, Juliette Olivré, Fabrice Priouzeau, Thamilla Zamoum, Pierre-Laurent Merle, and Paola Furla. 2013. Establishment of Primary Cell Culture from the Temperate Symbiotic Cnidarian, *Anemonia Viridis*. *Cytotechnology* 65: 697–704. [CrossRef]

Bedi, Supinder S., Ali Salim, Shanping Chen, and David L. Glanzman. 1998. Long-Term Effects of Axotomy on Excitability and Growth of Isolated Aplysia Sensory Neurons in Cell Culture: Potential Role of CAMP. *Journal of Neurophysiology* 79: 1371–83. [CrossRef]

Bely, Alexandra E., and Kevin G. Nyberg. 2010. Evolution of Animal Regeneration: Re-Emergence of a Field. *Trends in Ecology & Evolution* 25: 161–70. [CrossRef]

Bilej, Martin, Ludmila Tučková, and Pavel Rossmann. 1994. A New Approach to in Vitro Studies of Antigenic Response in Earthworms. *Developmental & Comparative Immunology* 18: 363–67.

Bodnar, Andrea G. 2009. Marine Invertebrates as Models for Aging Research. *Experimental Gerontology* 44: 477–84. [CrossRef] [PubMed]

Bookman, Richard J., and Yuan Liu. 1990. Analysis of Calcium Channel Properties in Cultured Leech Retzius Cells by Internal Perfusion, Voltage-Clamp and Single-Channel Recording. *The Journal of Experimental Biology* 149: 223–37. [CrossRef] [PubMed]

Boulo, Viviane, Jean Paul Cadoret, Françoise Le Marrec, Germaine Dorange, and Eric Miahle. 1996. Transient expression of luciferase reporter gene after lipofection in oyster (*Crassostrea gigas*) primary cell cultures. *Molecular Marine Biology and Biotechnology* 5: 167–74. [PubMed]

Brody, Michael D., and Ernest S. Chang. 1989. Development and Utilization of Crustacean Long-Term Primary Cell Cultures: Ecdysteroid Effects in Vitro. *Invertebrate Reproduction & Development* 16: 141–47.

Burighel, Paolo, Riccardo Brunetti, and Giovanna Zaniolo. 1976. Hibernation of the Colonial Ascidian *Botrylloides Leachi* (Savigny): Histological Observations. *Italian Journal of Zoology* 43: 293–301.

Cai, Xiaoqing, Huamin Wang, Linxuan Huang, Juntao Chen, Qinfen Zhang, and Yan Zhang. 2013. Establishing Primary Cell Cultures from *Branchiostoma belcheri Japanese*. *In Vitro Cellular & Developmental Biology—Animal* 49: 97–102. [CrossRef]

Cai, Xiaoqing, and Yan Zhang. 2014. Marine Invertebrate Cell Culture: A Decade of Development. *Journal of Oceanography* 70: 405–14. [CrossRef]

Cao, Asunción, Luis Mercado, Juan Ignacio Ramos-Martinez, and Ramiro Barcia. 2003. Primary Cultures of Hemocytes from *Mytilus Galloprovincialis* Lmk.: Expression of IL-2Rα Subunit. *Aquaculture* 216: 1–8. [CrossRef]

Chapman, Arthur D. 2009. Numbers of Living Species in Australia and the World. Available online: http://citeseerx.ist.psu.edu/viewdoc/download?doi=10.1.1.231.2819&rep=rep1& type=pdf (accessed on 14 October 2021).

Chen, S. N., and C. S. Wang. 1999. Establishment of Cell Lines Derived from Oyster, Crassostrea Gigas Thunberg and Hard Clam, Meretrix Lusoria Röding. *Methods in Cell Science* 21: 183–92. [CrossRef]

Chernogor, Lubov I., Natalia N. Denikina, Sergey I. Belikov, and Alexander V. Ereskovsky. 2011. Long-Term Cultivation of Primmorphs from Freshwater Baikal Sponges Lubomirskia Baikalensis. *Marine Biotechnology* 13: 782–92. [CrossRef]

Claydon, Kerry, and Leigh Owens. 2008. Attempts at Immortalization of Crustacean Primary Cell Cultures Using Human Cancer Genes. *In Vitro Cellular & Developmental Biology-Animal* 44: 451–57.

Coleman, Annette W., Arturo Suarez, and Lynda J. Goff. 1994. Molecular Delineation of Species and Syngens in Volvocacean Green Algae (Chlorophyta) 1. *Journal of Phycology* 30: 80–90. [CrossRef]

Conkling, Megan, Kylie Hesp, Stephanie Munroe, Kenneth Sandoval, Dirk E. Martens, Detmer Sipkema, Rene H. Wijffels, and Shirley A. Pomponi. 2019. Breakthrough in Marine Invertebrate Cell Culture: Sponge Cells Divide Rapidly in Improved Nutrient Medium. *Scientific Reports* 9: 1–10. [CrossRef] [PubMed]

Corporeau, Charlotte, Arnaud Huvet, Vianney Pichereau, Lizenn Delisle, Claudie Quéré, Christine Dubreuil, Sebastien Artigaud, Catherine Brenner, Monique Meyenberg Cunha-De Padua, and Nathalie Mazure. 2019. The Oyster Crassostrea Gigas, a New Model against Cancer. *M S—Medecine Sciences* 35: 463–66.

Corporeau, Charlotte, David Tamayo, Fabrice Pernet, Claudie Quéré, and Stéphanie Madec. 2014. Proteomic Signatures of the Oyster Metabolic Response to Herpesvirus OsHV-1 Mvar Infection. *Journal of Proteomics* 109: 176–87. [CrossRef]

Čuperlović-Culf, Miroslava, David A. Barnett, Adrian S. Culf, and Ian Chute. 2010. Cell Culture Metabolomics: Applications and Future Directions. *Drug Discovery Today* 15: 610–21. [CrossRef]

Custodio, Marcio R., Ivo Prokic, Renate Steffen, Claudia Koziol, Radovan Borojevic, Franz Brümmer, Michael Nickel, and Werner E. G. Müller. 1998. Primmorphs Generated from Dissociated Cells of the Sponge *Suberites domuncula*: A Model System for Studies of Cell Proliferation and Cell Death. *Mechanisms of Ageing and Development* 105: 45–59. [CrossRef]

Daugavet, Mariya A., and Miralda I. Blinova. 2015. Culture of Mussel (*Mytiuls edulis* L.) Mantle Cells. *Cell and Tissue Biology* 9: 233–43. [CrossRef]

De Mulder, Katrien, Georg Kuales, Daniela Pfister, Maxime Willems, Bernhard Egger, Willi Salvenmoser, Marlene Thaler, Anne-Kathrin Gorny, Martina Hrouda, Gaëtan Borgonie, and et al. 2009. Characterization of the Stem Cell System of the Acoel Isodiametra Pulchra. *BMC Developmental Biology* 9: 69. [CrossRef]

De Rosa, Salvatore, Salvatore De Caro, Giuseppina Tommonaro, Krasimir Slantchev, Kamen Stefanov, and Simeon Popov. 2001. Development in a Primary Cell Culture of the Marine Sponge *Ircinia muscarum* and Analysis of the Polar Compounds. *Marine Biotechnology* 3: 281–86. [CrossRef]

DeBerardinis, Ralph J., and Navdeep S. Chandel. 2020. We Need to Talk about the Warburg Effect. *Nature Metabolism* 2: 127–29. [CrossRef] [PubMed]

Dessai, Shanti Nilesh. 2012. Primary Culture of Mantle Cells of Bivalve Mollusc, Paphia Malabarica. *In Vitro Cellular & Developmental Biology. Animal* 48: 473–77. [CrossRef]

Domart-Coulon, Isabelle J., David C. Elbert, Erik P. Scully, Precilia S. Calimlim, and Gary K. Ostrander. 2001. Aragonite Crystallization in Primary Cell Cultures of Multicellular Isolates from a Hard Coral, *Pocillopora damicornis*. *Proceedings of the National Academy of Sciences of the United States of America* 98: 11885–90. [CrossRef]

Domart-Coulon, I. J., C. S. Sinclair, R. T. Hill, S. Tambutté, S. Puverel, and G. K. Ostrander. 2004. A Basidiomycete Isolated from the Skeleton of *Pocillopora damicornis* (Scleractinia) Selectively Stimulates Short-Term Survival of Coral Skeletogenic Cells. *Marine Biology* 144: 583–92. [CrossRef]

Domart-Coulon, Isabelle, Dominique Doumenc, Stephanie Auzoux-Bordenave, and Yann Le Fichant. 1994. Identification of Media Supplements That Improve the Viability of Primarily Cell Cultures of Crassostrea Gigas Oysters. *Cytotechnology* 16: 109–20. [CrossRef]

Dominici, M., K. Le Blanc, I. Mueller, I. Slaper-Cortenbach, F. C. Marini, D. S. Krause, R. J. Deans, A. Keating, D. J. Prockop, and E. M. Horwitz. 2006. Minimal Criteria for Defining Multipotent Mesenchymal Stromal Cells. The International Society for Cellular Therapy Position Statement. *Cytotherapy* 8: 315–17. [CrossRef]

Downs, Craig A., John E. Fauth, Virgil D. Downs, and Gary K. Ostrander. 2010. In Vitro Cell-Toxicity Screening as an Alternative Animal Model for Coral Toxicology: Effects of Heat Stress, Sulfide, Rotenone, Cyanide, and Cuprous Oxide on Cell Viability and Mitochondrial Function. *Ecotoxicology* 19: 171–84. [CrossRef]

Dyachuk, Vyacheslav A., Maria A. Maiorova, and Nelly A. Odintsova. 2015. Identification of β Integrin-like-and Fibronectin-like Proteins in the Bivalve Mollusk Mytilus Trossulus. *Development, Growth & Differentiation* 57: 515–28.

Earle, Wilton R., Edward L. Schilling, Thomas H. Stark, Nancy P. Straus, Mary F. Brown, and Emma Shelton. 1943. Production of Malignancy in Vitro. IV. The Mouse Fibroblast Cultures and Changes Seen in the Living Cells. *JNCI: Journal of the National Cancer Institute* 4: 165–212. [CrossRef]

Echalier, Guy. 1997. *Drosophila Cells in Culture*. Cambridge: Academic Press.

Ermakov, Artem M., Olga N. Ermakova, Andrei A. Kudravtsev, and Natalia D. Kreshchenko. 2012. Study of Planarian Stem Cell Proliferation by Means of Flow Cytometry. *Molecular Biology Reports* 39: 3073–80. [CrossRef]

Fliegerová, Katerina, K. Hoffmann, J. Mrázek, and K. Voigt. 2008. The Design of Oligonucleotide Primers for the Universal Amplification of the N-Acetylglucosaminidase Gene (Nag1) in Chytridiomycetes with Emphasis on the Anaerobic Neocallimastigales. *Folia Microbiologica* 53: 209–13. [CrossRef] [PubMed]

Foster, Christie, Neil Portman, Min Chen, and Jan Šlapeta. 2014. Increased Growth and Pigment Content of Chromera Velia in Mixotrophic Culture. *FEMS Microbiology Ecology* 88: 121–28. [CrossRef] [PubMed]

Frank, U., C. Rabinowitz, and B. Rinkevich. 1994. In Vitro Establishment of Continuous Cell Cultures and Cell Lines from Ten Colonial Cnidarians. *Marine Biology* 120: 491–99. [CrossRef]

Fujita, Yoko, Kunio Oishi, and Ko Aida. 1972. Aggregation of Dissociated Sea Urchin and Sponge Cells by Culture Fluids of Microorganisms Having Hemagglutination Activity. *The Journal of General and Applied Microbiology* 18: 77–79. [CrossRef]

Garcia Camacho, T. Chileh, M. C. Cerón García, A. Sánchez Mirón, E. H. Belarbi, A. Contreras Gómez, and E. Molina Grima. 2006. Sustained Growth of Explants from Mediterranean Sponge *Crambe crambe* Cultured In Vitro with Enriched RPMI 1640. *Biotechnology Progress* 22: 781–90. [CrossRef]

Gardner, George R. 1993. Chemically Induced Histopathology in Aquatic Invertebrates. In *Pathobiology of Marine and Estuarine Organisms*. Boca Raton: CRC Press.

Gardner, S. G., D. A. Nielsen, K. Petrou, A. W. D. Larkum, and P. J. Ralph. 2015. Characterisation of Coral Explants: A Model Organism for Cnidarian–Dinoflagellate Studies. *Coral Reefs* 34: 133–42. [CrossRef]

Gardner, Stephanie G., Jean-Baptiste Raina, Peter J. Ralph, and Katherina Petrou. 2017. Reactive Oxygen Species (ROS) and Dimethylated Sulphur Compounds in Coral Explants under Acute Thermal Stress. *Journal of Experimental Biology* 220: 1787–91. [CrossRef]

Gates, R. D., and L. Muscatine. 1992. Three Methods for Isolating Viable Anthozoan Endoderm Cells with Their Intracellular Symbiotic Dinoflagellates. *Coral Reefs* 11: 143–45. [CrossRef]

Giard, Wilfrid, Jean Marc Lebel, Eve Boucaud-Camou, and Pascal Favrel. 1998. Effects of vertebrate growth factors on digestive gland cells from the mollusc *Pecten maximus* L.: An in vitro study. *Journal of Comparative Physiology B* 168: 81–86. [CrossRef]

Gierer, A., A. Berking, H. Bode, C. N. David, K. Flick, G. Hansmann, C. Schaller, and E. Trenkner. 1972. Regeneration of hydra from reaggregated cells. *Nature New Biology* 239: 98–101. [CrossRef]

Girard, Dorothée, Frédéric Torossian, Estelle Oberlin, Kylie A. Alexander, Jules Gueguen, Hsu-Wen Tseng, François Genêt, Jean-Jacques Lataillade, Marjorie Salga, and Jean-Pierre Levesque. 2021. Neurogenic Heterotopic Ossifications Recapitulate Hematopoietic Stem Cell Niche Development Within an Adult Osteogenic Muscle Environment. *Frontiers in Cell and Developmental Biology* 9: 499. [CrossRef] [PubMed]

Grace, T. D. C. 1962. Establishment of Four Strains of Cells from Insect Tissues Grown in Vitro. *Nature* 195: 788–89. [CrossRef] [PubMed]

Grasela, James J., Shirley A. Pomponi, Baruch Rinkevich, and Jennifer Grima. 2012. Efforts to develop a cultured sponge cell line: Revisiting an intractable problem. *In Vitro Cellular & Developmental Biology—Animal* 48: 12–20. [CrossRef]

Grimaldi, Annalisa, Serena Banfi, Laura Gerosa, Gianluca Tettamanti, Douglas M. Noonan, Roberto Valvassori, and Magda de Eguileor. 2009. Identification, Isolation and Expansion of Myoendothelial Cells Involved in Leech Muscle Regeneration. *PLoS ONE* 4: 7652. [CrossRef]

Halanych, Kenneth M. 2004. The New View of Animal Phylogeny. *Annual Review of Ecology, Evolution, and Systematics* 35: 229–56. [CrossRef]

Han, Qian, Pengtao Li, Xiongbin Lu, Zijuan Guo, and Huarong Guo. 2013. Improved Primary Cell Culture and Subculture of Lymphoid Organs of the Greasyback Shrimp *Metapenaeus ensis*. *Aquaculture* 10: 101–13. [CrossRef]

Hansen, Eder L. 1979. Initiating a Cell Line from Embryos of the Snail Biomphalaria Glabrata. *Tissue Culture Association Manual* 5: 1009–14. [CrossRef]

Hayashi, Tetsutaro, Maki Asami, Sayaka Higuchi, Norito Shibata, and Kiyokazu Agata. 2006. Isolation of Planarian X-Ray-Sensitive Stem Cells by Fluorescence-Activated Cell Sorting. *Development, Growth & Differentiation* 48: 371–80. [CrossRef]

Hayashi, Tetsutaro, Norito Shibata, Ryo Okumura, Tomomi Kudome, Osamu Nishimura, Hiroshi Tarui, and Kiyokazu Agata. 2010. Single-Cell Gene Profiling of Planarian Stem Cells Using Fluorescent Activated Cell Sorting and Its 'Index Sorting' Function for Stem Cell Research. *Development, Growth & Differentiation* 52: 131–44. [CrossRef]

Helman, Yael, Frank Natale, Robert M. Sherrell, Michèle LaVigne, Valentin Starovoytov, Maxim Y. Gorbunov, and Paul G. Falkowski. 2008. Extracellular matrix production and calcium carbonate precipitation by coral cells in vitro. *Proceedings of the National Academy of Sciences of the United States of America* 105: 54–58. [CrossRef]

Hemmrich, Georg, Konstantin Khalturin, Anna-Marei Boehm, Malte Puchert, Friederike Anton-Erxleben, Jörg Wittlieb, Ulrich C. Klostermeier, Philip Rosenstiel, Hans-Heinrich Oberg, Tomislav Domazet-Lošo, and et al. 2012. Molecular Signatures of the Three Stem Cell Lineages in Hydra and the Emergence of Stem Cell Function at the Base of Multicellularity. *Molecular Biology and Evolution* 29: 3267–80. [CrossRef] [PubMed]

Hetrick, Frank M., Edwards Stephens, Nancy Lomax, and Kathleen Lutrell. 1981. Attempts to Develop a Marine Molluscan Cell Line. In *Sea Grant College Program Technical Report UM-SG-TS-81-06*. College Park: University of Maryland.

Heude-Berthelin, Clotilde, Bruno Fievet, Gaël Leclerc, Pierre Germain, Kristell Kellner, and Michel Mathieu. 2003. In vivo and in vitro approaches to the analysis of glycogen metabolism in the Pacific oyster *Crassostrea gigas*. *Journal of Shellfish Research* 22: 715–20.

Hira, Jonathan, Deanna Wolfson, Aaron John Christian Andersen, Tor Haug, and Klara Stensvåg. 2020. Autofluorescence Mediated Red Spherulocyte Sorting Provides Insights into the Source of Spinochromes in Sea Urchins. *Scientific Reports* 10: 1–9. [CrossRef] [PubMed]

Hoshino, Kouichi, Kazunori Ohnishi, Wataru Yoshida, and Takao Shinozawa. 1991. Analysis of Ploidy in a Planarian by Flow Cytometry. *Hydrobiologia* 227: 175–78. [CrossRef]

Hung, Shun-Pei, Jennifer H. Ho, Yu-Ru V. Shih, Ting Lo, and Oscar K. Lee. 2012. Hypoxia Promotes Proliferation and Osteogenic Differentiation Potentials of Human Mesenchymal Stem Cells. *Journal of Orthopaedic Research* 30: 260–66. [CrossRef]

Hyams, Yosef, Guy Paz, Claudette Rabinowitz, and Baruch Rinkevich. 2017. Insights into the Unique Torpor of Botrylloides Leachi, a Colonial Urochordate. *Developmental Biology* 428: 101–17. [CrossRef]

Januszyk, Michael, Robert C. Rennert, Michael Sorkin, Zeshaan N. Maan, Lisa K. Wong, Alexander J. Whittam, Arnetha Whitmore, Dominik Duscher, and Geoffrey C. Gurtner. 2015. Evaluating the Effect of Cell Culture on Gene Expression in Primary Tissue Samples Using Microfluidic-Based Single Cell Transcriptional Analysis. *Microarrays* 4: 540–50. [CrossRef]

Juliano, Celina E., Adrian Reich, Na Liu, Jessica Götzfried, Mei Zhong, Selen Uman, Robert A. Reenan, Gary M. Wessel, Robert E. Steele, and Haifan Lin. 2014. PIWI Proteins and PIWI-Interacting RNAs Function in Hydra Somatic Stem Cells. *Proceedings of the National Academy of Sciences of the United States of America* 111: 337–42. [CrossRef]

Kaczmarek, L. K., M. Finbow, J. P. Revel, and F. Strumwasser. 1979. The Morphology and Coupling of Aplysia Bag Cells within the Abdominal Ganglion and in Cell Culture. *Journal of Neurobiology* 10: 535–50. [CrossRef]

Kaneko, Hiroyuki, Yukio Kawahara, and Marina Dan-Sohkawa. 1995. Primary Culture of Mesodermal and Endodermal Cells of the Starfish Embryo. *Zoological Science* 12: 551–58. [CrossRef]

Kassmer, Susannah H., Adam D. Langenbacher, and Anthony W. De Tomaso. 2020. Integrin-Alpha-6+ Candidate Stem Cells Are Responsible for Whole Body Regeneration in the Invertebrate Chordate *Botrylloides Diegensis*. *Nature Communications* 11: 4435. [CrossRef] [PubMed]

Kawamura, Kazuo, and Shigeki Fujiwara. 1995. Establishment of cell lines from multipotent epithelial sheet in the budding tunicate, *Polyandrocarpa misakiensis*. *Cell Structure and Function* 20: 97–106. [CrossRef]

Kawamura, Kaz, Koki Nishitsuji, Eiichi Shoguchi, Shigeki Fujiwara, and Noriyuki Satoh. 2021. Establishing Sustainable Cell Lines of a Coral, Acropora Tenuis. *Marine Biotechnology* 23: 373–88. [CrossRef] [PubMed]

Khalesi, Mohammad K. 2008. Cell Cultures from the Symbiotic Soft Coral Sinularia Flexibilis. *In Vitro Cellular & Developmental Biology. Animal* 44: 330–38. [CrossRef]

Kingsley, Roni J., Asenath M. Bernhardt, Karl M. Wilbur, and Norimitsu Watabe. 1987. Scleroblast Cultures from the Gorgonian Leptogorgia Virgulata (Lamarck) (Coelenterata: Gorgonacea). *In Vitro Cellular & Developmental Biology* 23: 297–302. [CrossRef]

Kocot, Kevin M., Felipe Aguilera, Carmel McDougall, Daniel J. Jackson, and Bernard M. Degnan. 2016. Sea Shell Diversity and Rapidly Evolving Secretomes: Insights into the Evolution of Biomineralization. *Frontiers in Zoology* 13: 23. [CrossRef]

Kopecky, Elizabeth J., and Gary K. Ostrander. 1999. Isolation and Primary Culture of Viable Multicellular Endothelial Isolates from Hard Corals. *In Vitro Cellular & Developmental Biology—Animal* 35: 616–24. [CrossRef]

Koziol, Claudia, Radovan Borojevic, Renate Steffen, and Werner E. G. Müller. 1998. Sponges (Porifera) Model Systems to Study the Shift from Immortal to Senescent Somatic Cells: The Telomerase Activity in Somatic Cells. *Mechanisms of Ageing and Development* 100: 107–20. [CrossRef]

Kupper, Frithjof C., Ingo Maier, Dieter G. Müller, Susan Loiseaux-de Goer, and Laure Guillou. 2006. Phylogenetic Affinities of Two Eukaryotic Pathogens of Marine Macroalgae, Eurychasma Dicksonii (Wright) Magnus and Chytridium Polysiphoniae Cohn. *Cryptogamie-Algologie* 27: 165–84.

Lavrov, Andrey I., and Igor A. Kosevich. 2016. Sponge Cell Reaggregation: Cellular Structure and Morphogenetic Potencies of Multicellular Aggregates. *Journal of Experimental Zoology Part A: Ecological Genetics and Physiology* 325: 158–77. [CrossRef]

Lee, Aih Cheun, Boris Decourt, and Daniel Suter. 2008. Neuronal Cell Cultures from Aplysia for High-Resolution Imaging of Growth Cones. *The Journal of Visualized Experiments* 20: 662. [CrossRef] [PubMed]

Lee, Lucy E. J., Sung-Jae E. Cho, Daylan T. Pritchard, and Petra W. C. Lee. 2021. Initiation of Cell Cultures from Pandalid Shrimp and Effects of Astaxanthin. *In Vitro Cellular & Developmental Biology-Animal* 57: S44.

Lei, Kai, Sean A. McKinney, Eric J. Ross, Heng-Chi Lee, and Alejandro Sánchez Alvarado. 2019. Cultured Pluripotent Planarian Stem Cells Retain Potency and Express Proteins from Exogenously Introduced mRNAs. *BioRxiv.* [CrossRef]

Levy, Shani, Anamaria Elek, Xavier Grau-Bové, Simón Menéndez-Bravo, Marta Iglesias, Amos Tanay, Tali Mass, and Arnau Sebé-Pedrós. 2021. A Stony Coral Cell Atlas Illuminates the Molecular and Cellular Basis of Coral Symbiosis, Calcification, and Immunity. *Cell* 184: 2973–87.e18. [CrossRef] [PubMed]

Lopez, Jose V., C. L. Peterson, Robin Willoughby, A. E. Wright, E. Enright, S. Zoladz, John K. Reed, and Shirley A. Pomponi. 2002. Characterization of Genetic Markers for in Vitro Cell Line Identification of the Marine Sponge Axinella Corrugata. *Journal of Heredity* 93: 27–36. [CrossRef] [PubMed]

Lynn, Dwight E. 2001. Novel Techniques to Establish New Insect Cell Lines. *In Vitro Cellular & Developmental Biology—Animal* 37: 319–21. [CrossRef]

Magalhães, João Pedro de, Joana Costa, and George M. Church. 2007. An Analysis of the Relationship Between Metabolism, Developmental Schedules, and Longevity Using Phylogenetic Independent Contrasts. *The Journals of Gerontology: Series A* 62: 149–60. [CrossRef]

Magoski, Neil S., Naweed I. Syed, and Andrew GM Bulloch. 1994. A Neuronal Network from the Mollusc Lymnaea Stagnalis. *Brain Research* 645: 201–14. [CrossRef]

Maramorosch, Karl, and Jun Mitsuhashi. 1997. *Invertebrate Cell Culture: Novel Directions and Biotechnology Applications*. Boca Raton: Science Pub Incorporated.

Martinez, Pedro, Loriano Ballarin, Alexander V. Ereskovsky, Eve Gazave, Bert Hobmayer, Lucia Manni, Eric Rottinger, Simon G. Sprecher, Stefano Tiozzo, Ana Varela-Coelho, and et al. 2022. Articulating the "stem cell niche" paradigm through the lens of non-model aquatic invertebrates. *BMC Biology* 20: 23. [CrossRef]

Maselli, Valeria, Fenglian Xu, Naweed I. Syed, Gianluca Polese, and Anna Di Cosmo. 2018. A Novel Approach to Primary Cell Culture for Neurons. *Frontiers in Physiology* 9: 220. [CrossRef]

Masuda-Nakagawa, L. M., A. Walz, D. Brodbeck, M. D. Neely, and S. Grumbacher-Reinert. 1994. Substrate-dependent Interactions of Leech Microglial Cells and Neurons in Culture. *Journal of Neurobiology* 25: 83–91. [CrossRef] [PubMed]

Matsumoto, Jun, Chiaki Nakamoto, Shigeki Fujiwara, Toshitsugu Yubisui, and Kazuo Kawamura. 2001. A Novel C-Type Lectin Regulating Cell Growth, Cell Adhesion and Cell Differentiation of the Multipotent Epithelium in Budding Tunicates. *Development* 128: 3339–47. [CrossRef] [PubMed]

Millane, R. Cathriona, Justyna Kanska, David J. Duffy, Cathal Seoighe, Stephen Cunningham, Günter Plickert, and Uri Frank. 2011. Induced stem cell neoplasia in a cnidarian by ectopic expression of a POU domain transcription factor. *Development* 138: 2429–39. [CrossRef] [PubMed]

Mohmad-Saberi, Salfarina Ezrina, Yumi Zuhanis Has-Yun Hashim, Maizirwan Mel, Azura Amid, Raha Ahmad-Raus, and Vasila Packeer-Mohamed. 2013. Metabolomics Profiling of Extracellular Metabolites in CHO-K1 Cells Cultured in Different Types of Growth Media. *Cytotechnology* 65: 577–86. [CrossRef]

Moiseeva, Elisabeth, Claudette Rabinowitz, Irena Yankelevich, and Baruch Rinkevich. 2004. "Cup Cell Disease" in the Colonial Tunicate *Botryllus schlosseri*. *Diseases of Aquatic Organisms* 60: 77–84. [CrossRef]

Monnier, Z., and M. Bride. 1995. In Vitro Effects of Methionine-Enkephalin, Somatostatin and Insulin on Cultured Gonadal Cells of the Snail *Helix aspersa*. *Experientia* 51: 824–30. [CrossRef]

Montgomery, Michelle, Maria C. Messner, and Mark D. Kirk. 2002. Arterial Cells and CNS Sheath Cells from Aplysia Californica Produce Factors That Enhance Neurite Outgrowth in Co-Cultured Neurons. *Invertebrate Neuroscience* 4: 141–55. [CrossRef]

Moore, Robert B., Katherine M. Ferguson, William K. W. Loh, Ove Hoegh-Guldberg, and Dee A. Carter. 2003. Highly Organized Structure in the Non-Coding Region of the PsbA Minicircle from Clade C Symbiodinium. *International Journal of Systematic and Evolutionary Microbiology* 53: 1725–34. [CrossRef]

Mothersill, Carmen, A. L. Mulford, and B. Austin. 2000. Basic Methods and Media. In *Aquatic Invertebrate Cell Culture*. Chichester: Springer & Praxis Publishing, pp. 9–14.

Müller, Werner E. G., Matthias Wiens, Renato Batel, Renate Steffen, Heinz C. Schröder, Radovan Borojevic, and Marcio Reis Custodio. 1999. Establishment of a Primary Cell Culture from a Sponge: Primmorphs from Suberites Domuncula. *Marine Ecology Progress Series* 178: 205–19. [CrossRef]

Munroe, Stephanie, Kenneth Sandoval, Dirk E. Martens, Detmer Sipkema, and Shirley A. Pomponi. 2019. Genetic Algorithm as an Optimization Tool for the Development of Sponge Cell Culture Media. *In Vitro Cellular & Developmental Biology Animal* 55: 149–58. [CrossRef]

Nesa, Badrun, and Michio Hidaka. 2009. High Zooxanthella Density Shortens the Survival Time of Coral Cell Aggregates under Thermal Stress. *Journal of Experimental Marine Biology and Ecology* 368: 81–87. [CrossRef]

Odintsova, Nelly A., N. I. Belogortseva, A. V. Ermak, V. I. Molchanova, and P. A. Luk'yanov. 1999. Adhesive and Growth Properties of Lectin from the Ascidian Didemnum Ternatanum on Cultivated Marine Invertebrate Cells. *Biochimica et Biophysica Acta (BBA)-Molecular Cell Research* 1448: 381–89. [CrossRef]

Odintsova, Nelly A., I. Yu Dolmatov, and V. S. Mashanov. 2005. Regenerating Holothurian Tissues as a Source of Cells for Long-Term Cell Cultures. *Marine Biology* 146: 915–21. [CrossRef]

Odintsova, Nelly A., Ludmila N. Usheva, Konstantin V. Yakovlev, and Konstantin V. Kiselev. 2011. Naturally Occurring and Artificially Induced Tumor-like Formations in Marine Invertebrates: A Search for Permanent Cell Lines. *Journal of Experimental Marine Biology and Ecology* 407: 241–49. [CrossRef]

Odintsova, Nelly A., Natalya V. Ageenko, Yulia O. Kipryushina, Mariia A. Maiorova, and Andrey V. Boroda. 2015. Freezing Tolerance of Sea Urchin Embryonic Cells: Differentiation Commitment and Cytoskeletal Disturbances in Culture. *Cryobiology* 71: 54–63. [CrossRef] [PubMed]

Ootani, Akifumi, Xingnan Li, Eugenio Sangiorgi, Quoc T. Ho, Hiroo Ueno, Shuji Toda, Hajime Sugihara, Kazuma Fujimoto, Irving L. Weissman, Mario R. Capecchi, and et al. 2009. Sustained in vitro intestinal epithelial culture within a Wnt-dependent stem cell niche. *Nature Medicine* 15: 701–6. [CrossRef] [PubMed]

Oppenheimer, Steven B., and James T. Meyer. 1982. Isolation of Species-Specific and Stage-Specific Adhesion Promoting Component by Disaggregation of Intact Sea Urchin Embryo Cells. *Experimental Cell Research* 137: 472–76. [CrossRef]

Otero, Jorge, Anna Ulldemolins, Ramon Farré, and Isaac Almendros. 2021. Oxygen Biosensors and Control in 3D Physiomimetic Experimental Models. *Antioxidants* 10: 1165. [CrossRef]

Oviedo, Néstor J., and Wendy S. Beane. 2009. Regeneration: The origin of cancer or a possible cure? *Seminars in Cell & Developmental Biology* 20: 557–64. [CrossRef]

Park, Sang Eon, Hyeongseop Kim, Soojin Kwon, Suk-joo Choi, Soo-young Oh, Gyu Ha Ryu, Hong Bae Jeon, and Jong Wook Chang. 2020. Pressure Stimuli Improve the Proliferation of Wharton's Jelly-Derived Mesenchymal Stem Cells under Hypoxic Culture Conditions. *International Journal of Molecular Sciences* 21: 7092. [CrossRef]

Peters, Esther C. 2006. Appendix 1: Invertebrate Neoplasms. In *Invertebrate Medicine*. Hoboken: John Wiley & Sons, Ltd., pp. 297–99. [CrossRef]

Pichon, Delphine, Isabelle Domart-Coulon, and Renata Boucher-Rodoni. 2007. Cephalopod Bacterial Associations: Characterization and Isolation of the Symbiotic Complex in the Accessory Nidamental Glands. *Bollettino Malacologico* 43: 96.

Pomponi, Shirley A., Allison Jevitt, Jignasa Patel, and M. Cristina Diaz. 2013. Sponge Hybridomas: Applications and Implications. *Integrative and Comparative Biology* 53: 524–30. [CrossRef] [PubMed]

Prinz, Astrid A., and Peter Fromherz. 2000. Electrical Synapses by Guided Growth of Cultured Neurons from the Snail Lymnaea Stagnalis. *Biological Cybernetics* 82: L1–L5. [CrossRef] [PubMed]

Przysiezniak, Jan, and Andrew N. Spencer. 1989. Primary Culture of Identified Neurones from a Cnidarian. *Journal of Experimental Biology* 142: 97–113. [CrossRef]

Puverel, Sandrine, Eric Tambutte, Didier Zoccola, Isabelle Domart-Coulon, Andre Bouchot, Severine Lotto, Denis Allemand, and Sylvie Tambutte. 2005. Antibodies against the Organic Matrix in Scleractinians: A New Tool to Study Coral Biomineralization. *Coral Reefs* 24: 149–56. [CrossRef]

Rabinowitz, Claudette, and Baruch Rinkevich. 2004. In Vitro Delayed Senescence of Extirpated Buds from Zooids of the Colonial Tunicate Botryllus *schlosseri*. *Journal of Experimental Biology* 207: 1523–32. [CrossRef]

Rabinowitz, Claudette, and Baruch Rinkevich. 2011. De Novo Emerged Stemness Signatures in Epithelial Monolayers Developed from Extirpated Palleal Buds. *In Vitro Cellular & Developmental Biology —Animal* 47: 26–31. [CrossRef]

Rabinowitz, Claudette, Gilad Alfassi, and Baruch Rinkevich. 2009. Further Portrayal of Epithelial Monolayers Emergent de Novo from Extirpated Ascidians Palleal Buds. *In Vitro Cellular & Developmental Biology—Animal* 45: 334–42. [CrossRef]

Rabinowitz, Claudette, Elisabeth Moiseeva, and Baruch Rinkevich. 2016. In Vitro Cultures of Ectodermal Monolayers from the Model Sea Anemone Nematostella *vectensis*. *Cell and Tissue Research* 366: 693–705. [CrossRef]

Raftos, David A., and Edwin L. Cooper. 1991. Proliferation of Lymphocyte-like Cells from the Solitary Tunicate, Styela *clava*, in Response to Allogeneic Stimuli. *Journal of Experimental Zoology* 260: 391–400. [CrossRef]

Ramirez, M. A., E. Pericuesta, M. Yáñez-Mó, A. Palasz, and A. Gutiérrez-Adán. 2011. Effect of Long-term Culture of Mouse Embryonic Stem Cells under Low Oxygen Concentration as Well as on Glycosaminoglycan Hyaluronan on Cell Proliferation and Differentiation. *Cell Proliferation* 44: 75–85. [CrossRef] [PubMed]

Ren, Yuan, Yingpei He, Sherlene Brown, Erica Zbornik, Michael J. Mlodzianoski, Donghan Ma, Fang Huang, Seema Mattoo, and Daniel M. Suter. 2019. A Single Tyrosine Phosphorylation Site in Cortactin Is Important for Filopodia Formation in Neuronal Growth Cones. *Molecular Biology of the Cell* 30: 1817–33. [CrossRef] [PubMed]

Reverter, Miriam, Sven Rohde, Christelle Parchemin, Nathalie Tapissier-Bontemps, and Peter J. Schupp. 2020. Metabolomics and Marine Biotechnology: Coupling Metabolite Profiling and Organism Biology for the Discovery of New Compounds. *Frontiers in Marine Science* 7. [CrossRef]

Revilla-I-Domingo, Roger, Clara Schmidt, Clara Zifko, and Florian Raible. 2018. Establishment of Transgenesis in the Demosponge *Suberites domuncula*. *Genetics* 210: 435–43. [CrossRef] [PubMed]

Reyes-Bermudez, A., and D. J. Miller. 2009. In Vitro Culture of Cells Derived from Larvae of the Staghorn Coral Acropora Millepora. *Coral Reefs* 28: 859. [CrossRef]

Reyes-Bermudez, Alejandro, Michio Hidaka, and Alexander Mikheyev. 2021. Transcription Profiling of Cultured Acropora Digitifera Adult Cells Reveals the Existence of Ancestral Genome Regulatory Modules Underlying Pluripotency and Cell Differentiation in Cnidaria. *Genome Biology and Evolution* 13: evab008. [CrossRef]

Ribeiro, Raphaela Cantarino, Alexandra Caroline da Silva Veronez, Thaís Tristão Tovar, Serean Adams, Dayse Aline Bartolomeu, Clayton Peronico, and Tatiana Heid Furley. 2018. Cryopreservation: Extending the Viability of Biological Material from Sea Urchin (Echinometra *lucunter*) in Ecotoxicity Tests. *Cryobiology* 80: 139–43. [CrossRef]

Rinkevich, Baruch. 2005. Marine Invertebrate Cell Cultures: New Millennium Trends. *Marine Biotechnology* 7: 429–39. [CrossRef]

Rinkevich, Baruch. 2011. Cell Cultures from Marine Invertebrates: New Insights for Capturing Endless Stemness. *Marine Biotechnology* 13: 345–54. [CrossRef]

Rinkevich, Baruch, and Claudette Rabinowitz. 1993. In Vitro Culture of Blood Cells from the Colonial Protochordate *Botryllus schlosseri*. *In Vitro Cellular & Developmental Biology—Animal* 29: 79–85. [CrossRef]

Rinkevich, Baruch, and Claudette Rabinowitz. 1994. Acquiring embryo-derived cell cultures and aseptic metamorphosis of larvae from the colonial proto-chordate Botryllus schlosseri. *Invertebrate Reproduction & Development* 25: 59–72. [CrossRef]

Rinkevich, Baruch, and Claudette Rabinowitz. 1997. Initiation of Epithelial Cell Cultures from Palleal Buds of *Botryllus schlosseri*, a Colonial Tunicate. *In Vitro Cellular & Developmental Biology—Animal* 33: 422–24. [CrossRef]

Rinkevich, Yuval, Amalia Rosner, Claudette Rabinowitz, Ziva Lapidot, Elithabeth Moiseeva, and Buki Rinkevich. 2010. *Piwi* Positive Cells That Line the Vasculature Epithelium, Underlie Whole Body Regeneration in a Basal Chordate. *Developmental Biology* 345: 94–104. [CrossRef]

Rinkevich, B., L. Ballarin, P. Martinez, I. Somorjai, O. Ben-Hamo, I. Borisenko, E. Berezikov, A. Ereskovsky, E. Gazave, D. Khnykin, and et al. 2022. A pan-metazoan concept for adult stem cells: The wobbling Penrose landscape. *Biological Reviews* 97: 299–325. [CrossRef]

Robert, Jacques. 2010. Comparative study of tumorigenesis and tumor immunity in invertebrates and nonmammalian vertebrates. *Developmental & Comparative Immunology* 34: 915–25. [CrossRef]

Robinson, Jeffrey M. 2015. MicroRNA Expression during Demosponge Dissociation, Reaggregation, and Differentiation and a Evolutionarily Conserved Demosponge MiRNA Expression Profile. *Development Genes and Evolution* 225: 341–51. [CrossRef]

Rosental, Benyamin, Zhanna Kozhekbaeva, Nathaniel Fernhoff, Jonathan M. Tsai, and Nikki Traylor-Knowles. 2017. Coral Cell Separation and Isolation by Fluorescence-Activated Cell Sorting (FACS). *BMC Cell Biology* 18: 30. [CrossRef]

Rosental, Benyamin, Mark Kowarsky, Jun Seita, Daniel M. Corey, Katherine J. Ishizuka, Karla J. Palmeri, Shih-Yu Chen, Rahul Sinha, Jennifer Okamoto, Gary Mantalas, and et al. 2018. Complex Mammalian-like Haematopoietic System Found in a Colonial Chordate. *Nature* 564: 425–29. [CrossRef]

Rosner, Amalia, Jean Armengaud, Loriano Ballarin, Stéphanie Barnay-Verdier, Francesca Cima, Ana Varel Coelhoe, Isabell Domart-Coulon, Damjana Drobne, Anne-Marie Genevière, Anita Jemec Kokalj, and et al. 2021. Stem cells of aquatic invertebrates as an advanced tool for assessing ecotoxicological impacts. *Science of The Total Environment* 771: 144565. [CrossRef]

Russell, W. C., F. L. Graham, J. Smiley, and R. Nairn. 1977. Characteristics of a Human Cell Line Transformed by DNA from Human Adenovirus Type 5. *Journal of General Virology* 36: 59–72. [CrossRef]

Sakai, H. 1960. Studies on Sulfhydryl Groups during Cell Division of Sea Urchin Egg: II. Mass Isolation of the Egg Cortex and Change in Its—SH Groups during Cell Division. *The Journal of Cell Biology* 8: 603–7. [CrossRef] [PubMed]

Schacher, S., and E. Proshansky. 1983. Neurite Regeneration by Aplysia Neurons in Dissociated Cell Culture: Modulation by Aplysia Hemolymph and the Presence of the Initial Axonal Segment. *Journal of Neuroscience* 3: 2403–13. [CrossRef] [PubMed]

Scherer, William F. 1953. Studies on the Propagation in Vitro of Poliomyelitis Viruses: Iv. Viral Multiplication in a Stable Strain of Human Malignant Epithelial Cells (Strain Hela) Derived from an Epidermoid Carcinoma of the Cervix. *Journal of Experimental Medicine* 97: 695–710. [CrossRef]

Schippers, Klaske J., Dirk E. Martens, Shirley A. Pomponi, and René H. Wijffels. 2011. Cell Cycle Analysis of Primary Sponge Cell Cultures. *In Vitro Cellular & Developmental Biology-Animal* 47: 302–11. [CrossRef]

Schmid, Volker. 1992. Transdifferentiation in Medusae. *International Review of Cytology* 142: 213–61.

Schmid, Volker, and Hansjürg Alder. 1984. Isolated, Mononucleated, Striated Muscle Can Undergo Pluripotent Transdifferentiation and Form a Complex Regenerate. *Cell* 38: 801–9. [CrossRef]

Schmid, Volker, and Susanne Reber-Müller. 1995. Transdifferentiation of Isolated Striated Muscle of Jellyfish in vitro: The Initiation Process. In *Seminars in Cell Biology*. Amsterdam: Elsevier, pp. 109–16.

Schmid, Volker, Shin-Ichi Ono, and Susanne Reber-Müller. 1999. Cell-Substrate Interactions in Cnidaria. *Microscopy Research and Technique* 44: 254–68. [CrossRef]

Schürmann, Wolfgang, and Peter Roland. 2001. Planarian Cell Culture: A Comparative Review of Methods and an Improved Protocol for Primary Cultures of Neoblasts. *Belgian Journal of Zoology* 131: 123–30.

Sebé-Pedrós, Arnau, Baptiste Saudemont, Elad Chomsky, Flora Plessier, Marie-Pierre Mailhé, Justine Renno, Yann Loe-Mie, Aviezer Lifshitz, Zohar Mukamel, and Sandrine Schmutz. 2018. Cnidarian Cell Type Diversity and Regulation Revealed by Whole-Organism Single-Cell RNA-Seq. *Cell* 173: 1520–34.e20. [CrossRef]

Shinzato, Chuya, Yuki Yasuoka, Sutada Mungpakdee, Nana Arakaki, Manabu Fujie, Yuichi Nakajima, and Nori Satoh. 2014. Development of Novel, Cross-Species Microsatellite Markers for Acropora Corals Using next-Generation Sequencing Technology. *Frontiers in Marine Science* 1: 11. [CrossRef]

Siboni, Nachshon, Diana Rasoulouniriana, Eitan Ben-Dov, Esti Kramarsky-Winter, Alex Sivan, Yossi Loya, Ove Hoegh-Guldberg, and Ariel Kushmaro. 2010. Stramenopile Microorganisms Associated with the Massive Coral Favia Sp. *Journal of Eukaryotic Microbiology* 57: 236–44.

Siebert, Stefan, Jeffrey A. Farrell, Jack F. Cazet, Yashodara Abeykoon, Abby S. Primack, Christine E. Schnitzler, and Celina E. Juliano. 2019. Stem Cell Differentiation Trajectories in Hydra Resolved at Single-Cell Resolution. *Science* 365: 6451. [CrossRef] [PubMed]

Simpson, Tracy L. 1984. *The Cell Biology of Sponges*. New York: Springer. [CrossRef]

Slack, Jonathan M. W. 2017. Animal Regeneration: Ancestral Character or Evolutionary Novelty? *EMBO Reports* 18: 1497–508. [CrossRef] [PubMed]

Snyder, Grace A., Shir Eliachar, Michael T. Connelly, Shani Talice, Uzi Hadad, Orly Gershoni-Yahalom, William E. Browne, Caroline V. Palmer, Benyamin Rosental, and Nikki Traylor-Knowles. 2021. Functional Characterization of Hexacorallia Phagocytic Cells. *Frontiers in Immunology* 12: 662803. [CrossRef]

Sogin, Emilia M., Erik Puskás, Nicole Dubilier, and Manuel Liebeke. 2019. Marine Metabolomics: A Method for Nontargeted Measurement of Metabolites in Seawater by Gas Chromatography–Mass Spectrometry. *MSystems* 4. [CrossRef]

Song, Yuefan, Yi Qu, Xupeng Cao, Wei Zhang, Fuming Zhang, Robert J. Linhardt, and Qi Yang. 2021. Cultivation of Fractionated Cells from a Bioactive-Alkaloid-Bearing Marine Sponge Axinella Sp. *In Vitro Cellular & Developmental Biology—Animal* 57: 539–49. [CrossRef]

Stephens, Edward B., and Frank M. Hetrick. 1979. Decontamination of the American Oyster Tissues for Cell and Organ Culture. *TCA Manual/Tissue Culture Association* 5: 1029–31. [CrossRef]

Suja, C. P., N. Sukumaran, S. Dharmaraj, and Anu Meena. 2014. Effect of Depuration of Animals and Use of Antimicrobial Agents on Proliferation of Cells and Microbial Contamination in In-Vitro Mantle Explant Culture of the Abalone Haliotis Varia Linnaeus. *Indian Journal of Fisheries* 61: 93–98.

Sun, Liming, Yuefan Song, Yi Qu, Xingju Yu, and Wei Zhang. 2007. Purification and in Vitro Cultivation of Archaeocytes (Stem Cells) of the Marine Sponge Hymeniacidon Perleve (Demospongiae). *Cell and Tissue Research* 328: 223–37. [CrossRef]

Suter, Daniel M. 2011. Live Cell Imaging of Neuronal Growth Cone Motility and Guidance in Vitro. In *Cell Migration*. Berlin and Heidelberg: Springer, pp. 65–86.

Tascedda, Fabio, and Enzo Ottaviani. 2014. Tumors in Invertebrates. *Invertebrate Survival Journal* 11: 197–203.

Thammasorn, Thitiporn, Nozaki Reiko, Kondo Hidehiro, and Hirono Ikuo. 2020. Investigation of essential cell cycle regulator genes as candidates for immortalized shrimp cell line establishment based on the effect of in vitro culturing on gene expression of shrimp primary cells. *Aquaculture* 529: 735733. [CrossRef]

Tikhonenkov, Denis V., Elisabeth Hehenberger, Anton S. Esaulov, Olga I. Belyakova, Yuri A. Mazei, Alexander P. Mylnikov, and Patrick J. Keeling. 2020a. Insights into the Origin of Metazoan Multicellularity from Predatory Unicellular Relatives of Animals. *BMC Biology* 18: 1–24. [CrossRef] [PubMed]

Tikhonenkov, Denis V., Kirill V. Mikhailov, Elisabeth Hehenberger, Sergei A. Karpov, Kristina I. Prokina, Anton S. Esaulov, Olga I. Belyakova, Yuri A. Mazei, Alexander P. Mylnikov, Vladimir V. Aleoshin, and et al. 2020b. New Lineage of Microbial Predators Adds Complexity to Reconstructing the Evolutionary Origin of Animals. *Current Biology* 30: 4500–9. [CrossRef] [PubMed]

Tomczyk, Szymon, Kathleen Fischer, Steven Austad, and Brigitte Galliot. 2015. Hydra, a Powerful Model for Aging Studies. *Invertebrate Reproduction & Development* 59: 11–16. [CrossRef]

Toullec, J. Y., Y. Crozat, J. Patrois, and P. Porcheron. 1996. Development of Primary Cell Cultures from the Penaeid Shrimps Penaeus Vannamei and P. Indicus. *Journal of Crustacean Biology* 16: 643–49. [CrossRef]

Toullec, Jean-Yves. 1999. Crustacean Primary Cell Culture: A Technical Approach. *Methods in Cell Science* 21: 193–98. [CrossRef]

Urban-Gedamke, Elizabeth, Megan Conkling, Peter J. McCarthy, Paul S. Wills, and Shirley A. Pomponi. 2021. 3-D Culture of Marine Sponge Cells for Production of Bioactive Compounds. *Marine Drugs* 19: 569. [CrossRef]

Vago, Constantin. 2012. *Invertebrate Tissue Culture*. Amsterdam: Elsevier.

Vandepas, Lauren E., Kaitlyn J. Warren, Chris T. Amemiya, and William E. Browne. 2017. Establishing and Maintaining Primary Cell Cultures Derived from the Ctenophore Mnemiopsis Leidyi. *The Journal of Experimental Biology* 220: 1197–1201. [CrossRef]

Ventura, P., G. Toullec, C. Fricano, L. Chapron, V. Meunier, E. Röttinger, P. Furla, and S. Barnay-Verdier. 2018. Cnidarian Primary Cell Culture as a Tool to Investigate the Effect of Thermal Stress at Cellular Level. *Marine Biotechnology* 20: 144–54. [CrossRef]

Vincent, Mark. 2012. Cancer: A de-repression of a default survival program common to all cells? A life-history perspective on the nature of cancer. *Bioessays* 34: 72–82. [CrossRef]

Vizel, Maya, Yossi Loya, Craig A Downs, and Esti Kramarsky-Winter. 2011. A novel method for coral explant culture and micropropagation. *Marine Biotechnology* 13: 423–32. [CrossRef] [PubMed]

Vogt, Günter. 2008. How to minimize formation and growth of tumours: Potential benefits of decapod crustaceans for cancer research. *International Journal of Cancer* 123: 2727–34. [CrossRef] [PubMed]

Walcourt, Asikiya, and William Winlow. 2019. A Comparison of the Electrophysiological Characteristics of Identified Neurons of the Feeding System of Lymnaea Stagnalis (L.) in Situ and in Culture. *EC Neurol* 11: 323–33.

Walker, Charles, Stefanie Böttger, and Ben Low. 2006. Mortalin-Based Cytoplasmic Sequestration of P53 in a Nonmammalian Cancer Model. *The American Journal of Pathology* 168: 1526–30. [CrossRef]

Walker, Charles W., Benjamin E. Low, and S. Anne Böttger. 2013. Mortalin in invertebrates and the induction of apoptosis by wild-type p53 following defeat of mortalin-based cytoplasmic sequestration in cancerous clam hemocytes. In *Mortalin Biology: Life, Stress and Death*. Edited by Sunil Kaul and Renu Wadhwa. Netherlands: Springer, pp. 97–114.

Wang, Changliu, Shicui Zhang, Feng Su, Lei Wang, and Hongyan Li. 2009. Initiation of Primary Cell Culture from Amphioxus Branchiostoma Belcheri Tsingtauense. *Chinese Journal of Oceanology and Limnology* 27: 69. [CrossRef]

Wilson, D. E., and D. M. Reeder. 2011. Class Mammalia Linnaeus, 1758. In *Animal Biodiversity: An Outline of Higher-Level Classification and Survey of Taxonomic Richness*. Edited by Zhi-Qiang Zhang. Available online: https://biotaxa.org/Zootaxa/article/view/zootaxa. 3148.1.9 (accessed on 14 October 2021).

Xu, Xiaohui, Hu Duan, Yingli Shi, Shijun Xie, Zhan Song, Songjun Jin, Fuhua Li, and Jianhai Xiang. 2018. Development of a Primary Culture System for Haematopoietic Tissue Cells from Cherax Quadricarinatus and an Exploration of Transfection Methods. *Developmental and Comparative Immunology* 88: 45–54. [CrossRef]

Yang, Gong, Daniel G. Rosen, Imelda Mercado-Uribe, Justin A. Colacino, Gordon B. Mills, Robert C. Bast Jr., Chenyi Zhou, and Jinsong Liu. 2007. Knockdown of P53 Combined with Expression of the Catalytic Subunit of Telomerase Is Sufficient to Immortalize Primary Human Ovarian Surface Epithelial Cells. *Carcinogenesis* 28: 174–82. [CrossRef]

Yoshino, T. P., U. Bickham, and C. J. Bayne. 2013. Molluscan Cells in Culture: Primary Cell Cultures and Cell Lines. *Canadian Journal of Zoology* 91: 6. [CrossRef]

Zanetti, Laura, Filomena Ristoratore, Maria Francone, Stefania Piscopo, and Euan R. Brown. 2007. Primary Cultures of Nervous System Cells from the Larva of the Ascidian Ciona Intestinalis. *Journal of Neuroscience Methods* 165: 191–97. [CrossRef]

Zhang, Zhi-Qiang. 2013. Animal Biodiversity: An Update of Classification and Diversity in 2013. *Zootaxa* 3703: 5. [CrossRef]

Zhang, Xiaoying, Gaël Le Pennec, Renate Steffen, Wener E. G. Müller, and Wei Zhang. 2004. Application of a MTT Assay for Screening Nutritional Factors in Growth Media of Primary Sponge Cell Culture. *Biotechnology Progress* 20: 151–55. [CrossRef] [PubMed]

Zhang, Sihui, Diya Banerjee, and Jeffrey R. Kuhn. 2011. Isolation and Culture of Larval Cells from *C. elegans*. *PLoS ONE* 6: e19505. [CrossRef] [PubMed]

Zhao, Yali, Dan O. Wang, and Kelsey C. Martin. 2009. Preparation of Aplysia Sensory-Motor Neuronal Cell Cultures. *Journal of Visualized Experiments: JoVE* 28: 1355. [CrossRef] [PubMed]

Zhu, Ling-Ling, Li-Ying Wu, David Tai Yew, and Ming Fan. 2005. Effects of Hypoxia on the Proliferation and Differentiation of NSCs. *Molecular Neurobiology* 31: 231–42. [CrossRef]

Adult Stem Cells Host Intracellular Symbionts: The Poriferan Archetype

Alexander Ereskovsky, Baruch Rinkevich and Ildiko M. L. Somorjai

Abstract: Unlike vertebrates, adult stem cells (ASC) in a wide range of aquatic invertebrate phyla are morphologically diverse, exhibiting a wide range of differentiation states as well as somatic and germline physiognomies. They may arise de novo by trans-differentiation from somatic cells and above all represent phenotypes of specialized cells with multifunctionality. One unexpected phenomenon is the presence of intracellular symbionts in the ASCs of some invertebrates. Overviewing the literature on intracellular symbionts in sponge (Porifera) ASCs and in other aquatic invertebrates, we reveal that ASC intracellular prokaryotic and eukaryotic symbionts are restrictive to a single sponge class, the Demospongiae. The eukaryotic symbionts in sponges are exclusively unicellular photosynthetic algae, and are found only in pluripotent stem cells, most frequently in the archaeocytes; they are documented in five orders of Demospongiae. Bacteriocyte-like cells have been reported in sponges and three other phyla, indicative of their independent evolutionary origins. The results of this study add considerable insight into the establishment and maintenance of intracellular symbioses in ASCs of aquatic invertebrates, and provide new a understanding of the diversity of symbiotic associations across the tree of life.

1. Introduction

According to the prevailing dogma in cell biology, adult stem cells (ASC) in animals are committed lineage-specific cells, with tissue-/organ-restricted fates, and which are moreover capable of regeneration and repair of tissues and organs (Clevers and Watt 2018). Ordinarily, ASCs are undifferentiated cells that give rise to either daughter stem cells, non-self-renewing progenitors, or to lineage-specific differentiated cells (Clevers and Watt 2018; Raff 2003). Model ASCs (in vertebrates and insects) typically possess high nucleo-cytoplasmic ratios, are small in size compared to lineage-differentiated progenies, and are often rare. However, ASCs in many aquatic invertebrates are not only very common (up to one third of all animal cells), but are also morphologically highly diverse, and exhibit a wide range of differentiation states as well as somatic and germline characteristics, just to name some key biological properties (summarized in Rinkevich et al. 2022). Moreover, ASCs in aquatic invertebrates may arise de novo by trans-differentiation from somatic cells (Borisenko et al. 2015; Ferrario et al. 2020) and above all represent phenotypes of specialized cells with multifunctionality. Examples include the ecto-/endodermal epitheliomuscular cells in polyps of Cnidaria (Bosch et al. 2010; Hobmayer et al. 2012) or the archaeocytes and choanocytes in Porifera (Funayama 2018).

One unexpected and, as yet, little-explored phenomenon is the presence of intracellular symbionts in the ASCs of some invertebrates (for example, Bright and Giere 2005; Masuda 1990; Pflugfelder et al. 2009; Saller 1989), and the evidence that ASCs manipulate symbiont maintenance (Bosch et al. 2010; Dirks et al. 2012; Kovacevic 2012). Below, we review the literature on ASCs and their symbionts in sponges (Phylum: Porifera), which represent the best-known model case, as well as the few examples from other systems. We place this within the context of intracellular symbionts more generally, concluding with a discussion of how the application of modern methodologies in sponges to this problem may improve our understanding of this unusual symbiosis.

2. Symbiosis

2.1. What Is Endosymbiosis?

Symbiosis, an inter-dependent relationship between two species, is an important factor for ecological diversity and evolutionary novelty (Sitte and Eschbach 1992; Wernegreen 2012). The most comprehensive definition of symbiosis includes the full range of interaction modes, from harmful (parasitic) to beneficial (mutualistic). It applies not only to organisms living anywhere within the host body—such as within tissues (extracellular) or within cells (intracellular)—but also to cytosymbiosis, the intimate and long-lasting association of cells belonging to different taxa, and often considered as the most intricate partnership among living entities (Sitte and Eschbach 1992; Wernegreen 2012). Both parasitic and mutualistic symbiotic interactions can evolve into a state where there is a stable and permanent association between symbionts and hosts. In the case of intracellular mutualists, evolutionary processes may lead to cytosymbiosis through both morphological alterations as well as via physiological/molecular incorporation of the symbionts into the hosts' cellular environments, to the point where endosymbionts are no longer easily recognizable as foreign intrusions. Following such integrations, endosymbionts enhance the ability of hosts to succeed in diverse contexts, from unbalanced diets and nitrogen-poor soils, to hydrothermal vents and oligotrophic aquatic environments (Hinzke et al. 2021; Wernegreen 2012). Key functions performed by mutualistic, intracellular endosymbionts include harvesting energy from chemicals or light, to converting nitrogen into a usable form, and synthesizing nutrients that supplement the host's diet, to name just a few (Wernegreen 2012).

Cytosymbiotic associations can be organized within a graded series of cumulative morphological integrations, including the development of arrays of mechanisms targeting the interactions between host and symbiont (Bandi et al. 1995; Melo-Clavijo et al. 2018; Song et al. 2017). They can also be exposed to partner switching and rapid compensatory evolution (Sørensen et al. 2021). In cytosymbiosis, the interrelations between the partners of each specific symbiosis can be commensalic, parasitic, or mutualistic; but in every case, cytosymbiotic partnership leads to adaptive interaction of the partners or even to strict co-evolution (Sitte and Eschbach 1992). Intracellular symbiotic microorganisms commonly reside in specialized or non-specialized host cells, but not in ASCs; the property of "stemness" could be regarded as mutually exclusive to a highly differentiated and specialized cytosymbiotic state. The appearance of such an association, therefore, should be studied not only from functional but also from host/symbiont co-evolutionary perspectives, as unicellular symbionts have been associated with sponges (and their ASCs) since their initial evolution as multicellular animals (Ereskovsky 2010; Wilkinson 1983).

2.2. Porifera as Model Systems for ASC Cytosymbiosis

2.2.1. Overview of Characteristics of Organization and Cellular Plasticity

Sponges branch off basally in the metazoan phylogenetic tree and comprise four distinct classes: Demospongiae, Hexactinellida, Calcarea and Homoscleromorpha. Living sponges are found in all aquatic environments at all depths. A sponge is traditionally defined as "a sedentary, filter-feeding metazoan", and has no nerves, muscles, specialised digestive system or gonads (Borchiellini et al. 2021).

Sponges have two cell layers, the choanoderm and the pinacoderm (Figure 1), formed by choanocytes and pinacocytes, respectively. Choanocytes are flagellated collar cells lining the filtering cavities of the aquiferous system, the choanocyte chambers. Pinacocytes are flattened cells covering the outer parts of the body and lining the canals of the aquiferous system. The space between the external pinacocyte layer and the aquiferous system is filled by the mesohyl, a loose layer composed of collagen fibrils, skeletal elements, and up to ten cell types with different degrees of motility (Ereskovsky and Lavrov 2021; Harrison and De Vos 1991).

The tissues in sponges are simpler, both structurally and functionally, than in other Metazoa. In particular, sponge tissues tend to be highly multifunctional when compared to counterparts in more recent branching animal lineages, permitting a higher rate of cell migration and thus an almost constant reorganization of tissues. Moreover, the cells of sponge tissues possess a very high capacity for transdifferentiation into other cell types (Gaino et al. 1995; Nakanishi et al. 2014). In addition, sponges possess very high regenerative and reconstitutive abilities, culminating in the re-building of a functional body from dissociated cells (reviewed in Ereskovsky et al. 2015, 2020, 2021; Lavrov and Kosevich 2014; Simpson 1984).

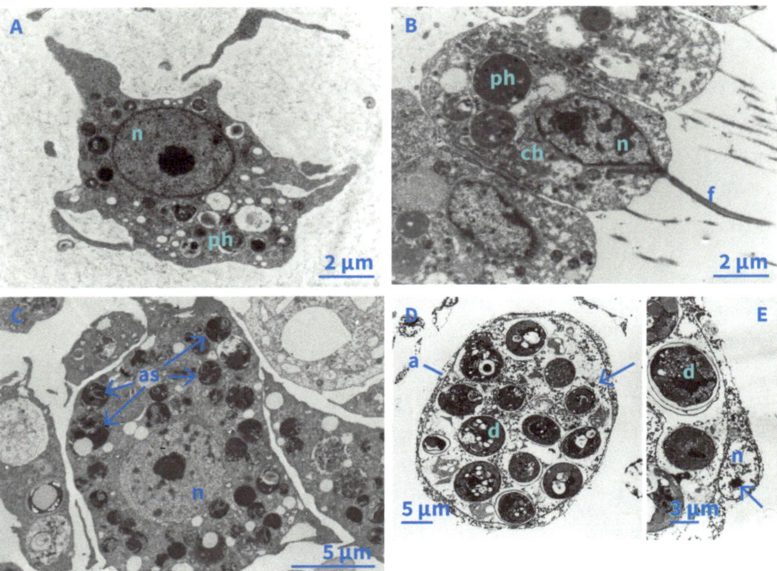

Figure 1. TEM images of principal pluripotent cells in sponges. (**A**)—archaeocyte of *Halisarca dujardinii* (Demospongiae); (**B**)—choanocytes of *Leucosolenia variabilis* (Calcarea); (**C**)—archaeocyte of the freshwater sponge *Lubomirskia baicalensis* (Demospongiae) with intra-cellular algal symbionts; (**D,E**)—archaeocytes of the marine sponge *Haliclona* sp. (Demospongiae) with intra-cellular dinoflagellate symbionts (showing fibrillar material between algae and archaeocyte (arrowed) (Modified from Garson et al. 1998). as—algal symbionts, ch—choanocyte, d—dinoflagellate, f—flagellum, n—nucleus, ph—phagosome. Source: Graphic by authors.

2.2.2. Sponge ASC Characteristics

As one of the most basal metazoan groups (Redmond and McLysaght 2021; Simion et al. 2017), sponges hold a key position to address stem cell origins.

Most research on stem cells in sponges has been conducted in demosponges, and until recently, consisted almost entirely of microscopic studies. However, in the past few years, molecular studies have provided new insights. According to the most recent investigations in Porifera, there are not only two (Funayama 2018), but rather at least four types of pluripotent ASC: the archaeocytes and choanocytes, as well as pinacocytes and particular amoeboid vacuolar cells (Ereskovsky et al. 2015; Fierro-Constaín et al. 2017; Lavrov et al. 2018).

Three main criteria are generally accepted as defining a stem cell: (1) the capacity for self-renewal, (2) differentiation (or transdifferentiation) of this cell type into others, and (3) contribution of this cell to the processes of homeostasis and regeneration (Melton 2014). The molecular evidence of their stemness includes the expression of genes encoding GMP (germline multipotency program) proteins (*piwi, vasa, bruno, pl-10,* and all the genes encoding Tudor domains, *ddx6,* and *mago-nashi*); the observation that genes encoding RNA helicase and proteins involved in mRNA splicing are elevated in the archaeocytes of the freshwater demosponge *Ephydatia fluviatilis* (Alié et al. 2015); and expression of the *EfPiwiA* and *EfPiwiB* genes detected in choanocytes (Funayama et al. 2010). GMP genes (*piwi, argonaute, vasa, nanos, pl10, tudor, pumillo, boule*) are expressed in the choanocytes of adult *Oscarella lobularis* (Homoscleromorpha) (Fierro-Constaín et al. 2017). *VasaB* and *SciPL10B* are also strongly expressed in the choanocytes of *Sycon ciliatum* (Calcarea) (Leininger et al. 2014).

Choanocytes are specialized epithelial cells responsible for water movement inside the sponge aquiferous system and food particle capture. These cells are characterized by apical–basal polarity and the presence of a flagellum surrounded by the collar of microvilli at the apical pole (Simpson 1984) (Figure 1B).

Archaeocytes are amoeboid cells of the mesohyl devoid of any polarity or specialized features, and are typical in Demospongiae. These cells manifest high polymorphism and multifunctionality. Up to now, no generally accepted characteristics of archaeocytes have been defined. Only general features exist, which are present in all archaeocyte descriptions: an amoeboid shape, a large nucleolated nucleus and the absence of specialized inclusions in the cytoplasm (Ereskovsky and Lavrov 2021; Simpson 1984), (Figure 1A,C). As for the function of demosponge archaeocytes, their role has been described in: (1) the transport of food particles and elimination of digestive products (Godefroy et al. 2019; Willenz and Van de Vyver 1984); (2) outgoing particulate organic matter (Maldonado 2016); (3) the burrowing processes in excavating sponges (Rützler and Rieger 1973); (4) spicules secretion (Funayama et al. 2005; Rozenfeld 1980); (5) immunity role (Fernàndez-Busquets 2008; Smith and Hildemann 1986); (6) gametogenesis (Ereskovsky 2010; Simpson 1984); (7) asexual reproduction (budding, gemmulogenesis, reduction body formation) (Ereskovsky et al. 2017; Harrison et al. 1975; Simpson 1984); (8) regeneration, somatic embryogenesis and growth (Buscema et al. 1980; Ereskovsky et al. 2020, 2021; Lavrov and Kosevich 2014). Thus, this sponge archaeocyte multifunctionality is unusual for the stem cells of Metazoa.

Notably, there is another unusual feature of archaeocytes in Demospongiae—the presence of intracellular photosynthetic algal symbionts. Freshwater sponges (order Spongillida) harbour Chlorophyta from the classes Trebouxiophyceae and Chlorophyceae (zoochlorella), and Ochrophyta from the class Eustigmatophytacea. Some marine demosponges (orders Haplosclerida and Clionaida) also harbour Dinoflagellata *Symbiodinium* spp (zooxanthella) (Table 1).

Table 1. Distribution of symbiotic intracellular unicellular algae in demosponge adult stem cells.

Demosponge Species	Adult Sponge Cell Type	Gemmule	Buds	Algal Species	Method	References
Order Spongillida						
Family Spongillidae						
Spongilla lacustris (geen)	Archaeocytes, choanocytes (Williamson)	Thesocytes of green gemmules	No	Phylum Chlorophyta Trebouxiophyceae, zoochlorellae	TEM	(Masuda 1990; Gilbert and Allen 1973; Williamson 1979)
Spongilla lacustris	Archaeocytes, choanocytes, pinacocytes	Thesocytes	No	Trebouxiophyceae, *Chlorella sp.* Germany	In vivo microscopy LM, TEM	(Saller 1989, 1991)
Spongilla lacustris	?	Thesocytes	No	Trebouxiophyceae, *Choricystis minor* - Japon	LM,	(Handa et al. 2006)
Spongilla lacustris	?	?	No	Trebouxiophyceae, *Lewiniosphaera symbiontica* USA	MB	(Pröschold et al. 2010)
Nudospongilla moorei	Amoebocytes	?	?	Zoochlorella	LM	(Brien and Govaert-Mallebranche 1958)
Radiospongilla sendai (geen)	Archaeocytes	Thesocytes of green gemmules	No	Trebouxiophyceae, (zoochlorellae) *Choricystis minor*	TEM	(Masuda 1990; Handa et al. 2006; Okuda et al. 2002)
Radiospongilla cerebellata (geen)	Archaeocytes	Thesocytes of green gemmules	No	Trebouxiophyceae, (zoochlorellae) *Choricystis minor* and *Chlorella*	TEM	(Handa et al. 2006; Masuda 1985, 1990)
Radiospongilla cerebellata	Archaeocytes, amoebocytes, choanocytes, pinacocytes	?	Archaeocytes, amoebocytes, choanocytes, pinacocytes	Trebouxiophyceae, *Chlorella*	LM, TEM	(Saller 1990)
Eunapius fragilis	?	Thesocytes	No	Trebouxiophyceae, *Choricystis minor*	LM, *In vivo* microscopy	(Handa et al. 2006)
Heteromeyenia slepanowii (geen)	Archaeocytes	No	No	Zoochlorellae	TEM	(Masuda 1990)
Ephydatia fluviatilis (geen)	Archaeocytes	Thesocytes	No	Trebouxiophyceae, *Chlorella* sp	LM, TEM	(Wilkinson 1980; Gaino et al. 2003)
Ephydatia fluviatilis (brown)	Archaeocytes	No	No	No algae	LM, TEM	(Gaino et al. 2003)
Ephydatia muelleri	Archaeocytes	Thesocytes	No	*Chlorella* sp.	LM, cell fractioning	(Hall et al. 2021)
Ephydatia muelleri	Archaeocytes	No	No	Trebouxiophyceae *Choricystis*, *Chlorella* sp.	MB, CM, TEM	(Masuda 1990; Gilbert and Allen 1973; Williamson 1979)

Table 1. *Cont.*

Demosponge Species	Adult Sponge Cell Type	Gemmule	Buds	Algal Species	Method	References
			Family Lubomirskiidae			
Lubomirskia baicalensis	Archaeocytes	No	No	Chlorophyceae *Mychonastes jurisii*	LM, TEM, SEM, MB	(Chernogor et al. 2013)
Lubomirskia baicalensis	Archaeocytes	No	No	Trebouxiophyceae Chlorophyceae *Mychonastes* sp.	TEM	(Ereskovsky et al. 2016)
Lubomirskia incrustans	Archaeocytes	No	No	Trebouxiophyceae *Choricystis parasitica*	LM	(Kulakova et al. 2014)
Lubomirskia abietina	Archaeocytes	No	No	Trebouxiophyceae *Choricystis krienitzii*	LM	(Kulakova et al. 2020)
Baikalospongia bacillifera	Archaeocytes	No	No	Trebouxiophyceae *Choricystis parasitica*	LM	(Kulakova et al. 2014)
Baikalospongia intermedia	Archaeocytes	No	No	Trebouxiophyceae *Choricystis krienitzii*	LM	(Kulakova et al. 2020)
			Family Metaniidae			
Corvomeyenia everetti	Archaeocytes	No	No	Phylum Ochrophyta Eustigmatophytacea	TEM	(Frost et al. 1997)
			Order Haplosclerida			
Haliclona sp.	Archaeocytes	No	No	Dinoflagellata, *Symbiodinium microadriaticum*	LM, TEM	(Garson et al. 1998)
			Order Clionaida			
Cliona viridis	Archaeocytes	?	Archaeocytes	Dinoflagellata *Symbiodinium*	LM, TEM	(Rosell 1993)
Cliona inconstans, C. orientalis	Archaeocytes	?	?	Zooxantellae	LM, TEM	(Vacelet 1981)
Cliona caribbaea, C. varians	Archaeocytes	?	?	Dinoflagellata *Gymnodinium microadriaticum*	LM, TEM	(Rützler 1990)
Cervicornia cuspidifera	Amoeboid cells	No	No	Dinoflagellata *Symbiodinium microadriaticum*	LM	(Rützler and Rieger 1973)
			Order Suberitida			
Suberites aurantiacus	Archaeocytes	No	No	Zooxantella	LM	(Cheng et al. 1968)
			Order Tetractinellida			
Cinachyra tarentina	Amoeboid cells	No	No	Dinoflagellata Zooxantella *Symbiodinium microadriaticum*	LM, TEM	(Scalera-Liaci et al. 1999)

CM—confocal microscopy; LM—light microscopy; MB—molecular biological data; no—absence; SEM—scanning electron microscopy; TEM—transmission electron microscopy; ?—no data.

2.2.3. Diversity of Intracellular Algal Symbionts

Intracellular algal symbionts were described for the first time by Brandt (1881, 1882—see Krueger 2016) in mesohylar cells of the freshwater demosponge *Spongilla* sp. Subsequently, thanks to progress in light and electron microscopy, intracellular algal symbionts were found in a number of different sponge species, but exclusively from the class Demospongiae (Rützler 1990; Sarà and Vacelet 1973; Sarà et al. 1998; Simpson 1984; Vacelet 1981; Wilkinson 1987). These symbionts include different species of the phylum Chlorophyta, the classes Trebouxiophyceae (genera *Chlorella, Zoochlorella, Choricystis, Lewiniosphaera*), Chlorophyceae *(Mychonastes)* (Masuda 1985, 1990; Pröschold and Darienko 2020; Saller 1990; Simpson 1984; Williamson 1979), dinoflagellates (Zooxanthellae) of the genera *Symbiodinium* and *Gymnodinium* (Annenkova et al. 2011; Garson et al. 1998; Hill 1996; Pang 1973; Rosell and Uriz 1992; Rützler 1990; Sarà and Liaci 1964; Scalera-Liaci et al. 1999; Vacelet 1981), cryptophytes, cryptomonads (Wilkinson 1992), diatoms (Cox and Larkum 1983), coccoid red algae (Lemloh et al. 2009) eustigmatophytes (Frost et al. 1997), and macroscopic algae (Price et al. 1984; Rützler 1990) (Table 1).

2.2.4. Distribution of Archaeocytes with/without Symbionts in the Sponge

Archaeocytes are the principal cells acing as hosts (Table 1), and the same archaeocyte can contain from one to several algal symbionts (Gaino et al. 2003; Masuda 1990; Saller 1989). In some freshwater sponges, green algal symbionts can also be found inside choanocytes and pinacocytes (Gilbert and Allen 1973; Saller 1990, 1991). This is also true for some marine demosponges. In *Haliclona* sp., algal cells of *Symbiodinium microadriaticum* are grouped together in clusters of 6 ± 10 cells and enclosed by sponge cells, rather than being randomly distributed throughout the mesohyl (Garson et al. 1998). In the boring sponges *Cliona inconstans* and *C. orientalis*, the Zooxanthellae are always intracellular and occur in individual vacuoles of archaeocytes (Figure 1D,E). Each cell contains several algae (Vacelet 1981). In *Cliona caribbaea* and *C. varians*, the symbiotic dinoflagellates *Gymnodinium microadriaticum* are intracellular, either fully embedded in a host archaeocyte vacuole or encircled by host cell filopodia (Rützler 1990).

The spatial distribution of cells harboring symbionts in the sponge body is not homogeneous. In *Cinachyra tarentina*, the majority of the zoochlorellae are concentrated in the cortical zone of the sponge (Scalera-Liaci et al. 1999). Archaeocytes of *Ephydatia fluviatilis* harbour *Chlorella* concentrated mainly in the uppermost regions of the sponge body; in the inner parts of the sponge body, cells do not host zoochlorellae (Gaino et al. 2003).

The intracellular position of algal cells occurs in the host cytoplasm within vacuoles. At least in more thoroughly studied systems such as protists, the cnidarian *Hydra viridis* and the sponge *Spongilla lacustris*, two types of vacuole are observed (Reisser and Wiessner 1984). The first, the perialgal vacuole, always harbours only one algal cell. The wall of this type of vacuole is attached to the vacuolar membrane of the host. A perialgal vacuole divides simultaneously with the enclosed alga and apparently protects it from host lytic enzyme action (Reisser and Wiessner 1984). The chlorellae are able to divide inside the perialgal vacuole of sponge cells in *Spongilla lacustris* (Saller 1990). The second, the food vacuole, contains algae in various stages of digestion and other material (Simpson 1984). This may allow the host cell to absorb nutrients from damaged or dying algae, or under particularly adverse conditions in which the symbiont can no longer be maintained.

2.2.5. Intracellular Symbiosis Is Facultative

Three lines of evidence indicate that intracellular symbiosis of algal and sponge cells is facultative: (1) geographic, (2) ecological and (3) ontogenetic. The best geographic evidence comes from *Spongilla lacustris.* These sponges are able to host different algae species in their archaeocytes, depending on the geographic region they inhabit: *Chlorella* sp. in Germany (Saller 1989), *Choricystis minor* in Japan (Handa et al. 2006), *Choricystis parasitica* and *Lewiniosphaera symbiontica* in Massachusetts (USA) (Pröschold et al. 2010). (2) Ecological evidence. With respect to ecological evidence, many populations of the same freshwater sponge species contain green, brownish, and white individuals as a result of temporal and/or spatial variation in light availability. Electron microscopy investigation revealed that green sponges harbour zoochlorellae, which absent in the brownish ones (Gaino et al. 2003). Sponges that have green colour with zoochlorellae will quickly turn white when shaded (Frost and Williamson 1980), as zoochlorellae were digested by their host (Williamson 1979). Other examples are fresh-water sponge species that live in dark habitats, such as underground caves (*Eunapius subterraneus* in Croatia (Bilandija et al. 2007); *Racekiela cavernicola* in Brasil (Volkmer-Ribeiro et al. 2010)), or at great depths in lakes that completely lack symbiotic eukaryotic algae (e.g., *Baikalospongia abyssalis* in Baikal (Itskovich et al. 2017)).

There is also experimental evidence. For example, Hall et al. (2021) infected young aposymbiotic sponges of *Ephydatia muelleri* that had hatched from gemmules with sponge-derived algae. Evidence of the establishment of intracellular position by the algae was manifested within 4 h of infection. At the 24-hour time point, many sponge host archaeocytes harboured multiple or single algae within a single cell.

In adult sponges the algae are transmitted among the sponge cells in a very particular way. After the donor and the recipient cell getting closer each another, the vacuole includes *Chlorella* inside bulges out, surrounded by cell processes of the recipient cell. The vacuole opens, while the donor cell retracts and the recipient cell closes around the alga. Finally, the alga is incorporated into the recipient cell (Masuda 1990; Saller 1991). No release of the algae into the intercellular mesenchyme was detected. Then, the chlorella cells divide inside the sponge cells.

2.2.6. Horizontal and Vertical Transmission of Intracellular Algal Symbionts

As we showed above, the sponge-algal symbiosis is facultative. Accordingly, the transmission of algal symbionts occurs horizontally during sexual reproduction. In any event, not a single study has so far shown the presence of algal symbionts in sponge larvae. As for asexual reproduction, the situation there is more complicated.

In sponges there are three types of asexual reproduction: fragmentation, budding and gemmule formation (Ereskovsky 2010).

During fragmentation, the sponge is divided into two or more parts, each consisting of all tissue types and the symbionts. In contrast, during the budding process, the vertical transmission of intracellular symbiotic algae has been documented for two species: in the fresh-water sponge *Radiospongilla cerebellata*, where bud cell archaeocytes, amoebocytes, choanocytes and pinacocytes included *Chlorella* sp. (Saller 1990); and in the marine boring sponge *Cliona viridis*, in which the archaeocytes of the buds harbour intracellular dinoflagellate symbionts (Rosell 1993).

Gemmules of demosponges are special dormant structures that are capable, under suitable conditions, of developing asexually into new adult animals (Simpson 1984). Gemmules develop from the pluripotent archaeocytes. Gemmule thesocytes (resulting from archaeocyte differentiation) of many freshwater sponges include four or five functional algal endosymbionts per cell (Gilbert and Allen 1973; Masuda 1990; Okuda et al. 2002; Williamson 1979). The ultrastructure of zoochlorellae inside of gemmules differs from the ultrastructure of active symbionts in adult, green sponges: the gemmular symbionts contain loosely packed membranes of the chloroplasts, they generally lack lipid granules, and they lack chloroplast starch grains (Masuda 1990; Williamson 1979). This modification in structure could be a result of the relative inactivity of the symbionts inside gemmules. However, it has been shown that the symbionts within thesocytes are photosynthetically active, and could pass some of their photosynthate to the sponge cells (Gilbert and Allen 1973).

Before hatching, symbiotic algae could be phagocyted, and thus the young sponges that develop from such gemmules would be aposymbiotic (Rasmont 1970), without signs of symbiotic algal propagation (Simpson 1984; Williamson 1979). Yet, under dark conditions, brown gemmules do not host symbiotic algae, or only possess them in very low numbers (Gilbert and Allen 1973; Jorgensen 1947; Simpson 1984). Therefore, the vertical transmission of intracellular algal symbionts during gemmulogenesis in sponges is facultative.

2.3. Cytosymbiosis in ASCs-Beyond Poriferans

In contrast to the demosponges, cytosymbiosis in ASCs is a rare situation in other multicellular organisms in general, and in marine invertebrates in particular (Figure 2). Nevertheless, several well documented cases attest to the importance of ASCs in coordinating and maintaining intracellular symbiosis. Examples include the deep-sea vestimentiferan tubeworms (Polychaeta; best known are *Lamellibrachia luymesi* and *Riftia pachyptila*), which live in symbiosis with intracellular bacteria housed in bacteriocyte host cells (considered to be "tissue-specific unipotent bacteriocyte stem cells"; (Pflugfelder et al. 2009)), located within a special organ, the trophosome (Bright and Giere 2005). These stem cells continuously proliferate to produce new bacteriocytes, a process leading to self-renewal of bacteriocyte and to a complex control of the symbiont population in these host cells. Similarly, the free-living symbiotic flatworm *Paracatenula galateia* possesses intracellular, sulphur-oxidizing bacteria (also called bacteriocytes): as for all other somatic cells in adult worms, the bacteriocytes originate solely from the pool of aposymbiotic neoblasts, the ASCs of flatworms (Dirks et al. 2012). In addition, in *Hydra*, the epithelial stem cells lineages, but not the interstitial cells, actively shape the microbial intracellular communities of epithelial cells (Fraune et al. 2009). However, the elimination of nerve cells and secretory gland cells, two important cell types derived from interstitial cells, had a significant influence on the structure of symbiotic microbiota. Further, in the branching coral species *Stylophora pistillata*, algal containing cells in the endodermal layer express "stemness" genes such as *Nanos* and *Tudor*, as well as Tubulins and genes involved in the cell cycle (Levy et al. 2021), indicating that these cells may carry stem cell properties. There is also some preliminary evidence for ASC-related cytosymbiosis in hibernating colonies of botryllid ascidians (Hyams et al. 2017). About 15% of the blood cell population in the vasculature of hibernating colonies was first identified as phagocytes. However, transmission electron microscope studies revealed specific facultative symbionts—*Endozoicomonas* bacteria—inside their phagosomes. This novel case of cytosymbiosis develops de novo and only during stress conditions, a phenomenon most probably controlled by circulating ASCs (B.R., unpublished data).

There are additional unique examples of the involvement of ASCs of terrestrial invertebrates in maintaining or controlling intracellular symbionts. In early developmental stages of the aphids Acyrthosiphon pisum and Megoura viciae and in the cockroach Periplaneta americana, studies revealed de novo bacteriocyte formation from aposymbiotic ASCs, followed in the cockroaches by postembryonic divisions of the bacteriocytes (Braendle et al. 2003; Chevalier et al. 2011; Lambiase et al. 1997; Maire et al. 2020; Miura et al. 2003), suggesting that insect and tubeworm bacteriocytes proliferate (Dirks et al. 2012). The same applies to haemocytes of the isopod Armadillidium vulgare, which host endosymbiotic Wolbachia cells, intracellular α-proteobacteria (Chevalier et al. 2011) that are considered parasites in many insects such as Drosophila, in which they colonize female germline stem cells (Ote and Yamamoto 2020). However, some strains also appear to confer protection against RNA viruses in flies and mosquitoes in the laboratory, indicating a mutualism, although it is still unclear if this antiviral effect exists in the wild (reviewed in (Pimentel et al. 2021)). Intracellular Wolbachia symbionts are not only the cytosymbiotic bacteria in insect stem cells, as germline cells can also be colonized by other microorganisms, such as the Gram-positive bacterium Spiroplasma in Drosophila (Hackett et al. 1986), or the Gram-negative bacterium Arsenophonus, which infects the Sulcia symbiont of the leafhopper Macrosteles laevis (Kobiałka et al. 2016). The aforementioned means of ASC control is further illustrated by bacteria from larval bacteriocytes in uninfected nuclei of putative stem cells, as assessed over the course of metamorphosis (Maire et al. 2020).

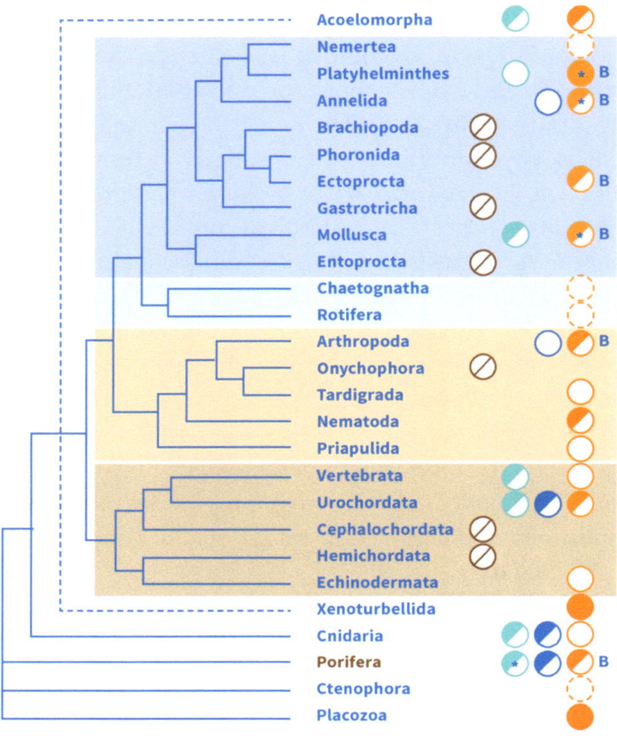

Figure 2. Distribution of algal, cyanobacterial and bacterial endosymbionts in metazoan phyla. To the left, a cladogram illustrates phylogenetic relationships among phyla; branch lengths are not proportional to evolutionary divergence. The position of sponges (Porifera) is highlighted in red. Coloured boxes indicate bilaterian lineages belonging to the Ecdysozoa (yellow), Gnathifera (green) Lophotrochozoa (blue), and Deuterostomia (pink). The positions of Acoelomorpha and Xenoturbellida are still debated and are indicated by dotted lines. To the right of each phylum, absence of endosymbionts (red symbols) as well as presence of algal (green symbols), cyanobacterial (blue symbols) and bacterial (orange symbols) endosymbionts are shown. Note that for many lineages, examples of both (mixed circles) intracellular (filled circles) and extracellular (empty circles) symbionts exist and where endosymbiosis is uncertain, dotted circles are used. Groups in which bacteriocytes have been reported are indicated by a "B" next to the bacterial endosymbiont column. An asterisk (*) denotes cases in which ASCs have been reported in the literature to contain endosymbionts. Sources: Acoelomorpha:

(Melo-Clavijo et al. 2018; Hikosaka-Katayama et al. 2012; Venn et al. 2008);
Nemertea: (McDermott 2006); Platyhelminthes: (Dirks et al. 2012; Dubilier et al.
2008; Gruber-Vodicka et al. 2011; Melo-Clavijo et al. 2018; Venn et al. 2008); Annelida:
(Dubilier et al. 2008); Ectoprocta: (Karagodina et al. 2018; Saffo 1992; Sharp et al.
2007); Gastrotricha: (Todaro et al. 2017); Mollusca: (Dubilier et al. 2008; Duperron
et al. 2006; Melo-Clavijo et al. 2018; Venn et al. 2008); Chaetognatha: (Thuesen and
Kogure 1989); Rotifera: (Selmi 2001); Arthropoda: (Dubilier et al. 2008; Lindquist
et al. 2005); Tardigrada: (Vecchi et al. 2016); Nematoda: (Dubilier et al. 2008);
Priapulida: (Kroer et al. 2016); Vertebrata: (Baker et al. 2019; Kerney et al. 2011;
Melo-Clavijo et al. 2018); Urochordata: (Melo-Clavijo et al. 2018; Mutalipassi
et al. 2021; Saffo 1992); Echinodermata: (Carrier and Reitzel 2020; Saffo 1992);
Xenoturbellida: (Kjeldsen et al. 2010); Cnidaria: (Melo-Clavijo et al. 2018; Venn
et al. 2008); Porifera: (Rützler 1990; Saller 1991; Sarà et al. 1998; Williamson 1979);
Ctenophora: (Daniels and Breitbart 2012; Hernandez and Ryan 2018); Placozoa:
(Gruber-Vodicka et al. 2019). Please see text for details.

The important interplay between ASCs and their intracellular symbionts has
also been recorded in vertebrates. For instance, the intracellular bacterial pathogen
Mycobacterium leprae has the capacity to alter the developmental reprogramming
of lineage committed host glial cells to progenitor/stem cell-like cells in mammals
(Hess and Rambukkana 2015). In addition, the host–pathogen symbiosis commonly
recorded between bacteria and stem cells of the intestine, where microbial products
can stimulate stem cell survival, trigger regeneration and provide protection against
stress (Nigro et al. 2014), or the ways in which Escherichia coli cells can mobilize
functional hematopoietic stem cells (Burberry et al. 2014), are but two of many
examples of what may be a widespread but poorly understood phenomenon in
animals.

3. Discussion and Future Perspectives

Here, we showed that the intracellular symbionts (either prokaryotic or
eukaryotic) of sponges are found only in representative species of Demospongiae,
one of the four Porifera classes (Demospongiae, Hexactinellida, Homoscleromorpha
and Calcarea; Table 2). Prokaryotic organisms are found in specialized
cells—bacteriocytes—in representatives of different orders of Demospongiae (Table 2).
Bacteriocyte-like cells have been reported in four phyla, indicative of their
independent evolutionary origins (Figure 2). Eukaryotic symbionts are exclusively
unicellular photosynthetic algae in sponges, and are found in pluripotent stem cells,
most frequently in the archaeocytes; they are documented in five orders: Spongillida,
Haplosclerida, Clionaida, Suberitida, and Tetractinellida (Table 1). It is interesting to
note that the representatives of the green algae from the phylum Chlorophyta were
found only in freshwater sponges of the order Spongillida.

Table 2. Distribution of bacteriocytes harboring intracellular symbiotic bacteria within Porifera.

Order	Species	Bacteriocyte Localization	Stem-Cell	References
		Class Demospongiae		
Biemnida	*Biemna ehrenbergi*	Mesohyl	No	(Ilan and Abelson 1995)
	Neofibularia irata	Mesohyl	No	(Wilkinson 1978)
Axinellida	*Cymbastella concentrica*	Mesohyl	No	(Nguyen et al. 2014)
Verongiida	*Aplysina cavernicola*	Mesohyl	No	(Vacelet 1975)
	Aplysina aerophoba	Mesohyl	No	(Vacelet 1975)
	Aplysina cauliformis	Mesohyl	No	(Gochfeld et al. 2019)
	Aplysina fistularis	Mesohyl	No	(Negandhi et al. 2010)
Haplosclerida	*Petrosia ficiformis*	Mesohyl	No	(Vacelet and Donaday 1977)
	Haliclona tubifera	Larva	No	(Woollacott 1993)
	Haliclona cnidata	Mesohyl	No	(Schellenberg et al. 2020)
	Haliclona sp.	Mesohyl	No	(Tianero et al. 2019)
	Oceanapia sagittaria	Mesohyl	No	(Salomon et al. 2001)
	Cribochalina	Mesohyl	No	(Rützler 1990)
Chondrosida	*Chondrosia reniformis*	Mesohyl	No	(Lévi and Lévi 1976)
Chondrillida	*Halisarca dujardinii*	Mesohyl	No	(Ereskovsky unpublished)
	Halisarca restingaensis	Mesohyl	No	(Alvizu et al. 2013)
	Chondrilla australiensis	Mesohyl and Larva	No	(Usher and Ereskovsky 2004)
Suberitida	*Suberites domuncula*	Mesohyl	No	(Bohm et al. 2001)
Tetractinellida	*Thoosa* sp., *Alectona* sp.	Mesohyl	No	(Garrone 1974)
	Jaspis stellifera	Mesohyl	No	(Wilkinson 1978)
Tethyida	*Tethya stolonifera*	Mesohyl	No	(Taylor et al. 2021)
Poecilosclerida	*Lycopodina hypogea*	Mesohyl	No	(Vacelet and Boury-Esnault 1996)
	Cladorhiza sp.	Mesohyl	No	(Vacelet et al. 1996)
	Crambe crambe	Mesohyl	No	(Maldonado 2007)
	Hymedesmia methanophila	Mesohyl	No	(Rubin-Blum et al. 2019)
Scopalinida	*Svenzea zeae*	Mesohyl and Larva	No	(Rützler et al. 2003)
	Scopalina ruetzleri	Mesohyl	No	(Rützler et al. 2003)
Agelasida	*Astrosclera willeyana*	Mesohyl	No	(Worheide 1998)
Demospongiae incertae sedis	*Myceliospongia araneosa*	Mesohyl	No	(Vacelet and Perez 1998)

It is generally accepted that all multicellular organisms actively coordinate somatic maintenance properties, including growth (in organisms with indeterminate growth -such as sponges, corals, and the immortal *Hydra*- throughout the organism´s life span; (Vogt 2012)); cell proliferation and cell death for tissue homeostasis; and for phenomena such as regeneration, with ASCs in some of these organisms acting as the building blocks for all needs (Biteau et al. 2011; Merrell and Stanger 2016; Rinkevich et al. 2022). The additional cellular homeostasis required for the management and coordination of intracellular symbiosis clearly presents a scenario in which non-traditional functions were imposed on ASC performance during evolution. In contrast to the sponge examples, most other ASC types do not possess intracellular symbionts and yet directly or indirectly influence cytosymbiosis in a wide range of marine and terrestrial taxa (but see *Wolbachia* infections in isopods that harbour these intracellular parasites not only inside haemocytes but also within ASCs of the hematopoietic system, or the *Wolbachia* that highjack the female germline of insects (Chevalier et al. 2011; Ote and Yamamoto 2020). Indeed, intracellular symbionts are rarely associated with ASCs, and to our knowledge—with the exception of sponges—only in the case of bacterial symbionts (Figure 2).

It is, therefore, of great interest to illuminate the mechanisms driving the highly coordinated behaviours of ASCs in specific symbioses, such as the unipotent bacteriocyte stem cells that continuously proliferate to produce new bacteriocytes in some annelids (Bright and Giere 2005); the maintenance of symbiosis during the continuous bacteriocyte formation from aposymbiotic neoblasts in adult paracatenulid flatworms (Dirks et al. 2012); the epithelial stem cells that actively shape the microbial intracellular communities in Hydra (Fraune et al. 2009); or the larval bacteriocytes that develop from uninfected putative stem cells in the rice weevil Sitophilus oryzae (Alvizu et al. 2013). Thus, cytosymbiosis-borne ASC phenomena are either established (in sponges) or supported (directly and indirectly; at least in Cnidaria, Platyhelminthes, Annelida, Arthropoda [insects and crustacean alike], Urochordata and Vertebrata). However, the most prominent examples of endosymbiotic ASCs come from the sponges.

Many challenges remain in studying symbioses at the mechanistic level. First, it should be possible to isolate and culture host and symbiont separately; this is rarely possible. Many symbioses have arisen in inhospitable environments (e.g., deep sea Bathymodiolus mussels and their sulphide- and methane-oxidizing bacterial symbionts, (Duperron et al. 2006), which cannot be easily recreated in the laboratory. Marine algae are particularly difficult to culture, and yet are the basis for many photosymbiotic associations. In addition, many symbioses are obligate, or transmitted vertically, making them near impossible to manipulate without killing host or symbiont, or affecting embryonic survival. It should be possible to generate aposymbiotic and symbiotic hosts at will to understand the metabolic and genetic changes directly caused by symbiosis. Such studies on sponges have recently been initiated (Geraghty et al. 2021; Hall et al. 2021). Finally, from a technical perspective, it is often difficult to separate host and symbiont genomes in intracellular symbioses. In particular, RNA sequencing of endosymbiotic host tissues en masse fails to adequately define transcriptional profiles at the fine resolution necessary to assess changes at the cellular level.

In spite of these many limitations, metagenomic approaches are now giving new insight into host–symbiont interactions. For instance, dual RNA-seq combined with imaging has allowed the time course of endosymbiont-embryonic host cellular interactions to be mapped during cereal weevil metamorphosis (Maire et al. 2020). RNAseq of aposymbiotic and symbiotic bobtail squid tissues as well as Vibrio both before and after venting from the light organ (Thompson et al. 2017) and hybridization chain reaction-fluoresencent in situ hybridization of both partners at the onset of symbiosis maps transcriptional changes in situ (Nikolakakis et al. 2015). A recent RNASeq analysis, combined with electron and confocal microscopy of fresh-water demosponge model Ephydatia muelleri, has revealed some of the genetic pathways involved in intracellular host/photosymbiont interactions, identifying putative genetic pathways involved with endosymbiosis establishment (Hall et al. 2021). RNASeq analysis and comparative analyses of the transcriptomes of aposymbiotic and symbiotic sponges have identified a suite of genes that are regulated at the early establishment stages of the stable symbiosis between E. muelleri and its native green algal symbionts (Geraghty et al. 2021). Authors have also begun to differentiate these genes from those involved in generalized phagocytosis events related to feeding and/or immunity. Single cell analyses are providing new avenues for understanding that might be well suited to tackling the ASC/endosymbiont–poriferan mutualism. As a case in point, recent work on the cnidarian coral Xenia has identified the cell lineage containing the Symbiodinium algal symbiont as originating as a pre-endosymbiotic progenitor pool (Hu et al. 2020). Similarly, Levy et al. (2021) simultaneously queried the transcriptomes of Symbiodinium-containing host cells and their symbionts, and compared with "free" Symbiodinium and non-symbiotic gastrodermal cells and in the stony coral Stylophora pistillata. They identified shared lipid metabolism pathways in algal hosting cells with those of Xenia (Hu et al. 2020) and Exaiptasia pallida (Hambleton et al. 2019), a symbiotic anemone, suggesting cnidarian-dinoflagellate photosymbioses may generate very particular constraints on physiologies despite their independent evolutionary origins. Similar efforts in sponges would thus add considerable insight into the establishment and maintenance of photosymbioses, and provide new insight into the diversity of symbiotic associations seen across the tree of life.

Author Contributions: A.E., I.M.L.S. and B.R. conceived the work, wrote the manuscript and approved the final version. All authors have read and agreed to the published version of the manuscript.

Funding: This research was funded by the European Cooperation in Science & Technology program (EU COST). Grant title: "Stem cells of marine/aquatic invertebrates: from basic research to innovative applications" (MARISTEM). The work of AE was conducted under the IDB RAS Government basic research program in 2021 No. 0088-2021-0009.

Conflicts of Interest: The authors declare no conflict of interest.

References

Alié, Alexandre, Tetsutaro Hayashi, Itsuro Sugimura, Michaël Manuel, Wakana Sugano, Akira Mano, Nori Satoh, Kiyokazu Agata, and Noriko Funayama. 2015. The ancestral gene repertoire of animal stem cells. *Proceedings of the National Academy of Sciences of the United States of America* 112: E7093–100. [CrossRef] [PubMed]

Alvizu, Adriana, Maria-Chistina Díaz, Christina Bastidas, Klaus Rützler, Rob Thacker, and Linda M. Márquez. 2013. A skeleton-less sponge of Caribbean mangroves: Invasive or undescribed? *Invertebrate Biology* 132: 81–94. [CrossRef]

Annenkova, Natalia, Dennis Lavrov, and Sergey Belikov. 2011. Dinoflagellates Associated with Freshwater Sponges from the Ancient Lake Baikal. *Protist* 162: 222–36. [CrossRef] [PubMed]

Baker, Lydia, Lindsay Freed, Cole Easson, Jose Lopez, Dante Fenolio, Tracey Sutton, Spencer V. Nyholm, and Tory A. Hendry. 2019. Diverse deep-sea anglerfishes share a genetically reduced luminous symbiont that is acquired from the environment. *Elife* 8: e47606. [CrossRef] [PubMed]

Bandi, Claudio, Massimo Sironi, Guiseppe Damiani, Lorenzo Magrassi, Christine Nalepa, Ugo Laudani, and Luciano Sacchi. 1995. The establishment of intracellular symbiosis in an ancestor of cockroaches and termites. *Proceedings of the Royal Society B: Biological Sciences* 259: 293–99. [PubMed]

Bohm, Markus, Ute Hentschel, Anja Friedrich, Lars Fieseler, Richard Stefen, Vera Gamulin, Isabel M. Muller, and Werner E.G. Muller. 2001. Molecular response of the sponge *Suberites domuncula* to bacterial infection. *Marine Biology* 139: 1037–45.

Bilandija, Helena, Jana Bedek, Branco Jalzic, and Sanja Gottstein. 2007. The morphological variability, distribution patterns and endangerment in the Ogulin cave sponge *Eunapius subterraneus* Sket & Velikonja, 1984 (Demospongiae, Spongillida). *Natura Croatica* 16: 1–17.

Biteau, Benoit, Christine Hochmuth, and Henrich Jasper. 2011. Maintaining Tissue Homeostasis: Dynamic Control of Somatic Stem Cell Activity. *Cell Stem Cell* 9: 402–11. [CrossRef]

Borchiellini, Carole, Kassandra de Pao-Mendonca, Amelie Vernale, Caroline Rocher, Alexander Ereskovsky, Jean Vacelet, and Emmanuelle Renard. 2021. Porifera (Sponges): Recent knowledge and new perspectives. *eLS* 2: 1–10. [CrossRef]

Borisenko, Ilya, Maya Adamska, Daria Tokina, and Alexander Ereskovsky. 2015. Transdifferentiation is a driving force of regeneration in *Halisarca dujardini* (Demospongiae, Porifera). *Peer J* 3: e1211. [CrossRef]

Bosch, Thomas C. G., Friederike Anton-Erxleben, Georg Hemmrich, and Konstantin Khalturin. 2010. The *Hydra* polyp: Nothing but an active stem cell community. *Development Growth and Differentiation* 52: 15–25. [CrossRef]

Braendle, Christian, Toru Miura, Ryan Bickel, Alexander Shingleton, Shrinivas Kambhampati, and David L. Stern. 2003. Developmental origin and evolution of bacteriocytes in the aphid-*Buchnera* symbiosis. *PLoS Biology* 1: 70–76. [CrossRef]

Brandt, Karl. 1881. *Ueber das Zusammenleben von Thieren und Algen. Verhandlungen der physiologischen Gesellschaft zu Berlin 1880–1881*. Berlin: Archiv für Anatomie und Physiologie—Physiologische Abtheilung. Physiologische Gesellschaft zu Berlin/Berlin Physiological Society, pp. 570–574.

Brien, Paul, and Denise Govaert-Mallebranche. 1958. A propos de deux éponges du Tanganika. *Académie Royale des Sciences Coloniales* 8: 1–43.

Bright, Monika, and Olav Giere. 2005. Microbial symbiosis in Annelida. *Symbiosis* 38: 1–45.

Burberry, Aaron, Melody Zeng, Lei Ding, Ian Wicks, Naohiro Inohara, Sean Morrison, and Gabriel Nunez. 2014. Infection mobilizes hematopoietic stem cells through cooperative NOD-like receptor and Toll-like receptor signalling. *Cell Host and Microbe* 15: 779–91. [CrossRef]

Buscema, Marco, Danielle De Sutter, and Gisele Van de Vyver. 1980. Ultrastructural study of differentiation processes during aggregation of purified sponge archaeocytes. *Wilhelm Roux's Archive of Developmental Biology* 53: 45–53. [CrossRef] [PubMed]

Carrier, Tyler, and Adam Reitzel. 2020. Symbiotic life of echinoderm larvae. *Frontiers in Ecology and Evolution* 7: 509. [CrossRef]

Cheng, Thomas C., Herbert W. F. Yee, and Erik Rifkin. 1968. Studies on the Internal Defense Mechanisms of Sponges I. The Cell Types Occurring in the Mesoglea of *Terpios zeteki* (de Laubenfels) (Porifera: Demospongiae). *Pacific Sciences* 22: 395–401.

Chernogor, Lubov, Natalia Denikina, Igor Kondratov, Ivan Solovarov, Igor Khanaev, Sergej Belikov, and Hermann Ehrlich. 2013. Isolation and identification of the microalgal symbiont from primmorphs of the endemic freshwater sponge *Lubomirskia baicalensis* (Lubomirskiidae, Porifera). *European Journal of Phycology* 48: 497–508. [CrossRef]

Chevalier, Frederic, Juline Herbiniére-Gaboreau, Joanne Bertaux, Maryline Raimond, Franck Morel, Didier Bouchon, Pierre Grève, and Christine Braquart-Varnier. 2011. The Immune Cellular Effectors of Terrestrial Isopod *Armadillidium vulgare*: Meeting with Their Invaders, *Wolbachia. PLoS ONE* 6: e18531. [CrossRef]

Clevers, Hans, and Fiona Watt. 2018. Defining adult stem cells by function, not by phenotype. *Annual Review Biochemistry* 87: 1015–27. [CrossRef]

Cox, Guy, and Anthony W. Larkum. 1983. A diatom apparently living in symbiosis with a sponge. *Bulletin of Marine Science* 33: 943–45.

Daniels, Camille, and Mya Breitbart. 2012. Bacterial communities associated with the ctenophores *Mnemiopsis leidyi* and *Beroe ovata. FEMS Microbiology Ecology* 82: 90–101. [CrossRef] [PubMed]

Dirks, Ulrich, Harald R. Gruber-Vodicka, Nikolaus Leisch, Silvia Bulgheresi, Bernhard Egger, Peter Ladurner, and Jörg A. Ott. 2012. Bacterial symbiosis maintenance in the asexually reproducing and regenerating flatworm *Paracatenula galateia. PLoS ONE* 7: e34709. [CrossRef]

Dubilier, Nivole, Claudia Bergin, and Christian Lott. 2008. Symbiotic diversity in marine animals: The art of harnessing chemosynthesis. *Nature Reviews Microbiology* 6: 725–40. [CrossRef]

Duperron, Sebastien, Claudia Bergin, Frank Zielinski, Anna Blazejak, Annelie Pernthaler, Zoe McKiness, and Nikole Dubilier. 2006. A dual symbiosis shared by two mussel species, Bathymodiolus azoricus and *Bathymodiolus puteoserpentis* (Bivalvia: Mytilidae), from hydrothermal vents along the northern Mid-Atlantic Ridge. *Environmental Microbiology* 8: 1441–47. [CrossRef]

Ereskovsky, Alexander, Alisia Geronimo, and Thierry Pérez. 2017. Asexual and puzzling sexual reproduction of the Mediterranean sponge *Haliclona fulva* (Demospongiae): Life cycle and cytological structures. *Invertebrate Biology* 136: 403–21. [CrossRef]

Ereskovsky, Alexander, and Andrey Lavrov. 2021. Porifera. In *Invertebrate Histology*. Edited by Elise E.B. LaDouceur. Hoboken: John Wiley & Sons, Inc., pp. 19–54.

Ereskovsky, Alexander, Daria B. Tokina, Damian Saidov, Stephen Baghdiguian, Emilie Le Goff, and Andrey Lavrov. 2020. Transdifferentiation and mesenchymal-to-epithelial transition during regeneration in Demospongiae (Porifera). *Journal of Experimental Zoology, Part B: Molecules, Development and Evolution* 334: 37–58. [CrossRef] [PubMed]

Ereskovsky, Alexander, Ilya E. Borisenko, Feodor V. Bolshakov, and Andrey I. Lavrov. 2021. Whole-body regeneration in sponges: Diversity, fine mechanisms and future prospects. *Genes* 12: 506. [CrossRef]

Ereskovsky, Alexander, Ilya E. Borisenko, Pascal Lapébie, Eve Gazave, Daria Tokina, and Carole Borchiellini. 2015. *Oscarella lobularis* (Homoscleromorpha, Porifera) regeneration: Epithelial morphogenesis and metaplasia. *PLoS ONE* 10: e0134566. [CrossRef]

Ereskovsky, Alexander, Lubov Chernogor, and Sergey Belikov. 2016. Ultrastructural description of development and cell composition of primmorphs in the endemic Baikal sponge *Lubomirskia baicalensis*. *Zoomorphology* 135: 1–17. [CrossRef]

Ereskovsky, Alexander. 2010. *The Comparative Embryology of Sponges*. Dordrecht: Springer, pp. 1–329.

Fernàndez-Busquets, Xavier. 2008. The sponge as a model of cellular recognition. In *Sourcebook of Models for Biomedical Research*. Edited by Philipp M. Conn. Totowa, NJ: Humana Press Inc., pp. 75–84.

Ferrario, Cinzia, Michela Sugni, Ildiko Somorjai, and Loriano Ballarin. 2020. Beyond adult stem cells: Dedifferentiation as a unifying mechanism underlying regeneration in invertebrate deuterostomes. *Frontiers in Cell and Developmental Biology* 8: 587320. [CrossRef]

Fierro-Constaín, Laura, Quentin Schenkelaars, Eve Gazave, Anne Haguenauer, Caroline Rocher, Alexander Ereskovsky, Carole Borchiellini, and Emmanuelle Renard. 2017. The conservation of the germline multipotency program, from sponges to vertebrates: A stepping stone to understanding the somatic and germline origins. *Genome Biology and Evolution* 9: 474–88. [CrossRef] [PubMed]

Fraune, Sebastian, Yuichi Abe, and Thomas Bosch. 2009. Disturbing epithelial homeostasis in the metazoan *Hydra* leads to drastic changes in associated microbiota. *Environmental Microbiology* 11: 2361–69. [CrossRef] [PubMed]

Frost, Thomas M., Linda E. Graham, Joan E. Elias, Mark J. Haase, Donald Kretchmer, and James Kranzfelder. 1997. A yellow-green algal symbiont in the freshwater sponge, *Corvomeyenia everetti*: Convergent evolution of symbiotic associations. *Freshwatar Biology* 38: 395–99. [CrossRef]

Frost, Thomas, and Craig Williamson. 1980. *In situ* determination of the effect of symbiotic algae on the growth of the freshwater sponge *Spongilla lacustris*. *Ecology* 61: 1361–70. [CrossRef]

Funayama, Noriko, Mikiko Nakatsukasa, Kurato Mohri, Yoshiki Masuda, and Kiyokazu Agata. 2010. *Piwi* expression in archeocytes and choanocytes in demosponges: Insights into the stem cell system in demosponges. *Evolution and Development* 12: 275–87. [CrossRef]

Funayama, Noriko, Mikiko Nakatsukasa, Tetsutaro Hayashi, Shigehiro Kuraku, Katsuaki Takechi, and Mikako Dohi. 2005. Isolation of Et silicatein and Et lectin as Molecular Markers for Sclerocytes and Cells Involved in Innate Immunity in the Freshwater Sponge *Ephydatia tluviatilis*. *Zoological Science* 22: 1113–22. [CrossRef]

Funayama, Noriko. 2018. The cellular and molecular bases of the sponge stem cell systems underlying reproduction, homeostasis and regeneration. *The International Journal of Developmental Biology* 62: 513–25. [CrossRef]

Gaino, Elda, Manuela Rebora, Carla Corallini, and Tisza Lancioni. 2003. The life-cycle of the sponge *Ephydatia fluviatilis* (L.) living on the reed *Phragmites australis* in an artificially regulated lake. *Hydrobiologia* 495: 127–42. [CrossRef]

Gaino, Elda, Renata Manconi, and Roberto Pronzato. 1995. Organizational plasticity as a successful conservative tactics in sponges. *Animal Biology* 4: 31–43.

Garrone, Robert. 1974. Ultrastructure d'une "gemmule armée" planctonique d'éponge Clionidae. *Inclusions fibrillaires et genèse du collagène, Arhives d'Anatomie Microscopique* 63: 163–82.

Garson, Mary J., Andrew E. Flowers, Richard I. Webb, Romila D. Charan, and Elizabeth J. McCaffrey. 1998. A sponge/dinoflagellate association in the haplosclerid sponge *Haliclona* sp.: Cellular origin of cytotoxic alkaloids by Percoll density gradient fractionation. *Cell and Tissue Researches* 293: 365–73. [CrossRef] [PubMed]

Geraghty, Sara, Koutsouveli Vasiliki, Hall Chelsea, Chang Lillian, Sacristan-Soriano Oriol, Malcolm Hill, Ana Riesgo, and April Hill. 2021. Establishment of Host–Algal Endosymbioses: Genetic response to symbiont versus prey in a sponge host. *Genome Biology Evolution* 13: evab252. [CrossRef] [PubMed]

Gilbert, John J., and Harold L. Allen. 1973. Chlorophyll and primary productivity of some green, freshwater sponges. *Internationale Revue der gesamten Hydrobiologie und Hydrographie* 58: 633–58. [CrossRef]

Gochfeld, Deborah, Maria-Christina Diaz, Abigail Renegar, and Julie Olson. 2019. Histological and ultrastructural features of Aplysina cauliformis affected by Aplysina red band syndrome. *Invertebrate Biology* 138. [CrossRef]

Godefroy, Nelly, Emilie Le Goff, Camille Martinand-Mari, Khalid Belkhir, Jean Vacelet, and Stephen Baghdiguian. 2019. Sponge digestive system diversity and evolution: Filter feeding to carnivory. *Cell and Tissue Researches* 377: 341–51. [CrossRef]

Gruber-Vodicka, Harald, Nikolaus Leisch, Manuel Kleiner, Tjorven Hinzke, Manuel Liebeke, Margaret McFall-Ngai, Michael G. Hadfield, and Nicole Dubilier. 2019. Two intracellular and cell type-specific bacterial symbionts in the placozoan *Trichoplax* H2. *Nature Microbiology* 4: 1465–74. [CrossRef]

Gruber-Vodicka, Harald, Ulrich Dirks, Nikolaus Leisch, Christian Baranyi, and Kilian Stoecker. 2011. *Paracatenula*, an ancient symbiosis between thiotrophic *Alphaproteobacteria* and catenulid flatworms. *Proceedings of the National Academy of Sciences of the United States of America* 108: 12078–83. [CrossRef]

Hackett, Kevin, Dwight Lynn, David Williamson, Annette Ginsberg, and Robert Whitcomb. 1986. Cultivation of the *Drosophila* sex-ratio spiroplasma. *Science* 232: 1253–55. [CrossRef]

Hall, Chelsea, Sara Camilli, Henry Dwaah, Benjamin Kornegay, Christie Lacy, Malcolm S. Hill, and April L. Hill. 2021. Freshwater sponge hosts and their green algae symbionts: A tractable model to understand intracellular symbiosis. *Ephydatia muelleri* Algal symbiosis transcriptomes. *PeerJ* 9: e10654. [CrossRef] [PubMed]

Hambleton, Elizabeth, Victor Jones, Ira Maegele, David Kvaskoff, Timo Sachsenheimer, and Annika Guse. 2019. Sterol transfer by atypical cholesterolbinding NPC2 proteins in coral-algal symbiosis. *eLife* 8: e43923. [CrossRef] [PubMed]

Handa, Shinji, Miro Nakahara, Hiromi Tsubota, Hironori Deguchi, Yoshiko Masuda, and Taketo Nakano. 2006. Choricystis minor (Trebouxiophyceae, Chlorophyta) as a symbiont of several species of freshwater sponge. *Hikobia* 14: 365–73.

Harrison, Frederic, and Leo De Vos. 1991. Porifera. In *Microscopic anatomy of invertebrates*. Edited by Frederic W. Harrison and John A. Westfall. New York: Wiley, Volume 2, pp. 29–89.

Harrison, Frederic W., Dana Dunkelberger, and Norimitsu Watabe. 1975. Cytological examination of reduction bodies of *Corvomeyenia carolinensis* Harrison (Porifera: Spongillidae). *Journal of Morphology* 145: 483–91. [CrossRef] [PubMed]

Hernandez, Alexandra, and Joseph Ryan. 2018. Horizontally transferred genes in the ctenophore *Mnemiopsis leidyi*. *PeerJ* 6: e5067. [CrossRef] [PubMed]

Hess, Samuel, and Anura Rambukkana. 2015. Bacterial-induced cell reprogramming to stem cell-like cells: New premise in host–pathogen interactions. *Current Opinion in Microbiology* 23: 179–88. [CrossRef] [PubMed]

Hikosaka-Katayama, Tomoe, Kanae Koike, Hiroshi Yamashita, Arira Hikosaka, and Kazuhiko Koike. 2012. Mechanisms of maternal inheritance of dinoflagellate symbionts in the acoelomorph worm *Waminoa litus*. *Zoological Sciences* 29: 559–67. [CrossRef]

Hill, Malcolm S. 1996. Symbiotic zooxanthellae enhance boring and growth rates of the tropical sponge *Anthosigmella varians* formavarians. *Marine Biology* 125: 649–54. [CrossRef]

Hinzke, Tjorven, Manuel Kleiner, Mareike Meister, Rabea Schlüter, Chrisyian Hentschker, Jan Pané-Farré, Petra Hildebrandt, Horst Felbeck, Stefan M. Sievert, Florian Bonn, and et al. 2021. Bacterial symbiont subpopulations have different roles in a deep-sea symbiosis. *eLife* 10: e58371. [CrossRef]

Hobmayer, Bert, Marcell Jenewein, Dominik Eder, Marie-Kristin Eder, and Stella Glasauer. 2012. Stemness in *Hydra*—A current perspective. *International Journal of Developmental Biology* 56: 509–17. [CrossRef]

Hu, Minjie, Xiaobin Zheng, Chen-Ming Fan, and Yixian Zheng. 2020. Lineage dynamics of the endosymbiotic cell type in the soft coral *Xenia*. *Nature* 582: 534–38. [CrossRef] [PubMed]

Hyams, Yosef, Guy Paz, Claudette Rabinowitz, and Baruh Rinkevich. 2017. Insights into the unique torpor of *Botrylloides leachi*, a colonial urochordate. *Developmental Biology* 428: 101–17. [CrossRef] [PubMed]

Ilan, Micha, and Avigdor Abelson. 1995. The life of a sponge in a sandy lagoon. *Biological Bulletin* 189: 363–69. [CrossRef]

Itskovich, Valeria B., Oxana V. Kaluzhnaya, Elena Veynberg, and Dirck Erpenbeck. 2017. Endemic Lake Baikal sponges from deep water 2: Taxonomy and Bathymetric Distribution. *Zootaxa* 4236: 335–42. [CrossRef]

Jorgensen, Barker. 1947. On the gemmules of *Spongilla lacustris* together with some remarks on the taxonomy of the species. *Det Kgl. Danske Videnskabernes Selskab, Biologiske Meddelelser* 20: 69–79.

Karagodina, Natalia, Andrey Vishnyakov, Olga Kotenko, Alina Maltseva, and Andrey Ostrovsky. 2018. Ultrastructural evidence for nutritional relationships between a marine colonial invertebrate (Bryozoa) and its bacterial symbionts. *Symbiosis* 75: 155–64. [CrossRef] [PubMed]

Kerney, Ryan, Eunsoo Kim, Roger Hangarter, Aaron Heiss, Cory Bishop, and Bryan Hall. 2011. Intracellular invasion of green algae in a salamander host. *Proceedings of the National Academy of Sciences of the United States of America* 108: 6497–502. [CrossRef] [PubMed]

Kjeldsen, Kasper, Mattias Obst, Hiroaki Nakano, Peter Funch, and Andreas Schramm. 2010. Two types of endosymbiotic bacteria in the enigmatic marine worm *Xenoturbella bocki*. *Applied and Environmental Microbiology* 76: 2657–62. [CrossRef]

Kobiałka, Michat, Anna Michalik, Marcin Walczak, Lukasz Junkiert, and Tereza Szklarzewicz. 2016. *Sulcia* symbiont of the leafhopper *Macrosteles laevis* (Ribaut, 1927) (Insecta, Hemiptera, Cicadellidae: Deltocephalinae) harbors *Arsenophonus* bacteria. *Protoplasma* 253: 903–12. [CrossRef]

Kovacevic, Goran. 2012. Value of the *Hydra* model system for studying symbiosis. *The International Journal of Developmental Biology* 56: 627–35. [CrossRef]

Kroer, Paul, Kasper Kjeldsen, Jens Nyengaard, Andreas Schramm, and Peter Funch. 2016. A novel extracellular gut symbiont in the marine worm *Priapulus caudatus* (Priapulida) reveals an alphaproteobacterial symbiont clade of the Ecdysozoa. *Frontiers in Microbiology* 7: 539. [CrossRef] [PubMed]

Krueger, Thomas. 2016. Concerning the cohabitation of animals and algae—An English translation of K. Brandt's 1881 presentation "Ueber das Zusammenleben von Thieren und Algen". *Symbiosis* 71: 167–74. [CrossRef]

Kulakova, Nina, Natalia Denikina, and Sergey Belikov. 2014. Diversity of Bacterial Photosymbionts in Lubomirskiidae Sponges from Lake Baikal. *International Journal of Biodiversity* 2014: 152097. [CrossRef]

Kulakova, Nina, Serrgey Kashin, and Yuriy Bukin. 2020. The genetic diversity and phylogeny of green microalgae in the genus *Choricystis* (Trebouxiophyceae, Chlorophyta) in Lake Baikal. *Limnology* 21: 15–24. [CrossRef]

Lambiase, Simonetta, Aldo Grigolo, Ugo Laudani, Luciano Sacchi, and Baccio Baccetti. 1997. Pattern ofbacteriocyte formation in *Periplaneta americana* (L.) (Blattaria: Blattidae). *International Journal of Insect Morphology and Embryology* 26: 9–19. [CrossRef]

Lavrov, Andrey I., Feodor V. Bolshakov, Daria Tokina, and Alexander Ereskovsky. 2018. Sewing up the wounds: The epithelial morphogenesis as a central mechanism of calcaronean sponge regeneration. *Journal of Experimental Zoology, Part B: Molecules, Development and Evolution* 330: 351–71. [CrossRef] [PubMed]

Lavrov, Andrey, and Igor Kosevich. 2014. Sponge cell reaggregation: Mechanisms and dynamics of the process. *Russian Journal of Developmental Biology* 45: 205–23. [CrossRef]

Leininger, Sven, Marcin Adamski, Brith Bergum, Corina Guder, Jing Liu, Mary Laplante, Jon Bråte, Friederike Hoffmann, Sofia Fortunato, Signe Jordal, and et al. 2014. Developmental gene expression provides clues to relationships between sponge and eumetazoan body plans. *Nature Communications* 5: 3905. [CrossRef] [PubMed]

Lemloh, Marie-Louise, Janny Fromont, Franz Brummer, and Kayley Usher. 2009. Diversity and abundance of photosynthetic sponges in temperate Western Australia. *BMC Ecology* 9: 4. [CrossRef]

Lévi, Claude, and Pierrete Lévi. 1976. Embryogenèse de *Chondrosia reniformis* (Nardo), démosponge ovipareé et transmission des bactéries symbiotiques. *Annales des Sciellces Naturelles, Zoologie* 18: 367–80.

Levy, Shani, Anamaria Elek, Xavier Grau-Bové, Simon Menéndez-Bravo, Marta Iglesias, Amos Tanay, Tali Mass, and Arnau Sebé-Pedrós. 2021. A stony coral cell atlas illuminates the molecular and cellular basis of coral symbiosis, calcification, and immunity. *Cell* 11: 2973–87. [CrossRef] [PubMed]

Lindquist, Neils, Paul Barber, and Jeremy Weisz. 2005. Episymbiotic microbes as food and defence for marine isopods: Unique symbioses in a hostile environment. *Proceedings of the Royal Society B: Biological Sciences* 272: 1209–16. [CrossRef]

Maire, Justin, Nicolas Parisot, Mariana Ferrarini, Agnes Vallier, Benjamin Gillet, Sandrine Hughes, Séverine Balmand, Carole Vincent-Monégat, Anna Zaidman-Rémy, and Abdelaziz Heddi. 2020. Spatial and morphological reorganization of endosymbiosis during metamorphosis accommodates adult metabolic requirements in a weevil. *Proceedings of the National Academy of Sciences of the United States of America* 117: 19347–58. [CrossRef] [PubMed]

Maldonado, Manuel. 2007. Intergenerational transmission of symbiotic bacteria in oviparous and viviparous demosponges, with emphasis on intracytoplasmically compartmented bacterial types. *Journal of the Marine Biological Association of the UK* 87: 1701–13. [CrossRef]

Maldonado, Manuel. 2016. Sponge waste that fuels marine oligotrophic food webs: A re-assessment of its origin and nature. *Marine Ecology* 37: 477–91. [CrossRef]

Masuda, Yoshiki. 1990. Electron microscopic study on the zoochlorellae of some freshwater sponges. In *New Perspectives in Sponge Biology*. Edited by Klais Rützler. Washington, DC: Smithsonian Institution Press Washington, pp. 467–71.

Masuda, Yoshiko. 1985. Electron Microscopic Study on the Zoochlorellae of Adult Green Sponges and Gemmules of *Radiospongilla cerebellata* (Bowerbank) (Porifera: Spongillidae). *Kawasaki Igakkai Shi Liberal Arts Sci Course* 11: 63–66.

McDermott, John. 2006. Nemerteans as hosts for symbionts: A review. *Journal of Natural History* 40: 1007–20. [CrossRef]

Melo-Clavijo, Jenny, Alexander Donath, Joao Serôdio, and Gregor Christa. 2018. Polymorphic adaptations in metazoans to establish and maintain photosymbioses. *Biological Reviews* 93: 2006–20. [CrossRef]

Melton, Douglas. 2014. 'Stemness' 'Stemness': Definitions, Criteria, and Standards. In *Essentials of Stem Cell Biology, 3d ed.* Edited by Ronald Lanza and Alex Atala. London: Academic Press, pp. 7–17.

Merrell, Allyson, and Ben Stanger. 2016. Adult cell plasticity in vivo: De-differentiation and transdifferentiation are back in style. *Nature Reviews Molecular Cell Biology* 17: 413–25. [CrossRef]

Miura, Toru, Christian Braendle, Alexander Shingleton, Geoffroy Sisk, Srinivas Kambhampati, and David Stern. 2003. A comparison of parthenogenetic and sexual embryogenesis of the pea aphid *Acyrthosiphon pisum* (Hemiptera: Aphidoidea). *Journal of Experimental Zoology Part B Molecular and Developmental Evolution* 295B: 59–81. [CrossRef]

Mutalipassi, Mirko, Gennaro Riccio, Valerio Mazzella, Christian Galasso, Emmanuele Somma, Antonia Chiarore, Donatella de Pascale, and Valerio Zupo. 2021. Symbioses of cyanobacteria in marine environments: Ecological insights and biotechnological perspectives. *Marine Drugs* 19: 227. [CrossRef]

Nakanishi, Nagayasu, Shunsuke Sogabe, and Bernard Degnan. 2014. Evolutionary origin of gastrulation: Insights from sponge development. *BMC Biology* 12: 26. [CrossRef] [PubMed]

Negandhi, Karita, Patricia Blackwelder, Alexander Ereskovsky, and Jose Lopez. 2010. Florida reef sponges harbor coral disease-associated microbes. *Symbiosis* 51: 117–29. [CrossRef]

Nguyen, Mary, Michael Liu, and Torsten Thomas. 2014. Ankyrin-repeat proteins from sponge symbionts modulate amoebal phagocytosis. *Molecular Ecology* 23: 1635–45. [CrossRef] [PubMed]

Nigro, Giulina, Raffaella Rossi, Pierre-Henri Commere, Philippe Jay, and Philippe Sansonetti. 2014. The cytosolic bacterial peptidoglycan sensor Nod2 affords stem cell protection and links microbes to gut epithelial regeneration. *Cell Hostand Microbe* 15: 792–98. [CrossRef]

Nikolakakis, Kiel, Erik Lehnert, Margaret McFall-Ngai, and Edward Ruby. 2015. Use of hybridization chain reaction-fluorescent in situ hybridization to track gene expression by both partners during initiation of symbiosis. *Applied and Environmental Microbiology* 81: 4728–35. [CrossRef]

Okuda, Naomi, Atsuko Yamamoto, Yuriko Satoh, Yoshihiro Fujimoto, and Yoshihisa Kamishima. 2002. Role of Symbiotic Algae on Gemmule Germination of a Freshwater Sponge, *Radiospongilla cerebellata*. *Chugokugakuen Journal* 1: 7–12.

Ote, Manabu, and Daisuke Yamamoto. 2020. Impact of *Wolbachia* infection on *Drosophila* female germline stem cells. *Current Opinion in Insect Science* 37: 8–15. [CrossRef]

Pang, Rosemary. 1973. The ecology of some Jamaican excavating sponges. *Bulletin Marine Sciences* 23: 227–43.

Pflugfelder, Bettina, Craig S. Cary, and Monika Bright. 2009. Dynamics of cell proliferation and apoptosis reflect different life strategies in hydrothermal vent and cold seep vestimentiferan tubeworms. *Cell Tissue Research* 337: 149–65. [CrossRef]

Pimentel, Andre C., Casia Cesar, Marcos Martins, and Rodrigo Cogni. 2021. The antiviral effects of the symbiont bacteria *Wolbachia* in insects. *Frontiers in Immunology* 11: 626329. [CrossRef] [PubMed]

Price, Ian, Richard Fricker, and Clive Wilkinson. 1984. *Ceratodictyon spongiosum* (Rhodophyta), the macroalgal partner in an alga- sponge symbiosis, grown in unialgal culture. *Journal of Phycology* 20: 156–58. [CrossRef]

Pröschold, Thomas, and Tatiana Darienko. 2020. *Choricystis* and *Lewiniosphaera* gen. nov. (Trebouxiophyceae Chlorophyta), two different green algal endosymbionts in freshwater sponges. *Symbiosis* 82: 175–88. [CrossRef] [PubMed]

Pröschold, Thomas, Tatyana Darienko, Paul Silva, Werner Reisser, and Lothar Krienitz. 2010. The systematics of *Zoochlorella* revisited employing an integrative approach. *Environmental Microbiology* 13: 350–64. [CrossRef] [PubMed]

Raff, Martin. 2003. Adult stem cell plasticity: Fact or artifact? *Annual Review of Cell and Developmental Biology* 19: 1–22. [CrossRef]

Rasmont, Robert. 1970. Some new aspects of the physiology of freshwater sponges. In *The Biology of the Porifera*. Edited by William G. Fry. London: Academic Press, pp. 415–22.

Redmond, Anthony K., and Aoife McLysaght. 2021. Evidence for sponges as sister to all other animals from partitioned phylogenomics with mixture models and recoding. *Nature Communication* 12: 1783. [CrossRef] [PubMed]

Reisser, Werner, and Werner Wiessner. 1984. Autotrophic Eukaryotic Freshwater Symbionts. In *Cellular Interactions. Encyclopedia of Plant Physiology*. Edited by Hans Linskens and John Heslop-Harrison. Berlin/Heidelberg: Springer, Volume 17, pp. 59–74.

Rinkevich, Baruh, Loriano Ballarin, Pedro Martinez, Ildiko Somorjai, Oshrat Ben-Hamo, Ilya Borisenko, Eugene Berezikov, Alexander Ereskovsky, Eve Gazave, Denis Khnykin, and et al. 2022. A pan-metazoan concept for adult stem cells: The wobbling Penrose landscape. *Biological Review* 97: 299–325. [CrossRef]

Rosell, Doloris, and Maria J. Uriz. 1992. Do associated zooxanthellae and the nature of the substratum affect survival, attachment and growth of *Cliona viridis* (Porifera, Hadromerida)—An experimental approach. *Marine Biology* 114: 503–7. [CrossRef]

Rosell, Doloris. 1993. Effects of reproduction in Cliona viridis (Hadromerida) on zooxanthellae. In *Recent advances in ecology and systematics of sponges*. Edited by Maria J. Uriz and Klaus Rützler. Barcelona: Instituto de Ciencias del mar, C.S.I.C., Volume 54, pp. 405–13.

Rozenfeld, Francine. 1980. Effects of puromycin on the differentiation of the freshwater sponge: *Ephydatia fluviatilis*. *Differentiation* 17: 193–98. [CrossRef]

Rubin-Blum, Maxim, Chakkitah Antony, Lisbeth Sayavedra, Clara Martínez-Pérez, Daniel Birgel, Jörn Peckmann, Yu-Chen Wu, Paco Cardenas, Ian MacDonald, Yann Marcon, and et al. 2019. Fueled by methane: Deep-sea sponges from asphalt seeps gain their nutrition from methane-oxidizing symbionts. *The ISME Journal* 13: 1209–25. [CrossRef]

Rützler, Klaus, and Gregor Rieger. 1973. Sponge burrowing: Fine structure of *Cliona lampa* penetrating calcareous substrata. *Marine Biology* 21: 144–6246. [CrossRef]

Rützler, Klaus, Rob W.B. van Soest, and Belinda Alvarez. 2003. *Svenzea zeai*, a Caribbean reef sponge with a giant larva, and *Scopalina ruetzleri*: A comparative fine-structural approach to classification (Demospongiae, Halichondrida, Dictyonellidae). *Invertebrate Biology* 122: 203–22. [CrossRef]

Rützler, Klaus. 1990. Associations between Caribbean sponges and photosynthetic organisms. In *New Perspectives in Sponge Biology*. Edited by Klaus Rützler. Washington, DC: Smithsonian Institution Press, pp. 455–66.

Saffo, Mary. 1992. Invertebrates in endosymbiotic associations. *American Zoologist* 32: 557–65. [CrossRef]

Saller, Uwe. 1989. Microscopical aspects on symbiosis of *Spongilla lacustris* (Porifera, Spongillidae) and green algae. *Zoomorphology* 108: 291–96. [CrossRef]

Saller, Uwe. 1990. Formation and construction of asexual buds of the freshwater sponge *Radiospongilla cerebellata* (Porifera, Spongillidae). *Zoomorphology* 109: 295–301. [CrossRef]

Saller, Uwe. 1991. Symbiosis of Spongilla lacustris (Spongillidae) and Green Algae. Algae Uptake, Distribution and Final Whereabouts. In *Fossil and Recent Sponges*. Edited by Joachim Reitner and Hans Keupp. Berlin: Springer, pp. 299–305.

Salomon, Christine E., Thomas Deerinck, Mark H. Ellisman, and David J. Faulkner. 2001. The cellular localization of dercitamide in the Palauan sponge *Oceanapia sagittaria*. *Marine Biology* 139: 313–19.

Sarà, Michele, and Jean Vacelet. 1973. Ecologie des Démosponges. In *Traité de Zoologie*. Edited by Pierre P. Grassé. Paris: Masson, Volume 3, pp. 462–576.

Sarà, Michele, and Lidia Liaci. 1964. Symbiotic association between Zooxanthellae and two Marine Sponges of the Genus *Cliona*. *Nature* 203: 321. [CrossRef]

Sarà, Michele, Gorgio Bavestrello, Riccardo Cattaneo-Vietti, and Carlo Cerrano. 1998. Endosymbiosis in Sponges—Relevance for Epigenesis and Evolution. *Symbiosis* 25: 57–70.

Scalera-Liaci, Lidia, Margaretta Sciscioli, Elena Lepore, and Elda Gaino. 1999. Symbiotic zooxanthellae in *Cinachyra tarentina*, a non-boring demosponge. *Endocytobiosis Cell Research* 13: 105–14.

Schellenberg, Johannes, Jessica Reichert, Martin Hardt, Ines Klingelhöfer, Gertrud Morlock, Patrick Schubert, Mina Bižić, Hans-Peter Grossart, Peter Kämpfer, Thomas Wilke, and et al. 2020. The Bacterial Microbiome of the Long-Term Aquarium Cultured High-Microbial Abundance Sponge *Haliclona cnidata*—Sustained Bioactivity Despite Community Shifts Under Detrimental Conditions. *Frontiers in Marine Sciences* 7: 266. [CrossRef]

Selmi, Carlo. 2001. Ectosymbiotic bacteria on ciliated cells of a rotifer. *Tissue and Cell* 33: 258–61. [CrossRef]

Sharp, Koty, Seana Davidson, and Margo Haygood. 2007. Localization of 'Candidatus Endobugula sertula' and the bryostatins throughout the life cycle of the bryozoan *Bugula neritina*. *The ISME Journal* 1: 693–702. [CrossRef] [PubMed]

Simion, Paule, Herve Philippe, Dennis Baurain, Muriel Jager, Daniel Richter, Arnaud Di Franco, B. Roure, N. Satoh, E. Quéinnec, A. Ereskovsky, and et al. 2017. Tackling the conundrum of metazoan phylogenomics: Sponges are sister to all other animals. *Current Biology* 27: 1–10. [CrossRef] [PubMed]

Simpson, Tracey. 1984. *The Cell Biology of Sponges*. New York: Springer, p. 678.

Sitte, Peter, and Stefan Eschbach. 1992. Cytosymbiosis and Its Significance in Cell Evolution. In *Progress in Botany*. Edited by H.-Dietmar Behnke, Karl Esser, Klaus Kubitzki, Michael Runge and Hubert Ziegler. Berlin: Springer, Volume 53, pp. 29–43.

Smith, Courtney, and William H. Hildemann. 1986. Allograft rejection, autograft fusion and inflammatory responses to injury in *Callyspongia diffusa* (Porifera; Demospongia). *Proceeding of Royal Society, London* 226: 445–64.

Song, Chihong, Kazuyoshi Murata, and Tochinobu Suzaki. 2017. Intracellular symbiosis of algae with possible involvement of mitochondrial dynamics. *Scientific Reports* 7: 1221. [CrossRef] [PubMed]

Sørensen, Megan E.S., Jamie A. Wood, Duncan D. Cameron, and Michael Brockhurst. 2021. Rapid compensatory evolution can rescue low fitness symbioses following partner switching. *Current Biology* 31: 3721–28. [CrossRef] [PubMed]

Taylor, Jessica, Giorgia Palladino, Bernd Wemheuer, Georg Steinert, Detmer Sipkema, Timothy J. Williams, and Torsten Thomas. 2021. Phylogeny resolved, metabolism revealed: Functional radiation within a widespread and divergent clade of sponge symbionts. *The ISME Journal* 15: 503–19. [CrossRef] [PubMed]

Thompson, Luke, Kiel Nikolakakis, Shu Pan, Jennifer Reed, Rob Knight, and Edward Ruby. 2017. Transcriptional characterization of *Vibrio fischeri* during colonization of juvenile *Euprymna scolopes*. *Environmental Microbiology* 19: 1845–56. [CrossRef] [PubMed]

Thuesen, Erik, and Kazuhiro Kogure. 1989. Bacterial production of tetrodotoxin in four species of Chaetognatha. *The Biological Bulletin* 176: 191–94. [CrossRef]

Tianero, Ma Diarey, Jared Balaich, and Mohamed Donia. 2019. Localized production of defence chemicals by intracellular symbionts of *Haliclona* sponges. *Nature Microbiology* 4: 1149–59. [CrossRef]

Todaro, Antonio, Matteo Dal Zotto, Sarah Bownes, and Robert Perissinotto. 2017. Two new interesting species of Macrodasyida (Gastrotricha) from KwaZulu-Natal (South Africa). *Proceedings of the Biological Society of Washington* 130: 140–55. [CrossRef]

Usher, Kayley, and Alexander Ereskovsky. 2004. Larval development, ultrastructure and metamorphosis in *Chondrilla australiensis* Carter, 1873 (Demospongiae, Chondrosida, Chondrillidae). *Invertebrate Reproduction and Development* 47: 51–62. [CrossRef]

Vacelet, Jean, Aline Fiala-Medioni, Christine R. Fisher, and Nicole Boury-Esnault. 1996. Symbiosis between methane-oxidizing bacteria and a deep-sea carnivorous cladorhizid sponge. *Marine Ecology, Progress Serie* 145: 77–85. [CrossRef]

Vacelet, Jean, and Claude Donaday. 1977. Electron microscope study of the association between some sponges and bacteria. *Journal of Experimental Marine Biology and Ecology* 30: 301–14. [CrossRef]

Vacelet, Jean, and Nicole Boury-Esnault. 1996. A new species of carnivorous sponge (Demospongiae: Cladorhizidae) from a Mediterranean cave. *Bulletin de L'institut Royal des Sciences Naturelles de Belgique* 66: 109–15.

Vacelet, Jean, and Thierry Perez. 1998. Two new genera and species of sponges without skeleton (Porifera, Demospongiae) from a Mediterranean cave. *Zoosystema* 20: 5–22.

Vacelet, Jean. 1975. Etude en microscopie electronique de l'association entre bacteries et spongiaires du genre *Verongia* (Dictyoceratida). *Journal Microscopie Biol Cell* 23: 271–88.

Vacelet, Jean. 1981. Algal-sponge symbioses in the coral reefs of New Caledonia: A morphological study. Paper presented at Fourth International Coral Reef Symposium, Manila, Philippines, 18–22 May 1981; pp. 713–19.

Vecchi, Matteo, Filipe Vicente, Roberto Guidetti, Roberto Bertolani, Lorena Rebecchi, and Michele Cesari. 2016. Interspecific relationships of tardigrades with bacteria, fungi and protozoans, with a focus on the phylogenetic position of *Pyxidium tardigradum* (Ciliophora). *Zoological Journal of the Linnean Society* 178: 846–55. [CrossRef]

Venn, A. Alexander, John E. Loram, and Angeka E. Douglas. 2008. Photosynthetic symbioses in animals. *Journal of Experimental Botany* 59: 1069–80. [CrossRef]

Vogt, Gunter. 2012. Hidden treasures in stem cells of indeterminately growing bilaterian invertebrates. *Stem Cell Reviews and Reports* 8: 305–17. [CrossRef]

Volkmer-Ribeiro, Cecille, Maria Bichuette, and Vanessa Machado. 2010. *Racekiela cavernicola* (Porifera: Demospongiae) new species and the first record of cave freshwater sponge from Brazil. *Neotropical Biology and Conservation* 5: 53–58. [CrossRef]

Wernegreen, Jennifer J. 2012. Endosymbiosis. *Current Biology* 22: R555–R561. [CrossRef]

Wilkinson, Clive. 1978. Microbial associations in sponges. III. Ultrastructure of the in situ associations in coral reef sponges. *Marine Biology* 49: 177–85. [CrossRef]

Wilkinson, Clive R. 1980. Nutrient translocation from green algal symbionts to the freshwater sponge *Ephydatia fluviatilis*. *Hydrobiologia* 75: 241–50. [CrossRef]

Wilkinson, Clive. 1983. Phylogeny of bacterial and cyanobacterial symbionts in marine sponges. *Endocytobiology* 2: 993–1002.

Wilkinson, Clive. 1987. Significance of microbial symbionts in sponge evolution and ecology. *Symbiosis* 4: 135–46.

Wilkinson, Clive. 1992. Symbiotic interactions between marine sponges and algae. In *Algae and Symbioses: Plants, Animals, Fungi and Viruses, Interactions Explored*. Edited by William Reisser. Bristol: Biopress Ltd., pp. 112–5187.

Willenz, Philippe, and Gisele Van de Vyver. 1984. Ultrastructural localization of lysosomal digestion in the freshwater sponge *Ephydatia fluviatilis*. *Journal of Ultrastructural Researcjes* 87: 13–22. [CrossRef]

Williamson, Craig E. 1979. An ultrastructural investigation of algal symbiosis in white and green *Spongilla lacustris* (L.) (Porifera: Spongillidae). *Transactions of the American Microscopical Society* 98: 59–77. [CrossRef]

Woollacott, Robert. 1993. Structure and swimming behavior of the larva of *Haliclona tubifera* (Porifera, Demospongiae). *Journal of Morphology* 218: 301–21. [CrossRef]

Worheide, Gerd. 1998. The reef cave dwelling ultraconservative coralline demosponge *Astrosclera willeyana* Lister 1900 from the Indo-Pacific. *Facies* 38: 1–88. [CrossRef]

Somatic Expression of Stemness Genes in Aquatic Invertebrates

Loriano Ballarin, Bert Hobmayer, Amalia Rosner and Baruch Rinkevich

Abstract: Adult stem cells (ASCs) of aquatic invertebrates are involved in important biological processes such as regeneration and asexual reproduction. Unlike in vertebrates, they share pluripotency and even totipotency, and do not reside in permanent niches. Aquatic invertebrates represent the widest phylogenetic animal radiation on earth, but until now, limited research data have been available on their ASCs. Although less studied than their vertebrate counterparts, aquatic invertebrate ASCs express orthologues of many vertebrate genes usually associated with stemness. With this review, we aim at providing a database for current and future studies on ASC properties through a comprehensive literature analysis of intra- and inter-phylum comparisons of gene expressions and their functions in aquatic invertebrate ASCs. We concentrate on major gene families where sufficient data are available; gaps in our results will be filled by future studies on ASCs of aquatic invertebrates.

1. Introduction

Aquatic invertebrates present the widest metazoan radiation, and by virtue of their intraphylum diversity, they form large assemblages of multicellular animals and represent many model species in a wide range of biological disciplines (Ballarin et al. 2018). They include sponges (phylum Porifera), diploblastic jellyfish, anemones and corals (phylum Cnidaria), and triploblastic animals, the latter further divided into protostomes and deuterostomes. Protostomes include Spiralia (e.g., phyla Platyhelminthes, Annelida, Mollusca) and Ecdysozoa (e.g., phyla Nematoda and Arthropoda), whereas deuterostomes have Echinodermata and the subphyla Cephalochordata and Tunicata of the phylum Chordata as the prominent representatives (Figure 1). However, only limited research data are available on adult stem cells (ASCs) in general and ASC characteristics in particular in aquatic invertebrates (Rinkevich and Matranga 2009; Ballarin et al. 2021; Rinkevich et al. 2022). This is in clear contrast to the fact that ASCs in aquatic invertebrates are key players in many biological processes, such as regeneration, asexual reproduction/budding, torpor phenomena and more (Rinkevich et al. 2022).

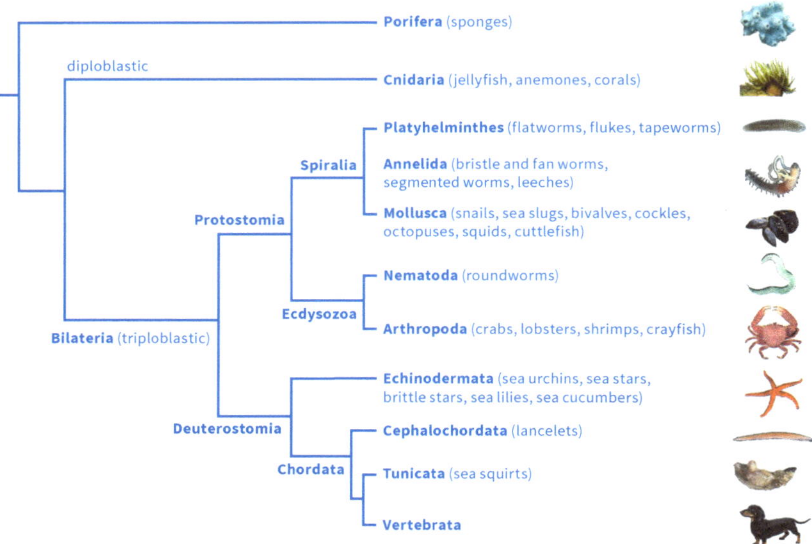

Figure 1. Phylogenetic relationship among the main invertebrate phyla. Graphic by authors.

The vertebrates possess ASCs homed to defined niches that are known to hold multipotency at best, representing lineage-specific self-renewing cells with tissue/organ-specific activities that generate limited numbers of daughter cell types (self-renewing progenitors and differentiated cells; Wagers and Weissman 2004; Clevers and Watt 2018). A careful examination of the animal phylogenetic tree (Rinkevich et al. 2022) reveals that ASCs have been studied only in a few metazoan phyla outside of the vertebrates, mostly taxa with high capabilities for asexual reproduction and regeneration (including whole-body regeneration) such as sponges, cnidarians, platyhelminthes, tunicates and echinoderms. These ASCs reveal dramatic disparities from vertebrate ASCs. Many of them are pluripotent and even totipotent, they do not follow the germline-sequestering model of the vertebrates, many exhibit morphologies of highly differentiated cells, they may generate the entire repertoire of cell types in adult animals, and some of them do not reside in stem cell niches (Rinkevich et al. 2022; Martinez et al. 2022).

It is thus of obvious importance to evaluate the properties of aquatic invertebrate ASCs using inter- and intra-phylum comparative and mechanistic analyses focusing on biological processes or specific properties. Indeed, aquatic invertebrate ASCs express orthologues of many vertebrate "stemness" genes (associated or disparate from the biological phenomena studied in the vertebrates), even though the molecular machinery by which these organisms hold sustainable ASC stocks along their lifespan is still unsolved (Conte et al. 2009; Rinkevich et al. 2022).

In the following sections, we provide a comprehensive literature analysis of intra- and inter-phylum comparisons of gene expressions and their functions in aquatic invertebrate ASCs. We concentrated on major gene families where sufficient data are available, and we aim at providing a database for current and future studies on ASC properties and at deducing general aspects of ASC gene-regulatory programs across the metazoan tree of life. The most important current models for the study of aquatic invertebrate ASCs are represented by sponges, cnidarians, flatworms, annelids, echinoderms and colonial ascidians, but limited data are available also for ctenophores (comb jellies), xenacelomorphs (acoel flatworms), hemichordates (acorn worms) and cephalochordates (lancelets). It must be noted that since most of the cited studies were not performed using the most cutting-edge methodological approaches and techniques (including using well-annotated genome assemblies and in-depth, single-cell-transcriptome data), there may be gaps in our results, for which future studies are required to resolve the presence or absence of stem cell gene expression.

2. RNA-Binding Proteins (RBPs)

One main commonality that characterizes the pluripotent ASCs of aquatic invertebrates is the abundance of expressed RBPs. They are represented by several families of proteins situated in the cytoplasm and/or the nucleus, and account for the major differences between transcribed mRNA and protein levels eventually synthesized by the cells. RBPs have a role in every aspect of mRNA post-transcriptional regulation (mRNAs biogenesis, stability, function, transport, structure and interactions with other RNAs and proteins). Specific RBPs and mRNA combinations in stem cells (ribonucleoprotein, RNPs) lead to mRNA alternative splicing, 3′ UTR cleavage and polyadenylation, mRNA sequence alteration at control regions, and altogether impact the specific function or stability of mRNAs in stems cells (Shigunov and Dallagiovanna 2015). Germ stem cells and early differentiating cells of a germline lineage contain condensed protein–mRNA complexes called nuage/pole plasm/germ plasm/germ granule/chromatoid bodies, similar in content to complexes found in some pluri/multipotent somatic cells (Juliano et al. 2010). The RBP families listed below are all part of the nuage-like structures and function in post-transcriptional regulation and in curtailing the activity of transposable elements (TEs) in order to assure genome integrity. These proteins are believed to be part of the "germline multi-potency program" (GMP), but are also expressed in adult somatic stem cells.

2.1. Argonaute (Ago)/Piwi Family

The ago/piwi family is composed of three subfamilies, two of which—ago and piwi—have vital functions in all multicellular organisms. Ago/piwi proteins form complexes with small noncoding RNAs. These complexes silence transposons and specific genes at various stages of RNA metabolism, perform chromatin modifications or inhibit mRNA translation. Piwi proteins associate with piwi-interacting RNAs (piRNAs) and are restricted to the germ lineage and pluripotent stem cells. Piwi proteins not only pair with piRNA, but also participate in piRNA biogenesis through the so-called "ping-pong" amplification that occurs in the nuage-like regions (Czech et al. 2018). The ago proteins are involved in the formation of the RNA-induced silencing complex (RISC), which specifically targets mRNA or DNA sequences in the genome and silences them. They bind to micro RNAs (miRNAs) and silence complementary transcripts by either destroying them or preventing their correct translation (Höck and Meister 2008; Meister 2013).

Expression of *ago* genes in ASCs has been reported in choanocytes, pinacocytes and type 1 vacuolar cells of the homoscleromorph sponge *Oscarella lobularis* (Fierro-Constaín et al. 2017). *ago* genes are also present in placozoan and cnidarian genomes (Grimson et al. 2008), although their expression was not studied in detail. An *ago2* gene is transcribed by neoblasts of the planarian *Dugesia japonica* (Rouhana et al. 2010), as well as by some somatic cells. Another ago protein is present in neoblasts of the fluke *Schistosoma mansoni* (Collins et al. 2013). No data on the presence of ago proteins in ASCs of coelomate metazoans are present in the literature.

As far as piwi proteins are concerned, they are expressed in ASCs of almost all the metazoans characterized by high regenerative power. Two *piwi* genes have been described in sponges, active in choanocytes and archeocytes of the demosponge *Ephydatia fluviatilis* (Funayama et al. 2010; Alié et al. 2015) and in choanocytes, pinacocytes and mesohyl type 1 vacuolar cells of the homoschleromorph sponge *O. lobularis* (Fierro-Constaín et al. 2017). No *piwi* genes are present in Placozoans (Grimson et al. 2008), whereas a *piwi* gene is present in the comb jelly *Pleurobrachia pileus*, expressed in progenitors of colloblasts, muscle cell, cells of the forming combs and of the aboral sense organ (Alié et al. 2011). In Cnidaria, two *piwi* orthologues (*hywy* and *hyli*) are actively transcribed by i-cells and epidermal cells of *Hydra vulgaris* and *Hydra magnipapillata*: in the former species, their mRNAs are also located in nematoblasts, the precursors of stinging cells (Juliano et al. 2014). A single *piwi* gene is transcribed in the i-cells of the hydroid *Hydractinia echinata* (Plickert et al. 2012) and the siphonophoran *Nanomia bijuga* (Siebert et al. 2015), in transdifferentiating epitheliomuscular cells of *Podocoryne carnea* medusae (Seipel et al. 2004), and in nematoblasts of the *Clytia hemisferica* medusae (Denker et al. 2008). Among bilateria, *piwi* expression has been demonstrated in the neoblasts of the acoelomorph worm *Isodiametra pulchra* (Egger et al. 2009) and of the planarians *Macrostomum lignano* (two piwi proteins; Pfister et al. 2007; Zhou et al. 2015), *Schmidtea mediterranea* (three active piwi genes; Reddien et al. 2005a, 2005b; Palakodeti et al. 2008; Rouhana et al. 2014) and *D. japonica* (six piwi genes; Rossi et al. 2006, 2007; Rouhana et al. 2010;

Shibata et al. 2016). Two piwi proteins are located in cells of the posterior growth zone and of the regeneration blastema of the polychaete annelids *Platynereis dumerilii* (Rebscher et al. 2007; Gazave et al. 2013; Planques et al. 2019), *Alitta virens* (Kozin and Kostyuchenko 2015) and *Capitella teleta* (Giani et al. 2011), whereas a *piwi* gene is transcribed by proliferating, undifferentiated cells of the growth zone and fission zone of the oligochaete worm *Pristina leidyi* undergoing asexual reproduction by paratomy (Özpolat and Bely 2015). Two expressed *piwi* genes have also been reported in the gastropod mollusk *Lymnaea stagnalis* and the bivalve *Crassostrea gigas*; in addition to the reproductive tract, their mRNAs are located in cells of the gills and lung, musculature, brain and labial palps (Jehn et al. 2018). In the sea slug *Aplysia californica*, a piwi protein is present in the central nervous system (CNS), where it is involved in the epigenetic control of memory-related synaptic plasticity (Rajasethupathy et al. 2012). Among nonchordate deuterostomes, *piwi* expression has been demonstrated in a series of somatic tissues, including coelomocytes, esophagus and tube feet epithelium, the epithelium of the spines, and the musculature of the sea urchins *Lytechinus variegatus*, *Strongylocentrotus purpuratus* and *Mesocentrotus franciscanus* (Reinardy et al. 2015; Bodnar and Coffman 2016). Anti-piwi-positive cells were also observed in the coelomic cells, the coelomic epithelium and the connective tissue of the sea cucumber *Eupentacta fraudatrix* (Dolmatov et al. 2021): these cells reach a maximum number within 4 h after evisceration and contribute to the regeneration of the intestine (Dolmatov 2021). In addition, *piwi* transcripts were also found in the adult nerve cord of the holothurian *Holothuria glaberrima* (Mashanov et al. 2015a). As for invertebrate chordates, data are limited to ascidians. In the solitary species *Ciona intestinalis*, two *piwi* orthologues are actively transcribed in cells inside the vessels of the branchial basket, where the lymph nodes, representing hematopoietic organs, are located. They are also expressed in the endostyle, gut epithelium, cells of the basal stalk, cell clusters of the siphon walls, and cells of the atrial epithelium. They assure the growth and the continuous turnover of cells of the body (Jeffery 2014). In colonial botryllid ascidians, one *piwi* gene is present in *Botryllus schlosseri*, *Botryllus primigenus*, *Botrylloides leachii* and *Botrylloides violaceus*. The protein is located in phagocytes near the endostyle, in tunic cells and in cells of the stomach of zooids of *B. schlosseri* along the ontogeny (Rosner et al. 2009; Rinkevich et al. 2010). The gene is also expressed by the epithelial monolayers developed from extirpated palleal buds and isolated floating buds in vitro (Rabinowitz and Rinkevich 2011). In *B. schlosseri* and *B. leachii*, the *piwi* gene is also transcribed by activated dormant cells lining the vasculature epithelium during whole-body regeneration (Rinkevich et al. 2010). At the onset of hibernation, in *B. leachii*, Hyams et al. (2017) observed high expression levels of *piwi* within the cell islands, the stem cell niches at both sides of the endostyle; in the advanced hibernation state, *piwi* was expressed in the multinucleated cells, the probable reservoir cells for the generation of new zooids at the end of the torpor. In *B.*

primigenus, piwi is expressed in coelomic cells (Kawamura and Sunanaga 2011). In *B. violaceus, piwi*-positive cells are found in the endostyle and hemocytes of adult zooids and in hemocytes and a few cells of the peribranchial epithelium of the developing bud (Brown et al. 2009).

2.2. Tudor Domain-Containing Proteins

The tudor proteins allow for the reading of protein methylations and include methylarginine- and methyllysine-binding proteins. Piwi proteins contain symmetrically dimethylated arginine (sDMA) in their N-termini and form tudor–piwi interactions that are required for the proper function of the piwi-piRNA pathway. Tudor proteins also participate in the proper assembly of the nuage and control of gametogenesis (Pek et al. 2012).

tudor genes, which are part of the piwi machinery, show high expression in archeocytes of the sponge *E. fluviatilis* (Alié et al. 2015). *oltudor1* is expressed in choanocytes, pinacocytes and type 1 vacuolar cells of the sponge *O. lobularis* (Fierro-Constaín et al. 2017). *tdrd9* of the cnidarian *H. magnipapillata* is associated with both *piwi* orthologues, *hywi* and *hyli*, at nuage perinuclear granules of i-cells, and contribute to piRNA biogenesis (Lim et al. 2014); *tdrd5* of *H. vulgaris* is also expressed in i-cells (Alié et al. 2015). In the flatworm *Schmidtea polychroa*, the Tudor domain-containing protein, Spoltud-1, was identified as a chromatoid body component of neoblasts, essential for proliferation and differentiation; it is also expressed in cells of the CNS (Solana et al. 2009); *smtdrd5* of the planarian *S. mediterranea* is expressed in neoblasts (Alié et al. 2015). In the annelid *Platynereis dumerilii* there are three *tudor* genes expressed in a very similar way in cells situated in the segment addition zone and in germ cells (Gazave et al. 2013).

2.3. DEAD and "DEAH-Box"-Containing Helicases

The "DEAD-box" helicases form a family of proteins present in all eukaryotic cells, and are characterized by the existence of a domain of 400 amino acids that can be further divided into 12 characteristic motifs, one of which—the Asp-Glu-Ala-Asp (DEAD) motif—confers the name to the family (Cordin et al. 2006; Rosner and Rinkevich 2007). The conserved domain serves as a binding site for ATP and RNA to facilitate helicase activities. The motifs participate in various interactions, endowing the proteins with multifunctionality in many aspects of RNA metabolism, from transcription to decay. Despite the high conservation between "DEAD-box" proteins, they participate in different processes, some of which having very specific roles. "DEAD-box" proteins often function within large multiprotein complexes such as the exon junction complex, and are involved in processes such as the export of mRNA and translation initiation (Gilman et al. 2017; Perčulija and Ouyang 2019). This is a large family of proteins and an individual genome may contain dozens of members of genes encoding for proteins of this family. *vasa* (*ddx4*) and *pl10* (*ddx3*) are among the most prominent and well-studied members of this family.

The vasa proteins were considered for many years as specific markers of the germline lineage; vasa proteins are among the most important components of the nuage localized at the nuclear envelope, transporting piRNA transcripts to the cytoplasmic piRNA machinery. Additionally, they function as translational initiating factors involved in the translation of several stem cell-specific mRNAs (Poon et al. 2006; Liu et al. 2009; Xiol et al. 2014).

Pl10 (ddx3) has been extensively studied in many organisms and has been shown to participate in transcription and translation regulation, mRNA maturation and mRNA export. Additionally, it may be associated with stress responses and stress granules, innate immune response, and regulation of apoptosis (Chang and Liu 2010). *ddx3* is the closest paralogue of *vasa*, both of them being highly expressed in germ cells and indispensable for their integrity. However, vasa is more restricted to germ lineages and pluripotent stem cells.

Ddx6, also known as me31b (in *Drosophila*) and dhh1 (in yeast), is one of the GMP proteins found in P-bodies and in stress granules that function in both gene translation inhibition and deadenylation-dependent mRNA degradation, by forming complexes with other proteins (Chang and Liu 2010).

In the sponge *E. fluviatilis*, all the three "DEAD box" genes, *efvasa*, *efpl10*, and *efddx6*, are continuously expressed in archeocytes (Alié et al. 2015). In the calcarean sponge *Sycon ciliatum*, there are two *vasa* and two *pl10* orthologues. In mature animals, only one *vasa* (*scivasab*) and one *pl10* (*scipl10b*) genes are expressed in choanocytes (Leininger et al. 2014). In the homoscleromorph sponge *O. lobularis*, *olvasa* and *olpl10* are both expressed in choanocytes, pinacocytes and mesohyl type 1 vacuolar cells (Fierro-Constaín et al. 2017). The hydrozoan *H. magnipapillata* has two *vasa* orthologues (*cnvas1* and *cnvas2*) that are actively transcribed in i-cells and epidermal cells (Mochizuki et al. 2001). The *pl10* of this animal (*cnpl10*) is expressed in i-cells and epidermal cells as well as in nematoblasts (Mochizuki et al. 2001). In the medusa of the hydrozoan *P. carnea*, *vasa* in transcribed in nematoblasts of the tentacle bulbs and manubrium (Plickert et al. 2012). In the siphonophoran *N. bijuga*, *vasa* and *pl10* are expressed in the i-cells of both the epidermis and the gastrodermis (Siebert et al. 2015). In the ctenophoran *P. pileus*, one *vasa* and one *pl10* gene have been reported, and they are transcribed in progenitors of colloblasts and muscle cells, cells of the forming combs and cells of the aboral sense organ (Alié et al. 2011). In the flatworm *M. lignano*, the *vasa* homologue, *macvasa*, is expressed in neoblasts: *macvasa* knockdown does not affect ASCs population but dramatically reduces the quantity of piRNAs, suggesting that *macvasa* functions in piRNA biogenesis (Pfister et al. 2008). In the planarian *D. japonica*, *djvas1* is transcribed in neoblasts and is required for regeneration and differentiation, but not for neoblast maintenance. *D. japonica djvlga* is expressed in neoblasts and the CNS, while *djvlgb* is detected in a limited fraction of neoblasts (Shibata et al. 1999; Rouhana et al. 2010; Wagner et al. 2012). In the same species, the *djcbc1* gene, the orthologue of *ddx6*, similarly to *djvlga*, is expressed in neoblasts and cells of the CNS (Rouhana et al. 2010; Juliano et al. 2014) and is abundant in chromatoid bodies, different from those in which the *piwi* orthologue *djpiwic* is detected (Kashima et al. 2016). In the planarian *S. mediterranea*, smedvasa1 and smedvasa2 proteins are upregulated in neoblasts; *smedvasa1* is essential for proliferation and for promoting differentiation (Shibata et al. 1999; Wagner et al. 2012). The *vasa* orthologues of *D. japonica* (phylum Platyhelminthes) are involved in regeneration but not in cell proliferation (Shibata et al. 1999; Rossi et al. 2007; Rouhana et al. 2010; Wagner et al. 2012). In the polychaete worms *P. dumerilii* and *A. virens*, *vasa* and *pl10* are actively transcribed in the proliferating, undifferentiated cells of the growth (via posterior elongation) zone in metamorphosing larvae and in growing adults, as well as in the blastema of regenerating animals (Rebscher et al. 2007; Gazave et al. 2013; Planques et al. 2019). A similar location was reported for the *vasa* mRNA of *C. teleta* (Dill and Seaver 2008). In the oligochaete worm *P. leidy*, in addition to the posterior growth zone of metamorphosing larvae and adults, *vasa* is expressed in the fission zone of animals undergoing asexual reproduction by fragmentation (Özpolat and Bely 2015). Analogously, in the oligochaete annelid *Enchytraeus japonensis*, *ejvlg2* mRNA is located in undifferentiated cells of mesodermal origin in the posterior surface of the septa during asexual reproduction by autotomy (Sugio et al. 2012).

Arthropods are animals with low regenerative power and no asexual reproduction. Usually in this taxon, genes of the nuage proteins are expressed only in the germ line. However, in the crab *Eriocheir sinensis*, *esddx6* has low expression in some somatic tissues, such as heart, stomach, muscle, hemocytes, and cells of the thorax and intestine (Li et al. 2015). Moreover, in the rhizocephalan cirripede *Polyascus polygenea*, *ppvlg* and *ppdrh1* are expressed in cells of the stolons and buds of *interna* that will give rise to both the germline and the soma (Shukalyuk et al. 2007).

As for deuterostomes, one *vasa* gene was reported in the echinoids *L. variegatus*, *S. purpuratus* and *M. franciscanus*—it is expressed in muscles, epithelium of the gut, tube feet and spines, CNS and coelomocytes (Bodnar and Coffman 2016), and its protein product is located in the epithelium of tube feet, spines and esophagus, as well as in neurons and a fraction of coelomocytes (Reinardy et al. 2015).

In the colonial ascidian *B. schlosseri*, *bsvasa* is expressed in germ cells (Rosner and Rinkevich 2011) and two additional cell populations: (i) cells, resembling the PGCs, that aggregate exterior to the developing gonads or in hemolymphatic vessels; (ii) phagocytes in the cell island adjacent to the endostyle. Anti-vasa antibodies also stain the epithelium of the stomach and the intestine. Bspl10 protein is present in

PGCs, many germ-cell types and somatic cells of the epithelium of the digestive tract. In addition, *pl10* is upregulated in differentiating bud and budlet tissues, and in a fraction of hemocytes, and has low expression in mature tissue. Knockdown of *pl10*, in addition to reducing the number of germ cells, causes malformation in the developing buds and the alteration of the morphology of adult zooid tissues; however, high expression of *pl10* can be detected even in malformed tissues (Rosner et al. 2006, 2009). In the same species, *bsddx1* is expressed in hemocytes of the cell islands along ontogeny and astogeny (Rosner et al. 2013). In the ascidian *B. violaceus*, 3–6% of circulating hemocytes and some tunic cells closely associated with the vasculature stain positively with the *vasa* antisense probe (Brown and Swalla 2007).

2.4. Nanos Family Proteins

Nanos is a zinc finger protein with two C2HC zinc finger motifs and represents an additional component of the nuage-like structures. Nanos acts as a translational repressor of specific mRNAs by forming a complex with pum2 proteins. The complex associates with the 3′-UTR of mRNA targets and inhibits their translation (De Keuckelaere et al. 2018). Nanos is mainly a regulator of the germ lineage in embryos and adults. In adult tissues, *nanos* is mainly expressed in the spermatogonia and other early differentiating germ cells, and in some invertebrates (sponges, cnidarians, flatworms, annelids, echinoderms, lancelet and ascidians) it is expressed in ASCs.

The *nanos* mRNA in *O. lobularis* (phylum Porifera) is located in choanocytes, pinacocytes and mesohyl type 1 vacuolar cells (Fierro-Constaín et al. 2017). The cnidarian *H. magnipapillata* contains two *nanos* genes: *cnnos1*, expressed in the multipotent i-cells and cells of the germline, and *cnnos2*, expressed in the gastrodermal cells and a subset of germ cells (Mochizuki et al. 2000). In the hydroid *H. echinata*, in addition to the i-cells, the *nanos-2* gene is actively transcribed in nematoblasts and maturing nematocytes (Kanska and Frank 2013). Two *nanos* genes, *nanos1* and *nanos2*, are transcribed in the i-cells located in the siphonosomal buds and young zooids of the siphonophoran *N. bijuga* (Siebert et al. 2015). In the planarian *D. japonica*, *djnos* is expressed in a subset of germline-committed neoblasts, and in the early differentiating oogonia and spermatogonia (Sato et al. 2006). Furthermore, in the planarian *S. mediterranea*, *smednos* is expressed in eye precursor cells during regeneration (Handberg-Thorsager and Saló 2007), whereas *smnanos2* mRNA of the fluke *S. mansoni* is located in neoblasts (Collins et al. 2013). In the annelid *P. dumerilii*, *pdunos* is expressed in the posterior zone of segmental growth and cells of the CNS, and is upregulated during regeneration (Rebscher et al. 2007; Gazave et al. 2013; Planques et al. 2019). A similar location was reported for the *capinanos* mRNA of the worm *C. teleta* (Dill and Seaver 2008). The *bpnos* of the colonial tunicate *Botryllus primigenus* is strongly expressed in immature and mature male germ cells, and to a lesser extent, in the multipotent epithelia of the buds and in a fraction of blood cells. Knockdown of *bpnos* strongly interferes with male germ-cell differentiation, but does not affect the formation of female germ cells (Sunanaga et al. 2008).

2.5. PUF Family Proteins

The PUF family proteins associate with nanos and regulate the translation of specific genes by binding to a specific Pumilio Response Element situated at their 3′ UTR (De Keuckelaere et al. 2018).

In the sponge *O. lobularis*, *pumilio* is expressed in choanocytes, pinacocytes and mesohyl type 1 vacuolar cells (Fierro-Constaín et al. 2017). The *djpum* orthologue of the flatworm *D. japonica* is predominantly expressed in neoblasts, and RNAi-mediated gene silencing of it causes loss of nematoblasts and reduced regeneration (Salvetti et al. 2005; Rouhana et al. 2010). Similar expression was reported for *smedpumilio* of the planarian *S. mediterranea* (Solana et al. 2009). *pumilio*, and its related genes *pufa* and *pufb*, are expressed both in soma (posterior growth zone) and germ stem cells in *P. dumerilii* (phylum Annelida); however, the *pumilio* genes are not upregulated during posterior elongation, and they are also expressed in the gut of untreated animals (Gazave et al. 2013).

2.6. Mago-Nashi (or Mago)

Originally identified in *Drosophila*, Mago-nashi has emerged as essential for germ plasm assembly. It is characterized by a specific domain localized at the 5′ end of the molecule, and it is an integral part of a protein complex that forms the exon junction complex (Kataoka et al. 2001). In *Drosophila*, Mago-nashi acts during germ stem cell differentiation and is required for the polarization of the oocyte and the formation of perpendicular axes (Micklem et al. 1997; Parma et al. 2007).

The *mago-nashi* orthologue of the demosponge *E. fluviatilis*, *efmago-nashi*, is transcribed to a higher extent in the totipotent archeocytes than in other cells, however, its functions in these cells has not yet been defined (Alié et al. 2015). In the freshwater sponge *Lubomirskia baicalensis*, a *mago-nashi* orthologue is expressed at the top of the branches that characterize the deep-water morphs (Wiens et al. 2006).

3. RNA Recognition Motif (RRM) Containing Proteins

RRM is a 90 amino acid domain consisting of three aromatic side chains located between two conserved motives: RNP1 (octamer) and RNP2 (hexamer). RRMs usually binds a variable number of nucleotides, ranging from two to eight, within a single-strand RNA (ssRNA), but it can interact with single-strand DNA (ssDNA) as well. The number of RRMs varies among different subfamilies of proteins. For example, Bruno has three RRMs and Bruli has two domains. Both function in pre-mRNA alternative splicing, mRNA translation and stability (Maris et al. 2005).

3.1. Mbnl, Bruli and Bruno

Alternative splicing occurs with the involvement of highly regulated RBPs that bind pre-mRNA at specific sequences and regions and modulate the inclusion or exclusion of exons. In stem cells and their descendants, this regulation is an interplay between two kinds of RBPs with opposing functions in controlling splicing: Mbml (muscleblind-like splicing regulator), Bruli and Bruno. The former is a CCCH zinc finger protein that represses gene isoforms active in stem cells and is upregulated in differentiating cells, while Bruli and Bruno assist in the expression of gene isoforms that are active and upregulated in stem cells.

The sponge *O. lobularis* contains two copies of *bruno* genes: *bruno* and *brunob*. Both are continuously expressed in choanocytes, pinacocytes and mesohyl type 1 vacuolar cells. Similar expression was reported for the *boule* gene (Fierro-Constaín et al. 2017), acting as regulator of the translation of specific mRNAs and required for meiotic entry and germline differentiation at the transition between G2 and M phases of meiosis I (Shah et al. 2010). The sponge *E. fluviatilis* also shows high expressions of *bruno* in archeocytes and low expression in other cells, while *mbnl* has exactly the opposite expression: low in archeocytes and high in other somatic cells (Alié et al. 2015). In the ctenophoran *P. pileus*, *ppibruno* is expressed in progenitors of various somatic cell lineages (Alié et al. 2011). In the planarian *S. mediterranea*, neoblasts contain *smedbruli* mRNA; depletion of *bruli* results in neoblast loss and lack of regeneration. *smedbruli* is also expressed in cells of the CNS; loss of *mbnl* function results in slower regeneration (Guo et al. 2006; Solana et al. 2016). In the same species, *smedmbnl1* and three *smedmbnl-like* genes are present: *smedmbnl1* mRNA is present in differentiated cells of the body parenchyma, whereas *smedmbnl-like1* and *smedmbnl-like2* are transcribed in differentiated cells of the epidermis and gut tissues; no data on *smedmbnl-like3* expression are present in the literature (Solana et al. 2016). *djbruli* of the planarian *D. japonica* is actively transcribed in neoblasts and cells of the CNS (Rouhana et al. 2010). *pdubruno* of *P. dumerilii* is expressed in proliferating, undifferentiated cells of the of the posterior growth zone (Gazave et al. 2013).

3.2. Musashi

The Musashi proteins contain two RRMs and are expressed in stem and in neural lineage cells, including neural stem cells. Musashi proteins are involved in stem cell self-renewal. They function through binding of the 3'UTR of target mRNAs that prevent their translation, and by inhibiting 80 S ribosome assembly (Park et al. 2014).

In the sponge *E. fluviatilis*, *eflmsia*, the *musashi* orthologue is transcribed in archeocytes and the protein product is localized in their nucleus. Based on *eflmsia* expression at M-phase, archeocytes can be divided into a group undergoing self-renewal and expressing high quantities of the *eflmsia*, and another one expressing 30–60% of the quantity of mRNA of the previous group and protein and committed to differentiation (Okamoto et al. 2012). In the starlet sea anemone *Nematostella vectensis*, *nvmsi* is expressed in precursor cells of the CNS (Marlow et al. 2009). In the flatworm *D. japonica*, *musashi-like genes -a, -b, -c* are expressed in differentiated neural cells, and therefore are not markers of stem cells, whereas *djdmlg* (DAZAP-like/musashi-like gene) is expressed in stem cells and additional types of soma cells (Higuchi et al. 2008). In the polychaete worm *P. dumerilii*, the *musashi* gene is upregulated in the posterior zone of segment formation and in cells of the nervous system, but it is not transcribed in germ cells nor upregulated during regeneration (Gazave et al. 2013). In the holothurian *H. glaberrima*, *msi1/2* mRNA is located in the outermost part of the adult radial nerve cord (Mashanov et al. 2015a). In the colonial ascidian *B. schlosseri*, *dazap1* is transcribed in buds and in differentiating tissues of both, germ line and soma (Gasparini et al. 2011). Two additional genes were described in ascidians, *hrmsi* from *Halocynthia roretzi* and *cimsi* from *C. intestinalis*; however, their expressions were only tested in embryos (Kawashima et al. 2000).

4. Signal Transduction Factors

4.1. Wnt

Wnts represent a family of secreted, lipid-modified signaling glycoproteins that are 350–400 amino acids in length. The lipid modification of Wnts is required to bind its carrier protein Wntless (WLS) and to be transported to the plasma membrane for secretion and binding to the receptor Frizzled. Three Wnt signaling pathways have been characterized: (i) the canonical Wnt pathway leading to regulation of gene transcription by nuclear localization of β-Catenin; (ii) the noncanonical planar cell polarity pathway that regulates the cytoskeleton and thereby modulates cell shape and migration; (iii) the noncanonical Wnt/calcium pathway that regulates intracellular calcium. All three pathways are activated by the binding of an Wnt-protein ligand to a Frizzled family receptor, which passes the biological signal to the Dishevelled protein inside the cell (Zhan et al. 2017). In the hydrozoan *H. magnipapillata*, *wnt* mRNA is located in the epidermis and gastrodermis of the hypostome, the oral end of the buds and the apical tip of regenerating animals (Hobmayer et al. 2000; Lengfeld et al. 2009). In embryos of the anthozoan *N. vectensis*, it is present around the blastopore and later in the oral end of growing polyps (Kusserow et al. 2005). Both the oral hypostome of hydrozoan polyps and the blastoporal region in anthozoan embryos represent the main inductive organizers for patterning the entire cnidarian oral–aboral body axis. Moreover, the canonical Wnt/β-Catenin pathway represents a core element of these inductive signaling centers. There is also accumulating evidence that Wnt/β-Catenin signaling is involved in self renewal in cnidarian and vertebrate ASCs. In fact, recent data show that global activation of Wnt/β-Catenin signaling along the major body axis in *H. vulgaris* enhances self-renewal in i-cells, strongly activates neurogenesis and inhibits the differentiation of nematocytes (stinging cells) (Khalturin et al. 2007; Hartl et al. 2019). Enhanced i-cell maintenance in *Hydra* is most likely mediated by β-Catenin regulating the activity of myc transcription factors (see below; see also the accompanying chapter in this book by Lechable et al. 2022). In the related colonial hydrozoan *H. echinata*, *wnt3a* is expressed in i-cells of the epidermis and gastrodermis, as well as in nematoblasts along the body column of the polyps (Müller et al. 2007). Furthermore, action of Wnt/β-Catenin signaling in global patterning and ASC decision making is reported in several bilaterians, particularly in flatworms, cephalochordates and tunicates. *wntA*, *wnt4*, *wnt6*, *wnt16*, *frizzled1/2/7*, *frizzled4* and *frizzled5/8* are overexpressed in somatic tissues of the holothurian *E. fraudatrix* during the regeneration of the internal organs after the induction of evisceration (Girich et al. 2017). A *wnt* orthologue is transcribed in cells of the tail-regenerating blastema of the lancelet *Branchiostoma lanceolatum* (Somorjai 2017), whereas two *wnt* orthologues, *wnt2b* and *wnt5a*, have been identified in the ascidian *B. schlosseri*: they are actively transcribed in all the tissues of the early buds (Di Maio et al. 2015). In the regular blastogenetic cycle of *B. schlosseri*, Wnt is an important signal transduction pathway, and the administration of both Wnt agonist and antagonist imposed significant alterations in the prosecution of the cycle and bud development (Rosner et al. 2014). In whole-body regeneration of the colonial ascidian *B. leachii*, differential gene expression analysis of the transcriptome revealed upregulation of genes involved in developmental signaling pathways including *wnt* (Blanchoud et al. 2018). In the congeneric species *Botrylloides diegensis*, *frizzled5/8*, *β-catenin* and *disheveled* increase their transcription during whole-body regeneration (Kassmer et al. 2020).

4.2. TGF-β/BMP

The transforming growth factor-beta (TGF-β)/bone morphogenetic protein (BMP) signaling pathway plays a fundamental role ni regulating cell development and growth through the activation of receptor serine/threonine kinases (Guo and Wang 2009). Smad (small mother against decapentaplegic) is the main signal transducer for receptors of the transforming growth factor beta (TGF-β) superfamily, which are critically important in cell proliferation and differentiation (Blobe et al. 2000).

Choanocytes of the calcarean sponge *S. ciliatum* weakly express two *smad* orthologues: *smad1/5* and *smad4* (Leininger et al. 2014). The same cell type in the demosponge *Chondrosia reniformis* expresses *tgf6* mRNA (Pozzolini et al. 2019). The gene orthologue *smed-smad-6/7* is expressed by neoblasts of *S. mediterranea* (Van Wolfswinkel et al. 2014), whereas *bssmad1/5/8* is expressed by a fraction of circulating phagocytes of the colonial ascidian *B. schlosseri* (Rosner et al. 2013). In the cephalochordate *B. lanceolatum*, *chordin* and *bmp2/4* increase their transcription in the regenerating tail (Somorjai et al. 2012; Somorjai 2017; Liang et al. 2019; Ferrario et al.

2020). TGF-β, together with Wnt/β-catenin and MAPK/ERK, emerged as important signal transduction pathways in organizing the colony of the ascidian *B. schlosseri* (Rosner et al. 2014).

5. Other Transcription Factors

Many transcription factors have been described as being expressed in ASCs of aquatic invertebrates, and most of them derive from studies in acoel flatworms and planarians, some from studies in sponges, cnidarians, cephalochordates and tunicates. *scibra1*, *scibra2*, *scigata*, *smad1/5* and *smad4* are expressed by choanocytes of the calcarean sponge *S. ciliatum* (Leininger et al. 2014). Among cnidarians, *hymyc1* and *hymyc2* are transcribed in proliferating i-cells and other proliferating somatic cells and gamete precursors of *H. vulgaris* (Hartl et al. 2010, 2014; Hobmayer et al. 2012). Interstitial (i-) cells of *H. vulgaris* also express *foxo* (Boehm et al. 2012). Functional interference suggests that the action of *myc* genes and *foxo* is directed to *Hydra* i-cell maintenance (Boehm et al. 2012; Hartl et al. 2019). In the stolon of the colonial hydroid *H. echinata*, i-cells express the POU protein Pln (Millane et al. 2011). In the acoel flatworm *I. pulchra*, *ipptx*, *ipsix1/2*, *ipfox a1*, *ipfox a2*, *ipfox c*, *ipgata4/5/6*, *iptwist1*, *iptwist2* are transcribed in the neoblasts and the musculature (Chiodin et al. 2013). In the planarian *S. mediterranea*, *smedprox-1*, *smedpbx1*, *smednkx2.2*, *smedsoxb1*, *smedsoxP1*, *smedsoxP3*, *smedgata4/5/6*, *smedtcf15*, *smedjunl1*, *smedzfmym1*, *smedzf2071*, *smedfhl1*, *smedzfp1*, *smedprog1*, *smedprog2* and *smedegr1* are transcribed by neoblasts, whereas *smedpax3/7* is expressed in differentiating sensory neurons (Wagner et al. 2012). The annelid *P. dumerilii* expresses *pdumyc*, *pduhes2*, *pduhes4*, *pduhes5*, *pduhes6*, *pduhes8*, *pduid*, *pduap2*, *pdugcm*, *pducdx*, *pduevx*, *pduhox3* in cells of its posterior growth zone and in the regenerating blastema (Gazave et al. 2013). In the sea cucumber *H. glaberrima*, *myc* is transcribed by progenitor cells in the adult radial nerve cord and by scattered cells of the neural parenchyma. The same cells also host the transcripts of a series of stem cell-associated genes, such as *soxb1*, *foxj1*, *hes*, *klf1/2/4* and *oct1/2/11*. Most of the cells are located in the outer layer of the adult radial nerve cord and correspond to the radial glial cells that also express a series of proneural genes (Mashanov et al. 2015a, 2015b). In the holothurian *E. fraudatrix*, *efsox9/10* and *efsox17* are actively transcribed by cells of the epithelia of the coelom and of the gut anlage during the regeneration of the internal organs consequent to evisceration (Dolmatov et al. 2021). A *pax3/7* orthologue is transcribed in the tail-regenerating blastema of *B. lanceolatum* (Somorjai et al. 2012; Somorjai 2017). In colonial ascidians, *myc* is expressed by cells of the peribranchial epithelium and fibroblast-like cells involved in organogenesis of the developing buds of *P. misakiensis* (Fujiwara et al. 2011), and by cells of the peribranchial epithelium of growing palleal and vascular buds as well as by some circulating hemocytes of *B. primigenus* (Kawamura et al. 2008). In the colonial species *B. schlosseri*, *bspitx* is transcribed by cells of the peribranchial epithelium of buds at early developmental stages, the inner wall of the oral siphon and the oral tentacles, the forming cerebral ganglion and the developing gut of young zooids (Tiozzo et al. 2005; Tiozzo and de Tomaso 2009); Bsoct4 protein is present in the epithelial cells of the branchial sac along ontogeny and astogeny (Rosner et al. 2009). In the same species, *bspou3* is expressed by few cells of the proximal side of the bud atrial epithelium (Ricci et al. 2016). In *B. schlosseri*, *bsgata4/5/6* is transcribed by cells of the posterior side of the new bud (Ricci et al. 2016).

6. Chromatin Modification/Cell Cycle/Differentiation

6.1. Proliferation Markers

Proliferating Cell Nuclear Antigen (PCNA) is a protein found in the nucleus and serves as an auxiliary protein of DNA polymerase delta. High expression levels of this molecule correlate with high rates of division.

In the demosponges *Hymeniacidon perleve* and *E. fluviatilis*, *pcna* is actively transcribed in archeocytes (Sun et al. 2007; Alié et al. 2015). mRNAs for HvPCNA of *H. vulgaris*, and DjPCNA of the planarian *D. japonica* are located in the i-cells and neoblasts, respectively (Orii et al. 2005; Alié et al. 2015). In addition, *smedpcna* of *S. mediterranea* is actively transcribed in neoblasts (Eisenhoffer et al. 2008; Onal et al. 2012; Alié et al. 2015). *pcna* of the annelid *P. dumerilii* is actively transcribed in the posterior growth zone (Gazave et al. 2013; Planques et al. 2019). In the enteropneust hemichordate *P. flava*, *pcna* is expressed in cells of the regeneration blastema (Rychel and Swalla 2008). In early regenerating fragments of the tunicate *B. leachii*, *pcna* predominantly stains piwi-positive cells attached to the vascular epithelium, directly involved in the formation of new buds, whereas later on, it also labels piwi-positive cells within the lumen of the colonial vasculature (Rinkevich et al. 2010). In the colonial ascidian *B. leachii*, at the onset of hibernation, leading to the resorption of the colonial zooids, a high expression level of *pcna* was recorded within the cell islands, the stem cell niches on both sides of the endostyle, whereas in a deep hibernation state, *pcna* was expressed in the multinucleated cells, probably a reservoir of cell types for fast regeneration of zooids from the circulation during arousal from torpor (Hyams et al. 2017). In the colonial ascidian *P. misakiensis*, *pcna* is transcribed in cells of the epithelia of the developing buds (Kawamura et al. 2012), whereas in the tunicate *B. violaceus*, *pcna* is expressed by clusters of hemocytes during whole-body regeneration (Brown et al. 2009).

Mini chromosome Maintenance Complex Component 2 (MCM2) forms a complex with additional proteins that function in the initiation of eukaryotic genome replication. The gene is transcribed by archeocytes of the demosponge *E. fluviatilis* (Alié et al. 2015), i-cells of the cnidarian *H. vulgaris* (Alié et al. 2015), and neoblasts of the planarians *D. japonica* and *S. mediterranea* (Salvetti et al. 2000; Rossi et al. 2007; Onal et al. 2012).

Cyclin B1 (CCNB1) is necessary for proper control of the G2/M transition phase of the cell cycle. The protein encoded by this gene, together with phospho-histone H3, is considered a good marker for cell proliferation. Its mRNA is present in archeocytes of the sponge *E. fluviatilis* (Alié et al. 2015), i-cells of the hydrozoan *H. vulgaris* (Alié et al. 2015) and neoblasts of the flatworm *S. mediterranea* (Reddien et al. 2005b; Eisenhoffer et al. 2008); in addition, two cyclin genes are expressed in proliferating, undifferentiated cells of the growth zone and regeneration blastema of the annelid *P. dumerilii* (Planques et al. 2019). A cyclin gene is over-transcribed during whole body regeneration in the colonial ascidian *B. diegensis* (Kassmer et al. 2020).

6.2. Genes for Chromatin Modifications

Many genes involved in the epigenetic modification of chromatin are actively transcribed in ASCs of aquatic invertebrates. Three transcriptional silencers are expressed by the planarian *S. mediterranea* neoblasts (Eisenhoffer et al. 2008; Resch et al. 2012; Trost et al. 2018); the same cells express various genes involved in methylation/demethylation and acetylation/deacetylation as well as histone modifications in the flatworms *S. mediterranea* (Reddien et al. 2005a, 2005b; Eisenhoffer et al. 2008; Onal et al. 2012; Cao et al. 2019) and *D. japonica* (Rossi et al. 2007; Bonuccelli et al. 2010; Cao et al. 2019).

6.3. Telomere Protection

Telomeres are guanine-rich DNA repeats ((TTAGGG)n) located at the termini of chromosomes, that stabilize and protect chromosome ends through a protein complex called shelterin, which also serves to recruit telomerase to the telomeres. In most mature cells, telomeres progressively shorten through each cell division and trigger DNA damage responses that, eventually, mediate cell cycle arrest or apoptosis. Homeostasis can be achieved via telomere lengthening by a telomerase, a process that occurs in germline and somatic stem cells. Telomerase is a cellular reverse transcriptase that synthesizes telomeric DNA directly onto chromosome ends. Although the gene encoding the telomerase has been detected in many metazoans, equivalent telomerase activity has not been demonstrated in many organisms. Other molecules associated with telomerase activity are Pot1 and RTEL1. Protection of telomeres 1 (Pot1), a component of shelterin, contributes to the suppression of unnecessary DNA damage response at the telomeres and their maintenance. RTEL1 is an ATP-dependent DNA helicase implicated in telomere-length regulation, DNA repair and the maintenance of genomic stability. It also regulates meiotic recombination and crossover homeostasis (Udroiu et al. 2017).

The archeocytes of the demosponge *H. perleve*, when cultured in vitro, show the presence of mRNAs for telomerase and telomerase reverse transcriptase (Sun et al. 2007). Similarly, the archeocytes of the sponge *E. fluviatilis* transcribe *efrtel1*, and the i-cells of the solitary cnidarian *H. vulgaris* contain mRNA for HvRTEL1 (Alié et al. 2015). The neoblasts of the flatworm *S. mediterranea* express *smrtel1* and the telomerase reverse transcriptase orthologue *smedtert*. Four different alternative splice isoforms are encoded by the latter gene, only one of which is coding for a full active enzyme. Following fission or regeneration, there is an increased expression of the genes for all four *smedtert* isoforms in ASCs of the asexual individuals, with an increase in the relative proportion of the full-length isoform. The combined result of about an eight-fold increase in telomerase mRNA may contribute to indefinite somatic telomerase activity in proliferating stem cells during regeneration or reproduction by fission (Tan et al. 2012). In the same species, *smedob1*, a *pot1* orthologue, is ubiquitously expressed, however its knockdown impairs homeostasis and regeneration (Yin et al. 2016). The same phenomenon occurs in the planarian *D. japonica* (Yin et al. 2016). A telomerase gene is also actively transcribed by undifferentiated cells of the posterior surface of the septa, and during asexual reproduction by autotomy in the oligochaete annelid *E. japonensis* (Sugio et al. 2012). *tert* is also transcribed in muscles, the esophagus, CNS and coelomocytes of the sea urchins *L. variegatus*, *S. purpuratus* and *M. franciscanus* (Bodnar and Coffman 2016). The multipotent bud epithelia of the colonial ascidian *B. schlosseri* express *bspot1* for telomere protection and an orthologous of telomerase (Laird and Weissman 2004; Ricci et al. 2016).

7. Discussion and Conclusions

ASC evolution is associated with the origin of multicellular animals. Beyond the obvious role in tissue homeostasis, ASCs in aquatic invertebrates often are key participants in sustaining important biological processes for indeterminate growth, regeneration, asexual reproduction (agametic cloning), torpor phenomena and more (Sköld et al. 2009; Vogt 2012; Rinkevich et al. 2022). The technologies for ASC isolation and growth under in vitro conditions (Odintsova 2009; Rinkevich 2011; Zahiri and Zahiri 2016), for the study of their expression repertoires, for proteomic metabolomics and bioinformatic approaches to the biology of ASC, are advancing rapidly (Ballarin et al. 2018). The application of the above techniques will offer the possibility to obtain, in the near future, unprecedented insights into the biology of ASCs from aquatic invertebrates.

A wide range of disparate characteristics have been found when comparing vertebrate and aquatic invertebrate ASCs, including features such as morphology, differentiation states and somatic/germ lineage characteristics. As these and numerous other important traits in aquatic invertebrates differ significantly from those recorded in the vertebrates' ASCs (Isaeva et al. 2009; Rinkevich et al. 2022), it is of great importance to find common shared characteristics, many of which are associated with stemness gene expressions. Stem cells in various marine taxa, including members of Porifera, Cnidaria, Ctenophora, Annelida, Acoela, Platyhelminthes, Echinodermata, Cephalochordata and Tunicata, express stemness genes and other key genes. As they exist throughout the lifespan of these organisms and are not rare (may contribute >25–30% of the total animals' cells), with high potency (pluripotency and totipotency), a comparative analysis of their gene expression may add relevant information to our understanding of their nature (e.g., Lai and Aboobaker 2018; Alié et al. 2015; Fierro-Constaín et al. 2017). Indeed, in order to clarify the biological phenomena evolved, there is a need to understand the nature, the biology and gene expressions of ASCs from aquatic invertebrates (Rinkevich et al. 2022). Such an analysis is of further importance since ASCs from aquatic invertebrates participate in biological phenomena not found in the vertebrates, such as whole-body regeneration, asexual budding and dormancy (Vogt 2012; Rinkevich et al. 2022).

The data collected in this report are summarized in Table 1; a detailed version can be found in Appendix A. From these data, we propose a few general conclusions. First, it is difficult to identify clear ASC gene-expression signatures across and even within phyla. This is in agreement with initial in-depth single-cell-transcriptome data sets established in cnidarians and flatworms, where it was not possible to characterize core programs of ASC-specific gene activation within the studied species (Fincher et al. 2018; Plass et al. 2018; Siebert et al. 2019). Based on this, it is difficult to define conserved molecular mechanisms for maintenance of long-term ASC stocks and to define conserved elements for communication between ASCs and their short- and long-range environment (Martinez et al. 2022). Much larger data sets at this level of resolution in many more species studied in the future may improve this dilemma.

Second, our analysis shows clear coexpression of somatic- and germ line-specific genes in ASCs across and within phyla, confirming earlier reports (Juliano et al. 2010; Alié et al. 2015; Fierro-Constaín et al. 2017). This indicates that an ancestral conserved multi- or toti-potency program may coordinate ASC dynamics during tissue homeostasis, (indefinite) asexual growth and sexual reproduction. Third, ASC transcriptional regulation, when compared to vertebrate ASC programs, is limited when we consider the activation of genes encoding for somatic transcription factors, but rich if referring to RNA regulatory genes known from pan-metazoan embryonic stem cells (Rinkevich et al. 2022).

Table 1. Main stem cell genes expressed by ASCs of aquatic invertebrates.

Protein Family	Porifera	Ctenophora	Cnidaria	Acoelomorpha	Platyhelminthes	Annelida	Mollusca	Arthropoda Rhizocephala	Echinodermata	Hemichordata	Cephalochordata	Tunicata
RNA-binding proteins												
ago/piwi family	✓	✓	✓	✓	✓	✓	✓		✓			✓
tudor domain-containing proteins	✓		✓		✓	✓						
DEAD and "DEAD-box"-containing helicases	✓	✓	✓	✓	✓	✓		✓	✓			✓
nanos family proteins	✓		✓		✓	✓						✓
PUF family proteins	✓				✓	✓						✓
mago-nashi	✓											
RRM-containing proteins												
mbnl, bruli and *bruno*	✓	✓			✓	✓						
musashi	✓		✓		✓	✓						✓
Signal transduction factors												
wnt			✓								✓	✓
tgf-β/bmp	✓				✓						✓	✓
Chromatin modification/cell cycle/differentiation												
pcna	✓	✓			✓	✓				✓		✓
mcm2	✓				✓							
cyclin b1	✓	✓			✓	✓						✓

Author Contributions: Concept and design: L.B., B.R., and B.H. Writing of the manuscript: L.B., B.R., A.R., and B.H. All authors have read and agreed to the published version of the manuscript.

Funding: This reserach was supported by the EC H2020 Marie Sklodowska-Curie COFUND research grant No 847681 "ARDRE—Ageing, Regeneration and Drug Research" to B.H., the NSF/BSF research grant No 2021650 "Somatic cell evolution towards immortalization in a marine tunicate" to B.R., and the EU-COST network 16203 "MARISTEM—Stem Cells of marine/aquatic invertebrates: From basic research to innovative applications".

Conflicts of Interest: The authors declare no conflict of interest.

Appendix A

On the basis of the codified proteins, genes were grouped as follows: RNA-binding proteins, RNA recognition motif (RRM)-containing proteins, signal transduction proteins, transcription factors, chromatin modification proteins, proteins involved in autophagy, cell-cycle proteins, control of differentiation proteins, niche interaction proteins and genes for miRNAs. Each category of genes (titles in boxes with green background) can contain various subgroups (titles in boxes with blue background). Within each subgroup, genes are listed according to the phylogenetic position of the organisms.

Table A1. Genes expressed in invertebrate ASCs and progenitor cells during potency state changes.

Species	Gene	Cell Types	Tissue/Organ	Interval of Expression	Identification Methods	References
			RNA-binding proteins			
			Argonaute family silencing genes			
Ephydatia fluviatilis (Porifera, Demospongiae)	*efpiwia* *efpiwib*	choanocytes archeocytes	choanocyte chambers mesohyl	throughout ontogeny	ISH, scRNAsec	Funayama et al. (2010); Alié et al. (2015)
Oscarella lobularis (Porifera, Homoscleromorpha)	*olpiwia*	choanocytes pinacocytes vacuolar cells type 1	choanocyte chambers pinacoderm mesohyl	throughout ontogeny during wound healing during wound healing	ISH	Fierro-Constaín et al. (2017)
	olpiwib *olago*	choanocytes pinacocytes vacuolar cells type 1	choanocyte chambers pinacoderm mesohyl	throughout ontogeny	ISH	Fierro-Constaín et al. (2017)
Podocoryne carnea (Cnidaria, Hydrozoa)	*cniwi*	epitheliomuscular cells	epiderm	throughout ontogeny	ISH	Seipel et al. (2004); Plickert et al. (2012)
Hydra vulgaris (Cnidaria, Hydrozoa)	*hywi* *hyli*	i-cells nematoblasts epidermal cells gastrodermal cells	body column	throughout ontogeny	ISH, IB	Juliano et al. (2014)
Hydra magnipapillata (Cnidaria, Hydrozoa)	*hywi* *hyli*	i-cells epidermal cells	body column	throughout ontogeny	ISH, IB	Juliano et al. (2014)

Table A1. *Cont.*

Species	Gene	Cell Types	Tissue/Organ	Interval of Expression	Identification Methods	References
Nanomia bijuga (Cnidaria, Hydrozoa, Siphonophora)	*piwi*	i-cells (epiderm and gastroderm)	siphonosomal horn buds young zooids	throughout ontogeny	ISH	Siebert et al. (2015)
Clytia hemisphaerica (Cnidaria, Hydrozoa)	*piwi*	nematoblasts	tentacle bulb	throughout ontogeny	ISH	Denker et al. (2008); Plickert et al. (2012)
Hydractinia echinata (Cnidaria, Hydrozoa)	*piwi*	i-cells	stolon and polyp	throughout ontogeny	ISH	Rebscher et al. (2008)
Pleurobrachia pileus (Ctenophora)	*ppipiwi1*	progenitors of colloblasts muscle cell progenitors cells of the forming combs cells of the aboral sense organ	tentacle roots tentacle roots comb rows aboral sense organ	throughout ontogeny	ISH	Alié et al. (2011)
Isodiametra pulchra (Acoelomorpha)	*ipiwi*	neoblasts	whole animals	throughout ontogeny	ISH	Egger et al. (2009)
Macrostomum lignano (Platyhelminthes, Rhabditophora)	*macpiwi1*	neoblasts	whole animals	throughout ontogeny	ISH	Pfister et al. (2007)
	macpiwi2	neoblasts γ-radiation resistant cells	whole animals regeneration blastema	throughout ontogeny during regeneration	ISH	Pfister et al. (2007); Zhou et al. (2015)

Table A1. *Cont.*

Species	Gene	Cell Types	Tissue/Organ	Interval of Expression	Identification Methods	References
Schmidtea mediterranea (Platyhelminthes, Rhabditophora)	*smedwi1*	neoblasts	whole animals	throughout ontogeny	ISH, FACS	Reddien et al. (2005a, 2005b)
	smedwi2	neoblasts	whole animals	throughout ontogeny	ISH, FACS,	Reddien et al. (2005a, 2005b); Palakodeti et al. (2008)
	smedwi3	neoblasts	whole animals	throughout ontogeny	ISH, FACS	Palakodeti et al. (2008)
Dugesia japonica (Platyhelminthes, Rhabditophora)	*djpiwi1*	neoblasts of the dorsal parenchyma	whole animals	throughout ontogeny	ISH, RNAseq	Rossi et al. (2006, 2007)
	djpiwi2 djpiwi3	neoblasts	whole animals	throughout ontogeny	ISH, RNAseq	Rossi et al. (2007)
	djpiwi-a (djpiwi4)	neoblasts and CNS (cytoplasm)	whole animals	throughout ontogeny	ISH	Rouhana et al. (2010); Shibata et al. (2016)
	djpiwi-b	neoblasts and their descendants (nucleus)	whole animals	throughout ontogeny	ICC, ISH	Shibata et al. (2016)
	djpiwi-c	neoblasts (cytoplasm)	whole animals	throughout ontogeny	ICC, ISH	Shibata et al. (2016)
	djago2	neoblasts, brain, intestine	whole animals	throughout ontogeny	ISH	Rouhana et al. (2010)

Table A1. *Cont.*

Species	Gene	Cell Types	Tissue/Organ	Interval of Expression	Identification Methods	References
Schistosoma mansoni (Platyhelminthes, Neodermata, Trematoda)	*smago2-1*	neoblasts	whole animals	throughout ontogeny	ISH, iRNA	Collins et al. (2013)
Platynereis dumerilii (Annelida, Polychaeta)	*piwia* *piwib*	proliferating, undifferentiated cells of the growth zone and blastema	posterior growth zone blastema	metamorphosing larva throughout ontogeny (posterior elongation) regeneration metamorphosing larva	ICC, ISH	Rebscher et al. (2007); Gazave et al. (2013); Planques et al. (2019)
Alitta virens (Annelida, Polychaeta)	*piwil1*	proliferating, undifferentiated cells of the growth zone and blastema	posterior growth zone blastema	metamorphosing larva throughout ontogeny (posterior elongation) regeneration	ISH	Kozin and Kostyuchenko (2015)
	avipiwil2	proliferating, undifferentiated cells of the blastema	blastema	regeneration	ISH	Kozin and Kostyuchenko (2015)
Capitella teleta (Annelida, Polychaeta)	*piwi1* *piwi2*	proliferating, undifferentiated cells of the growth zone and blastema	posterior growth zone blastema	metamorphosing larva throughout ontogeny (posterior elongation) regeneration	ISH	Giani et al. (2011)

Table A1. *Cont.*

Species	Gene	Cell Types	Tissue/Organ	Interval of Expression	Identification Methods	References
Pristina leidyi (Annelida, Oligochaeta)	*piwi1*	proliferating, undifferentiated cells of the growth zone and fission zone blastema	posterior growth zone fission zone blastema	metamorphosing larva throughout ontogeny (posterior elongation) asexual reproduction regeneration metamorphosing larva	ISH	Özpolat and Bely (2015)
Lymnaea stagnalis (Mollusca, Gastropoda)	*piwil1* *piwil2*	not reported	muscle lung brain	throughout ontogeny	qPCR	Jehn et al. (2018)
Aplysia californica (Mollusca. Gastropoda)	*piwi*	not reported	central nervous system, heart	throughout ontogeny	NB, ICC	Rajasethupathy et al. (2012)
Crassostrea gigas (Mollusca, Bivalvia)	*piwil1* *piwil2*	not reported	labial palps gills adductor muscle mantle	throughout ontogeny	qPCR	Jehn et al. (2018)
Lytechinus variegatus (Echinodermata, Echinoidea)	*piwi*	muscle, epithelia of the esophagus, spines, tube feet, spines, coelomocytes, radial nerve	tube feet spines coelomocytes central nervous system gut	throughout ontogeny	ICC, qRT PCR	Reinardy et al. (2015); Bodnar and Coffman (2016)

Table A1. *Cont.*

Species	Gene	Cell Types	Tissue/Organ	Interval of Expression	Identification Methods	References
Strongylocentrotus purpuratus (Echinodermata, Echinoidea)	*piwi*	muscles, epithelia of the esophagus, spines, tube feet, spines, coelomocytes radial nerve	tube feet spines coelomocytes central nervous system gut	throughout ontogeny	qRT PCR	Bodnar and Coffman (2016)
Mesocentrotus franciscanus (Echinodermata, Echinoidea)	*piwi*	muscle, epithelia of the esophagus, spines, tube feet, spines, radial nerve	tube feet spines central nervous system gut	throughout ontogeny	qRT PCR	Bodnar and Coffman (2016)
Eupentacta fraudatrix (Echinodermata, Holothoroidea)	*piwi*	coelomic cells, coelomic epithelium, connective tissue	coelomic cells, coelomic epithelium, connective tissue	throughout ontogeny	IHC, ISH	Dolmatov et al. (2021)
Holothuria glaberrima (Echinodermata, Holothoroidea)	*piwi*	nerve cord	neuroepithelium of the radial nerve cord cells of the neural parenchyma	adults	ISH	Mashanov et al. (2015a)
Botryllus schlosseri (Chordata, Tunicata)	*piwi*	phagocytes near the endostyle (Ab) tunic cells (Ab) stomach cells (Ab)	hemolymph, tunic	throughout ontogeny and astogeny	ISH, IHC; iRNA	Rosner et al. (2009)

Table A1. *Cont.*

Species	Gene	Cell Types	Tissue/Organ	Interval of Expression	Identification Methods	References
Botrylloides leachii (Chordata, Tunicata)	*piwi*	activated dormant cells lining the vasculature epithelium cell islands	colonial vasculature	during WBR	IHC; iRNA	Rinkevich et al. (2010)
Botrylloides violaceus (Chordata, Tunicata)	*piwi*	endostyle, hemocytes hemocytes, few epithelial cells	filtering zooids regenerating buds	throughout ontogeny buds from stage 3–6	IHC with commercial Abs	Brown et al. (2009)
Botrylloides diegensis (Chordata, Tunicata)	*piwi2*	hemoblasts	colonial vasculature	throughout ontogeny	ISH	Kassmer et al. (2020)
Ciona intestinalis (Chordata, Tunicata)	*piwi-like (1,2)*	stem cells	gut epithelium vessels of the branchial sac basal stalk cell clusters in the siphon walls lymph nodes in pharynx, endostyle, atrial epithelium	continuous turnover of cells forming body growth zone	IHC with commercial Abs	Jeffery (2014)
Styela plicata (Chordata, Tunicata)	*piwi*	hemoblasts	intestine submucosa	adults	IHC with commercial Abs	Jiménez-Merino et al. (2019)

Table A1. *Cont.*

Species	Gene	Cell Types	Tissue/Organ	Interval of Expression	Identification Methods	References
DEAD and "DEAH-box"-containing helicases						
Ephydatia fluviatilis (Porifera, Demospongiae)	*efvasa* *efpl10* *ef-ddx6*	archeocytes	mesohyl	throughout ontogeny	ISH, scRNAsec	Alié et al. (2015)
Sycon ciliatum (Porifera, Calcarea)	*scivasab* *scipl10b*	choanocytes	choanocyte chambers	throughout ontogeny	ISH	Leininger et al. (2014)
Oscarella lobularis (Porifera, Homoscleromorpha)	*olvasa* *olpl10*	choanocytes pinacocytes vacuolar cells type 1	choanocyte chambers pinacoderm mesohyl	throughout ontogeny during wound healing throughout ontogeny	ISH	Fierro-Constaín et al. (2017)
Hydra magnipapillata (Cnidaria, Hydrozoa)	*cnvas1* *cnvas2*	i-cells (+) epidermal cells	whole animals body column	throughout ontogeny	ISH	Mochizuki et al. (2001)
	cnpl10	i-cells nematoblasts epidermal cells	whole animals whole animals body column	throughout ontogeny	ISH	Mochizuki et al. (2001)
Hydractinia echinata (Cnidaria, Hydrozoa)	*vasa*	i-cells	epidermis of stolon and polyp	throughout ontogeny	ISH	Rebscher et al. (2008); Plickert et al. (2012)
Podocoryne carnea (Cnidaria, Hydrozoa)	*vasa*	nematoblasts	tentacle bulbs and manubrium	medusa	ISH	Plickert et al. (2012)
Nanomia bijuga (Cnidaria, Hydrozoa, Siphonophora)	*vasa1* *pl10*	i-cells (epiderm and gastroderm)	siphonosomal horn buds young zooids	throughout ontogeny	ISH	Siebert et al. (2015)

Table A1. *Cont.*

Species	Gene	Cell Types	Tissue/Organ	Interval of Expression	Identification Methods	References
Pleurobrachia pileus (Ctenophora)	*vasa* *pl10*	progenitors of colloblasts muscle cell progenitors cells of the forming combs cells of the aboral sense organ	tentacle roots tentacle roots comb rows aboral sense organ	whole, adult animals	ISH	Alié et al. (2011)
Macrostomum lignano (Platyhelminthes, Rhabditophora)	*macvasa*	neoblasts (Ab + ISH))	body parenchyma	whole animals	ICC, ISH	Pfister et al. (2008)
Dugesia japonica (Platyhelminthes, Rhabditophora)	*djvas1*	neoblasts	body parenchyma	whole animals	ISH	Rouhana et al. (2010); Wagner et al. (2012)
	djvlga	neoblasts and CNS	body parenchyma	regenerating animals	ISH, iRNA	Shibata et al. (1999); Rouhana et al. (2010); Wagner et al. (2012)
	djvlgb	neoblasts	body parenchyma	whole animals	ISH, RNAseq	Shibata et al. (1999); Rossi et al. (2007)
	djtud1	neoblasts	body parenchyma	ontogeny	ISH	Rouhana et al. (2010)
	djdhx-8c	neoblasts and CNS	body parenchyma and brain	whole animals	ISH	Rouhana et al. (2010)

Table A1. *Cont.*

Species	Gene	Cell Types	Tissue/Organ	Interval of Expression	Identification Methods	References
	djcbc1 (ddx6)	neoblasts and CNS	body parenchyma and brain	throughout ontogeny	ISH	Rouhana et al. (2010); Juliano et al. (2014)
Schmidtea mediterranea (Platyhelminthes, Rhabditophora)	*smedvasa1*	neoblasts	body parenchyma	adult animals	ICC, ISH, iRNA	Pfister et al. (2008); Wagner et al. (2012)
	smedvasa2	neoblasts	body parenchyma	adult animals	ISH, iRNA, NB	Shibata et al. (1999); Wagner et al. (2012)
	smedtud1a smedtud1b	neoblasts	body parenchyma	throughout ontogeny	ISH	Solana et al. (2009)
	smedtdrd1l2	neoblasts	body parenchyma	throughout ontogeny	ISH, scqPCR, cell transplantation	Van Wolfswinkel et al. (2014)
Platynereis dumerilii (Annelida, Polychaeta)	*vasa pl10*	proliferating, undifferentiated cells of the growth zone and blastema	posterior growth zone blastema	metamorphosing larva throughout ontogeny (posterior elongation) regeneration	ICC, ISH	Rebscher et al. (2007); Gazave et al. (2013); Planques et al. (2019)
Alitta virens (Annelida, Polychaeta)	*vasa pl10*	proliferating, undifferentiated cells of the growth zone and blastema	posterior growth zone blastema	metamorphosing larva throughout ontogeny (posterior elongation) regeneration	ISH	Kozin and Kostyuchenko (2015)

Table A1. *Cont.*

Species	Gene	Cell Types	Tissue/Organ	Interval of Expression	Identification Methods	References
Capitella teleta (Annelida, Polychaeta)	*vasa*	proliferating, undifferentiated cells of the growth zone and blastema	posterior growth zone blastema	metamorphosing larva throughout ontogeny (posterior elongation) regeneration metamorphosing larva	ISH	Dill and Seaver (2008)
Pristina leidyi (Annelida, Oligochaeta)	*vasa*	proliferating, undifferentiated cells of the growth zone and fission zone blastema	posterior growth zone fission zone	metamorphosing larva throughout ontogeny (posterior elongation) asexual reproduction regeneration metamorphosing larva	ISH	Özpolat and Bely (2015)
Enchytraeus japonensis (Anellida, Oligochaeta)	*ejvlg2*	neoblasts and N-cells (only for mesoderm)	posterior surface of septa (N-cells dorsal to neoblasts)	during asexual reproduction by autotomy throughout	ISH	Sugio et al. (2012)
Polyascus polygenea (Cirripedia, Rhizocephala)	*ppvlg*	stem cells	stolons and buds of the asexual organism	asexual reproduction	ISH	Shukalyuk et al. (2007)
	ppdrh1	stem cells	stolons and buds of *interna*	*interna* asexual reproduction	ISH	Shukalyuk et al. (2007)
Eriocheris sinensis (Crustacea, Decapoda)	*esddx6*	cells of various tissues	heart, stomach muscle, hemocytes, thorax, intestine	throughout ontogeny	IHC with commercial Abs	Li et al. (2015)

Table A1. *Cont.*

Species	Gene	Cell Types	Tissue/Organ	Interval of Expression	Identification Methods	References
Lytechinus variegatus (Echinodermata, Echinoidea)	*vasa*	muscle, epithelia of the esophagus, spines, tube feet, spines, coelomocytes, radial nerve	tube feet spines coelomocytes central nervous system gut	throughout ontogeny	ICC, qPCR	Reinardy et al. (2015); Bodnar and Coffman (2016)
Strongylocentrotus purpuratus (Echinodermata, Echinoidea)	*vasa*	muscles, epithelia of the esophagus, spines, tube feet, spines, coelomocytes radial nerve	tube feet spines coelomocytes central nervous system gut	throughout ontogeny	qRT PCR	Bodnar and Coffman (2016)
Mesocentrotus franciscanus (Echinodermata, Echinoidea)	*vasa*	muscle, epithelia of the esophagus, spines, tube feet, spines, radial nerve	tube feet spines central nervous system gut	throughout ontogeny	qRT PCR	Bodnar and Coffman (2016)
Botryllus schlosseri (Chordata, Tunicata)	*bsvasa*	epithelial cells phagocytes in cell islands (Ab) stomach cells (Ab)	bud tissues hemolymph tissues of filtering zooids	throughout astogeny throughout ontology and astogeny	ISH, IHC	Rosner et al. (2009)

Table A1. *Cont.*

Species	Gene	Cell Types	Tissue/Organ	Interval of Expression	Identification Methods	References
	bspl10	epithelial cells some blood cells, phagocytes in cell islands (Ab) stomach cells (Ab)	bud tissues hemolymph tissues of filtering zooids	throughout astogeny throughout ontology and astogeny	ISH, IHC, ICC	Rosner et al. (2006, 2009)
	bsddx1	cell islands	hemolymph	throughout ontogeny and astogeny	IHC, ISH	Rosner et al. (2013)
Botrylloides violaceus (Chordata, Tunicata)	*bvvasa*	hemocytes in vasculature (3–6%) cells outside the vessels in proximity of the tunic	hemolymph	throughout ontogeny and astogeny	ISH	Brown and Swalla (2007)
Botrylloides diegensis (Chordata, Tunicata)	*vasa*	hemoblasts	colonial vasculature	throughout ontogeny	ISH	Kassmer et al. (2020)
tudor domain-containing proteins						
Ephydatia fluviatilis (Porifera, Demospongiae)	*eftudor9* *eftdrkh* *eftdrd1* *eftdrd5*	archeocytes	mesohyl	throughout ontogeny	ISH, scRNAsec	Alié et al. (2015)
Oscarella lobularis (Porifera, Homoscleromorpha)	*oltudor1*	choanocytes pinacocytes vacuolar cells type 1	choanocyte chambers pinacoderm mesohyl	throughout ontogeny	ISH	Fierro-Constaín et al. (2017)

Table A1. *Cont.*

Species	Gene	Cell Types	Tissue/Organ	Interval of Expression	Identification Methods	References
Hydra magnipapillata (Cnidaria, Hydrozoa)	*tdrd9*	i-cells	body column	throughout ontogeny	ICC	Lim et al. (2014)
Hydra vulgaris (Cnidaria, Hydrozoa)	*tdrd5*	i-cells	body column	throughout ontogeny	ISH, scRNAsec	Alié et al. (2015)
Schmidtea mediterranea (Platyhelminthes, Rhabditophora)	*smtdrd5*	neoblasts	whole animals	throughout ontogeny	ISH, scRNAsec	Alié et al. (2015)
Schmidtea polychroa (Platyhelminthes, Rhabditophora)	*spoltud-1*	neoblasts central nervous system	whole animals	throughout ontogeny	ISH	Solana et al. (2009)
Platynereis dumerilii (Annelida, Polychaeta)	*tdrd1* *tdrd2* *tdrd3*	proliferating, undifferentiated cells of the growth zone	posterior growth zone	metamorphosing larva throughout ontogeny (posterior elongation)	ISH	Gazave et al. (2013)
			PUF family proteins			
Oscarella lobularis (Porifera, Homoscleromorpha)	*pumilio*	choanocytes pinacocytes vacuolar cells type 1	choanocyte chambers pinacoderm mesohyl	throughout ontogeny throughout ontogeny throughout ontogeny	ISH	Fierro-Constaín et al. (2017)
Dugesia japonica (Platyhelminthes, Rhabditophora)	*djpum*	neoblasts	whole animals	throughout ontogeny	ISH	Salvetti et al. (2005) Rouhana et al. (2010)

127

Table A1. *Cont.*

Species	Gene	Cell Types	Tissue/Organ	Interval of Expression	Identification Methods	References
Schmidtea mediterranea (Platyhelminthes, Rhabditophora)	*smed-pumilio*	neoblasts	whole animals	throughout ontogeny	ISH	Solana et al. (2009)
Platynereis dumerilii (Annelida, Polychaeta)	*pumilio* *pufa* *pufb*	proliferating, undifferentiated cells of the growth zone	posterior growth zone	metamorphosing larva throughout ontogeny (posterior elongation)	ISH	Gazave et al. (2013)
nanos family proteins						
Oscarella lobularis (Porifera, Homoscleromorpha)	*nanos*	choanocytes pinacocytes vacuolar cells type 1	choanocyte chambers pinacoderm mesohyl	throughout ontogeny throughout ontogeny throughout ontogeny	ISH	Fierro-Constaín et al. (2017)
.5 *Hydra magnipapillata* (Cnidaria, Hydrozoa)	*cnnos1*	i-cells	whole animals	throughout ontogeny	ISH	Mochizuki et al. (2000)
	cnnos2	endodermal epithelial cells	hypostome	throughout ontogeny	ISH, ICC	Mochizuki et al. (2000); Kanska and Frank (2013)
Hydractinia echinata (Cnidaria, Hydrozoa)	*nanos2*	nematoblasts, maturing nematocytes i-cells	whole animals	throughout ontogeny	ISH, ICC	Kanska and Frank (2013)
Nanomia bijuga (Cnidaria, Hydrozoa, Siphonophora)	*nanos1* *nanos2*	i-cells (epiderm and gastroderm)	siphonosomal horn buds young zooids	throughout ontogeny	ISH	Siebert et al. (2015)

Table A1. *Cont.*

Species	Gene	Cell Types	Tissue/Organ	Interval of Expression	Identification Methods	References
Dugesia japonica (Platyhelminthes, Rhabditophora)	*djnos*	neoblasts	asexual and sexual individuals mesenchyme	throughout ontogeny	ISH	Sato et al. (2006)
Schmidtea mediterranea (Platyhelminthes, Rhabditophora)	*smednos*	eye precursor cells	body parenchyma	during regeneration	ISH	Handberg-Thorsager and Saló (2007)
Schistosoma mansoni (Platyhelminthes, Neodermata, Trematoda)	*smnanos2*	neoblasts	body parenchyma	throughout ontogeny	ISH, iRNA	Collins et al. (2013)
Platynereis dumerilii (Annelida, Polychaeta)	*pdunos*	proliferating, undifferentiated cells of the growth zone and blastema	posterior growth zone blastema	metamorphosing larva throughout ontogeny (posterior elongation) regeneration	ICC, ISH	Rebscher et al. (2007); Gazave et al. (2013); Planques et al. (2019)
Capitella teleta (Annelida, Polychaeta)	*capinanos*	proliferating, undifferentiated cells of the growth zone and blastema	posterior growth zone blastema	metamorphosing larva throughout ontogeny (posterior elongation) regeneration metamorphosing larva	ISH	Dill and Seaver (2008)
Botryllus primigenus (Chordata, Tunicata)	*bpnos*	weak staining of pharyngeal epithelia of the developing budlets	pharyngeal epithelia	blastogenesis (budlets stages 1–6)	ISH, IHC	Sunanaga et al. (2008)

Table A1. *Cont.*

Species	Gene	Cell Types	Tissue/Organ	Interval of Expression	Identification Methods	References
Mago nashi family proteins						
Ephydatia fluviatilis (Porifera, Demospongiae)	*efmago-nashi*	archeocytes	mesohyl	throughout ontogeny	ISH, scRNAsec	Alié et al. (2015)
Lubomirskia baicalensis (Porifera, Demospongiae)	*mago-nashi*	?	top of the branches	throughout ontogeny	ISH	Wiens et al. (2006)
RNA recognition motif (RRM)-containing proteins						
Ephydatia fluviatilis (Porifera, Demospongiae)	*bruno*	archeocytes	mesohyl	throughout ontogeny	ISH, ICC	Alié et al. (2015)
	msia	archeocytes	mesohyl	throughout ontogeny	ISH, scRNAsec	Okamoto et al. (2012)
Chondrosia reniformis	*msi1*	archeocytes	mesohyl	throughout ontogeny	ICC	Pozzolini et al. (2019)
Oscarella lobularis (Porifera, Homoscleromorpha)	*boule* *bruno* *brunob*	choanocytes pinacocytes vacuolar cells type 1	choanocyte chambers pinacoderm mesohyl	throughout ontogeny during wound healing throughout ontogeny	ISH	Fierro-Constaín et al. (2017)
Nematostella vectensis (Cnidaria, Anthozoa)	*nvmsi*	neuronal progenitors	tentacle epidermis	young polyp	ISH	Marlow et al. (2009)

Table A1. *Cont.*

Species	Gene	Cell Types	Tissue/Organ	Interval of Expression	Identification Methods	References
Pleurobrachia pileus (Ctenophora)	*ppibruno*	progenitors of colloblasts muscle cell progenitors cells of the forming combs cells of the aboral sense organ	tentacle roots tentacle roots comb rows aboral sense organ	whole animals	ISH	Alié et al. (2011)
Schmidtea mediterranea (Platyhelminthes, Rhabditophora)	*smedbruli*	neoblasts, CNS	body parenchyma and brain	throughout ontogeny	ISH	Guo et al. (2006); Solana et al. (2016)
	smedkhd1 *smedcip29*	neoblasts	body parenchyma	throughout ontogeny	ISH	Mochizuki et al. (2001)
	smedsmb	neoblasts	body parenchyma	throughout ontogeny	ISH, iRNA, FACS	Fernandéz-Taboada et al. (2010)
	smedmbnl1	neoblasts differentiated cells	body parenchyma	throughout ontogeny	ISH	Okamoto et al. (2012)
	smedmbnl-like1 *smedmbnl-like2* *smedmbnl-like3*	epidermis gut tissues	whole animals	throughout ontogeny	ISH	Solana et al. (2016)
Dugesia japonica (Platyhelminthes, Rhabditophora)	*djtial1* *djtial2* *djtial3* *djtial4* *djtial5*	neoblasts	body parenchyma	throughout ontogeny	ISH	Rouhana et al. (2010)

131

Table A1. *Cont.*

Species	Gene	Cell Types	Tissue/Organ	Interval of Expression	Identification Methods	References
	djbruli *djpabpc2* *djedc4* *djcnot6* *djgemin5* *djdicer1* *djlsm14* *djsgm7* *djrbm18* *djfmrp1*	neoblasts and CNS	body parenchyma and brain	throughout ontogeny	ISH	Rouhana et al. (2010)
	djxrn1 *djg3bp* *djcnot7* *djdcp1l* *djupf1*	neoblasts, brain, intestine	whole animals	throughout ontogeny	ISH	Rouhana et al. (2010)
	djdmlg	neoblasts neural precursors and additional cell types	body parenchyma	throughout ontogeny	ISH, FACS, scPCR	Higuchi et al. (2008)
Platynereis dumerilii (Annelida, Polychaeta)	*pdubruno* *pdusmb*	proliferating, undifferentiated cells of the growth zone	posterior growth zone	metamorphosing larva throughout ontogeny (posterior elongation)	ISH	Gazave et al. (2013)

Table A1. *Cont.*

Species	Gene	Cell Types	Tissue/Organ	Interval of Expression	Identification Methods	References
	pdumusashi	proliferating, undifferentiated cells of the growth zone ventral nerve cord brain	posterior growth zone	metamorphosing larva throughout ontogeny (posterior elongation)	ISH	Gazave et al. (2013)
Holothuria glaberrima (Echinodermata, Holothuroidea)	*msi1/2*	radial nerve cord	radial nerve cord	adult	ISH	Mashanov et al. (2015a, 2015b)
Botryllus schlosseri (Chordata, Tunicata)	*bsdazap1*	all tissues	buds	during blastogenesis	ISH	Gasparini et al. (2011)
Signal transduction pathways						
Helix–Loop–Helix-domain-containing proteins						
Isodiametra pulchra (Acoelomorpha)	*iptwist1* *iptwist2*	neoblasts musculature	whole animals	throughout ontogeny	ISH	Chiodin et al. (2013)
Sycon ciliatum (Porifera, Calcarea)	*scibra1*	choanocytes	choanocyte chambers	throughout ontogeny	ISH	Leininger et al. (2014)
Hydra magnipapillata (Cnidaria, Hydrozoa)	*wnt*	epidermis and gastrodermis	hypostome apical end of buds apical end of regenerating animals	throughout ontogeny budding during regeneration	ISH	Lengfeld et al. (2009)
Hydractinia echinata (Cnidaria, Hydrozoa)	*wnt3*	i-cells, nematoblasts	epidermis and gastrodermis	polyp	ISH	Müller et al. (2007)

Table A1. *Cont.*

Species	Gene	Cell Types	Tissue/Organ	Interval of Expression	Identification Methods	References
Nematostella vectensis (Cnidaria, Anthozoa)	*wnt*	epidermis and gastrodermis	apical end of the animal	throughout ontogeny	ISH	Kusserow et al. (2005)
Schmidtea mediterranea (Platyhelminthes, Rhabditophora)	*smedjunl1 smedtcf15*	neoblasts	body parenchyma	throughout ontogeny	ISH	Wagner et al. (2012)
Eupentacta fraudatrix (Echinodermata, Holothuroidea)	*wntA wnt4 wnt6 wnt16,*	somatic tissues	somatic tissues	during the regeneration of the internal organs	qRT PCR	Girich et al. (2017)
Eupentacta fraudatrix (Echinodermata, Holothuroidea)	*frizzled1/2/7, frizzled4 frizzled5/8*	somatic tissues	somatic tissues	during the regeneration of the internal organs	qRT PCR	Girich et al. (2017)
Branchiostoma lanceolatum (Chordata, Cephalochordata)	*wnt5*	cells of the blastema	regenerating tail	throughout ontogeny, regeneration	ISH	Somorjai (2017)
Botryllus schlosseri (Chordata, Tunicata)	*wnt2b*	all the tissues	secondary buds	stages 1–3	ISH	Di Maio et al. (2015)
	wnt5a	mesenchymal cells	developing gonads	primary buds	ISH	Di Maio et al. (2015)
	wnt9a	all the tissues	secondary buds	emerging secondary buds	ISH	Di Maio et al. (2015)
Botrylloides diegensis (Chordata, Tunicata)	*frizzled5/8 β-catenin dishevelled*	cycling hemoblasts	colonial vasculature	during WBR	ISH	Kassmer et al. (2020)

Table A1. *Cont.*

Species	Gene	Cell Types	Tissue/Organ	Interval of Expression	Identification Methods	References
			TGF-β/BMP			
Sycon ciliatum (Porifera, Calcarea)	*smad1/5*	choanocytes (weak)	choanocyte chambers	throughout ontogeny	ISH	Leininger et al. (2014)
	smad4	choanocytes (weak) mesohyl cells (weak)	choanocyte chambers mesohyl	throughout ontogeny throughout ontogeny	ISH	Leininger et al. (2014)
Chondrosia reniformis (Porifera, Demospongiae)	*tgf6*	choanocytes	choanocyte chambers	during regeneration	ISH	Pozzolini et al. (2019)
Schmidtea mediterranea (Platyhelminthes, Rhabditophora)	*smedsmad6/7*	neoblasts	body parenchyma	throughout ontogeny	ISH, scqPCR, cell transplantation	Van Wolfswinkel et al. (2014)
Branchiostoma lanceolatum (Chordata, Cephalochordata)	*chordin*	cells of the regenerating notochord	regenerating tail	throughout ontogeny, regeneration	ISH	Somorjai et al. (2012); Di Maio et al. (2015); Somorjai (2017)
Branchiostoma japonicum (Chordata, Cephalochordata)	*bmp2/4*	cells around wound edge	regenerating tail	throughout ontogeny, regeneration	ISH, IHC, RNASeq	Ferrario et al. (2020); Liang et al. (2019); Kaneto and Wada (2011)
Botryllus schlosseri (Chordata, Tunicata)	*bssmad1/5/8*	phagocytes	hemocytes	throughout ontogeny	IHC, ISH	Rosner et al. (2013)

Table A1. *Cont.*

Species	Gene	Cell Types	Tissue/Organ	Interval of Expression	Identification Methods	References
			Notch			
Platynereis dumerilii (Annelida, Polychaeta)	*pdudelta* *pdunotch*	proliferating, undifferentiated cells of the growth zone	posterior growth zone	metamorphosing larva	ISH	Gazave et al. (2017)
	pduhes4 *pduhes5* *pduhes6* *pduhes8*	proliferating, undifferentiated cells of the growth zone	posterior growth zone	metamorphosing larva throughout ontogeny (posterior elongation)	ISH	Gazave et al. (2014)
Botrylloides diegensis (Chordata, Tunicata)	*notch1* *notch2* *hes1*	cycling hemoblasts	colonial vasculature	during WBR	ISH	Kassmer et al. (2020)
			Hedgehog			
Sycon ciliatum (Porifera, Calcarea)	*scigli*	choanocytes	choanocyte chambers	throughout ontogeny	IHC; iRNA	Rinkevich et al. (2010)
			Kinases			
Schmidtea mediterranea (Platyhelminthes, Rhabditophora)	*smednlk1* *smedfgfr1* *smedfgfr4*	neoblasts	body parenchyma	throughout ontogeny	ISH	Wagner et al. (2012)
Schistosoma mansoni (Platyhelminthes, Neodermata, Trematoda)	*smfgfra*	neoblasts	body parenchyma	throughout ontogeny	ISH, iRNA	Collins et al. (2013)

Table A1. *Cont.*

Species	Gene	Cell Types	Tissue/Organ	Interval of Expression	Identification Methods	References
Polyandrocarpa misakiensis (Chordata, Tunicata)	*pmrack1*	atrial epithelium undifferentiated mesenchymal cells associated with epidermis pharynx epithelium	developing buds whole zooids	during dedifferentiation throughout ontogeny	ISH, IHC	Tatzuke et al. (2012)
Pair rule and segment polarity genes						
Platynereis dumerilii (Annelida, Polychaeta)	*pduhunchback*	proliferating, undifferentiated cells of the growth zone and blastema	posterior growth zone	metamorphosing larva throughout ontogeny (posterior elongation)	ISH	Gazave et al. (2013)
	runt	proliferating, undifferentiated cells of the growth zone and blastema	posterior growth zone blastema	metamorphosing larva throughout ontogeny (posterior elongation) regeneration	ISH	Gazave et al. (2013); Planques et al. (2019)
Branchiostoma japonicum (Chordata, Cephalochordata)	*runx*	distal cells of the regenerating oral cirrus	regenerating cirrus	throughout ontogeny, regeneration	ISH	Kaneto and Wada (2011)
Other transcription factors						
Homeobox-containing proteins						
Hydractinia echinata (Cnidaria, Hydrozoa)	*pln (pou protein)*	i-cells	stolons	throughout ontogeny	ISH	Millane et al. (2011)

Table A1. *Cont.*

Species	Gene	Cell Types	Tissue/Organ	Interval of Expression	Identification Methods	References
Isodiametra pulchra (Acoelomorpha)	*ippitx* *ipsix1/2*	neoblasts musculature	whole animals	throughout ontogeny	ISH	Chiodin et al. (2013)
Schmidtea mediterranea (Platyhelminthes, Rhabditophora)	*smedprox1*	neoblasts	body parenchyma	throughout ontogeny	ICC, ISH	Pfister et al. (2008)
	smedpbx1 *smednkx2.2*	neoblasts	body parenchyma	throughout ontogeny	ISH	Lim et al. (2014)
	smedpax3/7	differentiating sensory neurons			ISH	Lim et al. (2014)
Platynereis dumerilii (Annelida, Polychaeta)	*pducdx* *pduhox3*	proliferating, undifferentiated cells of the growth zone and blastema	posterior growth zone	metamorphosing larva throughout ontogeny (posterior elongation)	ISH	Gazave et al. (2013)
	pduevx	proliferating, undifferentiated cells of the growth zone and blastema	posterior growth zone	metamorphosing larva throughout ontogeny (posterior elongation)	ISH	Planques et al. (2019)
Branchiostoma lanceolatum (Chordata, Cephalochordata)	*pax3/7*	cells of the blastema and regenerating nerve cord	tail regenerate	throughout ontogeny, regeneration	ISH, IHC	Somorjai et al. (2012); Somorjai (2017)
	msx	cells of the blastema	regenerating tail	throughout ontogeny, regeneration	ISH	Somorjai et al. (2012); Somorjai (2017)

Table A1. *Cont.*

Species	Gene	Cell Types	Tissue/Organ	Interval of Expression	Identification Methods	References
Botryllus schlosseri (Chordata, Tunicata)	*bspitx*	cells of the peribranchial epithelium inner wall of the oral siphon and tentacles; forming cerebral ganglion, developing gut	budlets (stage 1–3); left peribranchial chamber (stage 4–6) zooids at stage 8–9	during budding and blastogenesis	ISH, qPCR	Tiozzo et al. (2005); Tiozzo and de Tomaso (2009)
	bsoct4	epithelial cells (Ab)	branchial sac	throughout ontogeny and astogeny	ISH, IHC	Rosner et al. (2009)
	bspou3	few cells of the proximal side of the bud	atrial epithelium	bud at stage 3	ISH	Ricci et al. (2016)
Botrylloides diegensis (Chordata, Tunicata)	*pou3*	hemoblasts	colonial vasculature	throughout ontogeny	ISH	Kassmer et al. (2020)
Sox family proteins						
Clytia hemisphaerica (Cnidaria, Hydrozoa)	*chesox1, chesox3, chesox10 chesox12*	i-cells	tentacle bulb	medusa	ISH	Jager et al. (2011)
Schmidtea mediterranea (Platyhelminthes, Rhabditophora)	*smedsoxb1*	neoblasts	body parenchyma	throughout ontogeny	IHC, ISH	Onal et al. (2012)
	smedsoxp1	neoblasts	body parenchyma	throughout ontogeny	ISH	Wagner et al. (2012)

Table A1. *Cont.*

Species	Gene	Cell Types	Tissue/Organ	Interval of Expression	Identification Methods	References
	smedsoxp3	neoblasts	body parenchyma	throughout ontogeny	ISH	Wagner et al. (2012)
Holothuria glaberrima (Echinodfermata, Holothuroidea)	*soxb1*	progenitor cells in the radial nerve cord and by scattered cells of the neural parenchyma	radial nerve cord	adult	ISH	Mashanov et al. (2015a, 2015b)
E. fraudatrix (Echinodfermata, Holothuroidea)	*efsox9/10* *efsox17*	cells of the epithelia of the coelom and of the gut anlage	coelomic epithelium gut	during regeneration of internal organs after evisceratio	ISH	Dolmatov et al. (2021)
Branchiostoma lanceolatum (Chordata, Cephalochordata)	*soxb2*	cells of the regenerating nerve cord	tail regenerate	throughout ontogeny, regeneration	ISH	Somorjai et al. (2012); Somorjai (2017)
Branchiostoma japonicum (Chordata, Cephalochordata)	*soxe*	distal cells of the regenerating oral cirrus	cirrus regenerate	throughout ontogeny, regeneration	ISH	Kaneto and Wada (2011); Ferrario et al. (2020)
Fox family proteins						
Hydra vulgaris (Cnidaria, Hydrozoa)	*foxo*	i-cells	polyp	polyp	ISH	Boehm et al. (2012)
Isodiametra pulchra (Acoelomorpha)	*ipfoxa1* *ipfoxa2* *ipfoxc*	neoblasts musculature	whole animals	throughout ontogeny	ISH	Chiodin et al. (2013)

140

Table A1. *Cont.*

Species	Gene	Cell Types	Tissue/Organ	Interval of Expression	Identification Methods	References
Holothuria glaberrima (Echinodermata, Holothuroidea)	*foxj1*	progenitor cells in the radial nerve cord and by scattered cells of the neural parenchyma	radial nerve cord	adult	ISH	Mashanov et al. (2015a, 2015b)
Zinc finger proteins						
Sycon ciliatum (Porifera, Calcarea)	*scigata*	choanocytes	choanocyte chambers	throughout ontogeny	ISH	Leininger et al. (2014)
Hydra (Cnidaria, Hydrozoa)	*hymyc1*	proliferating i-cells nematoblast gland cells	whole animals	polyp	ISH	Hobmayer et al. (2012); Hartl et al. (2014)
	hymyc2	proliferating i-cells (+++) epidermal cells (+) gastrodermal cells (+)	whole animals	polyp	ISH	Hobmayer et al. (2012); Hartl et al. (2014)
Hydractinia echinata (Cnidaria, Hydrozoa)	*myc2*	i-cells	stolon	polyp	ISH	Plickert et al. (2012)
Isodiametra pulchra (Acoelomorpha)	*ipgata456*	neoblasts musculature	whole animals	throughout ontogeny	ISH	Chiodin et al. (2013)

Table A1. *Cont.*

Species	Gene	Cell Types	Tissue/Organ	Interval of Expression	Identification Methods	References
Schmidtea mediterranea (Platyhelminthes, Rhabditophora)	*smedgata4/5/6*	neoblasts	body parenchyma	throughout ontogeny	IHC, ISH	Flores et al. (2016)
	smedzfmym1 smedzf2071 smedfhl1 smedzfp1 smedegr1	neoblasts	body parenchyma	throughout ontogeny	ISH	Wagner et al. (2012)
Platynereis dumerilii (Annelida, Polychaeta)	*pdumyc*	proliferating, undifferentiated cells of the growth zone	posterior growth zone	metamorphosing larva throughout ontogeny (posterior elongation)	ISH	Gazave et al. (2013)
Holothuria glaberrima (Echinodermata, Holothuroidea)	*myc*	coelomic epithelium, intestinal cells, neuroepithelial and glial cells	regenerating intestine and radial nerve cord	during regeneration	ISH	Mashanov et al. (2015b)
Botryllus schlosseri (Chordata, Tunicata)	*bsgata4/5/6*	atrial epithelium of the bud	posterior side of the budlet	bud at stage 3	ISH	Ricci et al. (2016)
Botryllus primigenus (Chordata, Tunicata)	*myc*	cells of the branchial epithelia circulating hemocytes	growing palleal and vascular buds hemocoel of developing zooids	Strong signal during blastogenesis, in budlets (stages 1–6) and weak signal in the early primary buds.	ISH	Kawamura et al. (2008); Kawamura and Sunanaga (2011)

Table A1. *Cont.*

Species	Gene	Cell Types	Tissue/Organ	Interval of Expression	Identification Methods	References
Polyandrocarpa misakiensis (Chordata, Tunicata)	*myc*	cells of the atrial epithelium and fibroblast-like cells involved in organogenesis	developing bud	more than one day before the dedifferentiation	ISH	Fujiwara et al. (2011)
Helix–Loop–Helix-domain-containing proteins						
Isodiametra pulchra (Acoelomorpha)	*iptwist1* *iptwist2*	neoblasts musculature	whole animals	throughout ontogeny	ISH	Chiodin et al. (2013)
Initiation factors						
Dugesia japonica (Platyhelminthes, Rhabditophora)	*djeif2a* *djeif3a* *djeif4e* *djeif4g* *djeif5a* *djeif4a3*	neoblasts	body parenchyma	throughout ontogeny	ISH	Rouhana et al. (2010)
T-box proteins						
Sycon ciliatum (Porifera, Calcarea)	*scibra2*	choanocytes	choanocyte chambers	throughout ontogeny	ISH	Leininger et al. (2014)
Other transcription factors						
Dugesia japonica (Platyhelminthes, Rhabditophora)	*djekf2a*	neoblasts	body parenchyma	throughout ontogeny	ISH	Rouhana et al. (2010)
	djelk3 *djprohibitin2* *djctbp1*	neoblasts	body parenchyma	throughout ontogeny	ISH, RNAseq	Rossi et al. (2007)

Table A1. *Cont.*

Species	Gene	Cell Types	Tissue/Organ	Interval of Expression	Identification Methods	References
Platynereis dumerilii (Annelida, Polychaeta)	*pduid* *pdugcm*	proliferating, undifferentiated cells of the growth zone	posterior growth zone	metamorphosing larva throughout ontogeny (posterior elongation)	ISH	Gazave et al. (2013)
	pduap2	proliferating, undifferentiated cells of the growth zone and blastema	posterior growth zone blastema	metamorphosing larva throughout ontogeny (posterior elongation) regeneration	ISH	Gazave et al. (2013); Planques et al. (2019)
Holothuria glaberrima (Echinodfermata, Holothuroidea)	*hes* *klf1/2/4* *oct1/2/11*	progenitor cells in the radial nerve cord and by scattered cells of the neural parenchyma	radial nerve cord	adult	ISH	Mashanov et al. (2015a, 2015b)
Chromatin modification/cell cycle/differentiation						
Transcriptional silencers						
Schmidtea mediterranea (Platyhelminthes, Rhabditophora)	*smedbcl11a*	neoblasts	body parenchyma	throughout ontogeny	iRNA, FACS ISH	Resch et al. (2012); Trost et al. (2018)
	smedsirt	neoblasts	body parenchyma	throughout ontogeny	ISH	Ziman et al. (2020)
	smedcbx1	neoblasts	body parenchyma	throughout ontogeny	ISH, FACS	Eisenhoffer et al. (2008)

Table A1. *Cont.*

Species	Gene	Cell Types	Tissue/Organ	Interval of Expression	Identification Methods	References
Proteins involved in methylation-demethylation						
Dugesia japonica (Platyhelminthes, Rhabditophora)	*djhrjda* *djhrjda*	neoblasts differentiated cells	whole animals	throughout ontogeny	ISH	Cao et al. (2019)
	smedsedt8	neoblasts	body parenchyma	throughout ontogeny	ISH	Onal et al. (2012)
	smhrjda *smhrjdb*	neoblasts differentiated cells	whole animals	throughout ontogeny	ISH	Cao et al. (2019)
Schmidtea mediterranea (Platyhelminthes, Rhabditophora)	*smedsetd81*	neoblasts	body parenchyma	throughout ontogeny	ISH	Wagner et al. (2012); Torre et al. (2017)
	smednsd1 *smedmrg1* *smedrbbp41*	neoblasts	body parenchyma	throughout ontogeny	ISH	Wagner et al. (2012)
	smedbrg1l	neoblasts	body parenchyma	throughout ontogeny	ISH	Onal et al. (2012)
Proteins involved in acetylation–deacetylation						
Dugesia japonica (Platyhelminthes, Rhabditophora)	*djtaf1β*	neoblasts	body parenchyma	throughout ontogeny	ISH, RNAseq	Rossi et al. (2007)
	djbap48	neoblasts	body parenchyma	throughout ontogeny	ISH, RNAseq	Rossi et al. (2007); Bonuccelli et al. (2010)
Schmidtea mediterranea (Platyhelminthes, Rhabditophora)	*smedash2l* *smedprmt5*	neoblasts	body parenchyma	throughout ontogeny	ISH	Onal et al. (2012)
	smedhdac1	neoblasts	body parenchyma	throughout ontogeny	ISH, FACS	Eisenhoffer et al. (2008)

Table A1. *Cont.*

Species	Gene	Cell Types	Tissue/Organ	Interval of Expression	Identification Methods	References
			Histones			
Dugesia japonica (Platyhelminthes, Rhabditophora)	*djh2az djrbp4 djcip-29 djhp1*	neoblasts	body parenchyma	throughout ontogeny	ISH, RNAseq	Rossi et al. (2007)
Schmidtea mediterranea (Platyhelminthes, Rhabditophora)	*smedxrn1 smedsmarcc2 smedssrp1*	neoblasts	body parenchyma	throughout ontogeny	ISH	Onal et al. (2012)
Branchiostoma lanceolatum (Chordata, Cephalochordata)	*ph3*	Cells of the blastema and regenerating nerve cord and notochord (many pax3/7+)	Tail regenerate	Throughout ontogeny, regeneration	IHC	Somorjai et al. (2012)
Branchiostoma japonicum (Chordata, Cephalochordata)	*ph3*	Isolated cells in the regenerating oral cirrus	Oral cirrus regenerate	regeneration	IHC	Kaneto and Wada (2011)
Botryllus schlosseri (Chordata, Tunicata)	*ph3*	Budlet and primary buds Zooidal stomach	adults, buds	throughout ontogeny and astogeny	IHC with commercial antibodies	Rosner et al. (2014)
Botrylloides diegensis (Chordata, Tunicata)	*ph3*	hemoblasts	colonial vasculature	throughout ontogeny	ISH	Kassmer et al. (2020)
Styela plicata (Chordata, Tunicata)	*ph3*	hemoblasts	intestine submucosa	adults	putative ASC	Jiménez-Merino et al. (2019)

Table A1. *Cont.*

Species	Gene	Cell Types	Tissue/Organ	Interval of Expression	Identification Methods	References
			Polycomb group proteins			
Schmidtea mediterranea (Platyhelminthes, Rhabditophora)	*smedezh*	neoblasts	body parenchyma	throughout ontogeny	ISH	Wagner et al. (2012)
	smedezh2	neoblasts	body parenchyma	throughout ontogeny	ISH, IHC	Onal et al. (2012)
	smedsz12-1	neoblasts	body parenchyma	throughout ontogeny	ISH	Wagner et al. (2012)
	smedeed-1	neoblasts	body parenchyma	throughout ontogeny	ISH	Wagner et al. (2012); Onal et al. (2012)
	smedbmi1 *smedrnf2* *smedsuz12*	neoblasts	body parenchyma	throughout ontogeny	ISH	Onal et al. (2012)
			Control of transcription			
Schmidtea mediterranea (Platyhelminthes, Rhabditophora)	*smedhcf1* *smedleo1* *smedctr9*	neoblasts	body parenchyma	throughout ontogeny	ISH	Onal et al. (2012)
	smedthoc4	neoblasts	body parenchyma	throughout ontogeny	ISH	Eisenhoffer et al. (2008); Bonuccelli et al. (2010)
	smedrrm2-1	neoblasts	body parenchyma	throughout ontogeny	ISH	Eisenhoffer et al. (2008)
	smedchd4	neoblasts	body parenchyma	throughout ontogeny	ISH, IHC	Scimone et al. (2010)

Table A1. *Cont.*

Species	Gene	Cell Types	Tissue/Organ	Interval of Expression	Identification Methods	References
			Proliferation markers			
Hymeniacidon perleve (Porifera, Demospongiae)	*pcna*	archeocytes	mesohyl	cultured cells	ICC with commercial Ab	Sun et al. (2007)
Ephydatia fluviatilis (Porifera, Demospongiae)	*pcna* *efmcm2* *efccnb1*	archeocytes	mesohyl	throughout ontogeny	ISH, scRNAsec	Alié et al. (2015)
Hydra vulgaris (Cnidaria, Hydrozoa)	*hvpcna* *hvmcm2* *hvccnb1*	i-cells	body column	throughout ontogeny	ISH, scRNAsec	Alié et al. (2015)
Dugesia japonica (Platyhelminthes, Rhabditophora)	*djmcm2*	neoblasts polymorphic large cells	whole animals macerates of tissues excised just below the wound	proliferating cells after X-ray irradiation proliferating cells of intact and regenerating planaria 30–60 min after wound infliction	ISH, RNAseq	Salvetti et al. (2000); Rossi et al. (2007)
	djpcna	neoblasts	body parenchyma	proliferating cells after X-ray irradiation	IHC	Orii et al. (2005)
	djkif3a *djkif3b* *djkif19b*	neoblasts and CNS	body parenchyma and brain	throughout ontogeny	ISH	Rouhana et al. (2010)
.5*Schmidtea mediterranea* (Platyhelminthes, Rhabditophora)	*smedrb*	neoblasts	body parenchyma	throughout ontogeny	IHC, ISH	Onal et al. (2012)
	smedpcna *smedmcm2*	neoblasts	body parenchyma	throughout ontogeny	ISH, scRNAsec	Onal et al. (2012); Alié et al. (2015)

Table A1. *Cont.*

Species	Gene	Cell Types	Tissue/Organ	Interval of Expression	Identification Methods	References
	smedcyclinb	neoblasts	body parenchyma	throughout ontogeny	ISH	Eisenhoffer et al. (2008); Alié et al. (2015)
	smedp53	neoblasts	body parenchyma	throughout ontogeny	IHC, ISH	Onal et al. (2012)
	smedpp32a smedprohibitin-1	neoblasts	body parenchyma	throughout ontogeny	ISH	Eisenhoffer et al. (2008)
Macrostomum lignano (Platyhelminthes, Rhabditophora)	*duf2366/tim29*	neoblasts	body parenchyma		ISH, iRNA, RNAseq	Mouton et al. (2021)
Platynereis dumerilii (Annelida, Polychaeta)	*pdupcna*	proliferating, undifferentiated cells of the growth zone and blastema	posterior growth zone blastema	metamorphosing larva throughout ontogeny (posterior elongation) regeneration	ISH	Gazave et al. (2013); Planques et al. (2019)
	pducycb1 pducycb3	proliferating, undifferentiated cells of the growth zone and blastema	posterior growth zone blastema	metamorphosing larva throughout ontogeny (posterior elongation) regeneration	ISH	Planques et al. (2019)
Ptychodera flava (Hemichordata, Emteropneusta)	*pcna*	cells of blastemal (Ab)	regeneration blastema	throughout ontogeny	IHC with commercial Abs	Rychel and Swalla (2008)

Table A1. *Cont.*

Species	Gene	Cell Types	Tissue/Organ	Interval of Expression	Identification Methods	References
Polyandrocarpa misakiensis (Chordata, Tunicata)	*pmpcna*	cells of the atrial epithelium cells associated with epithelia	developing buds	during dedifferentiation	IHC	Kawamura et al. (2012)
Botrylloides violaceus (Chordata, Tunicata)	*pcna*	hemocytes	regenerating buds, during WBR	buds from stage 4	IHC with commercial Abs	Brown et al. (2009)
Botrylloides diegenesis (Chordata, Tunicata)	*cyclin b*	hemoblasts	colonial vasculature	during WBR	ISH	Kassmer et al. (2020)
Cytostactic proteins						
Polyandrocarpa misakiensis (Chordata, Tunicata)	*tc14-1 (lectin)*	atrial epithelial cells hemoblasts	bud	primordial bud stage growing buds	IHC	Kawamura et al. (1991)
	tc14-3 (lectin)	hemocytes, atrial epithelium	adults, buds	throughout ontogeny and astogeny in hemocytes, only in growing buds for epithelium	IHC	Matsumoto et al. (2001)
Regulators of mitochondrial dynamics						
Dugesia japonica (Platyhelminthes, Rhabditophora)	*djsam68*	neoblasts	body parenchyma	throughout ontogeny	ISH, RNAseq	Rossi et al. (2006)
Schmidtea mediterranea (Platyhelminthes, Rhabditophora)	*smedarmc1*	neoblasts	body parenchyma	throughout ontogeny	ISH	Wagner et al. (2012)

Table A1. *Cont.*

Species	Gene	Cell Types	Tissue/Organ	Interval of Expression	Identification Methods	References
			Telomere protection			
Hymeniacidon perleve (Porifera, Demospongiae)	*telomerase*	archeocytes	mesohyl	cultured cells	PCR	Sun et al. (2007)
	tert	archeocytes	mesohyl	cultured cells	activity	Sun et al. (2007)
Ephydatia fluviatilis (Porifera, Demospongiae)	*efrtel1*	archeocytes	mesohyl	throughout ontogeny	ISH, scRNAsec	Alié et al. (2015)
Hydra vulgaris (Cnidaria, Hydrozoa)	*hvrtel1*	i-cells	body column	throughout ontogeny	ISH, scRNAsec	Alié et al. (2015)
Schmidtea mediterranea (Platyhelminthes, Rhabditophora)	*smrtel1*	neoblasts	body parenchyma	throughout ontogeny	ISH, scRNAsec	Alié et al. (2015)
	smedtert	neoblasts	body parenchyma	throughout ontogeny	ISH	Tan et al. (2012)
	smedob1	all cells	Whole body	throughout ontogeny	ISH	Yin et al. (2016)
Enchytraeus japonensis (Anellida, Oligochaeta)	*telomerase*	neoblasts and N-cells (only for mesoderm)	posterior surface of the septa	during asexual reproduction by autotomy	ISH	Sugio et al. (2012)
Litechinus variegatus (Echinodermata, Echinoidea)	*tert*	muscles, esophagus, CNS, coelomocytes	muscles, esophagus, CNS, coelomocytes	adult	qRT PCR	Bodnar and Coffman (2016)
Strongylocentrotus purpuratus (Echinodermata, Echinoidea)	*tert*	muscles, esophagus, CNS, coelomocytes	muscles, esophagus, CNS, coelomocytes	adult	qRT PCR	Bodnar and Coffman (2016)

Table A1. *Cont.*

Species	Gene	Cell Types	Tissue/Organ	Interval of Expression	Identification Methods	References
Mesocentrotus franciscanus (Echinodermata, Echinoidea)	*tert*	muscles, esophagus, CNS, coelomocytes	muscles, esophagus, CNS, coelomocytes	adult	qRT PCR	Bodnar and Coffman (2016)
Botryllus schlosseri (Chordata, Tunicata)	*pot1*	multipotent epithelia	budlets	throughout blastogenesis	ISH	Tiozzo and de Tomaso (2009)
	telomerase	earliest asexual bud	budlets	throughout blastogenesis	ISH	Laird and Weissman (2004)
Other nucleic acid binding-proteins						
Schmidtea mediterranea (Platyhelminthes, Rhabditophora)	*smedpairbp1* *smedhmg1* *smedhmg2*	neoblasts	body parenchyma	throughout ontogeny	ISH	Eisenhoffer et al. (2008)
Markers of postmitotic cells						
Schmidtea mediterranea (Platyhelminthes, Rhabditophora)	*smedprog1* *smedprog2* *smedporcna* *smedmhc1*	neoblasts	body parenchyma	throughout ontogeny	ISH	Wagner et al. (2012); Van Wolfswinkel et al. (2014)
Other genes						
Schmidtea mediterranea (Platyhelminthes, Rhabditophora)	*smedhnf4*	neoblasts	body parenchyma	throughout ontogeny	IHC, ISH	Onal et al. (2012)

Table A1. *Cont.*

Species	Gene	Cell Types	Tissue/Organ	Interval of Expression	Identification Methods	References
Proteins involved in autophagy						
Polyandrocarpa misakiensis (Chordata, Tunicata)	*pmatg7*	atrial epithelium	developing buds	during dedifferentiation	ISH	Kawamura et al. (2018)
Control of differentiation						
Dugesia japonica (Platyhelminthes, Rhabditophora)	*djjy1*	neoblasts, neurons	body parenchyma and brain	throughout ontogeny	ISH	Rouhana et al. (2010)
	dhsp60 djahnak	neoblasts	body parenchyma	throughout ontogeny	ISH, RNAseq	Rossi et al. (2007)
Schmidtea mediterranea (Platyhelminthes, Rhabditophora)	*smedmhc1*	neoblasts	body parenchyma	throughout ontogeny	ISH	Van Wolfswinkel et al. (2014)
	cbp-2	ubiquitous	body tissues	throughout ontogeny	ISH, iRNA	Fraguas et al. (2021)
Botrylloides leachii (Chordata, Tunicata)	*blraldh*	circulating phagocytes	hemolymph	throughout ontogeny and astogeny	ISH	Rinkevich et al. (2007)
	if-b	atrial epithelium of buds	budlets	buds at stage 1–3	ISH	Ricci et al. (2016)
	bsraldh	inner epithelium of the bud	posterior side of the budlet	bud at stage 3	ISH	Ricci et al. (2016)
myocyte enhancer factor-2 (Mef2) genes						
Isodiametra pulchra (Acoelomorpha)	*ipmef2*	neoblasts, musculature	whole animals	throughout ontogeny	ISH	Chiodin et al. (2013)

Table A1. *Cont.*

Species	Gene	Cell Types	Tissue/Organ	Interval of Expression	Identification Methods	References
			Niche interaction			
Ephydatia fluviatilis (Porifera)	*efannexin*	choanocytes	choanocyte chambers	throughout ontogeny	ISH	Funayama et al. (2010)
Schmidtea mediterranea (Platyhelminthes, Rhabditophora)	*smedinx13*	neoblasts	body parenchyma	throughout ontogeny	IHC, ISH	Van Wolfswinkel et al. (2014)
Botryllus schlosseri (Chordata, Tunicata)	*bscadherin*	aggregates of hemoblasts aggregates of phagocytes near the endostyle bud epithelia	ampullae cell islands buds	throughout ontogeny and astogeny buds at stage 1–5	ISH, IHC	Rosner et al. (2007)
	bscd133	budding ampullar epithelium some hemocytes	circulation and vasculature	during vasculature regeneration	ISH, FACS	Braden et al. (2014)
Botrylloides diegensis	*integrin α6*	hemoblasts	colonial vasculature	throughout ontogeny	ISH	Kassmer et al. (2020)
			Others			
Polyandrocarpa misakiensis (Chordata, Tunicata)	*pmpumpa*	atrial epithelium	developing buds	during dedifferentiation	ISH	Kawamura et al. (2018)

Table A1. *Cont.*

Species	Gene	Cell Types	Tissue/Organ	Interval of Expression	Identification Methods	References
			miRNA			
Schmidtea mediterranea (Platyhelminthes, Rhabditophora)	*let7a* *mir71b* *mir756* *mir13* *mir752*	neoblasts	body parenchyma	throughout ontogeny	differential expression after irradiation	Lu et al. (2009); Friedländer et al. (2009)
	let7b *mir2160*	neoblasts	body parenchyma	throughout ontogeny	differential expression after irradiation	Lu et al. (2009)
	mir36b *mir2a* *mir2d*	neoblasts	body parenchyma	throughout ontogeny	differential expression after irradiation	Friedländer et al. (2009)

FACS: fluorescence-activated cell sorter; IB: immunoblot; ICC: immunocytochemistry; IHC: immunohistochemistry; iRNA: RNA interference; ISH: in situ hybridization; NB: Northern blot; qPCR: quantitative RT PCR; RNA seq: RNA sequencing; scRNAseq: single-cell RNA sequencing; scqPCR: single-cell qRT PCR.

References

Alié, Alexandre, Lucas Leclère, Muriel Jager, Cyrielle Dayraud, Patrick Chang, Hervé Le Guyader, Eric Quéinnec, and Michaël Manuel. 2011. Somatic stem cells express Piwi and Vasa genes in an adult ctenophore: Ancient association of "germline genes" with stemness. *Developmental Biology* 350: 183–97. [CrossRef] [PubMed]

Alié, Alexandre, Tetsutaro Hayashi, Itsuro Sugimura, Michaël Manuel, Wakana Sugano, Akira Mano, Nori Satoh, Kiyokazu Agata, and Noriko Funayama. 2015. The ancestral gene repertoire of animal stem cells. *Proceedings of the National Academy of Sciences* 112: E7093–E7100. [CrossRef] [PubMed]

Ballarin, Loriano, Arzu Karahan, Alessandra Salvetti, Leonardo Rossi, Lucia Manni, Baruch Rinkevich, Amalia Rosner, Ayelet Voskoboynik, Benyamin Rosental, Laura Canesi, and et al. 2021. Stem cells and innate immunity in aquatic invertebrates: Bridging two seemingly disparate disciplines for new discoveries in biology. *Frontiers in Immunology* 12: 688106. [CrossRef] [PubMed]

Ballarin, Loriano, Baruch Rinkevich, Kerstin Bartscherer, Artur Burzynski, Sebastien Cambier, Matteo Cammarata, Isabelle Domart-Coulon, Damjana Drobne, Juanma Encinas, Uri Frank, and et al. 2018. Maristem—stem cells of marine/aquatic invertebrates: From basic research to innovative applications. *Sustainability* 10: 526. [CrossRef]

Blanchoud, Simon, Buki Rinkevich, and Megan J. Wilson. 2018. Whole-body regeneration in the colonial tunicate *Botrylloides leachii*. *Results and Problems in Cell Differentiation* 65: 337–55. [CrossRef]

Blobe, Gerard C., William P. Schiemann, and Harvey F. Lodish. 2000. Role of transforming growth factor beta in human disease. *New England Journal of Medicine* 342: 1350–58. [CrossRef]

Bodnar, Andrea G., and James A. Coffman. 2016. Maintenance of somatic tissue regeneration with age in short- and long-lived species of sea urchins. *Aging Cell* 15: 778–87. [CrossRef]

Boehm, Anna-Marei, Konstantin Khalturin, Friederike Anton-Erxleben, Georg Hemmrich, Ulrich C. Klostermeier, Javier A. Lopez-Quintero, Hans-Heinrich Oberg, Malte Puchert, Philip Rosenstiel, Jörg Wittlieb, and et al. 2012. FoxO is a critical regulator of stem cell maintenance in immortal *Hydra*. *Proceedings of the National Academy of Sciences* 109: 19697–702. [CrossRef]

Bonuccelli, Lucia, Leonardo Rossi, Annalisa Lena, Vittoria Scarcelli, Giuseppe Rainaldi, Monica Evangelista, Paola Iacopetti, Vittorio Gremigni, and Alessandra Salvetti. 2010. An RbAp48-like gene regulates adult stem cells in planarians. *Journal of Cell Science* 123: 690–98. [CrossRef]

Braden, Brian P., Daryl A. Taketa, James D. Pierce, Susannah Kassmer, Daniel D. Lewis, and Anthony W. De Tomaso. 2014. Vascular regeneration in a basal chordate is due to the presence of immobile, bi-functional cells. *PLoS ONE* 9: e95460. [CrossRef]

Brown, Federico D., and Billie J. Swalla. 2007. Vasa expression in a colonial ascidian, *Botrylloides violaceus*. *Evolution & Development* 9: 165–77. [CrossRef]

Brown, Federico D., Elena L. Keeling, Anna D. Le, and Billie J. Swalla. 2009. Whole body regeneration in a colonial ascidian, *Botrylloides violaceus*. *Journal of Experimental Zoology Part B: Molecular and Developmental Evolution* 312B: 885–900. [CrossRef]

Cao, Ping-Lin, Nobuyoshi Kumagai, Takeshi Inoue, Kiyokazu Agata, and Takashi Makino. 2019. JmjC domain-encoding genes are conserved in highly regenerative metazoans and are associated with planarian whole-body regeneration. *Genome Biology and Evolution* 11: 552–564. [CrossRef] [PubMed]

Chang, Ti-Cheng, and Wan-Sheng Liu. 2010. The molecular evolution of PL10 homologs. *BMC Evolutionary Biology* 10: 127. [CrossRef] [PubMed]

Chiodin, Marta, Aina Børve, Eugene Berezikov, Peter Ladurner, Pedro Martinez, and Andreas Hejnol. 2013. Mesodermal gene expression in the acoel *Isodiametra pulchra* indicates a low number of mesodermal cell types and the endomesodermal origin of the gonads. *PLoS ONE* 8: e55499. [CrossRef]

Clevers, Hans, and Fiona M. Watt. 2018. Defining adult stem cells by function, not by phenotype. *Annual Review of Biochemistry* 87: 1015–27. [CrossRef]

Collins, James J., III, Bo Wang, Bramwell G. Lambrus, Marla E. Tharp, Harini Iyer, and Phillip A. Newmark. 2013. Adult somatic stem cells in the human parasite *Schistosoma mansoni*. *Nature* 494: 476–79. [CrossRef]

Conte, Maria, Paolo Deri, Maria Emilia Isolani, Linda Mannini, and Renata Batistoni. 2009. A mortalin-like gene is crucial for planarian stem cell viability. *Developmental Biology* 334: 109–18. [CrossRef]

Cordin, Olivier, Josette Banroques, N. Kyle Tanner, and Patrick Linder. 2006. The DEAD-box protein family of RNA helicases. *Gene* 367: 17–37. [CrossRef]

Czech, Benjamin, Marzia Munafò, Filippo Ciabrelli, Evelyn L. Eastwood, Martin H. Fabry, Emma Kneuss, and Gregory J. Hannon. 2018. piRNA-guided genome defense: From biogenesis to silencing. *Annual Review of Genetics* 52: 131–57. [CrossRef]

De Keuckelaere, Evi, Paco Hulpiau, Yvan Saeys, Geert Berx, and Frans van Roy. 2018. Nanos genes and their role in development and beyond. *Cellular and Molecular Life Sciences* 75: 1929–46. [CrossRef] [PubMed]

Denker, Elsa, Michaël Manuel, Lucas Leclère, Hervé Le Guyader, and Nicolas Rabet. 2008. Ordered progression of nematogenesis from stem cells through differentiation stages in the tentacle bulb of *Clytia hemisphaerica* (Hydrozoa, Cnidaria). *Developmental Biology* 315: 99–113. [CrossRef] [PubMed]

Di Maio, Alessandro, Leah Setar, Stefano Tiozzo, and Anthony W. De Tomaso. 2015. Wnt affects symmetry and morphogenesis during post-embryonic development in colonial chordates. *EvoDevo* 6: 17. [CrossRef] [PubMed]

Dill, Kariena K., and Elaine C. Seaver. 2008. Vasa and nanos are coexpressed in somatic and germ line tissue from early embryonic cleavage stages through adulthood in the polychaete *Capitella* sp. I. *Development Genes and Evolution* 218: 453–63. [CrossRef]

Dolmatov, Igor Yu, Nadezhda V. Kalacheva, Ekaterina S. Tkacheva, Alena P. Shulga, Eugenia G. Zavalnaya, Ekaterina V. Shamshurina, Alexander S. Girich, Alexey V. Boyko, and Marina G. Eliseikina. 2021. Expression of Piwi, MMP, TIMP, and Sox during gut regeneration in holothurian *Eupentacta fraudatrix* (Holothuroidea, Dendrochirotida). *Genes* 12: 1292. [CrossRef]

Dolmatov, Igor Yu. 2021. Molecular aspects of regeneration mechanisms in holothurians. *Genes* 12: 250. [CrossRef]

Egger, Bernhard, Dirk Steinke, Hiroshi Tarui, Katrien De Mulder, Detlev Arendt, Gaëtan Borgonie, Noriko Funayama, Robert Gschwentner, Volker Hartenstein, Bert Hobmayer, and et al. 2009. To be or not to be a flatworm: The acoel controversy. *PLoS ONE* 4: e5502. [CrossRef]

Eisenhoffer, George T., Hara Kang, and Alejandro Sánchez Alvarado. 2008. Molecular analysis of stem cells and their descendents during cell turnover and regeneration in the planarian *Schmidtea mediterranea*. *Cell Stem Cell* 3: 327–39. [CrossRef]

Fernandéz-Taboada, Enrique, Sören Moritz, Dagmar Zeuschner, Martin Stehling, Hans R. Schöler, Emili Saló, and Luca Gentile. 2010. Smed-SmB, a member of the LSm protein superfamily, is essential for chromatoid body organization and planarian stem cell proliferation. *Development* 137: 1055–65. [CrossRef]

Ferrario, Cinzia, Michela Sugni, Ildiko M. L. Somorjai, and Loriano Ballarin. 2020. Beyond adult stem cells: Dedifferentiation as a unifying mechanism underlying regeneration in invertebrate deuterostomes. *Frontiers in Cell and Developmental Biology* 8: 587320. [CrossRef]

Fierro-Constaín, Laura, Quentin Schenkelaars, Eve Gazave, Anne Haguenauer, Caroline Rocher, Alexander Ereskovsky, Carole Borchiellini, and Emmanuelle Renard. 2017. The conservation of the germline multipotency program, from sponges to vertebrates: A stepping stone to understanding the somatic and germline origins. *Genome Biology and Evolution* 9: 474–88. [CrossRef] [PubMed]

Fincher, Christopher T., Omri Wurtzel, Thom de Hoog, Kellie M. Kravarik, and Peter W. Reddien. 2018. Cell type transcriptome atlas for the planarian *Schmidtea mediterranea*. *Science* 360: eaaq1736. [CrossRef] [PubMed]

Flores, Natasha M., Néstor J. Oviedo, and Julien Sage. 2016. Essential role for the planarian intestinal GATA transcription factor in stem cells and regeneration. *Developmental Biology* 418: 179–188. [CrossRef] [PubMed]

Fraguas, Susanna, Sheila Cárcel, Coral Vivancos, Ma Dolores Molina, Jordi Ginés, Judith Mazariegos, Thileepan Sekaran, Kerstin Bartscherer, Rafael Romero, and Francesc Cebrià. 2021. CREB-binding protein (CBP) gene family regulates planarian survival and stem cell differentiation. *Developmental Biology* 476: 53–67. [CrossRef] [PubMed]

Friedländer, Marc R., Catherine Adamidi, Ting Han, Svetlana Lebedeva, Thomas A. Isenbarger, Martin Hirst, Marco Marra, Chad Nusbaum, William L. Lee, James C. Jenkin, and et al. 2009. High-resolution profiling and discovery of planarian small RNAs. *Proceedings of the National Academy of Sciences USA* 106: 11546–11551. [CrossRef]

Fujiwara, Shigeki, Takaomi Isozaki, Kyoko Mori, and Kazuo Kawamura. 2011. Expression and function of myc during asexual reproduction of the budding ascidian *Polyandrocarpa misakiensis*. *Development, Growth & Differentiation* 53: 1004–14. [CrossRef]

Funayama, Noriko, Mikiko Nakatsukasa, Kurato Mohri, Yoshiki Masuda, and Kiyokazu Agata. 2010. Piwi expression in archaeocytes and choanocytes in demosponges: Insights into the stem cell system in demosponges. *Evolution & Development* 12: 275–87. [CrossRef]

Gasparini, Fabio, Sebastian M. Shimeld, Elena Ruffoni, Paolo Burighel, and Lucia Manni. 2011. Expression of a Musashi-like gene in sexual and asexual development of the colonial chordate *Botryllus schlosseri* and phylogenetic analysis of the protein group. *Journal of Experimental Zoology Part B: Molecular and Developmental Evolution* 316B: 562–73. [CrossRef]

Gazave, Eve, Aurélien Guillou, and Guillaume Balavoine. 2014. History of a prolific family: The Hes/Hey-related genes of the annelid *Platynereis*. *Evodevo* 5: 29. [CrossRef]

Gazave, Eve, Julien Béhague, Lucie Laplane, Aurélien Guillou, Laetitia Préau, Adrien Demilly, Guillaume Balavoine, and Michel Vervoort. 2013. Posterior elongation in the annelid *Platynereis dumerilii* involves stem cells molecularly related to primordial germ cells. *Developmental Biology* 382: 246–67. [CrossRef]

Gazave, Eve, Quentin I. B. Lemaître, and Guillaume Balavoine. 2017. The Notch pathway in the annelid *Platynereis*: Insights into chaetogenesis and neurogenesis processes. *Open Biology* 7: 160242. [CrossRef] [PubMed]

Giani, Vincent C., Jr., Emi Yamaguchi, Michael J. Boyle, and Elaine C. Seaver. 2011. Somatic and germline expression of piwi during development and regeneration in the marine polychaete annelid *Capitella teleta*. *EvoDevo* 2: 10. [CrossRef] [PubMed]

Gilman, Benjamin, Pilar Tijerina, and Rick Russell. 2017. Distinct RNA-unwinding mechanisms of DEAD-box and DEAH-box RNA helicase proteins in remodeling structured RNAs and RNPs. *Biochemical Society Transactions* 45: 1313–21. [CrossRef] [PubMed]

Girich, Alexander S., Marina P. Isaeva, and Igor Yu Dolmatov. 2017. *Wnt* and *frizzled* expression during regeneration of internal organs in the holothurian *Eupentacta fraudatrix*. *Wound Repair and Regeneration* 25: 828–35. [CrossRef] [PubMed]

Grimson, Andrew, Mansi Srivastava, Bryony Fahey, Ben J. Woodcroft, Hyojin Rosaria Chiang, Nicole King, Bernard M. Degnan, Daniel S. Rokhsar, and David P. Bartel. 2008. Early origins and evolution of microRNAs and Piwi-interacting RNAs in animals. *Nature* 455: 1193–97. [CrossRef]

Guo, Tingxia, Antoine H. F. M. Peters, and Phillip A. Newmark. 2006. A Bruno-like gene is required for stem cell maintenance in planarians. *Developmental Cell* 11: 159–69. [CrossRef]

Guo, Xing, and Xiao-Fan Wang. 2009. Signaling cross-talk between TGF-beta/BMP and other pathways. *Cell Research* 19: 71–88. [CrossRef]

Handberg-Thorsager, Mette, and Emili Saló. 2007. The planarian nanos-like gene Smednos is expressed in germline and eye precursor cells during development and regeneration. *Development Genes and Evolution* 217: 403–11. [CrossRef]

Hartl, Markus, Anna-Maria Mitterstiller, Taras Valovka, Kathrin Breuker, Bert Hobmayer, and Klaus Bister. 2010. Stem cell-specific activation of an ancestral myc protooncogene with conserved basic functions in the early metazoan *Hydra*. *Proceedings of the National Academy of Sciences* 107: 4051–56. [CrossRef]

Hartl, Markus, Stella Glasauer, Sabine Gufler, Andrea Raffeiner, Kane Puglisi, Kathrin Breuker, Klaus Bister, and Bert Hobmayer. 2019. Differential regulation of myc homologs by Wnt/β-Catenin signaling in the early metazoan *Hydra*. *The FEBS Journal* 286: 2295–310. [CrossRef]

Hartl, Markus, Stella Glasauer, Taras Valovka, Kathrin Breuker, Bert Hobmayer, and Klaus Bister. 2014. *Hydra myc2*, a unique pre-bilaterian member of the *myc* gene family, is activated in cell proliferation and gametogenesis. *Biology Open* 3: 397–407. [CrossRef] [PubMed]

Higuchi, Sayaka, Tetsutaro Hayashi, Hiroshi Tarui, Osamu Nishimura, Kaneyasu Nishimura, Norito Shibata, Hiroshi Sakamoto, and Kiyokazu Agata. 2008. Expression and functional analysis of musashi-like genes in planarian CNS regeneration. *Mechanisms of Development* 125: 631–45. [CrossRef] [PubMed]

Hobmayer, Bert, Fabian Rentzsch, Kerstin Kuhn, Christoph M. Happel, Christoph Cramer von Laue, Petra Snyder, Ute Rothbächer, and Thomas W. Holstein. 2000. Wnt signalling molecules act in axis formation in the diploblastic metazoan *Hydra*. *Nature* 407: 186–89. [CrossRef] [PubMed]

Hobmayer, Bert, Marcell Jenewein, Dominik Eder, Marie-Kristin Eder, Stella Glasauer, Sabine Gufler, Markus Hartl, and Willi Salvenmoser. 2012. Stemness in *Hydra*—A current perspective. *The International Journal of Developmental Biology* 56: 509–17. [CrossRef] [PubMed]

Höck, Julia, and Gunter Meister. 2008. The Argonaute protein family. *Genome Biology* 9: 210. [CrossRef]

Hyams, Yosef, Guy Paz, Claudette Rabinowitz, and Baruch Rinkevich. 2017. Insights into the unique torpor of *Botrylloides leachi*, a colonial urochordate. *Developmental Biology* 428: 101–17. [CrossRef]

Isaeva, Valeria V., Anna V. Akhmadieva, Ia N. Aleksandrova, and Andrey I. Shukalyuk. 2009. Morphofunctional organization of reserve stem cells providing for asexual and sexual reproduction of invertebrates. *Russian Journal of Developmental Biology* 40: 57–68. [CrossRef]

Jager, Muriel, Eric Quéinnec, Hervé Le Guyader, and Michaël Manuel. 2011. Multiple Sox genes are expressed in stem cells or in differentiating neuro-sensory cells in the hydrozoan Clytia hemisphaerica. *EvoDevo* 2: 12. [CrossRef]

Jeffery, William R. 2014. Distal regeneration involves the age dependent activity of branchial sac stem cells in the ascidian *Ciona intestinalis*. *Regeneration* 2: 1–18. [CrossRef]

Jehn, Julia, Daniel Gebert, Frank Pipilescu, Sarah Stern, Julian Simon Thilo Kiefer, Charlotte Hewel, and David Rosenkranz. 2018. PIWI genes and piRNAs are ubiquitously expressed in mollusks and show patterns of lineage-specific adaptation. *Communications Biology* 1: 137. [CrossRef]

Jiménez-Merino, Juan, Isadora Santos de Abreu, Laurel S. Hiebert, Silvana Allodi, Stefano Tiozzo, Cintia M. De Barros, and Federico D. Brown. 2019. Putative stem cells in the hemolymph and in the intestinal submucosa of the solitary ascidian *Styela plicata*. *EvoDevo* 10: 31. [CrossRef] [PubMed]

Juliano, Celina E., Adrian Reich, Na Liu, Jessica Götzfried, Mei Zhong, Selen Uman, Robert A. Reenan, Gary M. Wessel, Robert E. Steele, and Haifan Lin. 2014. PIWI proteins and PIWI-interacting RNAs function in Hydra somatic stem cells. *Proceedings of the National Academy of Sciences* 111: 337–42. [CrossRef] [PubMed]

Juliano, Celina E., S. Zachary Swartz, and Gary M. Wessel. 2010. A conserved germline multipotency program. *Development* 137: 4113–26. [CrossRef] [PubMed]

Kaneto, Satoshi, and Hiroshi Wada. 2011. Regeneration of amphioxus oral cirri and its skeletal rods: Implications for the origin of the vertebrate skeleton. *Journal of Experimental Zoology* 316B: 409–17. [CrossRef] [PubMed]

Kanska, J., and Uri Frank. 2013. New roles for Nanos in neural cell fate determination revealed by studies in a cnidarian. *Journal of Cell Science* 126: 3192–203. [CrossRef] [PubMed]

Kashima, Makoto, Nobuyoshi Kumagai, Kiyokazu Agata, and Norito Shibata. 2016. Heterogeneity of chromatoid bodies in adult pluripotent stem cells of planarian *Dugesia japonica*. *Development, Growth & Differentiation* 58: 225–37. [CrossRef]

Kassmer, Susannah H., Adam D. Langenbacher, and Anthony W. De Tomaso. 2020. Integrin-alpha-6+ Candidate stem cells are responsible for whole body regeneration in the invertebrate chordate *Botrylloides diegensis*. *Nature Communications* 11: 4435. [CrossRef] [PubMed]

Kataoka, Naoyuki, Michael D. Diem, V. Narry Kim, Jeongsik Yong, and Gideon Dreyfuss. 2001. Magoh, a human homolog of *Drosophila* mago nashi protein, is a component of the splicing-dependent exon-exon junction complex. *The EMBO Journal* 20: 6424–33. [CrossRef] [PubMed]

Kawamura, Kazuo, and Takeshi Sunanaga. 2011. Role of Vasa, Piwi, and Myc-expressing coelomic cells in gonad regeneration of the colonial tunicate, *Botryllus primigenus*. *Mechanisms of Development* 128: 457–70. [CrossRef] [PubMed]

Kawamura, Kazuo, Miki Tachibana, and Takeshi Sunanaga. 2008. Cell proliferation dynamics of somatic and germline tissues during zooidal life span in the colonial tunicate *Botryllus primigenus*. *Developmental dynamics: An Official Publication of the American Association of Anatomists* 237: 1812–25. [CrossRef]

Kawamura, Kazuo, Seigo Kitamura, Satoko Sekida, Masayuki Tsuda, and Takeshi Sunanaga. 2012. Molecular anatomy of tunicate senescence: Reversible function of mitochondrial and nuclear genes associated with budding cycles. *Development* 139: 4083–93. [CrossRef] [PubMed]

Kawamura, Kazuo, Shigeki Fujiwara, and Yasuo Sugino. 1991. Budding-specific lectin induced in epithelial cells is an extracellular matrix component for stem cell aggregation in tunicates. *Development* 113: 995–1005. [CrossRef] [PubMed]

Kawamura, Kazuo, Takuto Yoshida, and Satoko Sekida. 2018. Autophagic dedifferentiation induced by cooperation between TOR inhibitor and retinoic acid signals in budding tunicates. *Developmental Biology* 433: 384–393. [CrossRef] [PubMed]

Kawashima, Takeshi, Aya R. Murakami, Michio Ogasawara, Kimio J. Tanaka, Rieko Isoda, Yasunori Sasakura, Takahito Nishikata, and Kazuhiro W. Makabe. 2000. Expression patterns of musashi homologs of the ascidians, *Halocynthia roretzi* and *Ciona intestinalis*. *Development Genes and Evolution* 210: 162–65. [CrossRef]

Khalturin, Konstantin, Friederike Anton-Erxleben, Sabine Milde, Christine Plötz, Jörg Wittlieb, Georg Hemmrich, and Thomas C. G. Bosch. 2007. Transgenic stem cells in *Hydra* reveal an early evolutionary origin for key elements controlling self-renewal and differentiation. *Developmental Biology* 309: 32–44. [CrossRef] [PubMed]

Kozin, Vitaly, and Roman P. Kostyuchenko. 2015. Vasa, pl10, and piwi gene expression during caudal regeneration of the polychaete annelid *Alitta virens*. *Development Genes and Evolution* 225: 129–38. [CrossRef] [PubMed]

Kusserow, Arne, Kevin Pang, Carsten Sturm, Martina Hrouda, Jan Lentfer, Heiko A. Schmidt, Ulrich Technau, Arndt von Haeseler, Bert Hobmayer, Mark Q. Martindale, and et al. 2005. Unexpected complexity of the Wnt gene family in a sea anemone. *Nature* 433: 156–60. [CrossRef]

Lai, Alvina G., and A. Aziz Aboobaker. 2018. EvoRegen in animals: Time to uncover deep conservation or convergence of adult stem cell evolution and regenerative processes. *Developmental Biology* 433: 118–31. [CrossRef]

Laird, Diana J., and Irving L. Weissman. 2004. Telomerase maintained in self-renewing tissues during serial regeneration of the urochordate *Botryllus schlosseri*. *Developmental Biology* 273: 185–94. [CrossRef]

Lechable, Marion, Matthias Achrainer, Maren Kruus, Willi Salvenmoser, and Bert Hobmayer. 2022. Molecular regulation of decision making in the interstitial stem cell lineage of Hydra revisited. In *Advances in Aquatic Invertebrate Stem Cell Research: From Basic Research to Innovative Applications*. Edited by Loriano Ballarin, Baruch Rinkevich and Bert Hobmayer. Basel: MDPI, pp. 201–219.

Leininger, Sven, Marcin Adamski, Brith Bergum, Corina Guder, Jing Liu, Mary Laplante, Jon Bråte, Friederike Hoffmann, Sofia Fortunato, Signe Jordal, and et al. 2014. Developmental gene expression provides clues to relationships between sponge and eumetazoan body plans. *Nature Communications* 5: 3905. [CrossRef]

Lengfeld, Tobias, Hiroshi Watanabe, Oleg Simakov, Dirk Lindgens, Lydia Gee, Lee Law, Heiko A Schmidt, Suat Ozbek, Hans Bode, and Thomas W. Holstein. 2009. Multiple Wnts are involved in *Hydra* organizer formation and regeneration. *Developmental Biology* 330: 186–99. [CrossRef]

Li, Qing, Yuanli L. Wang, Jia-Wei Xie, Wen-Juan Sun, Ming Zhu, Lin He, and Qun Wang. 2015. Characterization and expression of DDX6 during gametogenesis in the Chinese mitten crab *Eriocheir sinensis*. *Genetics and Molecular Research* 14: 4420–37. [CrossRef] [PubMed]

Liang, Yujun, Delima Rathnayake, Shibo Huang, Anjalika Pathirana, Qiyu Xu, and Shicui Zhang. 2019. BMP signaling is required for amphioxus tail regeneration. *Development*, 146. [CrossRef] [PubMed]

Lim, Robyn S. M., Amit Anand, Chiemi Nishimiya-Fujisawa, Satoru Kobayashi, and Toshie Kai. 2014. Analysis of *Hydra* PIWI proteins and piRNAs uncover early evolutionary origins of the piRNA pathway. *Developmental Biology* 386: 237–51. [CrossRef] [PubMed]

Liu, Niankun, Hong Han, and Paul Lasko. 2009. Vasa promotes *Drosophila* germline stem cell differentiation by activating mei-P26 translation by directly interacting with a (U)-rich motif in its 3′ UTR. *Genes & Development* 23: 2742–52. [CrossRef]

Lu, Yi-Chien, Magda Smielewska, Dasaradhi Palakodeti, Michael T. Lovci, Stefan Aigner, Gene W. Yeo, and Brenton R. Graveley. 2009. Deep sequencing identifies new and regulated microRNAs in *Schmidtea mediterranea*. *RNA* 15: 1483–91. [CrossRef]

Maris, Christophe, Cyril Dominguez, and Frédéric H.-T. Allain. 2005. The RNA recognition motif, a plastic RNA-binding platform to regulate post-transcriptional gene expression. *The FEBS Journal* 272: 2118–31. [CrossRef]

Marlow, Heather Q., Mansi Srivastava, David Q. Matus, Daniel Rokhsar, and Mark Q. Martindale. 2009. Anatomy and development of the nervous system of *Nematostella vectensis*, an anthozoan cnidarian. *Developmental Neurobiology* 69: 235–54. [CrossRef]

Martinez, Pedro, Loriano Ballarin, Alexander V. Ereskovsky, Eve Gazave Bert Hobmayer, Lucia Manni, Eric Rottinger, Simon G. Sprecher, Stefano Tiozzo, A. Varela-Coelho, and B. Rinkevich. 2022. Articulating the "stem cell niche" paradigm through the lens of non-model aquatic invertebrates. *BMC Biology* 20: 23. [CrossRef] [PubMed]

Mashanov, Vladimir S., Olga R. Zueva, and José E. García-Arrarás. 2015a. Heterogeneous generation of new cells in the adult echinoderm nervous system. *Frontiers in Neuroanatomy* 9: 123. [CrossRef]

Mashanov, Vladimir S., Olga R. Zueva, and José E. García-Arrarás. 2015b. Expression of pluripotency factors in echinoderm regeneration. *Cell & Tissue Research* 359: 521–536. [CrossRef]

Matsumoto, Junko, Chiaki Nakamoto, Shigeki Fujiwara, Toshitsugu Yubisui, and Kazuo Kawamura. 2001. A novel C-type lectin regulating cell growth, cell adhesion and cell differentiation of the multipotent epithelium in budding tunicates. *Development* 128: 3339–3347. [CrossRef] [PubMed]

Meister, Gunter. 2013. Argonaute proteins: Functional insights and emerging roles. *Nature Reviews Genetics* 14: 447–59. [CrossRef]

Micklem, David R., Ramanuj Dasgupta, Heather Elliott, Fanni Gergely, Catherine Davidson, Andrea Brand, Acaimo González-Reyes, and Daniel St Johnston. 1997. The mago nashi gene is required for the polarisation of the oocyte and the formation of perpendicular axes in *Drosophila*. *Current Biology* 7: 468–78. [CrossRef]

Millane, R. Cathriona, Justyna Kanska, David J. Duffy, Cathal Seoighe, Stephen Cunningham, Günter Plickert, and Uri Frank. 2011. Induced stem cell neoplasia in a cnidarian by ectopic expression of a POU domain transcription factor. *Development* 138: 2429–39. [CrossRef] [PubMed]

Mochizuki, Kazufumi, Chiemi Nishimiya-Fujisawa, and Toshitaka Fujisawa. 2001. Universal occurrence of the vasa-related genes among metazoans and their germline expression in *Hydra*. *Development Genes and Evolution* 211: 299–308. [CrossRef]

Mochizuki, Kazufumi, Hiroko Sano, Satoru Kobayashi, Chiemi Nishimiya-Fujisawa, and Toshitaka Fujisawa. 2000. Expression and evolutionary conservation of nanos-related genes in *Hydra*. *Development Genes and Evolution* 210: 591–602. [CrossRef]

Mouton, Stijn, Kirill Ustyantsev, Frank Beltman, Lisa Glazenburg, and Eugene Berezikov. 2021. TIM29 is required for enhanced stem cell activity during regeneration in the flatworm *Macrostomum lignano*. *Scientific Reports* 11: 1166. [CrossRef]

Müller, Werner, Uri Frank, Regina Teo, Ofer Mokady, Christina Guette, and Günter Plickert. 2007. Wnt signaling in hydroid development: Ectopic heads and giant buds induced by GSK-3beta inhibitors. *The International Journal of Developmental Biology* 51: 211–20. [CrossRef]

Odintsova, Nelly A. 2009. Stem cells of marine invertebrates: Regulation of proliferation and induction of differentiation in vitro. *Cell and Tissue Biology* 3: 403–8. [CrossRef]

Okamoto, Kazuko, Mikiko Nakatsukasa, Alexandre Alié, Yoshiki Masuda, Kiyokazu Agata, and Noriko Funayama. 2012. The active stem cell specific expression of sponge Musashi homolog EflMsiA suggests its involvement in maintainingthe stem cell state. *Mechanisms of Development* 129: 24–37. [CrossRef] [PubMed]

Onal, Pinar, Dominic Grün, Catherine Adamidi, Agnieszka Rybak, Jordi Solana, Guido Mastrobuoni, Yongbo Wang, Hans-Peter Rahn, Wei Chen, Stefan Kempa, and et al. 2012. Gene expression of pluripotency determinants is conserved between mammalian and planarian stem cells. *The EMBO Journal* 31: 2755–69. [CrossRef] [PubMed]

Orii, Hidefumi, Takashige Sakurai, and Kenji Watanabe. 2005. Distribution of the stem cells (neoblasts) in the planarian *Dugesia japonica*. *Development Genes and Evolution* 215: 143–57. [CrossRef] [PubMed]

Özpolat, B. Duygu, and Alexandra E. Bely. 2015. Gonad establishment during asexual reproduction in the annelid *Pristina leidyi*. *Developmental Biology* 405: 123–36. [CrossRef] [PubMed]

Palakodeti, Dasaradhi, Magda Smielewska, Yi-Chien Lu, Gene W. Yeo, and Brenton R. Graveley. 2008. The PIWI proteins SMEDWI-2 and SMEDWI-3 are required for stem cell function and piRNA expression in planarians. *RNA* 14: 1174–86. [CrossRef] [PubMed]

Park, Sun-Mi, Raquel P. Deering, Yuheng Lu, Patrick Tivnan, Steve Lianoglou, Fatima Al-Shahrour, Benjamin L. Ebert, Nir Hacohen, Christina Leslie, George Q Daley, and et al. 2014. Musashi-2 controls cell fate, lineage bias, and TGF-β signaling in HSCs. *Journal of Experimental Medicine* 211: 71–87. [CrossRef] [PubMed]

Parma, David H., Paul E. Bennett Jr., and Boswell Robert E. 2007. Mago Nashi and Tsunagi/Y14, respectively, regulate *Drosophila* germline stem cell differentiation and oocyte specification. *Developmental Biology* 308: 507–19. [CrossRef]

Pek, Jun Wei, Amit Anand, and Toshie Kai. 2012. Tudor domain proteins in development. *Development* 139: 2255–66. [CrossRef]

Perčulija, Vanja, and Songying Ouyang. 2019. Diverse roles of DEAD/DEAH-box helicases in innate immunity and diseases. In *Helicases from All Domains of Life*. Edited by Renu Tuteja. Cambridge: Academic Press, pp. 141–71. [CrossRef]

Pfister, Daniela, Katrien De Mulder, Isabelle Philipp, Georg Kuales, Martina Hrouda, Paul Eichberger, Gaetan Borgonie, Volker Hartenstein, and Peter Ladurner. 2007. The exceptional stem cell system of *Macrostomum lignano*: Screening for gene expression and studying cell proliferation by hydroxyurea treatment and irradiation. *Frontiers in Zoology* 4: 9. [CrossRef]

Pfister, Daniela, Katrien De Mulder, Volker Hartenstein, Georg Kuales, Gaetan Borgonie, Florentine Marx, Joshua Morris, and Peter Ladurner. 2008. Flatworm stem cells and the germ line: Developmental and evolutionary implications of *macvasa* expression in *Macrostomum lignano*. *Developmental Biology* 319: 146–59. [CrossRef] [PubMed]

Planques, Anabelle, Julien Malem, Julio Parapar, Michel Vervoort, and Eve Gazave. 2019. Morphological, cellular and molecular characterization of posterior regeneration in the marine annelid *Platynereis dumerilii*. *Developmental Biology* 445: 189–210. [CrossRef] [PubMed]

Plass, Mireya, Jordi Solana, F. Alexander Wolf, Salah Ayoub, Aristotelis Misios, Petar Glažar, Benedikt Obermayer, Fabian J. Theis, Christine Kocks, and Nikolaus Rajewsky. 2018. Cell type atlas and lineage tree of a whole complex animal by single-cell transcriptomics. *Science*, 360. [CrossRef] [PubMed]

Plickert, Günter, Uri Frank, and Werner A. Müller. 2012. *Hydractinia*, a pioneering model for stem cell biology and reprogramming somatic cells to pluripotency. *The International Journal of Developmental Biology* 56: 519–34. [CrossRef]

Poon, Jessica, Gary M. Wessel, and Mamiko Yajima. 2006. An unregulated regulator: Vasa expression in the development of somatic cells and in tumorigenesis. *Developmental Biology* 415: 24–32. [CrossRef]

Pozzolini, Marina, Lorenzo Gallus, Stefano Ghignone, Sara Ferrando, Simona Candiani, Matteo Bozzo, Marco Bertolino, Gabriele Costa, Giorgio Bavestrello, and Sonia Scarfi. 2019. Insights into the evolution of metazoan regenerative mechanisms: Roles of TGF superfamily members in tissue regeneration of the marine sponge *Chondrosia reniformis*. *Journal of Experimental Biology*, 222. [CrossRef]

Rabinowitz, Claudette, and Baruch Rinkevich. 2011. De novo emerged stemness signatures in epithelial monolayers developed from extirpated palleal buds. *In Vitro Cellular & Developmental Biology-Animal* 47: 26–31. [CrossRef]

Rajasethupathy, Priyamvada, Igor Antonov, Robert Sheridan, Sebastian Frey, Chris Sander, Thomas Tuschl, and Eric R. Kandel. 2012. A role for neuronal piRNAs in the epigenetic control of memory-related synaptic plasticity. *Cell* 149: 693–707. [CrossRef]

Rebscher, Nicole, Fabiola Zelada-González, Torsten U. Banisch, Florian Raible, and Detlev Arendt. 2007. Vasa unveils a common origin of germ cells and of somatic stem cells from the posterior growth zone in the polychaete *Platynereis dumerilii*. *Developmental Biology* 306: 599–611. [CrossRef]

Rebscher, Nicole, Christopher Volk, Regina Teo, and Gunter Plickert. 2008. The germ plasm component Vasa allows tracing of the interstitial stem cells in the cnidarian *Hydractinia echinata*. *Developmental Dynamics* 237: 1736. [CrossRef]

Reddien, Peter W., Adam L. Bermange, Kenneth J. Murfitt, Joya R. Jennings, and Alejandro Sánchez Alvarado. 2005a. Identification of genes needed for regeneration, stem cell function, and tissue homeostasis by systematic gene perturbation in planaria. *Developmental Cell* 8: 635–49. [CrossRef] [PubMed]

Reddien, Peter W., Néstor J. Oviedo, Joya R. Jennings, James C. Jenkin, and Alejandro Sánchez Alvarado. 2005b. SMEDWI-2 is a PIWI-like protein that regulates planarian stem cells. *Science* 310: 1327–30. [CrossRef] [PubMed]

Reinardy, Helena C., Chloe E. Emerson, Jason M. Manley, and Andrea G. Bodnar. 2015. Tissue regeneration and biomineralization in sea urchins: Role of Notch signaling and presence of stem cell markers. *PLoS ONE* 10: e0133860. [CrossRef] [PubMed]

Resch, Alissa M., Dasaradhi Palakodeti, Yi-Chien Lu, Michael Horowitz, and Brenton R. Graveley. 2012. Transcriptome analysis reveals strain-specific and conserved stemness genes in *Schmidtea mediterranea*. *PLoS ONE* 7: e34447. [CrossRef]

Ricci, Lorenzo, Ankita Chaurasia, Pascal Lapébie, Philippe Dru, Rebecca R. Helm, Richard R. Copley, and Stefano Tiozzo. 2016. Identification of differentially expressed genes from multipotent epithelia at the onset of an asexual development. *Scientific Reports* 6: 27357. [CrossRef] [PubMed]

Rinkevich, Baruch, and Valeria Matranga, eds. 2009. *Stem Cells in Marine Organisms*. Berlin: Springer, p. 369.

Rinkevich, Baruch, Loriano Ballarin, Pedro Martinez, Ildiko Somorjai, Oshrat Ben-Hamo, Ilya Borisenko, Eugene Berezikov, Alexander Ereskovsky, Eve Gazave, Denis Khnykin, and et al. 2022. A pan-metazoan concept for adult stem cells: The wobbling Penrose landscape. *Biological Reviews* 97: 299–325. [CrossRef]

Rinkevich, Baruch. 2011. Cell cultures from marine invertebrates: New insights for capturing endless stemness. *Marine Biotechnology* 13: 345–54. [CrossRef]

Rinkevich, Yuval, Amalia Rosner, Claudette Rabinowitz, Ziva Lapidot, Elithabeth Moiseeva, and Buki Rinkevich. 2010. Piwi positive cells that line the vasculature epithelium, underlie whole body regeneration in a basal chordate. *Developmental Biology* 345: 94–104. [CrossRef]

Rinkevich, Yuval, Guy Paz, Baruch Rinkevich, and Ram Reshef. 2007. Systemic bud induction and retinoic acid signaling underlie whole body regeneration in the urochordate *Botrylloides leachi*. *PLoS Biology* 5: e71. [CrossRef]

Rosner, Amalia, and Baruch Rinkevich. 2007. The DDX3 subfamily of the DEAD Box helicases: Divergent roles as unveiled by studying different organisms and in vitro assays. *Current Medicinal Chemistry* 14: 2517–25. [CrossRef]

Rosner, Amalia, and Baruch Rinkevich. 2011. VASA as a specific marker for germ cells lineage: In light of evolution. *Trends in Comparative Biochemistry & Physiology* 15: 1–15.

Rosner, Amalia, Claudette Rabinowitz, Elizabeth Moiseeva, Ayelet Voskoboynik, and Baruch Rinkevich. 2007. BS-cadherin in the colonial urochordate *Botryllus schlosseri*: One protein, many functions. *Developmental Biology* 304: 687–700. [CrossRef] [PubMed]

Rosner, Amalia, Elizabeth Moiseeva, Yuval Rinkevich, Ziva Lapidot, and Baruch Rinkevich. 2009. Vasa and the germ line lineage in a colonial urochordate. *Developmental Biology* 331: 113–28. [CrossRef] [PubMed]

Rosner, Amalia, Elizabeth Moiseeva, Claudette Rabinowitz, and Baruch Rinkevich. 2013. Germ lineage properties in the urochordate *Botryllus schlosseri*—from markers to temporal niches. *Developmental Biology* 384: 356–74. [CrossRef] [PubMed]

Rosner, Amalia, Gilad Alfassi, Elizabeth Moiseeva, Guy Paz, Claudette Rabinowitz, Ziva Lapidot, Jacob Douek, Abraham Haim, and Baruch Rinkevich. 2014. The involvement of three signal transduction pathways in botryllid ascidians' astogeny, as revealed by expression patterns of representative genes. *The International Journal of Developmental Biology* 58: 677–92. [CrossRef]

Rosner, Amalia, Guy Paz, and Baruch Rinkevich. 2006. Divergent roles of the DEAD-box protein BS-PL10, the urochordate homologue of human DDX3 and DDX3Y proteins, in colony astogeny and ontogeny. *Developmental Dynamics: An Official Publication of the American Association of Anatomists* 235: 1508–21. [CrossRef]

Rossi, Leonardo, Alessandra Salvetti, Annalisa Lena, Renata Batistoni, Paolo Deri, Claudio Pugliesi, Elena Loreti, and Vittorio Gremigni. 2006. DjPiwi-1, a member of the PAZ-Piwi gene family, defines a subpopulation of planarian stem cells. *Development Genes and Evolution* 216: 335–46. [CrossRef]

Rossi, Leonardo, Alessandra Salvetti, Francesco M. Marincola, Annalisa Lena, Paolo Deri, Linda Mannini, Renata Batistoni, Ena Wang, and Vittorio Gremigni. 2007. Deciphering the molecular machinery of stem cells: A look at the neoblast gene expression profile. *Genome Biology* 8: R62. [CrossRef]

Rouhana, Labib, Jennifer A. Weiss, Ryan S. King, and Phillip A. Newmark. 2014. PIWI homologs mediate histone H4 mRNA localization to planarian chromatoid bodies. *Development* 141: 2592–601. [CrossRef]

Rouhana, Labib, Norito Shibata, Osamu Nishimura, and Kiyokazu Agata. 2010. Different requirements for conserved post-transcriptional regulators in planarian regeneration and stem cell maintenance. *Developmental Biology* 341: 429–43. [CrossRef]

Rychel, Amanda L., and Billie J. Swalla. 2008. Anterior regeneration in the hemichordate *Ptychodera flava*. *Developmental Dynamics: An Official Publication of the American Association of Anatomists* 237: 3222–32. [CrossRef] [PubMed]

Salvetti, Alessandra, Leonardo Rossi, Annalisa Lena, Renata Batistoni, Paolo Deri, Giuseppe Rainaldi, Maria Teresa Locci, Monica Evangelista, and Vittorio Gremigni. 2005. DjPum, a homologue of *Drosophila* Pumilio, is essential to planarian stem cell maintenance. *Development* 132: 1863–74. [CrossRef] [PubMed]

Salvetti, Alessandra, Leonardo Rossi, Paolo Deri, and Renata Batistoni. 2000. An MCM2-related gene is expressed in proliferating cells of intact and regenerating planarians. *Developmental Dynamics: An Official Publication of the American Association of Anatomists* 218: 603–14. [CrossRef]

Sato, Kimihiro, Norito Shibata, Hidefumi Orii, Reiko Amikura, Takashige Sakurai, Kiyokazu Agata, Satoru Kobayashi, and Kenji Watanabe. 2006. Identification and origin of the germline stem cells as revealed by the expression of nanos-related gene in planarians. *Development, Growth & Differentiation* 48: 615–28. [CrossRef]

Scimone, M. Lucila, Joshua Meisel, and Peter W. Reddien. 2010. The Mi-2-like Smed-CHD4 gene is required for stem cell differentiation in the planarian *Schmidtea mediterranea*. *Development* 137: 1231–41. [CrossRef]

Seipel, Katja, Nathalie Yanze, and Volker Schmid. 2004. The germ line and somatic stem cell gene Cniwi in the jellyfish *Podocoryne carnea*. *The International Journal of Developmental Biology* 48: 1–7. [CrossRef]

Shah, Chirag, Michael J. W. Vangompel, Villian Naeem, Yanmei Chen, Terrance Lee, Nicholas Angeloni, Yin Wang, and Eugene Yujun Xu. 2010. Widespread presence of human BOULE homologs among animals and conservation of their ancient reproductive function. *PLoS Genetics* 6: e1001022. [CrossRef]

Shibata, Norito, Makoto Kashima, Taisuke Ishiko, Osamu Nishimura, Labib Rouhana, Kazuyo Misaki, Shigenobu Yonemura, Kuniaki Saito, Haruhiko Siomi, Mikiko C. Siomi, and et al. 2016. Inheritance of a nuclear PIWI from pluripotent stem cells by somatic descendants ensures differentiation by silencing transposons in planarian. *Developmental Cell* 7: 226–37. [CrossRef]

Shibata, Norito, Yoshihiko Umesono, Hidefumi Orii, Takashige Sakurai, Kenji Watanabe, and Kiyokazu Agata. 1999. Expression of vasa(vas)-related genes in germline cells and totipotent somatic stem cells of planarians. *Developmental Biology* 206: 73–87. [CrossRef]

Shigunov, Patrícia, and Bruno Dallagiovanna. 2015. Stem cell ribonomics: RNA-binding proteins and gene networks in stem cell differentiation. *Frontiers in Molecular Biosciences* 2: 74. [CrossRef]

Shukalyuk, Andrey I., Kseniya A. Golovnina, Sergei I. Baiborodin, Konstantin V. Gunbin, Alexander G. Blinov, and Valeria V. Isaeva. 2007. Vasa-related genes and their expression in stem cells of colonial parasitic rhizocephalan barnacle *Polyascus polygenea* (Arthropoda: Crustacea: Cirripedia: Rhizocephala). *Cell Biology International* 31: 97–108. [CrossRef] [PubMed]

Siebert, Stefan, Freya E. Goetz, Samuel H. Church, Pathikrit Bhattacharyya, Felipe Zapata, Steven H. D. Haddock, and Casey W. Dunn. 2015. Stem cells in *Nanomia bijuga* (Siphonophora), a colonial animal with localized growth zones. *Evodevo* 6: 22. [CrossRef] [PubMed]

Siebert, Stefan, Jeffrey A. Farrell, Jack F. Cazet, Yashodara Abeykoon, Abby S. Primack, Christine E. Schnitzler, and Celina E. Juliano. 2019. Stem cell differentiation trajectories in *Hydra* resolved at single-cell resolution. *Science*, 365. [CrossRef] [PubMed]

Sköld, Helen Nilsson, Matthias Obst, Mattias Sköld, and Bertil Åkesson. 2009. Stem cells in asexual reproduction of marine invertebrates. In *Stem Cells in Marine Organisms*. Edited by Baruch Rinkevich and Valeria Matranga. Dordrecht: Springer, pp. 105–37. [CrossRef]

Solana, Jordi, Manuel Irimia, Salah Ayoub, Marta Rodriguez Orejuela, Vera Zywitza, Marvin Jens, Javier Tapial, Debashish Ray, Quaid Morris, Timothy R. Hughes, and et al. 2016. Conserved functional antagonism of CELF and MBNL proteins controls stem cell-specific alternative splicing in planarians. *Elife* 5: e16797. [CrossRef]

Solana, Jordi, Paul Lasko, and Rafael Romero. 2009. Spoltud-1 is a chromatoid body component required for planarian long-term stem cell self-renewal. *Developmental Biology* 328: 410–21. [CrossRef]

Somorjai, Ildikó M. L. 2017. Amphioxus regeneration: Evolutionary and biomedical implications. *The International Journal of Developmental Biology* 61: 689–96. [CrossRef]

Somorjai, Ildikó M. L., Rajmund L. Somorjai, Jordi Garcia-Fernàndez, and Hector Escrivà. 2012. Vertebrate-like regeneration in the invertebrate chordate amphioxus. *Proceedings of the National Academy of Sciences* 109: 517–22. [CrossRef]

Sugio, Mutsumi, Chikako Yoshida-Noro, Kaname Ozawa, and Shin Tochinai. 2012. Stem cells in asexual reproduction of *Enchytraeus japonensis* (Oligochaeta, Annelid): Proliferation and migration of neoblasts. *Development, Growth & Differentiation* 54: 439–50. [CrossRef]

Sun, Liming, Yuefan Song, Yi Qu, Xingju Yu, and Wei Zhang. 2007. Purification and in vitro cultivation of archaeocytes (stem cells) of the marine sponge *Hymeniacidon perleve* (Demospongiae). *Cell and Tissue Research* 328: 223–37. [CrossRef]

Sunanaga, Takeshi, Miho Satoh, and Kazuo Kawamura. 2008. The role of Nanos homologue in gametogenesis and blastogenesis with special reference to male germ cell formation in the colonial ascidian, *Botryllus primigenus*. *Developmental Biology* 324: 31–40. [CrossRef] [PubMed]

Tan, Thomas C. J., Ruman Rahman, Farah Jaber-Hijazi, Daniel A. Felix, Chen Chen, Edward J. Louis, and Aziz Aboobaker. 2012. Telomere maintenance and telomerase activity are differentially regulated in asexual and sexual worms. *Proceedings of the National Academy of Sciences* 109: 4209–14. [CrossRef] [PubMed]

Tatzuke, Yuki, Takeshi Sunanaga, Shigeki Fujiwara, and Kaz Kawamura. 2012. RACK1 regulates mesenchymal cell recruitment during sexual and asexual reproduction of budding tunicates. *Developmental Biology* 368: 393–403. [CrossRef]

Tiozzo, Stefano, and Antony W. de Tomaso. 2009. Functional analysis of pitx during asexual regeneration in a basal chordate. *Evolution & Development* 11: 152–62. [CrossRef]

Tiozzo, Stefano, Lionel Christiaen, Carole Deyts, Lucia Manni, Jean-Stéphane Joly, and Paolo Burighel. 2005. Embryonic versus blastogenetic development in the compound ascidian *Botryllus schlosseri*: Insights from Pitx expression patterns. *Developmental Dynamics: An Official Publication of the American Association of Anatomists* 232: 468–78. [CrossRef] [PubMed]

Torre, Cedric, Prasad Abnave, Landry Laure Tsoumtsa, Giovanna Mottola, Catherine Lepolard, Virginie Trouplin, Gregory Gimenez, Julie Desrousseaux, Stephanie Gempp, Anthony Levasseur, and et al. 2017. *Staphylococcus aureus* promotes smed-pgrp-2/smed-setd8-1 methyltransferase signalling in planarian neoblasts to sensitize anti-bacterial gene responses during re-infection. *EBioMedicine* 20: 150–160. [CrossRef]

Trost, Toria, Jessica Haines, Austin Dillon, Brittany Mersman, Mallory Robbins, Peyton Thomas, and Amy Hubert. 2018. Characterizing the role of SWI/SNF-related chromatin remodeling complexes in planarian regeneration and stem cell function. *Stem Cell Research* 32: 91–103. [CrossRef]

Udroiu, Ion, Valeria Russo, Tiziana Persichini, Marco Colasanti, and Antonella Sgura. 2017. Telomeres and telomerase in basal Metazoa. *ISJ-Invertebrate Survival Journal* 14: 233–40. [CrossRef]

Van Wolfswinkel, Josien C., Daniel E. Wagner, and Peter W. Reddien. 2014. Single-cell analysis reveals functionally distinct classes within the planarian stem cell compartment. *Cell Stem Cell* 15: 326–39. [CrossRef]

Vogt, Günter. 2012. Hidden treasures in stem cells of indeterminately growing bilaterian invertebrates. *Stem Cell Reviews and Reports* 8: 305–17. [CrossRef]

Wagers, Amy J., and Irving L. Weissman. 2004. Plasticity of adult stem cells. *Cell* 116: 639–48. [CrossRef]

Wagner, Daniel E., Jaclyn J. Ho, and Peter W. Reddien. 2012. Genetic regulators of a pluripotent adult stem cell system in planarians identified by RNAi and clonal analysis. *Cell Stem Cell* 10: 299–311. [CrossRef] [PubMed]

Wiens, Matthias, Sergey I. Belikov, Oxana V. Kaluzhnaya, Anatoli Krasko, Heinz C. Schröder, Sanja Perovic-Ottstadt, and Werner E. G. Müller. 2006. Molecular control of serial module formation along the apical–basal axis in the sponge *Lubomirskia baicalensis*: Silicateins, mannose-binding lectin and mago nashi. *Development Genes and Evolution* 216: 229–42. [CrossRef] [PubMed]

Xiol, Jordi, Pietro Spinelli, Maike A. Laussmann, David Homolka, Zhaolin Yang, Elisa Cora, Yohann Couté, Simon Conn, Jan Kadlec, Ravi Sachidanandam, and et al. 2014. RNA clamping by Vasa assembles a piRNA amplifier complex on transposon transcripts. *Cell Stem Cell* 10: 299–311. [CrossRef]

Yin, Shanshan, Yan Huang, Yingnan Zhangfang, Xiaoqin Zhong, Pengqing Li, Junjiu Huang, Dan Liu, and Zhou Songyang. 2016. SmedOB1 is required for planarian homeostasis and regeneration. *Scientific Reports* 6: 34013. [CrossRef]

Zahiri, Reyhane, and Mariya Zahiri. 2016. Marine invertebrate's stem cell culture: Biotechnology prospects of marine stem cells. *ISMJ* 19: 912–30. [CrossRef]

Zhan, Tailan, Niklas Rindtorff, and Michael Boutros. 2017. Wnt signaling in cancer. *Oncogene* 36: 1461–73. [CrossRef]

Zhou, Xin, Giorgia Battistoni, Osama El Demerdash, James Gurtowski, Julia Wunderer, Ilaria Falciatori, Peter Ladurner, Michael C. Schatz, Gregory J. Hannon, and Kaja A. Wasik. 2015. Dual functions of Macpiwi1 in transposon silencing and stem cell maintenance in the flatworm *Macrostomum lignano*. *RNA* 21: 1885–97. [CrossRef]

Ziman, Ben, Peter Karabinis, Paul Barghouth, and Néstor J. Oviedo. 2020. Sirtuin-1 regulates organismal growth by altering feeding behavior and intestinal morphology in planarians. *Journal of Cell Science*, 133. [CrossRef]

Oxylipins: Role in Stem Cell Biology

Helike Lõhelaid and Tarvi Teder

Abstract: Oxylipins, oxygenated fatty acid derivatives, are well-established stress mediators acting in auto- and paracrine manner. Eicosanoids, the most studied branch of oxylipins, are produced from twenty carbon polyunsaturated fatty acids (PUFAs). In vertebrates, they are synthesized mainly by lipoxygenase (LOX), cyclooxygenase (COX) and cytochrome P450-type monooxygenases. In corals, besides COX and LOX enzymes, the oxidation of arachidonic acid (AA) is catalyzed by natural fusion proteins, comprised of a LOX domain and a catalase related peroxidase domain, allene oxide synthase (AOS) or hydroperoxide lyase (HPL). Although oxylipins are well studied in vertebrate stem cells, their role in stem cells originating from marine invertebrates remains unexplored. Here, we present an overview of major oxylipin pathways in vertebrates and marine invertebrates, and discuss their potential role in invertebrate stem cells.

1. Introduction

There is a growing interest in invertebrate stem cells (SCs) due to their high toti- and pluripotency which makes them suitable model systems to investigate fundamental biological processes, such as cell fate, senescence, regeneration and cell reprogramming (Ballarin et al. 2018). Due to the simplicity of marine invertebrates, it is easier to track the expression of genes, test different compounds on differentiation/regeneration and discover underlying mechanisms of SCs (Manni et al. 2019). For instance, colonial ascidians are ideal organisms for the study of tissue regeneration and development because of their diverse reproductive strategies, relatively short lifespan, simple morphological and genomic organization, and easy experimental use. In addition, the high diversity of invertebrates creates an opportunity to use them as a source of novel natural products, including bioactive lipid mediators, which can be used to treat cancer, infections, autoimmune and inflammatory-related diseases, and can potentially be implemented in regenerative medicine (Palanisamy et al. 2017).

A group of bioactive oxylipins derived from arachidonic acid (AA), eicosanoids, are identified as important auto- and paracrine mediators of tissue repair and regeneration that act by regulating the stem cell biochemistry in vertebrates. Only a limited number of oxylipin studies have been conducted on invertebrates used for SC research (Kassmer et al. 2020). Screening and targeting of oxylipins from invertebrates would provide novel insights into the molecular mechanisms necessary

"

for either stemness or differentiation of SCs in marine invertebrates. For instance, profiling of oxylipins and tracking their secretion to surrounding tissues would reveal spatio-temporal distribution of oxylipins and their regulatory role in self-renewal and/or differentiation of SCs. This knowledge can be beneficial in the future studies of SCs across different species.

This review summarizes the status of oxylipin studies in invertebrate SC model systems and focuses on corals as the most studied model of oxylipin biosynthesis in invertebrates.

2. Oxylipin Pathways in Animals

Eicosanoids are the main group of oxylipins in animals synthesized from AA (C20: 4ω6) and other C20 polyunsaturated fatty acids (PUFAs) by fatty acid dioxygenases, e.g., lipoxygenase (LOX) and cyclooxygenase (COX), or monooxygenases, such as cytochrome P450 epoxygenases, respectively (Figure 1) (Brash 1999; Rouzer and Marnett 2003; Nelson et al. 2013).

In mammals, eicosanoids and other bioactive lipids are highly potent short-lived molecules that initiate signaling cascades and gene expression by binding to their corresponding receptors or being ligands for transcription factors. Activation of gene expression regulates cellular events, including cell proliferation and differentiation, and different physiological and pathological processes, e.g., inflammatory-related diseases and cancer. In addition to AA, other PUFAs, such as eicosapentaenoic acid (EPA, C20: 5ω3) and docosahexaenoic acid (DHA, C22: 6ω3), are the precursors for important bioactive lipids, e.g., resolvins and protectins (Serhan et al. 2002, 2008), which mediate the resolution of inflammation in animals.

The complexity of eicosanoid pathways is necessary to modulate cellular processes in a cell type and a metabolic state manner. On the other hand, the high variability of eicosanoids and sophisticated regulatory networks makes eicosanoid research challenging.

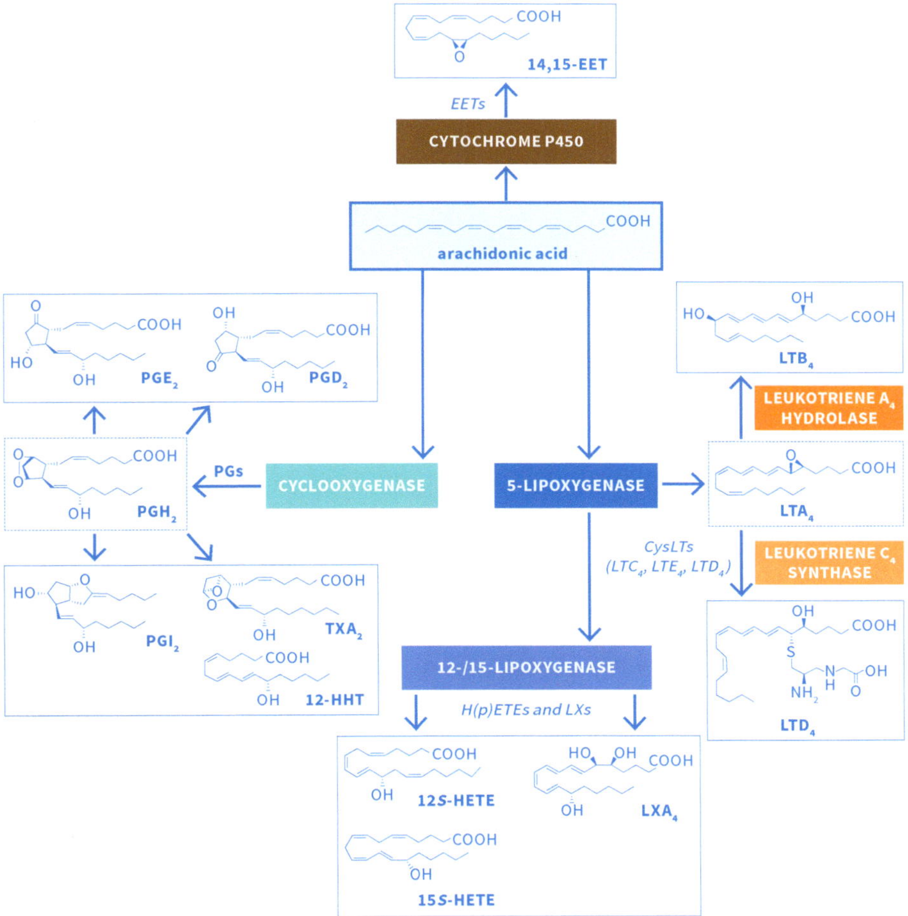

Figure 1. Biosynthetic routes of eicosanoids in animals. Arachidonic acid (AA) is released from cellular membranes by phospholipases in response to a variety of stimuli and converted to eicosanoids by cyclooxygenase (COX), lipoxygenase (LOX) and cytochrome P450 (CYP450) monooxygenase pathways. The COX pathway gives rise to prostaglandins (PGs), the LOX pathway produces hydroxy-eicosatetraenoic acids (HETEs), lipoxins (LXs) and leukotrienes (LTs), and CYP450 synthesizes epoxy-eicosatetraenoic acids (EETs). CysLT—cysteinyl leukotrienes; EET—epoxy-eicosatetraenoic acid; HETE—hydroxy-eicosatetraenoic acid; HHT—hydroxy-heptadecatrienoic acid; LT—leukotriene; LX—lipoxin; PG—prostaglandin; TX—thromboxane. Source: Graphic by authors.

2.1. Lipoxygenase

LOXs (E.C. 1.13.11.-) are non-heme iron containing dioxygenases that catalyze the regio- and stereo-specific peroxidation of PUFAs containing at least one *cis,cis*-1,4-pentadiene system to form biologically active mediators (Brash 1999). LOXs are classified in terms of their positional specificity. Animal LOXs are arachidonate 5-, 8-, 11-, 12- and 15-LOXs that catalyze the conversion of AA into corresponding 5-, 8-, 11-, 12- and 15-hydroperoxy-eicosatetraenoic acids (HpETEs) (Brash 1999). Depending on the species and cell type-specific expression of enzymes, the content and distribution of eicosanoids vary. Thus far, the LOX with $11R$-specificity has been identified only in marine invertebrates, such as hydra (Di Marzo et al. 1993), sea urchins (Hawkins and Brash 1987) and corals (Di Marzo et al. 1996; Varvas et al. 1999; Mortimer et al. 2006). In terrestrial organisms, the prevalent stereo-configuration of LOX products is S, while R stereospecificity is more pronounced in marine invertebrates.

HpETEs or their reduced derivatives, hydroxy-eicosatetraenoic acids (HETEs), are potent pro- or anti-tumorigenic agents and mediate cell migration due to their chemotactic properties and also. For example, 5- and 12-HETEs synthesized by 5- and 12-LOX, and 13-hydroxy-octadecadienoic acid formed by 15-LOX, respectively, are involved in the proliferation and inhibition of apoptosis, angiogenesis, cancer invasion and metastasis, while 15- and 8-HETE formed by 15-LOX-2 and 8-LOX are involved in the differentiation, growth arrest and induction of apoptosis (Pidgeon et al. 2007; Moreno 2009). In addition, lipid mediators generated in 5-LOX pathway mediate atherosclerosis and allergic inflammation (Haeggström 2018). Most importantly, HpETEs are precursors of many downstream biosynthetic routes, such as the leukotriene and lipoxin pathways (Figure 1), which are involved in the initiation and resolution of inflammation, respectively (Funk 2001; Serhan et al. 2002; Haeggström and Funk 2011).

2.2. Cyclooxygenase

Cyclooxygenases (COXs), also known as prostaglandin endoperoxide synthases (E.C. 1.14.99.1), are another oxygenation route converting AA to prostaglandins (PGs). All vertebrates have two COX isozymes, a constitutively expressed COX-1 and an inducible COX-2 (Funk 2001). Both COXs catalyze the formation of PGG_2 via cyclooxygenase activity and its reduction to PGH_2 via peroxidase activity (Rouzer and Marnett 2003; Schneider et al. 2007). The main differences between COX-1 and COX-2 are their genetic regulation and function (Rouzer and Marnett 2005; Blobaum and Marnett 2007). The formation of PGH_2 by COXs is a rate-limiting step in its downstream conversion to prostaglandin E_2 (PGE_2), $PGF_{2\alpha}$, and PGD_2, as well as the conversion to prostacyclin (PGI_2) and thromboxane A_2 (TXA_2) by corresponding isomerases or synthases (Figure 1) (Rouzer and Marnett 2009). Prostanoids are involved in inflammatory processes, wound healing, tissue regeneration and cardiovascular processes. Therefore, the inhibition of COX results in reduced inflammation, pain and fever (Flower 2006). Non-steroidal anti-inflammatory drugs (NSAIDs) have anti-inflammatory and pro-resolving effects through the inhibition of COX-2 (Vane and Botting 1998). In conferring their biological function, e.g., evoking an inflammatory response after injury, PGs have opposite effects. For example, depending on the timing and course of inflammation, they can either induce vasoconstriction ($PGF_{2\alpha}$, TXA_2, TXB_2) or vasodilation (PGE_1, PGE_2, PGI_2), inhibition of platelet aggregation (PGD_2, TXA_1, PGE_1, PGI_2) (Murakami 2011; Ricciotti and FitzGerald 2011) or aggregation of platelets (PGE_2) (Howie et al. 1973; Kobzar et al. 1997). Elevated levels of PGE_2 sensitize spinal neurons, which results in an increased sense of pain (Grace et al. 2014), causing fever via the hypothalamus-mediated manner (Coceani and Akarsu 1998), and are involved in the complex process of labor (Kelly et al. 2009).

3. Coral Eicosanoids

Corals are invertebrate animals (Kingdom *Animalia*; phylum *Cnidaria*; class *Anthozoa*) (Hyman 1940) that are divided into two major subclasses: reef-building *Hexacorallia* and soft corals *Octocorallia* (Zhang 2011), both comprised of azooxanthellate or zooxanthellate, the latter living in symbiosis with unicellular algae, *Symbiodinium sp.* species.

Coral oxylipin research started with the detection of large quantities of PGs and PG-esters (2–3% of dry weight) in the soft coral *Plexaura homomalla* (Weinheimer and Spraggins 1969). Thereafter, a plethora of eicosanoids have been discovered, which vary depending on the species and location (Corey et al. 1973, 1987, 1988; Varvas et al. 1993, 1999; Brash et al. 1987). In soft corals, AA is an abundant fatty acid (10–25%), being the primary precursor of eicosanoids (Imbs et al. 2006; Imbs and Yakovleva 2011). To a lesser degree (3–10%), AA also contributes to the fatty acid content of stony corals (Latyshev et al. 1991; Dunn et al. 2012; Figueiredo et al. 2012; Funk 2001). Released AA is metabolized by COX (Varvas et al. 1994; Koljak et al. 2001; Valmsen et al. 2001) or LOX (Mortimer et al. 2006; Brash et al. 1996) into PGs or H(p)ETEs, respectively (Figure 2). In addition to 11*R*-LOX (Eek et al. 2012; Mortimer et al. 2006; Järving et al. 2012), corals contain catalase-related allene oxide synthase-8*R*-lipoxygenase (AOS-LOX) and hydroperoxide lyase-8*R*-lipoxygenase (HPL-LOX) fusion protein pathways (Koljak et al. 1997).

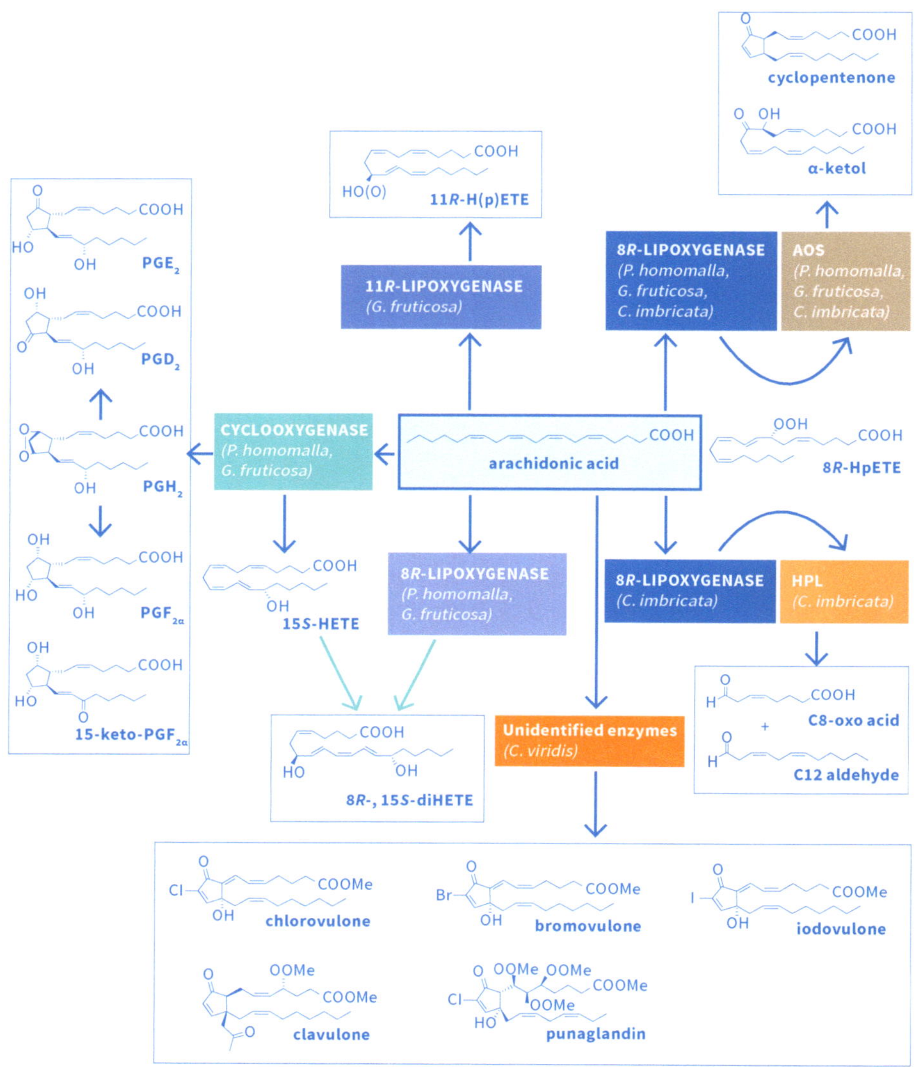

Figure 2. The eicosanoid pathways identified in soft corals (Varvas et al. 1993, 1999). AOS, allene oxide synthase; H(p)ETE—hydro(pero)xyeiocosatetraenoic acid; HPL—hydroperoxide lyase; PG—prostaglandin. *C. imbricata*—*Capnella imbricata*; *C. viridis*—*Clavularia viridis*; *G. fruticosa*—*Gersemia fruticosa*; *P. homomalla*—*Plexaura homomalla*. Source: Graphic by authors.

In principle, the coral AOS-LOX and HPL-LOX pathways are similar to the plant LOX pathways, except the fact that separately expressed and structurally distinct plant LOX, P450-type AOS and HPL metabolize only C18 PUFAs, e.g., linoleic acid

(Wasternack 2007). Initially, the cyclopentenone synthesized by coral AOS was thought to be the precursor of coral PGs, but the cloning and characterization of functional coral COXs indicated the existence of parallel oxygenation routes (Koljak et al. 2001; Valmsen et al. 2001). Even though *P. homomalla* contains a considerable amount of PGs, incubations with the tissue homogenate and exogenous AA do not produce PGs (Corey et al. 1973, 1988). In contrast, homogenates of *G. fruticosa* give rise to optically active PGs in vitro (Varvas et al. 1993, 1999). In addition, the soft coral *Clavularia viridis* converts AA to different cyclopentenone-type compounds, such as clavulones (preclavulone A) (Corey et al. 1987), bromovulones and iodovulones (Figure 2) (Honda et al. 1987; Watanabe et al. 2001). Although AOS-LOXs are not involved in the biosynthesis of coral PGs, they still might contribute to the production of clavulone-like derivatives. For today, the AOS-LOX pathway is identified in soft corals *P. homomalla*, *G. fruticosa*, and *C. imbricata*, while 11R-LOX is expressed only in *G. fruticosa*. In addition, no COX activity and PGs have been detected from *C. imbricata*. Although the sequence data implies the presence of AOS-LOX in soft and stony corals (Lõhelaid and Samel 2018), the fusion protein with the lyase activity is identified only in *C. imbricata*. Altogether, this data is indicative of species-specific eicosanoid biosynthesis.

The current literature on coral eicosanoids contains data on the identification of naturally occurring compounds (Corey et al. 1973, 1985; Varvas et al. 1993, 1994), the elucidation of metabolic pathways involved in their biosynthesis (Brash et al. 1987; Corey et al. 1987; Koljak et al. 1997, 2001; Varvas et al. 1999), and the effects of lipid extracts or isolated compounds on other systems (Hashimoto et al. 2003). For today, only the role of PGs in the defense of the coral *P. homomalla* against predators has been proposed (Pawlik et al. 1987; Gerhart 1991; O'Neal and Pawlik 2002; Whalen et al. 2010). In regard to the LOX activity in other marine invertebrates, it was demonstrated that 8R-HETE induces the maturation of starfish oocytes (Meijer et al. 1986) and 11R-HpETE is involved in the regeneration and bud formation of *Hydra vulgaris* (Di Marzo et al. 1993). In spite of the wide occurrence of different oxylipins (hydroxy fatty acids, PGs and their derivatives, etc.) in invertebrates (Rowley et al. 2005; Brash et al. 1987), their exact functions in those organisms remain unclear.

Coral Fusion Proteins in the Arachidonic Acid Pathway

In the arachidonate metabolism of corals, fusion proteins comprised of N-terminal catalase-like allene oxide synthase (AOS) or hydroperoxide lyase (HPL) and C-terminal 8R-LOX domains catalyze the conversion of AA via 8R-HpETE to allene oxide (Koljak et al. 1997; Lõhelaid et al. 2008, 2014a) or short-chain aldehydes (Teder et al. 2015), respectively (Figure 2). The 3D structure of the AOS-LOX fusion protein (Gilbert et al. 2008), as well as separately expressed AOS and LOX domains (Oldham et al. 2005a, 2005b; Neau et al. 2009), have been determined. Even though the structure of HPL-LOX has not been resolved, the differences in the substrate specificity and catalytic properties between HPL and AOS (Teder et al. 2017, 2019) indicate distinct regulation and roles of corresponding fusion proteins in vivo.

Several transcriptomic studies of stony corals have reported the increased expression of the *AOS-LOX* gene in response to white band disease (Libro et al. 2013), elevated UV radiation (Aranda et al. 2011) and temperature (Polato et al. 2013). However, transcriptomes lack information about expressed proteins and their activity. A targeted study with the soft coral *C. imbricata* demonstrated the elevated levels of AOS-LOX metabolites and increased gene expression in response to wounding (Lõhelaid et al. 2014a) and temperature (Lõhelaid et al. 2014b). In parallel, the levels of HPL-LOX mRNA and metabolites remained stable or even decreased. To date, involvement of the AOS-LOX pathway in the stress response of corals is evident, however, the biological importance of HPL-LOX remains elusive. Short-chain aldehydes also known as "green leaf volatiles" play an essential part in the communication and stress signaling of plants. In addition, aldehydes have antibacterial and antifungal properties due to their molecular attributes. Therefore, HPL-LOX-derived aldehydes may serve a housekeeping role, including defense against biotic stressors.

4. Eicosanoids in Stem Cells

SCs are undifferentiated progenitor cells with the ability to differentiate into specialized cell types and regenerate. Eicosanoids are best known for their inflammatory and immune-modulating properties, however, their ability to affect the cell fate has increased their importance in SC biology. Eicosanoids act in an auto- and paracrine manner to promote proliferation, migration, and differentiation of SCs which contribute to the tissue repair, regeneration and other cellular processes. For example, eicosanoids mediate the differentiation of SCs at each step of wound healing (Berry et al. 2017). Due to the diversity of bioactive lipids and other regulators, the role of eicosanoids in determining the fate of SCs is not very well understood.

The roles of PUFAs and eicosanoids have been studied in mammalian mesenchymal stem cells (MSCs) (Jang et al. 2012; Yun et al. 2009b, 2011; Ern et al. 2019; Kim et al. 2009b; Rinkevich et al. 2009), hematopoietic stem cells (HSCs) (Hoggatt and Pelus 2010), embryonic stem cells (ESCs) (Liou et al. 2007; Yanes et al. 2010; Yun et al. 2009b; Rajasingh and Bright 2006; Kim et al. 2009a), neural stem cells (NSCs) (Katura et al. 2010; Wada et al. 2006; Wiszniewska et al. 2011; Katakura et al. 2009, 2013; Beltz et al. 2007; He et al. 2009; Sakayori et al. 2011; Kawakita et al. 2006; Jung et al. 2006; Goncalves et al. 2010; Sasaki et al. 2003), endothelial progenitor stem cells (EPC) (Kawabe et al. 2010; Herrler et al. 2009) and others (Table 1). For instance, MSCs constitutively express COX, PGE_2 synthase (PGES) (Jang et al. 2012; Kleiveland et al. 2008), 5-LOX, and 12-LOX (Fang et al. 2015), giving rise to PGs, LTs and LXs, respectively. In addition, MSCs express different PG receptors, EP1-EP3, FP, and IP (Rinkevich et al. 2009). MSCs and eicosanoids are studied due to their involvement in immune-modulating and inflammatory-related processes (Bernardo and Fibbe 2013). In addition to MSCs, human periodontal ligament stem cells (hPDLSCs) produce PGE_2, PGD_2 and $PGF_{2\alpha}$ as well as specialized pro-resolving mediators (SPMs), e.g., different resolvins, protectin D1, maresins, and LXB_4 (Berry et al. 2017).

Table 1. Bioactions of eicosanoids in stem cells.

Fatty Acid or Eicosanoid	Stem Cell Type	Effects	References
Linoleic acid	Embryonic stem cells	Enhanced proliferation	(Kim et al. 2009a)
Arachidonic acid	Neuronal stem cells	Enhanced proliferation	(Vaca et al. 2008; He et al. 2009; Sakayori et al. 2011; Kawakita et al. 2006; Sakamoto et al. 2007)
Eicosapentaenoic acid	Neuronal stem cells	Improved differentiation	(Katakura et al. 2009)
Docosahexaenoic acid	Neuronal stem cells	Improved differentiation, increased proliferation	(Beltz et al. 2007; Katakura et al. 2009, 2013; Kan et al. 2007)
Prostaglandin E_1, E_2	Hematopoietic stem cells	Inhibited proliferation	(Gidali and Feher 1977; Kurland et al. 1978; Motomura and Dexter 1980)
	Embryonic stem cells	Enhanced proliferation, inhibited apoptosis	(Yun et al. 2009b; Liou et al. 2007; Hou et al. 2013)
	Human umbilical cord blood-derived mesenchymal stem cells	Enhanced proliferation	(Yun et al. 2011; Jang et al. 2012)
	Neuronal stem cells	Enhanced proliferation	(Jung et al. 2006; Goncalves et al. 2010; Sasaki et al. 2003)
	Bone marrow-derived cells	Improved endothelial differentiation	(Zhu et al. 2011)
	Tendon stem cells	Improved osteogenic differentiation	(Liu et al. 2013)
Δ12,14-prostaglandin J_2	Embryonic stem cells	Inhibited proliferation	(Rajasingh and Bright 2006)
15d-prostaglandin J_2	Neuronal stem cells	Regulation of proliferation	(Katura et al. 2010)
Leukotriene B_4	Neuronal stem cells	Regulation of proliferation, promoted differentiation to neurons	(Wada et al. 2006; Wiszniewska et al. 2011)
	Hematopoietic stem cells	Enhanced proliferation, inhibited apoptosis	(Chung et al. 2005)
Leukotriene D_4	Embryonic stem cells	Enhanced proliferation	(Kim et al. 2010)
Lipoxin A_4	Neuronal stem cells	Inhibited proliferation	(Wada et al. 2006)
	Human periodontal ligament stem cells	Enhanced proliferation, migration and wound healing	(Berry et al. 2017)
	Human dental apical papilla	Immunomodulation, proliferation, wound healing. Attenuated chemokine and growth factor secretion	(Gaudin et al. 2018)
	Bone marrow-derived mesenchymal stem cells	Resolution of inflammation and injury, bacterial clearance, increased SC growth.	(Fang et al. 2015; Tsoyi et al. 2016)
Lipoxin B_4	Bone marrow-derived mesenchymal stem cells	Radioprotection	(Walden 1988)
Neuroprotectin D_1	Embryonic stem cells	Improved neuronal and cardiac differentiation	(Yanes et al. 2010)
Thromboxane A_2	Adipose tissue-derived mesenchymal stem cells	Enhanced proliferation, promote differentiation to smooth-muscle-like cells	(Yun et al. 2009a; Kim et al. 2009b)

Table adapted from (Kang et al. 2014) and modified accordingly.

The role of eicosanoids has been extensively studied in tissue repair and generation. Overall, LTs and PGD_2 have a negative regulatory effect on tissue repair, while other lipid mediators, such as other PGs and LXs, promote healing (Esser-von Bieren 2019). It should be noted that the same type of lipid mediators may be differently regulated during proliferation, differentiation and migration of SCs (Rinkevich et al. 2009).

Modulation of eicosanoid pathways has an impact on the fate of SCs. For instance, the inhibition of COX and LOX pathways manifests in the pluripotency of ESC (Yanes et al. 2010). In contrast, supplementation of fatty acids and their derivatives promote proliferation and differentiation of mouse ESC (Yanes et al. 2010; Kim et al. 2009a). It is also known that SPMs lose their therapeutic effect when 5-LOX, 12-LOX and 15-LOX activities are attenuated (Romano et al. 2019).

4.1. Roles of Eicosanoids in Vertebrate Stem Cell Biology

4.1.1. The LOX Pathway in Stem Cells

The expression of 5-LOX and biosynthesis of LTs are increased in differentiated ESCs. Inhibition of the 5-LOX pathway results in impaired vasculogenesis by ESCs (Finkensieper et al. 2010). A downstream lipid mediator of the 5-LO pathway, LTB_4, induces the differentiation and anti-apoptotic effects of CD34+ HSCs and the inhibition of LTA_4H and its receptor, BLT2, resulted in self-renewal of HSCs (Chung et al. 2005). In addition, 12/15-LOX and its products, 12-HpETE and 15-HpETE, play important role in skin wound healing (Hong et al. 2014).

4.1.2. The COX Pathway in Stem Cells

The impact of PGs on the proliferation of HSCs was reported back in the 1970s (Feher and Gidali 1974; Gidali and Feher 1977). It was shown that PGE_2 released by monocytes or macrophages suppresses the proliferation of myeloid SCs in vitro. In addition, the presence of PGE_2 and higher expression of its receptors are linked to stimulation of angiogenesis and early state of inflammation (Ern et al. 2019). MSCs secrete different bioactive molecules, including PGE_2, that guide the polarization of pro-inflammatory to anti-inflammatory macrophages, resulting in lowered levels of inflammation (Prockop 2013). PGE_1 promotes the differentiation of HSCs to mature granulocytes and attenuates the production of macrophages. Similarly to PGE_2, PGI_2 is necessary to angiogenesis and the inhibition of PGI_2 synthase results in impaired wound healing (He et al. 2008). A short-term stimulation with PGE_2 enhances the proliferation of MSCs, while longer treatments inhibit growth. In contrast, PGD_2 has a growth-inhibitory effect in spite of the duration of the incubation (Ern et al. 2019). The development of human smooth muscle-like cells from adipose tissue-derived MSCs is controlled by another prostaglandin, TXA_2 (Yun et al. 2009a). Overall, inhibition of COX pathways by NSAIDS, e.g., aspirin (Liu et al. 2014) and ibuprofen (Goren et al. 2017), results in lower levels of TXs and PGs which delay the wound healing and self-renewal.

4.1.3. Pro-Resolving Mediators in Stem Cells

SPMs are formed in the cross-play between COXs, LOXs and other pathways or in the presence of drugs. For instance, LXA$_4$ can be formed cooperatively via 5-LOX and 12-/15-LOX pathways (Figure 1). It is evident that different SCs contain the biosynthetic machinery to produce different SPMs which can be potentially involved in the immune-modulating and anti-inflammatory properties of SCs (Romano et al. 2019). For example, MSCs secrete LXA$_4$ which regulates anti-inflammatory and pro-resolving processes (Rinkevich et al. 2009; Tsoyi et al. 2016). In fact, exogenous or MSC-derived LXA$_4$ contribute to the recovery from acute lung injury (Fang et al. 2015). Moreover, LXA$_4$ significantly enhances the wound healing capacity of hPDLSCs (Berry et al. 2017) and regulates the proliferation and differentiation of NSCs (Wada et al. 2006). Protectin D1 (also known as neuroprotection D1) promotes cardiac and neuronal differentiation and is essential in the regeneration of nerve cells (Yanes et al. 2010).

4.2. Model Systems for Marine Invertebrates

There are four main invertebrate adult SC models—the "big four": *Porifera*, *Cnidaria*, *Platyhelminthes* (flatworm), and *Tunicata* (Rinkevich et al. 2021). The PUFAs and eicosanoid pathways present in *Cnidaria* were discussed in detail above (see 3. Coral Eicosanoids). Although more than 250 fatty acids are determined in *Porifera*, there are no higher PUFAs, thus no traditional eicosanoids are present (Rod'kina 2005; Monroig et al. 2013) (Figure 3). In comparison, the main substrate PUFAs in *Platyhelminthes* (Angerer et al. 2019; Makhutova et al. 2009) and *Tunicates* are EPA and DHA, however, only trace amounts of AA are found (Mimura et al. 1986). It should be noted, that as in *Cnidarians*, there might be high variance in PUFA content between different species. In parallel, also the presence of LOXs varies between invertebrate species. For instance, no LOX sequences have been found in *Porifera* (Horn et al. 2015).

Dugasia tigrina was used as a planarian (*Platyhelminthes*) model to study regeneration by DHA and DHA-derived oxylipins from vertebrates (Serhan et al. 2012) (Figure 3). The ability to enhance the tissue regeneration by a lipid mediator, macrophage mediator in resolving inflammation (MaR1), indicates conserved regulatory roles and pathways of DHA-derived mediators. Inhibition of 12-LOX resulted in attenuated regeneration and formation of MaR1, suggesting that the 12-LOX pathway may play important role in *D. tigrina* (Figure 3). In addition, the genome of *Schistosoma japonicum* revealed conserved sequences of LOX, LTA$_4$H and putative receptors for LTB$_4$, cysteinyl-LTs, PGE$_2$ and PGF$_2$, indicating that these pathways may play a role in the physiology of planarian (Zhou et al. 2009). However, *Schmidtea mediterranea* does not contain any similar sequences to COX or LOX known in animals based on the PlanMine sequence database (Rozanski et al. 2019) (personal data).

	Stem cells	Whole animal				
	Vertebrates	**Invertebrates**				
		Porifera	*Cnidaria*		*Platyhelminthes*	*Tunicates*
			Hydra	*Coral*		
PUFA substrates	AA, EPA, DHA	No substrate	AA, EPA, DHA		EPA, DHA	EPA, DHA
Dioxygenases	5S-LOX 12/15-LOX COX	*n.d.*	11R-LOX	8R-LOX COX AOS-LOX	LOX*	LOX*, COX*
Detected metabolites	LTB$_4$, 12-, and 15-HpETE PGE$_2$, PGE$_1$, PGI$_2$, PGD$_2$, TXA$_2$, LXA$_4$, Neuroprotectin D1	No tradicional eicosanoids	11R-HETE	8R-, and 11R-HpETEs, cyclopentenone, PGs	Mar1	*n.d.*

Figure 3. The PUFA-dependent oxylipin pathways in vertebrate stem cells and in model organisms of invertebrates. * Predicted based on the gene sequence; *n.d.*—not determined. Source: Graphic by authors.

The PUFA composition of tunicates reveals that the most abundant PUFA substrates are EPA and DHA (Carballeira et al. 1995; Hou et al. 2021). Even though coral COX-like sequences exist in tunicates (Järving et al. 2004), it remains unknown if they encode functional dioxygenases and what is their catalytic specificity. Recently, it was shown that the germ cell migration and chemotaxis in *Botryllus schlosseri* is 12S-HETE-dependent (Kassmer et al. 2020). Unfortunately, only 12S-HETE was in the focus of their study and other HETEs remained untested. Furthermore, a *B. schlosseri* LOX sequence was described with a sequence identity of around 50% positives to human 5-LOX, 12-LOX and 15-LOX (Kassmer et al. 2020). The genome of closely related *Botrylloides diegensis* supports the presence of a single LOX gene in both species (Voskoboynik et al. 2013; Blanchoud et al. 2018). The sequence of *B. schlosseri* LOX contains conserved iron-coordinating amino acids and the amino acid determinant of regiospecificity (either S or R) suggests the presence of LOX with the S-specificity. However, only the end of the C-terminal domain without the N-terminal PLAT and part of the catalytic domains was present in the sequence (personal data). Thus, the presence of catalytically functional LOX in *B. schlosseri* needs to be confirmed by future studies.

Although major advances have been made in sequencing invertebrate genomes and transcriptomes, the prediction of bioactive metabolites only based on sequence data is not accurate due to highly conserved domains between dioxygenases with different catalytical specificities and biological roles, such as LOXs (Lõhelaid and Samel 2018). Additional experiments with dioxygenases need to be performed to supplement the sequence data. In conclusion, despite the progress in the field, very little is known about oxylipin biosynthesis or metabolites in invertebrate model systems.

Common precursor PUFAs for the oxylipin synthesis in vertebrate and invertebrate systems demonstrate the evolutionary requirement of lipid mediators in the physiology of animals (Figure 3). As in vertebrates the effect of different eicosanoids on the fate of SCs are clearly demonstrated (Table 1), it is likely that these processes in marine invertebrates are driven by ancestor genes and similar mediators.

4.3. Potential Role of Eicosanoids in the Stem Cells of Marine Invertebrates

In contrast to vertebrates, SCs in marine invertebrates are disseminated throughout the organism and instead of uni- or oligopotency, they possess pluri- and totipotent capabilities. Another unique property of invertebrate SCs is their ability to trans-differentiate from one cell type to other (Rinkevich et al. 2009). It occurs when a significant amount of SCs is needed, specifically during budding, regeneration and in response to severe abiotic or biotic stress (Rinkevich et al. 2009).

In all species studied to date, lipid mediators mediate important adaptation responses to cellular stress. Organisms continuously sense and respond to environmental conditions to maintain their homeostasis under changing conditions and survive. Biological stress can be defined as an adverse condition or force which disturbs the homeostasis and normal functioning of an organism (Jones et al. 2010). Overall, external stressors may be biotic, such as pathogens, or physical, such as temperature, salinity, water, nutrient deprivation, chemicals and pollutants, oxidative stress, mechanical stress and radiation.

The initial wound response in animals aims for rapid and efficient isolation of the wound to minimize both the loss of vital fluids and environmental challenges (Proksch et al. 2008; Rodriguez et al. 2008; Ariel and Timor 2013; Palmer et al. 2011; Maffei et al. 2007). In multicellular organisms, regeneration involves the repair of tissues/organs after injury and homeostatic renewal. The spatio-temporal immune cell activation is essential in regenerative response and its adequate regulation defines the regenerative success. The initial step in response to the incision in marine invertebrates, including corals, aims for rapid and efficient provisional plugging of the wound, similar to vertebrates (Palmer et al. 2011). On a cellular level, the wound repair in vertebrates has four phases: (1) hemostasis/coagulation, (2) inflammation, (3) proliferation and (4) remodeling (Singer and Clark 1999; Schultz et al. 2011; Maderna and Godson 2009). The same wound repair phases are observed in *Cnidarians* (Reitzel et al. 2008; Olano and Bigger 2000; Palmer et al. 2008). Coral wound response includes the recruitment of granular amoebocytes (Mydlarz et al. 2008; Palmer et al. 2008), which are important in pathogen clearance. Acting cooperatively, eicosanoids mediate the initial stages of wound response and the onset and end of the inflammatory phase of wound repair, promoting cell migration and modulating the central signal pathways involved in cell cycle control (Moreno 2009). Oxylipins are also involved in coral wound response (Lõhelaid et al. 2014a), but their effect on marine invertebrate stem cells is not known. Furthermore, innate immune response and regeneration are inter-connected processes during tissue repair (Aurora and Olson 2014). As pointed out before, 11*R*-HETE enhanced the tentacle regeneration and bud formation of decapitated *Hydra vulgaris* (Di Marzo et al. 1993) indicating its direct cellular regulator effect. The distribution of stem cells and molecular regulation of stemness in *Hydra* is complex (Hobmayer et al. 2012). Unfortunately, it is not known which cells are responding to this biomolecule and what is the underlying molecular mechanism.

In addition, the levels and production of eicosanoids in vertebrates are low and tightly controlled (Dennis and Norris 2015; Serhan and Chiang 2008), whereas corals contain an enormous amount of various oxylipins (Weinheimer and Spraggins 1969). Thus, the high production of oxylipins, such as PGE_2 in *P. homomalla*, could contribute to the differentiation of SCs and also increase the regenerative capacity of invertebrates.

4.4. Challenges in the Stem Cell Biology of Marine Invertebrates

Currently, we lack basic knowledge about oxylipins and oxylipin-mediated processes in marine invertebrates and their distribution in different cell populations, including stem cells. The main practical limitations for efficient studies are the absence of (1) SC definition in invertebrates, (2) adequate biomarkers to distinct cell populations, (3) developed protocols for SC isolation, and (4) proper knowledge of how to culture SCs and create SC lines. In addition, there are well-established protocols for extraction and analysis of different lipid subclasses (Hou et al. 2021), however, specific know-how, equipment and a certain amount of SCs for the proper detection are still required. Apart from the identification and profiling of oxylipins, it is challenging to determine the role of each of the individual oxylipins on the stem cells due to the high number of oxylipin derivatives and complexity of intracellular oxylipin pathways. Nevertheless, constantly improving state-of-the-art technology and methodology as well as greater networking opportunities contribute to the advancement of SC research.

5. Conclusions

Oxylipins, including eicosanoids, are short-lived lipid mediators, they act locally in an auto- and paracrine manner to control proliferation, migration, and differentiation of vertebrate SCs which contribute to tissue repair, regeneration and other cellular processes. Based on current knowledge, we propose that oxylipins are also involved in the renewal, proliferation and differentiation of marine invertebrate SCs. Still, due to a variety of lipid mediators and other regulators, and lack of studies, the role of eicosanoids in determining the fate of marine invertebrate SCs is far from being clear. For example, it is difficult to translate the function if there is a high variation in oxylipin content between different species and the regio- and stereoisomers of lipid mediators might have different or even opposite effects. Studies on marine invertebrate genomes and transcriptomes are able to give some clues, but they are insufficient to predict the specificity nor functionality of dioxygenases. To date, sequence data from different organisms are emerging, however, we lack systematic studies in different marine invertebrate species. For instance, profiling of oxylipin pathways and biological actions of PUFAs and oxylipins on model organisms and their SCs should be performed. Thus, only basic research on invertebrate SCs is able to define the compounds produced in model systems and the role of applied eicosanoids.

Author Contributions: H.L. and T.T. wrote the chapter and prepared figures. All authors have read and agreed to the published version of the manuscript.

Funding: This research was funded by the Academy of Finland grant number 322757 and the Estonian Research Council grant number PUTJD1046.

Conflicts of Interest: The authors declare no conflict of interest.

References

Angerer, Tina B., Neil Chakravarty, Michael J. Taylor, Carrie D. Nicora, Daniel J. Graham, Christopher R. Anderton, Eric H. Chudler, and Lara J. Gamble. 2019. Insights into the histology of planarian flatworm *Phagocata gracilis* based on location specific, intact lipid information provided by GCIB-ToF-SIMS imaging. *Biochimica et Biophysica Acta (BBA)-Molecular and Cell Biology of Lipids* 1864: 733–43. [CrossRef] [PubMed]

Aranda, Manuel, Anastazia T. Banaszak, Till Bayer, James R. Luyten, Monica Medina, and Christian R. Voolstra. 2011. Differential sensitivity of coral larvae to natural levels of ultraviolet radiation during the onset of larval competence. *Molecular Ecology* 20: 2955–72. [CrossRef] [PubMed]

Ariel, Amiram, and Orly Timor. 2013. Hanging in the balance: Endogenous anti-inflammatory mechanisms in tissue repair and fibrosis. *The Journal of Pathology* 229: 250–63. [CrossRef] [PubMed]

Aurora, Arin B., and Eric N. Olson. 2014. Immune modulation of stem cells and regeneration. *Cell Stem Cell* 15: 14–25. [CrossRef]

Ballarin, Loriano, Baruch Rinkevich, Kerstin Bartscherer, Artur Burzynski, Sebastien Cambier, Matteo Cammarata, Isabelle Domart-Coulon, Damjana Drobne, Juanma Encinas, Uri Frank, and et al. 2018. Maristem—Stem cells of marine/aquatic invertebrates: From basic research to innovative applications. *Sustainability* 10: 526. [CrossRef]

Beltz, Barbara S., Michael F. Tlusty, Jeanne L. Benton, and David C. Sandeman. 2007. Omega-3 fatty acids upregulate adult neurogenesis. *Neuroscience Letters* 415: 154–58. [CrossRef]

Bernardo, Maria E., and Willem E. Fibbe. 2013. Mesenchymal stromal cells: Sensors and switchers of inflammation. *Cell Stem Cell* 13: 392–402. [CrossRef]

Berry, Elizabeth, Yanzhou Liu, Li Chen, and Austin M. Guo. 2017. Eicosanoids: Emerging contributors in stem cell-mediated wound healing. *Prostaglandins Other Lipid Mediat* 132: 17–24. [CrossRef]

Blanchoud, Simon, Kim Rutherford, Lisa Zondag, Neil J. Gemmell, and Megan J. Wilson. 2018. *De novo* draft assembly of the *Botrylloides leachii* genome provides further insight into tunicate evolution. *Scientific Reports* 8: 5518. [CrossRef]

Blobaum, Anna L., and Lawrence J. Marnett. 2007. Structural and functional Basis of Cyclooxygenase Inhibition. *Journal of Medicinal Chemistry* 50: 1425–41. [CrossRef]

Brash, Alan R. 1999. Lipoxygenases: Occurrence, functions, catalysis, and acquisition of substrate. *Journal of Biological Chemistry* 274: 23679–82. [CrossRef] [PubMed]

Brash, Alan R., Steven W. Baertschi, Christiana D. Ingram, and Thomas M. Harris. 1987. On non-cyclooxygenase prostaglandin synthesis in the sea whip coral, *Plexaura homomalla*: An 8(R)-lipoxygenase pathway leads to formation of an alpha-ketol and a racemic prostanoid. *Journal of Biological Chemistry* 262: 15829–39. [CrossRef]

Brash, Alan R., William E. Boeglin, Min S. Chang, and Bih-Hwa Shieh. 1996. Purification and molecular cloning of an 8R-lipoxygenase from the coral *Plexaura homomalla* reveal the related primary structures of R- and S-lipoxygenases. *Journal of Biological Chemistry* 271: 20949–57. [CrossRef] [PubMed]

Carballeira, Nestor M., Fathi Shalabi, Kamen Stefanov, Krassimir Dimitrov, Simeon Popov, Athanas Kujumgiev, and Stoitze Andreev. 1995. Comparison of the fatty acids of the tunicate *Botryllus schlosseri* from the Black Sea with two associated bacterial strains. *Lipids* 30: 677–79. [CrossRef] [PubMed]

Chung, Jin W., Geun-Young Kim, Yeung-Chul Mun, Ji-Young Ahn, Chu-Myong Seong, and Jae-Hong Kim. 2005. Leukotriene B$_4$ pathway regulates the fate of the hematopoietic stem cells. *Experimental & Molecular Medicine* 37: 45–50.

Coceani, Flavio, and Eyup S. Akarsu. 1998. Prostaglandin E$_2$ in the pathogenesis of fever: An update. *Annals of the New York Academy of Sciences* 856: 76–82. [CrossRef] [PubMed]

Corey, Elias J., William N. Washburn, and Jong C. Chen. 1973. Studies on the prostaglandin A$_2$ synthetase complex from *Plexaura homomalla*. *Journal of the American Chemical Society* 95: 2054–55. [CrossRef]

Corey, Elias J., Peter T. Lansbury, and Yasuji Yamada. 1985. Identification of a new eicosanoid from in vitro biosynthetic experiments with *Clavularia Viridis*—Implications for the biosynthesis of clavulones. *Tetrahedron Letters* 26: 4171–74. [CrossRef]

Corey, Elias J., Marc Dalarcao, Seiichi P. T. Matsuda, Peter T. Lansbury, and Yasuji Yamada. 1987. Intermediacy of 8-(R)-Hpete in the conversion of arachidonic acid to pre-clavulone-A by *Clavularia Viridis*—Implications for the biosynthesis of marine prostanoids. *Journal of the American Chemical Society* 109: 289–90. [CrossRef]

Corey, Elias J., Seiichi P. T. Matsuda, Riu Nagata, and Martin B. Cleaver. 1988. Biosynthesis of 8-*R*-Hpete and preclavulone-A from arachidonate in several species of Caribbean coral—a widespread route to marine prostanoids. *Tetrahedron Letters* 29: 2555–58. [CrossRef]

Dennis, Edward A., and Paul C. Norris. 2015. Eicosanoid storm in infection and inflammation. *Nature Reviews Immunology* 15: 511–23. [CrossRef] [PubMed]

Di Marzo, Vincenzo, Luciano De Petrocellis, Carmen Gianfrani, and Guido Cimino. 1993. Biosynthesis, structure and biological activity of hydroxyeicosatetraenoic acids in *Hydra vulgaris*. *Biochemical Journal* 295: 23–29. [CrossRef] [PubMed]

Di Marzo, Vincenzo, Mariacarla Ventriglia, Ernesto Mollo, Mariarosaria Mosca, and Guido Cimino. 1996. Occurrence and biosynthesis of 11(R)-hydroxy-eicosatetraenoic acid (11-*R*-HETE) in the Caribbean soft coral. *Plexaurella dichotoma. Experientia* 52: 834–38. [CrossRef]

Dunn, Simon R., Michael C. Thomas, Geoffrey W. Nette, and Sophie. G. Dove. 2012. A lipidomic approach to understanding free fatty acid lipogenesis derived from dissolved inorganic carbon within *Cnidarian-Dinoflagellate* symbiosis. *PLoS ONE* 7: e46801. [CrossRef] [PubMed]

Eek, Priit, Reet Järving, Iivar Järving, Nathaniel C. Gilbert, Marcia E. Newcomer, and Nigulas Samel. 2012. Structure of a calcium-dependent 11*R*-lipoxygenase suggests a mechanism for Ca^{2+} regulation. *Journal of Biological Chemistry* 287: 22377–86. [CrossRef] [PubMed]

Ern, Christina, Iris Frasheri, Timo Berger, Hans-Georg Kirchner, Richard Heym, Reinhard Hickel, and Matthias Folwaczny. 2019. Effects of prostaglandin E$_2$ and D$_2$ on cell proliferation and osteogenic capacity of human mesenchymal stem cells. *Prostaglandins Leukot Essent Fatty Acids* 151: 1–7. [CrossRef] [PubMed]

Esser-von Bieren, Julia. 2019. Eicosanoids in tissue repair. *Immunology and Cell Biology* 97: 279–88. [CrossRef]

Fang, Xiaohui, Jason Abbott, Linda Cheng, Jennifer K. Colby, Jae W. Lee, Bruce D. Levy, and Michael A. Matthay. 2015. Human mesenchymal stem (stromal) cells promote the resolution of acute lung injury in part through lipoxin A$_4$. *The Journal of Immunology* 195: 875–81. [CrossRef]

Feher, Imre, and Julia Gidali. 1974. Prostaglandin E$_2$ as stimulator of haemopoietic stem cell proliferation. *Nature* 247: 550–51. [CrossRef]

Figueiredo, Joana, Andrew H. Baird, Michael F. Cohen, Jean-Francois Flot, Takayuki Kamiki, Tarik Meziane, Makoto Tsuchiya, and Hayata Yamasaki. 2012. Ontogenetic change in the lipid and fatty acid composition of scleractinian coral larvae. *Coral Reefs* 31: 613–19. [CrossRef]

Finkensieper, Andreas, Sophia Kieser, Mohamed M. Bekhite, Madeleine Richter, Joerg P. Mueller, Rolf Graebner, Hans-Reiner Figulla, Heinrich Sauer, and Maria Wartenberg. 2010. The 5-lipoxygenase pathway regulates vasculogenesis in differentiating mouse embryonic stem cells. *Cardiovascular Research* 86: 37–44. [CrossRef] [PubMed]

Flower, Roderick J. 2006. Prostaglandins, bioassay and inflammation. *British Journal of Pharmacology* 147: S182–S192. [CrossRef] [PubMed]

Funk, Colin D. 2001. Prostaglandins and leukotrienes: Advances in eicosanoid biology. *Science* 294: 1871–75. [CrossRef] [PubMed]

Gaudin, Alexis, Miroslav Tolar, and Ove A. Peters. 2018. Lipoxin A$_4$ Attenuates the inflammatory response in stem cells of the apical papilla via ALX/FPR2. *Scientific Reports* 8: 8921. [CrossRef]

Gerhart, Donald J. 1991. Emesis, learned aversion, and chemical defense in Octocorals—a central role for prostaglandins. *American Journal of Physiology* 260: R839–R843. [CrossRef]

Gidali, Julia, and Imre Feher. 1977. The effect of E type prostaglandins on the proliferation of haemopoietic stem cells in vivo. *Cell Tissue Kinet* 10: 365–73. [CrossRef]

Gilbert, Nathaniel C., Marc Niebuhr, Hiro Tsuruta, Tee Bordelon, Oswin Ridderbusch, Adam Dassey, Alan R. Brash, Sue G. Bartlett, and Marcia E. Newcomer. 2008. A covalent linker allows for membrane targeting of an oxylipin biosynthetic complex. *Biochemistry* 47: 10665–76. [CrossRef]

Goncalves, Maria B., Emma-Jane Williams, Ping Yip, Rafael J. Yanez-Munoz, Gareth Williams, and Patrick Doherty. 2010. The COX-2 inhibitors, meloxicam and nimesulide, suppress neurogenesis in the adult mouse brain. *British Journal of Pharmacology* 159: 1118–25. [CrossRef]

Goren, Itamar, Seo-Youn Lee, Damian Maucher, Rolf Nüsing, Thomas Schlich, Josef Pfeilschifter, and Stefan Frank. 2017. Inhibition of cyclooxygenase-1 and -2 activity in keratinocytes inhibits PGE$_2$ formation and impairs vascular endothelial growth factor release and neovascularisation in skin wounds. *International Wound Journal* 14: 53–63. [CrossRef]

Grace, Peter M., Mark R. Hutchinson, Steven F. Maier, and Linda R. Watkins. 2014. Pathological pain and the neuroimmune interface. *Nature Reviews Immunology* 14: 217–31. [CrossRef]

Haeggström, Jesper Z. 2018. Leukotriene biosynthetic enzymes as therapeutic targets. *Journal of Clinical Investigation* 128: 2680–90. [CrossRef] [PubMed]

Haeggström, Jesper Z., and Colin D. Funk. 2011. Lipoxygenase and leukotriene pathways: Biochemistry, biology, and roles in disease. *Chemical Reviews* 111: 5866–98. [CrossRef] [PubMed]

Hashimoto, Naoko, Shoko Fujiwara, Kinzo Watanabe, Kazuo Iguchi, and Mikio Tsuzuki. 2003. Localization of clavulones, prostanoids with antitumor activity, within the Okinawan soft coral *Clavularia viridis* (*Alcyonacea*, *Clavulariidae*): Preparation of a high-purity *Symbiodinium* fraction using a protease and a detergent. *Lipids* 38: 991–97. [CrossRef] [PubMed]

Hawkins, Dan J., and Alan R. Brash. 1987. Eggs of the sea urchin, *Strongylocentrotus purpuratus*, contain a prominent (11R) and (12R) lipoxygenase activity. *Journal of Biological Chemistry* 262: 7629–34. [CrossRef]

He, Tongrong, Tong Lu, Livius V. dUscio, Chen-Fuh Lam, Hon-Chi Lee, and Zvonimir S. Katusic. 2008. Angiogenic function of prostacyclin biosynthesis in human endothelial progenitor cells. *Circulation Research* 103: 80–88. [CrossRef]

He, Chengwei, Xiying Qu, Libin Cui, Jingdong Wang, and Jing X. Kang. 2009. Improved spatial learning performance of fat-1 mice is associated with enhanced neurogenesis and neuritogenesis by docosahexaenoic acid. *Proceedings of the National Academy of Sciences of the United States of America* 106: 11370–75. [CrossRef]

Herrler, Tanja, Simon F. Leicht, Stephan Huber, Patrick C. Hermann, Theresa M. Schwarz, Reinhard Kopp, and Christopher Heeschen. 2009. Prostaglandin E positively modulates endothelial progenitor cell homeostasis: An advanced treatment modality for autologous cell therapy. *Journal of Vascular Research* 46: 333–46. [CrossRef]

Hobmayer, Bert, Marcell Jenewein, Dominik Eder, Marie-Kristin Eder, Stella Glasauer, Sabine Gufler, Markus Hartl, and Willi Salvenmoser. 2012. Stemness in *Hydra*—A current perspective. *The International Journal of Developmental Biology* 56: 509–17. [CrossRef]

Hoggatt, Jonathan, and Louis M. Pelus. 2010. Eicosanoid regulation of hematopoiesis and hematopoietic stem and progenitor trafficking. *Leukemia* 24: 1993–2002. [CrossRef]

Honda, Atushi, Yo Mori, Kazuo Iguchi, and Yasuji Yamada. 1987. Antiproliferative and cytotoxic effects of newly discovered halogenated coral prostanoids from the Japanese stolonifer *Clavularia viridis* on human myeloid leukemia cells in culture. *Molecular Pharmacology* 32: 530–35.

Hong, Song, Bhagwat V. Alapure, Yan Lu, Haibin Tian, and Quansheng Wang. 2014. 12/15-Lipoxygenase deficiency reduces densities of mesenchymal stem cells in the dermis of wounded and unwounded skin. *British Journal of Dermatology* 171: 30–38. [CrossRef] [PubMed]

Horn, Thomas, Susan Adel, Ralf Schumann, Saubashya Sur, Kkumar Kakularam, Aparoy Polamarasetty, Pallu Redanna, Hartmut Kühn, and Dagmar Heydeck. 2015. Evolutionary aspects of lipoxygenases and genetic diversity of human leukotriene signaling. *Progress in Lipid Research* 57: 13–39. [CrossRef] [PubMed]

Hou, Pingping, Yanqin Li, Xu Zhang, Chun Liu, Jingyang Guan, Honggang Li, Ting Zhao, Junqing Ye, Weifeng Yang, Kang Liu, and et al. 2013. Pluripotent stem cells induced from mouse somatic cells by small-molecule compounds. *Science* 341: 651–54. [CrossRef] [PubMed]

Hou, Qing, Yuting Huang, Linghong Jiang, Kai Zhong, Yina Huang, Hong Gao, and Qian Bu. 2021. Evaluation of lipid profiles in three species of ascidians using UPLC-ESI-Q-TOF-MS-based lipidomic study. *Food Research International* 146: 110454. [CrossRef]

Howie, Peter W., Andrew A. Calder, Charles D. Forbes, and Colin R. Prentice. 1973. Effect of intravenous prostaglandin E_2 on platelet function, coagulation, and fibrinolysis. *Journal of Clinical Pathology* 26: 354–58. [CrossRef]

Hyman, Libbie Henrietta. 1940. *The Invertebrates*, 1st ed. McGraw-Hill Publications in the Zoological Sciences. New York: McGraw-Hill.

Imbs, Andrey B., and Irina M. Yakovleva. 2011. Dynamics of lipid and fatty acid composition of shallow-water corals under thermal stress: An experimental approach. *Coral Reefs* 31: 41–53. [CrossRef]

Imbs, Andrey B., Olga A. Demina, and Darja A. Demidkova. 2006. Lipid class and fatty acid composition of the boreal soft coral. *Gersemia rubiformis*. *Lipids* 41: 721–25. [CrossRef]

Jang, Min W., Seung P. Yun, Jae H. Park, Jung M. Ryu, Jang H. Lee, and Ho J. Han. 2012. Cooperation of Epac1/Rap1/Akt and PKA in prostaglandin E(2) -induced proliferation of human umbilical cord blood derived mesenchymal stem cells: Involvement of c-Myc and VEGF expression. *Journal of Cellular Physiology* 227: 3756–67. [CrossRef]

Järving, Reet, Ivar Järving, Reet Kurg, Alan R. Brash, and Nigulas Samel. 2004. On the evolutionary origin of cyclooxygenase (COX) isozymes: Characterization of marine invertebrate COX genes points to independent duplication events in vertebrate and invertebrate lineages. *Journal of Biological Chemistry* 279: 13624–33. [CrossRef]

Järving, Reet, Aivar Lõokene, Reet Kurg, Liina Siimon, Ivar Järving, and Nigulas Samel. 2012. Activation of 11*R*-lipoxygenase is fully Ca(2+)-dependent and controlled by the phospholipid composition of the target membrane. *Biochemistry* 51: 3310–20. [CrossRef]

Jones, Hamlyn G., Timothy J. Flowers, and Michael B. Jones. 2010. *Plants under Stress, Society for Experimental Biology Seminar Series*. Cambridge: Cambridge University Press.

Jung, Keun-Hwa, Kon Chu, Soon-Tae Lee, Juhyun Kim, Dong-In Sinn, Jeong-Min Kim, Dong-Kyu Park, Jung-Ju Lee, Seung U. Kim, Manho Kim, and et al. 2006. Cyclooxygenase-2 inhibitor, celecoxib, inhibits the altered hippocampal neurogenesis with attenuation of spontaneous recurrent seizures following pilocarpine-induced status epilepticus. *Neurobiology of Disease* 23: 237–46. [CrossRef] [PubMed]

Kan, Inna, Eldad Melamed, Daniel Offen, and Pnina Green. 2007. Docosahexaenoic acid and arachidonic acid are fundamental supplements for the induction of neuronal differentiation. *Journal of Lipid Research* 48: 513–17. [CrossRef] [PubMed]

Kang, Jing X., Jian-Bo Wan, and Chengwei He. 2014. Concise review: Regulation of stem cell proliferation and differentiation by essential fatty acids and their metabolites. *Stem Cells* 32: 1092–98. [CrossRef] [PubMed]

Kassmer, Susannah H., Delany Rodriguez, and Anthony W. De Tomaso. 2020. Evidence that ABC-transporter-mediated autocrine export of an eicosanoid signaling molecule enhances germ cell chemotaxis in the colonial tunicate *Botryllus schlosseri*. *Development* 147: dev184663. [CrossRef]

Katakura, Masanori, Michio Hashimoto, Hossain M. Shahdat, Shuji Gamoh, Tomoko Okui, Kentaro Matsuzaki, and Osamu Shido. 2009. Docosahexaenoic acid promotes neuronal differentiation by regulating basic helix-loop-helix transcription factors and cell cycle in neural stem cells. *Neuroscience* 160: 651–60. [CrossRef]

Katakura, Masanori, Michio Hashimoto, Tomoko Okui, Hossain M. Shahdat, Kentaro Matsuzaki, and Osamu Shido. 2013. Omega-3 polyunsaturated fatty acids enhance neuronal differentiation in cultured rat neural stem cells. *Stem Cells International* 2013: 490476. [CrossRef]

Katura, Takashi, Takahiro Moriya, and Norimichi Nakahata. 2010. 15-Deoxy-delta 12,14-prostaglandin J$_2$ biphasically regulates the proliferation of mouse hippocampal neural progenitor cells by modulating the redox state. *Molecular Pharmacology* 77: 601–11. [CrossRef]

Kawabe, Jun-Ichi, Koh-Ichi Yuhki, Motoi Okada, Takayasu Kanno, Atsushi Yamauchi, Naohiko Tashiro, Takaaki Sasaki, Shunsuke Okumura, Naoki Nakagawa, Youko Aburakawa, and et al. 2010. Prostaglandin I_2 promotes recruitment of endothelial progenitor cells and limits vascular remodeling. *Arteriosclerosis, Thrombosis, and Vascular Biology* 30: 464–70. [CrossRef]

Kawakita, Eisuke, Michio Hashimoto, and Osamu Shido. 2006. Docosahexaenoic acid promotes neurogenesis in vitro and in vivo. *Neuroscience* 139: 991–97. [CrossRef]

Kelly, Anthony J., Sidra Malik, Lee Smith, Josephine Kavanagh, and Jane Thomas. 2009. Vaginal prostaglandin (PGE_2 and PGF_{2a}) for induction of labour at term. *Cochrane Database of Systematic Reviews* 4: CD003101.

Kim, Min H., Mi O. Kim, Yun H. Kim, Jin S. Kim, and Ho J. Han. 2009a. Linoleic acid induces mouse embryonic stem cell proliferation via Ca^{2+}/PKC, PI3K/Akt, and MAPKs. *Cellular Physiology and Biochemistry* 23: 53–64. [CrossRef] [PubMed]

Kim, Mi R., Eun S. Jeon, Young M. Kim, Jung S. Lee, and Jae H. Kim. 2009b. Thromboxane A_2 induces differentiation of human mesenchymal stem cells to smooth muscle-like cells. *Stem Cells* 27: 191–99. [CrossRef] [PubMed]

Kim, Min H., Yu J. Lee, Mi O. Kim, Jin S. Kim, and Ho J. Han. 2010. Effect of leukotriene D4 on mouse embryonic stem cell migration and proliferation: Involvement of PI3K/Akt as well as GSK-3beta/beta-catenin signaling pathways. *Journal of Cellular Biochemistry* 111: 686–98. [CrossRef] [PubMed]

Kleiveland, Charlotte R., Moustapha Kassem, and Tor Lea. 2008. Human mesenchymal stem cell proliferation is regulated by PGE_2 through differential activation of cAMP-dependent protein kinase isoforms. *Experimental Cell Research* 314: 1831–38. [CrossRef]

Kobzar, Gennadi, Vilja Mardla, Ivar Järving, Nigulas Samel, and Madis Lõhmus. 1997. Modulatory effect of 8-iso-PGE2 on platelets. *General Pharmacology: The Vascular System* 28: 317–21. [CrossRef]

Koljak, Reet, Olivier Boutaud, Bih-Hwa Shieh, Nigulas Samel, and Alan R. Brash. 1997. Identification of a naturally occurring peroxidase-lipoxygenase fusion protein. *Science* 277: 1994–96. [CrossRef] [PubMed]

Koljak, Reet, Ivar Järving, Reet Kurg, William E. Boeglin, Külliki Varvas, Karin Valmsen, Mart Ustav, Alan R. Brash, and Nigulas Samel. 2001. The basis of prostaglandin synthesis in coral: Molecular cloning and expression of a cyclooxygenase from the Arctic soft coral *Gersemia fruticosa*. *Journal of Biological Chemistry* 276: 7033–40. [CrossRef]

Kurland, Jeffrey I., Hal E. Broxmeyer, Louis M. Pelus, Richard S. Bockman, and Malcolm A. Moore. 1978. Role for monocyte-macrophage-derived colony-stimulating factor and prostaglandin E in the positive and negative feedback control of myeloid stem cell proliferation. *Blood* 52: 388–407. [CrossRef]

Latyshev, Nikolay A., Nikolay V. Naumenko, Vasily I. Svetashev, and Yurii Y. Latypov. 1991. Fatty-acids of reef-building corals. *Marine Ecology Progress Series* 76: 295–301. [CrossRef]

Libro, Silvia, Stefan T. Kaluziak, and Steven V. Vollmer. 2013. RNA-seq profiles of immune related genes in the Staghorn coral *Acropora cervicornis* infected with white band disease. *PLoS ONE* 8: e81821. [CrossRef]

Liou, Jun-Yang, David P. Ellent, Sang Lee, Jennifer Goldsby, Bor-Sheng Ko, Nena Matijevic, Jaou-Chen Huang, and Kenneth K. Wu. 2007. Cyclooxygenase-2-derived prostaglandin E_2 protects mouse embryonic stem cells from apoptosis. *Stem Cells* 25: 1096–103. [CrossRef] [PubMed]

Liu, Junpeng, Lei Chen, Xu Tao, and Kanglai Tang. 2013. Phosphoinositide 3-kinase/Akt signaling is essential for prostaglandin E_2-induced osteogenic differentiation of rat tendon stem cells. *Biochemical and Biophysical Research Communications* 435: 514–9. [CrossRef] [PubMed]

Liu, Min, Kazuko Saeki, Takehiko Matsunobu, Toshiaki Okuno, Tomoaki Koga, Yukihiko Sugimoto, Chieko Yokoyama, Satoshi Nakamizo, Kenji Kabashima, Shuh Narumiya, and et al. 2014. 12-Hydroxyheptadecatrienoic acid promotes epidermal wound healing by accelerating keratinocyte migration via the BLT2 receptor. *Journal of Experimental Medicine* 211: 1063–78. [CrossRef]

Lõhelaid, Helike, and Nigulas Samel. 2018. Eicosanoid diversity of stony corals. *Marine Drugs* 16: 10. [CrossRef] [PubMed]

Lõhelaid, Helike, Reet Järving, Karin Valmsen, Külliki Varvas, Malle Kreen, Ivar Järving, and Nigulas Samel. 2008. Identification of a functional allene oxide synthase-lipoxygenase fusion protein in the soft coral *Gersemia fruticosa* suggests the generality of this pathway in octocorals. *Biochimica et Biophysica Acta* 1780: 315–21. [CrossRef] [PubMed]

Lõhelaid, Helike, Tarvi Teder, and Nigulas Samel. 2014a. Lipoxygenase-allene oxide synthase pathway in octocoral thermal stress response. *Coral Reefs* 34: 143–54. [CrossRef]

Lõhelaid, Helike, Tarvi Teder, Kadri Tõldsepp, Merrick Ekins, and Nigulas Samel. 2014b. Up-regulated expression of AOS-LOXa and increased eicosanoid synthesis in response to coral wounding. *PLoS ONE* 9: e89215. [CrossRef]

Maderna, Paola, and Catherine Godson. 2009. Lipoxins: Resolutionary road. *British Journal of Pharmacology* 158: 947–59. [CrossRef]

Maffei, Massimo E., Axel Mithöfer, and Wilhelm Boland. 2007. Before gene expression: Early events in plant–insect interaction. *Trends in Plant Science* 12: 310–16. [CrossRef]

Makhutova, Olesia N., Nadezhda N. Sushchik, Galina S. Kalachova, and Alexander V. Ageev. 2009. Fatty acid content and composition of freshwater planaria *Dendrocoelopsis* sp. (Planariidae, *Turbellaria*, *Platyhelminthes*) from the Yenisei River. *Journal of Siberian Federal University. Biology* 2: 135–44.

Manni, Lucia, Chiara Anselmi, Francesca Cima, Fabio Gasparini, Ayelet Voskoboynik, Margherita Martini, Anna Peronato, Paolo Burighel, Giovanna Zaniolo, and Loriano Ballarin. 2019. Sixty years of experimental studies on the blastogenesis of the colonial tunicate *Botryllus schlosseri*. *Developmental Biology* 448: 293–308. [CrossRef] [PubMed]

Meijer, Laurent, Alan R. Brash, Robert W. Bryant, Kwokei Ng, Jacques Maclouf, and Howard Sprecher. 1986. Stereospecific induction of starfish oocyte maturation by (8R)-hydroxyeicosatetraenoic acid. *Journal of Biological Chemistry* 261: 17040–47. [CrossRef]

Mimura, Tsutomu, Masaru Okabe, Mikio Satake, Tsutomu Nakanishi, Akira Inada, Yoshinori Fujimoto, Fumito Hata, Yasuko Matsumura, and Nobuo Ikekawa. 1986. Fatty acids and sterols of the tunicate, *Salpa thompsoni*, from the antarctic ocean: Chemical composition and hemolytic activity. *Chemical and Pharmaceutical Bulletin* 34: 4562–68. [CrossRef] [PubMed]

Monroig, Óscar, Douglas R. Tocher, and Juan C. Navarro. 2013. Biosynthesis of polyunsaturated fatty acids in marine invertebrates: Recent advances in molecular mechanisms. *Marine Drugs* 11: 3998–4018. [CrossRef] [PubMed]

Moreno, Juan J. 2009. New aspects of the role of hydroxyeicosatetraenoic acids in cell growth and cancer development. *Biochemical Pharmacology* 77: 1–10. [CrossRef] [PubMed]

Mortimer, Monika, Reet Järving, Alan R. Brash, Nigulas Samel, and Ivar Järving. 2006. Identification and characterization of an arachidonate 11R-lipoxygenase. *Archives of Biochemistry and Biophysics* 445: 147–55. [CrossRef]

Motomura, Seiji, and Michael T. Dexter. 1980. The effect of prostaglandin E_1 on hemopoiesis in long-term bone marrow cultures. *Experimental Hematology* 8: 298–303.

Murakami, Makoto. 2011. Lipid mediators in life science. *Experimental Animals* 60: 7–20. [CrossRef] [PubMed]

Mydlarz, Laura D., Sally F. Holthouse, Esther C. Peters, and Drew C. Harvell. 2008. Cellular responses in sea fan corals: Granular amoebocytes react to pathogen and climate stressors. *PLoS ONE* 3: e1811. [CrossRef]

Neau, David B., Nathaniel C. Gilbert, Sue G. Bartlett, William Boeglin, Alan R. Brash, and Marcia E. Newcomer. 2009. The 1.85 A structure of an 8R-lipoxygenase suggests a general model for lipoxygenase product specificity. *Biochemistry* 48: 7906–15. [CrossRef]

Nelson, David R., Jared V. Goldstone, and John J. Stegeman. 2013. The cytochrome P450 genesis locus: The origin and evolution of animal cytochrome P450s. *Philosophical Transactions of the Royal Society B* 368: 20120474. [CrossRef] [PubMed]

O'Neal, Will, and Joseph R. Pawlik. 2002. A reappraisal of the chemical and physical defenses of Caribbean gorgonian corals against predatory fishes. *Marine Ecology Progress Series* 240: 117–26. [CrossRef]

Olano, Cecile T., and Charles H. Bigger. 2000. Phagocytic activities of the gorgonian coral *Swiftia exserta*. *Journal of Invertebrate Pathology* 76: 176–84. [CrossRef] [PubMed]

Oldham, Michael L., Alan R. Brash, and Marcia E. Newcomer. 2005a. Insights from the X-ray crystal structure of coral 8R-lipoxygenase: Calcium activation via a C2-like domain and a structural basis of product chirality. *Journal of Biological Chemistry* 280: 39545–52. [CrossRef] [PubMed]

Oldham, Michael L., Alan R. Brash, and Marcia E. Newcomer. 2005b. The structure of coral allene oxide synthase reveals a catalase adapted for metabolism of a fatty acid hydroperoxide. *Proceedings of the National Academy of Sciences of the United States of America* 102: 297–302. [CrossRef] [PubMed]

Palanisamy, Satheesh K., Nadesan M. Rajendran, and Angela Marino. 2017. Natural products diversity of marine ascidians (*Tunicates*; *Ascidiacea*) and successful drugs in clinical development. *Natural Products and Bioprospecting* 7: 1–111. [CrossRef] [PubMed]

Palmer, Caroline V., Laura D. Mydlarz, and Bette L. Willis. 2008. Evidence of an inflammatory-like response in non-normally pigmented tissues of two scleractinian corals. *Proceedings of the Royal Society* 275: 2687–93. [CrossRef]

Palmer, Caroline V., Nikki G. Traylor-Knowles, Bette L. Willis, and John C. Bythell. 2011. Corals use similar immune cells and wound-healing processes as those of higher organisms. *PLoS ONE* 6: e23992. [CrossRef]

Pawlik, Joseph R., Mark T. Burch, and William Fenical. 1987. Patterns of chemical defense among caribbean gorgonian corals—a preliminary survey. *Journal of Experimental Marine Biology and Ecology* 108: 55–66. [CrossRef]

Pidgeon, Graham P., Joanne Lysaght, Sriram Krishnamoorthy, John V. Reynolds, Ken O'Byrne, Daotai Nie, and Kenneth V. Honn. 2007. Lipoxygenase metabolism: Roles in tumor progression and survival. *Cancer and Metastasis Reviews* 26: 503–24. [CrossRef]

Polato, Nicholas R., Naomi S. Altman, and Iliana B. Baums. 2013. Variation in the transcriptional response of threatened coral larvae to elevated temperatures. *Molecular Ecology* 22: 1366–82. [CrossRef] [PubMed]

Prockop, Darwin J. 2013. Concise review: Two negative feedback loops place mesenchymal stem/stromal cells at the center of early regulators of inflammation. *Stem Cells* 31: 2042–46. [CrossRef] [PubMed]

Proksch, Ehrhardt, Johanna M. Brandner, and Jens-Michael Jensen. 2008. The skin: An indispensable barrier. *Experimental Dermatology* 17: 1063–72. [CrossRef] [PubMed]

Rajasingh, Johnson, and John J. Bright. 2006. 15-Deoxy-delta12,14-prostaglandin J$_2$ regulates leukemia inhibitory factor signaling through JAK-STAT pathway in mouse embryonic stem cells. *Experimental Cell Research* 312: 2538–46. [CrossRef]

Reitzel, Adam M., James C. Sullivan, Nikki Traylor-Knowles, and John R. Finnerty. 2008. Genomic survey of candidate stress-response genes in the estuarine anemone. *Nematostella vectensis. Biological Bulletin* 214: 233–54. [CrossRef]

Ricciotti, Emanuela, and Garret A. FitzGerald. 2011. Prostaglandins and inflammation. *Arteriosclerosis, Thrombosis, and Vascular Biology* 31: 986–1000. [CrossRef]

Rinkevich, Yuval, Valeria Matranga, and Baruch Rinkevich. 2009. Stem cells in aquatic invertebrates: Common premises and emerging unique themes. *Stem Cells in Marine Organisms* 2009: 61–103. [CrossRef]

Rinkevich, Baruch, Loriano Ballarin, Pedro Martinez, Ildiko Somorjai, Oshrat Ben-Hamo, Ilya Borisenko, Eugene Berezikov, Alexander Ereskovsky, Eva Gazave, Denis Khnykin, and et al. 2021. A pan-metazoan concept for adult stem cells: The wobbling Penrose landscape. *Biological Reviews* 97: 299–325.

Rod'kina, Svetlana A. 2005. Fatty acids and other lipids of marine sponges. *Russian Journal of Marine Biology* 31: S49–S60. [CrossRef]

Rodriguez, Paola G., Frances N. Felix, David T. Woodley, and Elisabeth K. Shim. 2008. The role of oxygen in wound healing: A review of the literature. *Dermatologic Surgery* 34: 1159–69. [CrossRef]

Romano, Mario, Sara Patruno, Antonella Pomilio, and Antonio Recchiuti. 2019. Proresolving lipid mediators and receptors in stem cell biology: Concise review. *Stem Cells Translational Medicine* 8: 992–98. [CrossRef]

Rouzer, Carol A., and Lawrence J. Marnett. 2003. Mechanism of free radical oxygenation of polyunsaturated fatty acids by cyclooxygenases. *Chemical Reviews* 103: 2239–304. [CrossRef] [PubMed]

Rouzer, Carol A., and Lawrence J. Marnett. 2005. Structural and functional differences between cyclooxygenases: Fatty acid oxygenases with a critical role in cell signaling. *Biochemical and Biophysical Research Communications* 338: 34–44. [CrossRef] [PubMed]

Rouzer, Carol A., and Lawrence J. Marnett. 2009. Cyclooxygenases: Structural and functional insights. *Journal of Lipid Research* 50: S29–S34. [CrossRef] [PubMed]

Rowley, Andrew F., Claire L. Vogan, Graham W. Taylor, and Anthony S. Clare. 2005. Prostaglandins in non-insectan invertebrates: Recent insights and unsolved problems. *The Journal of Experimental Biology* 208: 3–14. [CrossRef] [PubMed]

Rozanski, Andrei, HongKee Moon, Holger Brandl, Jose M. Martin-Duran, Markus A. Grohme, Katja Huttner, Kerstin Bartscherer, Ian Henry, and Jochen C. Rink. 2019. PlanMine 3.0-improvements to a mineable resource of flatworm biology and biodiversity. *Nucleic Acids Research* 47: D812–D820. [CrossRef] [PubMed]

Sakamoto, Toshimasa, Mehmet Cansev, and Richard J. Wurtman. 2007. Oral supplementation with docosahexaenoic acid and uridine-5′-monophosphate increases dendritic spine density in adult gerbil hippocampus. *Brain Research* 1182: 50–59. [CrossRef] [PubMed]

Sakayori, Nobuyuki, Motoko Maekawa, Keiko Numayama-Tsuruta, Takashi Katura, Takahiro Moriya, and Noriko Osumi. 2011. Distinctive effects of arachidonic acid and docosahexaenoic acid on neural stem/progenitor cells. *Genes Cells* 16: 778–90. [CrossRef] [PubMed]

Sasaki, Tsutomu, Kazuo Kitagawa, Shiro Sugiura, Emi Omura-Matsuoka, Shigeru Tanaka, Yoshiki Yagita, Hideyuki Okano, Masayasu Matsumoto, and Masatsugu Hori. 2003. Implication of cyclooxygenase-2 on enhanced proliferation of neural progenitor cells in the adult mouse hippocampus after ischemia. *Journal of Neuroscience Research* 72: 461–71. [CrossRef] [PubMed]

Schneider, Claus, Derek A. Pratt, Ned A. Porter, and Alan R. Brash. 2007. Control of oxygenation in lipoxygenase and cyclooxygenase catalysis. *Chemical Biology* 14: 473–88. [CrossRef] [PubMed]

Schultz, Gregory S., Jeffrey M. Davidson, Robert S. Kirsner, Paul Bornstein, and Ira M. Herman. 2011. Dynamic reciprocity in the wound microenvironment. *Wound Repair and Regeneration* 19: 134–48. [CrossRef]

Serhan, Charles N., and Nan Chiang. 2008. Endogenous pro-resolving and anti-inflammatory lipid mediators: A new pharmacologic genus. *British Journal of Pharmacology* 153: S200–S215. [CrossRef] [PubMed]

Serhan, Charles N., Song Hong, Karsten Gronert, Sean P. Colgan, Pallavi R. Devchand, Gudrun Mirick, and Rose-Laure Moussignac. 2002. Resolvins: A family of bioactive products of omega-3 fatty acid transformation circuits initiated by aspirin treatment that counter proinflammation signals. *Journal of Experimental Medicine* 196: 1025–37. [CrossRef] [PubMed]

Serhan, Charles N., Nan Chiang, and Thomas E. Van Dyke. 2008. Resolving inflammation: Dual anti-inflammatory and pro-resolution lipid mediators. *Nature Reviews Immunology* 8: 349–61. [CrossRef]

Serhan, Charles N., Jesmond Dalli, Sergey Karamnov, Alexander Choi, Chul-Kyu Park, Zhen-Zhong Xu, Ru-Rong Ji, Min Zhu, and Nicos A. Petasis. 2012. Macrophage proresolving mediator maresin 1 stimulates tissue regeneration and controls pain. *The FASEB Journal* 26: 1755–65. [CrossRef] [PubMed]

Singer, Adam J., and Richard A. Clark. 1999. Cutaneous wound healing. *The New England Journal of Medicine* 341: 738–46. [CrossRef] [PubMed]

Teder, Tarvi, Helike Lõhelaid, William E. Boeglin, Wade M. Calcutt, Alan R. Brash, and Nigulas Samel. 2015. A catalase-related hemoprotein in coral is specialized for synthesis of short-chain aldehydes. *Journal of Biological Chemistry* 290: 19823–32. [CrossRef]

Teder, Tarvi, Helike Lõhelaid, and Nigulas Samel. 2017. Structural and functional insights into the reaction specificity of catalase-related hydroperoxide lyase: A shift from lyase activity to allene oxide synthase by site-directed mutagenesis. *PLoS ONE* 12: e0185291. [CrossRef]

Teder, Tarvi, Nigulas Samel, and Helike Lõhelaid. 2019. Distinct characteristics of the substrate binding between highly homologous catalase-related allene oxide synthase and hydroperoxide lyase. *Archives of Biochemistry and Biophysics* 676: 108126. [CrossRef]

Tsoyi, Konstantin, Sean R. Hall, Jesmond Dalli, Romain A. Colas, Sailaja Ghanta, Bonna Ith, Anna Coronata, Laura E. Fredenburgh, Rebecca M. Baron, Augustine M. Choi, and et al. 2016. Carbon monoxide improves efficacy of mesenchymal stromal cells during sepsis by production of specialized proresolving lipid mediators. *Critical Care Medicine* 44: e1236–e1245. [CrossRef]

Vaca, Pilar, Genoveva Berna, Raquel Araujo, Everardo M. Carneiro, Francisco J. Bedoya, Bernat Soria, and Franz Martin. 2008. Nicotinamide induces differentiation of embryonic stem cells into insulin-secreting cells. *Experimental Cell Research* 314: 969–74. [CrossRef]

Valmsen, Karin, Ivar Järving, William E. Boeglin, Külliki Varvas, Reet Koljak, Tõnis Pehk, Alan R. Brash, and Nigulas Samel. 2001. The origin of 15R-prostaglandins in the Caribbean coral *Plexaura homomalla*: Molecular cloning and expression of a novel cyclooxygenase. *Proceedings of the National Academy of Sciences of the United States of America* 98: 7700–5. [CrossRef]

Vane, John R., and Regina M. Botting. 1998. Anti-inflammatory drugs and their mechanism of action. *Inflammation Research* 47: S78–S87. [CrossRef] [PubMed]

Varvas, Külliki, Ivar Järving, Reet Koljak, Aino Vahemets, Tõnis Pehk, Aleksander-Mati Müürisepp, Ülo Lille, and Nigulas Samel. 1993. In vitro biosynthesis of prostaglandins in the white sea soft coral *Gersemia Fruticosa*—Formation of optically-active Pgd2, Pge2, Pgf2-Alpha and 15-Keto-Pgf2-Alpha from arachidonic acid. *Tetrahedron Letters* 34: 3643–46. [CrossRef]

Varvas, Külliki, Reet Koljak, Ivar Järving, Tõnis Pehk, and Nigulas Samel. 1994. Endoperoxide pathway in prostaglandin biosynthesis in the soft coral *Gersemia Fruticosa*. *Tetrahedron Letters* 35: 8267–70. [CrossRef]

Varvas, Külliki, Ivar Järving, Reet Koljak, Karin Valmsen, Alan R. Brash, and Nigulas Samel. 1999. Evidence of a cyclooxygenase-related prostaglandin synthesis in coral. The allene oxide pathway is not involved in prostaglandin biosynthesis. *Journal of Biological Chemistry* 274: 9923–29. [CrossRef]

Voskoboynik, Ayelet, Norma F. Neff, Debashis Sahoo, Aaron M. Newman, Dmitry Pushkarev, Winston Koh, Benedetto Passarelli, Christina H. Fan, Gary L. Mantalas, Karla J. Palmeri, and et al. 2013. The genome sequence of the colonial chordate, *Botryllus schlosseri*. *Elife* 2: e00569. [CrossRef]

Wada, Koichiro, Makoto Arita, Atsushi Nakajima, Kazufumi Katayama, Chiho Kudo, Yoshinori Kamisaki, and Charles N. Serhan. 2006. Leukotriene B_4 and lipoxin A_4 are regulatory signals for neural stem cell proliferation and differentiation. *The FASEB Journal* 20: 1785–92. [CrossRef]

Walden, Thomas L., Jr. 1988. Radioprotection of mouse hematopoietic stem cells by leukotriene A_4 and lipoxin B_4. *Journal of Radiation Research* 29: 255–60. [CrossRef]

Wasternack, Claus. 2007. Jasmonates: An update on biosynthesis, signal transduction and action in plant stress response, growth and development. *Annals of Botany* 100: 681–97. [CrossRef]

Watanabe, Kinzo, Miyuki Sekine, Haruko Takahashi, and Kazuo Iguchi. 2001. New halogenated marine prostanoids with cytotoxic activity from the Okinawan soft coral *Clavularia viridis*. *Journal of Natural Products* 64: 1421–25. [CrossRef]

Weinheimer, Alfred J., and Robert L. Spraggins. 1969. The occurrence of two new prostaglandin derivatives (15-epi-PGA$_2$ and its acetate, methyl ester) in the gorgonian *Plexaura homomalla* chemistry of coelenterates. XV. *Tetrahedron Letters* 10: 5185–88. [CrossRef]

Whalen, Kristen E., Victoria R. Starczak, David R. Nelson, Jared V. Goldstone, and Mark E. Hahn. 2010. Cytochrome P450 diversity and induction by gorgonian allelochemicals in the marine gastropod *Cyphoma gibbosum*. *BMC Ecology* 10: 24. [CrossRef] [PubMed]

Wiszniewska, Malgorzata, Maciej Niewada, and Anna Czlonkowska. 2011. Sex differences in risk factor distribution, severity, and outcome of ischemic stroke. *Acta Clinica Croatica* 50: 21–28. [PubMed]

Yanes, Oscar, Julie Clark, Diana M. Wong, Gary J. Patti, Antonio Sanchez-Ruiz, Paul H. Benton, Sunia A. Trauger, Caroline Desponts, Sheng Ding, and Gary Siuzdak. 2010. Metabolic oxidation regulates embryonic stem cell differentiation. *Nature Chemical Biology* 6: 411–17. [CrossRef] [PubMed]

Yun, Doo H., Hae Y. Song, Mi J. Lee, Mi R. Kim, Min Y. Kim, Jung S. Lee, and Jae H. Kim. 2009a. Thromboxane A_2 modulates migration, proliferation, and differentiation of adipose tissue-derived mesenchymal stem cells. *Experimental & Molecular Medicine* 41: 17–24.

Yun, Seung P., Min Y. Lee, Jung M. Ryu, and Ho J. Han. 2009b. Interaction between PGE_2 and EGF receptor through MAPKs in mouse embryonic stem cell proliferation. *Cellular and Molecular Life Sciences* 66: 1603–16. [CrossRef] [PubMed]

Yun, Seung P., Jung M. Ryu, Min W. Jang, and Ho J. Han. 2011. Interaction of profilin-1 and F-actin via a beta-arrestin-1/JNK signaling pathway involved in prostaglandin E(2)-induced human mesenchymal stem cells migration and proliferation. *Journal of Cellular Physiology* 226: 559–71. [CrossRef] [PubMed]

Zhang, Zhi-Qiang. 2011. Animal biodiversity: An outline of higher-level classification and survey of taxonomic richness. *Zootaxa* 3148: 1–237. [CrossRef]

Zhou, Yan, Huajun Zheng, Yangyi Chen, Lei Zhang, Kai Wang, Jing Guo, Zhen Huang, Bo Zhang, Wei Huang, Ke Jin, and et al. 2009. The *Schistosoma japonicum* genome reveals features of host–parasite interplay. *Nature* 460: 345–51.

Zhu, Zhenjiu, Chenglai Fu, Xiaoxia Li, Yimeng Song, Chenghong Li, Minghui Zou, Youfei Guan, and Yi Zhu. 2011. Prostaglandin E_2 promotes endothelial differentiation from bone marrow-derived cells through AMPK activation. *PLoS ONE* 6: e23554. [CrossRef]

Molecular Regulation of Decision Making in the Interstitial Stem Cell Lineage of *Hydra* Revisited

Marion Lechable, Matthias Achrainer, Maren Kruus, Willi Salvenmoser and Bert Hobmayer

Abstract: Multipotent interstitial stem cells in the freshwater polyp *Hydra* define one of the best-studied pre-bilaterian adult cell lineages. Most of them represent a population of small, fast-cycling cells that give rise to three somatic differentiation products (neurons, nematocytes, and gland cells) under conditions of continuous asexual growth and reproduction, and they also form the gametes when sexual reproduction is initiated. Few proliferate with a longer cell cycle. Interstitial stem cells in *Hydra* and other marine hydrozoans have been studied intensively using sophisticated cellular and molecular methods over several decades. Here, we discuss the properties of interstitial stem cells in *Hydra* and the known feedback control mechanisms maintaining tissue homeostasis and spatial distribution of interstitial cells along the polyp's major body axis. We summarize the current state of knowledge about molecular regulation of self-renewal and somatic differentiation and put particular emphasis on those molecular factors that have been shown to affect decision making using methods of functional interference.

1. Introduction

Interstitial cells (ICs) were discovered in the late 19th century by August Weismann and described as putative migratory germline precursor cells in several colonial marine hydrozoans, laying a basis for his theory of the germline published nearly ten years later (Weismann 1883). Labeling techniques and tissue manipulations revealed the lineage relationships and cellular dynamics of the various types of ICs in hydrozoans, especially in the freshwater polyp *Hydra* (Tardent 1954; Müller 1967; for review also see David et al. 1987; Bode 1996; Plickert et al. 2012). These studies showed that all ICs belong to a single adult stem cell lineage with three somatic cell types—neurons, nematocytes (stinging cells), gland cells—as well as the two types of gametes as differentiation products (Figure 1A).

Classic studies using *Hydra* as a model determined the probabilities for self-renewal and for differentiation into the different somatic interstitial cell types, and they precisely defined cell cycle and differentiation times (Campbell and David 1974; David and Gierer 1974; Schmidt and David 1986). Thus, there is a detailed quantitative understanding of IC lineage dynamics. More recently, omics approaches, transgenic *Hydra* polyps and genetic up- and down-regulation have provided an advanced understanding of the diversity and plasticity of sub-populations of cells within the IC lineage (Siebert et al. 2008, 2019; Chapman et al. 2010; Hemmrich et al. 2012; Buzgariu et al. 2015). An unexpected, modified model for nerve and gland cell differentiation arose from single-cell transcriptome analysis in *Hydra,* suggesting that these two differentiated cell types arise from a common precursor (Figure 1B; Siebert et al. 2019). Altogether, *Hydra* interstitial stem cells (ISCs), often referred to as "i-cells" in *Hydra* and other hydrozoans, and their differentiation products represent probably the best-studied pre-bilaterian adult stem cell system, and a number of comprehensive reviews have discussed its various features (Bosch 2008; Watanabe et al. 2009; Bosch et al. 2010; Hobmayer et al. 2012; David 2012; Nishimiya-Fujisawa and Kobayashi 2012). Here, after addressing major ISC properties in *Hydra,* we summarize the current state of knowledge about known molecular factors acting in lineage decision making in its asexual reproduction mode, and we focus on those factors shown to be active in functional interference assays. More detailed information about expression, putative function and functional validation of these factors is listed in Table 1. Table 1 also includes some regulators proposed to act in IC decision making on the basis of their cell type-specific gene expression.

Table 1. Selected molecular factors acting in the *Hydra* interstitial cell lineage based on available functional interference data and/or cell type-specific gene expression.

Factor (References)	Cellular Function	*Hydra* Genome Protein Model (Augustus)	Expression Pattern	Experimental Validation
		ISC self-renewal		
HyGSK-3β (Khalturin et al. 2007; Broun et al. 2005)	signal transduction	Sc4wPfr_488.g29970.t1	ISC	Smi: alsterpaullone
Hyβ-Catenin (Gee et al. 2010; Hartl et al. 2019)	signal transduction	Sc4wPfr_975.g7262.t1	ISC	transgenesis, Smi: alsterpaullone
Hy-I-cell1 (Siebert et al. 2019)	unknown	Sc4wPfr_559.g509.t1	ISC	scRNAseq
FoxO (Boehm et al. 2012)	transcription factor	Sc4wPfr_909.g33493.t2	ISC	RNAi, transgenesis
HyMyc1 (Ambrosone et al. 2012; Hartl et al. 2010, 2019)	transcription factor	Sc4wPfr_73.g11571.t1	ISC	RNAi, smi: 10058-F4
HyMyc2 (Hartl et al. 2014, 2019)	transcription factor	Sc4wPfr_850.1.g5732.t1	ISC	WISH

Factor (References)	Cellular Function	*Hydra* Genome Protein Model (Augustus)	Expression Pattern	Experimental Validation
Hywi, Hyli (Juliano et al. 2014; Teefy et al. 2020)	RNA binding protein	Sc4wPfr_597.2.g14333.t1, Sc4wPfr_661.g19809.t1	ISC	transgenesis
Nerve cell differentiation				
Cnash (Grens et al. 1995)	transcription factor	Sc4wPfr_147.g8607.t1	ISC, sensory neurons	WISH
Myb (Siebert et al. 2019)	transcription factor	Sc4wPfr_423.g13448.t1	neuronal progenitor cells	scRNAseq
HvSoxC (Siebert et al. 2019)	transcription factor	Sc4wPfr_351.g11299.t1	neuronal progenitor cells	scRNAseq
Myc3 (Siebert et al. 2019)	transcription factor	Sc4wPfr_199.g28684.t1	neuronal progenitor cells	scRNAseq
Cnox2 (Miljkovic-Licina et al. 2007)	transcription factor	Sc4wPfr_165.g10051.t1	apical neurons	RNAi
Head activator (Fenger et al. 1994)	signal peptide	-	-	peptide treatment
Hym-355 (Takahashi et al. 2000)	neuropeptide	Sc4wPfr_635.g14708.t1	neurons	peptide treatment, WISH
Hym33H (Takahashi et al. 1997)	neuropeptide	Sc4wPfr_59.2.g12471.t1	neurons	peptide treatment, WISH
NDA-1 (Augustin et al. 2017; Siebert et al. 2019)	neuropeptide	Sc4wPfr_824.g11313.t1	neurons	transgenic overexpression and knock-down
prdl-a (Miljkovic-Licina et al. 2007)	transcription factor	Sc4wPfr_1080.g15226.t1	nerve cells (ectoderm)	WISH
prdl-b (Miljkovic-Licina et al. 2007)	transcription factor	Sc4wPfr_372.g27997.t1	nematocyte, nerve cells	WISH
msh (Miljkovic-Licina et al. 2007)	transcription factor	Sc4wPfr_87.g16557.t1	nerve cells (ectoderm)	WISH
COUP-TF (Miljkovic-Licina et al. 2007)	transcription factor	Sc4wPfr_17.g15881.t1	nerve cells, nematocytes	WISH
Nematocyte differentiation				
HyZic (Lindgens et al. 2004)	transcription factor	Sc4wPfr_252.1.g15359.t1 Sc4wPfr_237.2.g16165.t1	early proliferating nematoblast nests (2-8)	BrdU, WISH
HvNotch (Käsbauer et al. 2007)	signal transduction receptor	Sc4wPfr_326.g15645.t1	early nematoblast differentiation	BrdU, smi: DAPT
GSK-3β (Khalturin et al. 2007)	phospho-kinase	Sc4wPfr_488.g29970.t1	early nematoblast differentiation	Smi: alsterpaullone

Table 1. Cont.

Factor (References)	Cellular Function	*Hydra* Genome Protein Model (Augustus)	Expression Pattern	Experimental Validation
Cnash (Grens et al. 1995)	transcription factor	Sc4wPfr_147.g8607.t1	nematoblast differentiation (8 and 16 cells)	WISH
HyEED co-expressed with HyEZH2 (Khalturin et al. 2007)	epigenetic regulator	Sc4wPfr_804.g24124.t1	nematoblast—ISC	transgenesis
Myc1 (Ambrosone et al. 2012)	transcription factor	Sc4wPfr_73.g11571.t1	nematoblast—ISC	RNAi, smi: 10058-F4
Dkk3 (Fedders et al. 2004)	secreted wnt modulator	Sc4wPfr_259_g33632.t1	differentiating nematocytes	WISH
Gland cell differentiation				
Myb (Siebert et al. 2019)	transcription factor	Sc4wPfr_839.g4024.t1	precursor gland cells	scRNAseq
Dkk 1/2/4 A-C (Augustin et al. 2006; Guder et al. 2006)	secreted wnt modulator	Sc4wPfr_134_g20117.t1	endodermal gland cells in gastric region	WISH
Gametogenesis				
HvNotch (Käsbauer et al. 2007)	signal transduction receptor	Sc4wPfr_326.g15645.t1	oocyte	smi-DAPT
Cnvas1 and Cnvas2 (Mochizuki et al. 2001)	germ-line factor	Sc4wPfr_861.g31120.t1 Sc4wPfr_2009.g19353.t1	germline—ISC	WISH
Cnnos1 and Cnnos2 (Mochizuki et al. 2000)	germ-line factor	Sc4wPfr_366.g23802.t1 Sc4wPfr_169.g29161.t1	germline—ISC	WISH
Hywi (Juliano et al. 2014; Teefy et al. 2020)	RNA-binding protein	Sc4wPfr_597.2.g14333.t1	germline—ISC	immunocytochemistry
Pumilio (Siebert et al. 2019)	RNA-binding protein	Sc4wPfr_112.1.g5130.t1	body column	scRNAseq
HyEED (Genikhovich et al. 2006)	epigenetic regulator	Sc4wPfr_6.g19136.t1	spermatogonia	WISH
HyMyc2 (Hartl et al. 2014)	transcription factor	Sc4wPfr_850.1.g5732.t1	spermatogenesis, oogenesis	WISH

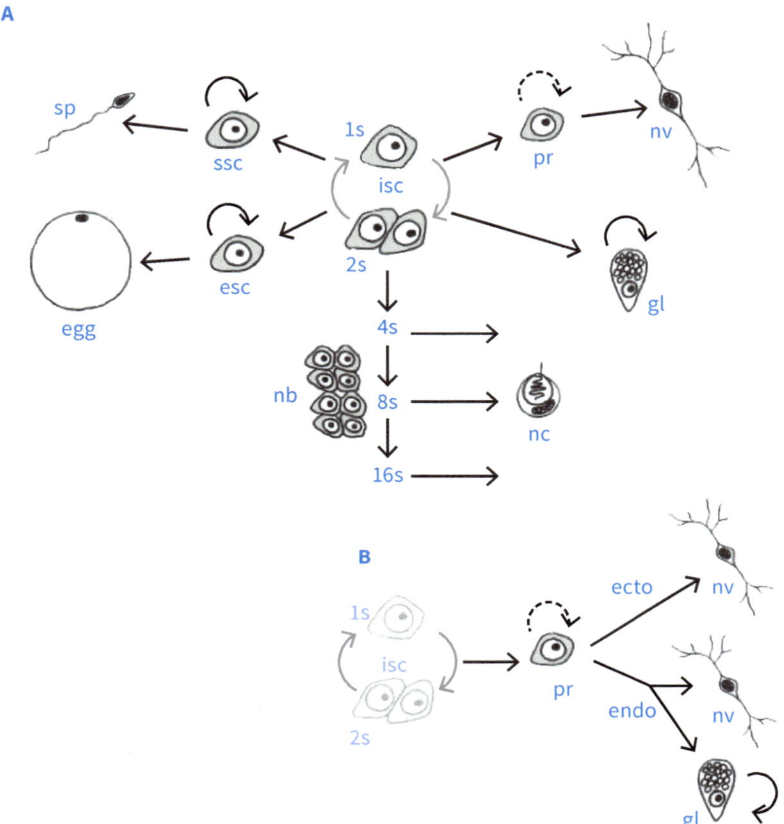

Figure 1. Schematics of the *Hydra* interstitial stem cell (ISC) lineage. (**A**) In the classic model, an ISC gives rise to somatic nerve cells (nv), gland cells (gl), and nematocytes (nc). ISCs also form sperm- and egg-restricted stem cells (ssc, esc), which can differentiate mature sperm cells (sp) and eggs during sexual reproduction. Nematocyte differentiation starts with the formation of nematoblast (nb) nests, which differentiate mature nematocytes after going through terminal mitosis. The committed precursor (pr) for nerve cell differentiation has a limited capacity for proliferation. (**B**) Results from single-cell transcriptome analysis suggest a modified model, in which nerve and gland cells derive from a common precursor (pr), whose capacity for proliferation is yet not clear. Source: Graphic by authors.

2. Interstitial Stem Cell (ISC) Properties in *Hydra*

2.1. ISC Self-Renewal and Stochastic Decision Making

ISCs represent small, undifferentiated cells appearing as single cells or cell pairs and exhibiting a large nuclear–cytoplasmic ratio, a de-condensed chromatin with conspicuous nucleoli, a poly-ribosome- and mitochondria-rich cytoplasm, and multiple chromatoid bodies (nuage) associated with the nuclear membrane and mitochondria or isolated within the cytoplasm without connections to other organelles (Figure 2; Hobmayer et al. 2012). In vivo stem cell cloning experiments using IC-free host tissue demonstrated the multipotency of *Hydra* ISCs and their capacity to differentiate into somatic cells and gametes (David and Murphy 1977; Bosch and David 1987; Nishimiya-Fujisawa and Sugiyama 1993). ISCs reside in the ectodermal epithelial layer throughout the gastric region. In intact, asexually growing polyps, ISCs continuously grow in contiguous patches and migrate only small distances at most (Bosch and David 1990; Boehm and Bosch 2012). Nearly all ISCs are fast-cycling cells with a cell cycle length of 18–30 h (Campbell and David 1974) and a probability for self-renewal (P_s) of around 0.6 (David and Gierer 1974). Notably, after keeping clonal lab strains under conditions of fast and indefinite growth over decades, ISCs do not show any sign of cellular senescence, indicating that they have evolved mechanisms counteracting the known limits to expanded stem cell division such as telomere reduction, mitochondrial dysfunction, DNA damage, etc. (Sun et al. 2020; Tomczyk et al. 2020). There is also a tiny population of slower cycling ISCs showing an expanded cell cycle length of several days, which can be activated to proliferate faster by regeneration signals (Govindasamy et al. 2014). Finally, tracking of DiI vitally labelled ISCs revealed the full capacity for decision making, in which both daughter cells of a stem cell can remain stem cells or become differentiation precursors, or in which asymmetric division yields one stem cell and one differentiation precursor (David 2012). This type of flexible decision making involves communication of ISCs with their environment and rather complex processing of incoming short- and long-range signals.

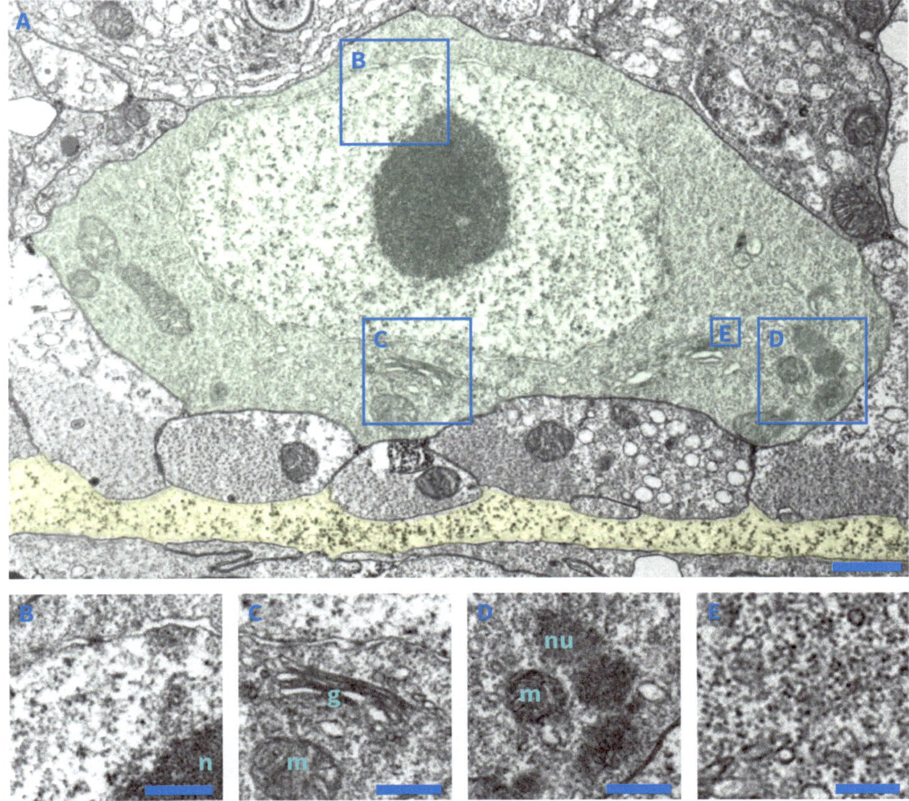

Figure 2. Ultrastructure of a *Hydra* interstitial stem cell (ISC). (**A**) ISC (green) positioned apically to the basal epithelial muscle fibers and the underlying mesoglea (yellow). Representative organelles of ISCs are depicted at higher magnification: (**B**) nuclear membrane with a nuclear pore, de-condensed chromatin and part of the nucleolus (n); (**C**) Golgi apparatus (g) and mitochondrium (m); (**D**) chromatoid body/nuage (nu) associated with mitochondria (m); (**E**) abundant ribosomes and chain-like poly-ribosomes in the cytoplasm. Bars: 1000 nm in (**A**), 500 nm in (**B–D**), 250 nm in (**E**). Source: Graphic by authors.

2.2. A Putative ISC Niche

Hydra ISCs reside in the interstitial spaces between ectodermal epithelial cells usually at the basal level of the epidermal layer close to the muscle fibers. The microenvironment of stem cells (the "niche") is commonly regarded as an important regulatory entity for stem cell decision making and for providing structural, trophic and physiological support. Thus, these interstitial spaces represent distinct niches for ISCs. They may create a communication space for maintaining the multipotent stem cell state or for becoming a committed precursor cell. However, none of the signals used to communicate has yet been isolated and characterized by now. Light and electron microscopic images reveal the direct contact of ISCs over almost their entire membrane surfaces with the membranes of surrounding epithelial cells (Figure 2). There has been speculation that classic cadherin, as in bilaterian stem cell niches, is involved in ISC–niche interactions (Bosch et al. 2010). The *Hydra* genome indeed encodes one large classic cadherin protein (Chapman et al. 2010), and this gene is transcriptionally activated in ectodermal epithelial cells and ISCs (*Hydra* single-cell transcriptome data available at the Broad Institute Single Cell Portal; Hobmayer lab, unpublished data). Functional analysis is required to validate this view. Furthermore, direct contact between ISCs and the mesoglea, *Hydra*'s extracellular matrix, has been discussed (Bosch et al. 2010), but also here a more detailed analysis using advanced imaging and molecular methods is needed to validate this idea.

2.3. Known Feedback Regulation through Signaling from Beyond the Niche

Two aspects of continuously growing asexual mass cultures of *Hydra* clearly suggest that ISC behavior must be under tight control of complex feedback signaling and global patterning mechanisms. First, ICs exhibit a defined distribution pattern along the polyp's major head–foot body axis. ISCs are restricted to the gastric region, and they do not occur in the differentiated head and foot areas. This was first demonstrated by David and Plotnick (1980) by analyzing the axial origin of self-renewing and clone-forming interstitial cells in host aggregates. Later, it was confirmed using ISC-specific antibody staining and stable transgenic polyps expressing GFP in ISCs (David et al. 1987; Wittlieb et al. 2006). The boundaries to the head and foot areas of differentiation are sharp, raising the question of how such sharp boundaries are maintained under conditions where the entire tissue is constantly growing and cells are permanently changing positions. Positional information provided by the primary axial patterning system was suggested to shift ISC decision making from self-renewal to differentiation at the gastric region-head and gastric region-foot boundaries (Bosch 2008). Wnt/beta-Catenin signaling plays a central role in the *Hydra* head organizer, the polyp's major signaling center for axial patterning and setting up positional information (Hobmayer et al. 2000; Broun et al. 2005). Furthermore, accumulating evidence as discussed below in more detail shows the effects of beta-Catenin on ISC maintenance, as well as on nerve and nematocyte differentiation.

Second, asexual polyp growth strictly follows the rules of homeostasis. All cell types maintain their numbers relative to each other. During permanent tissue growth, they increase in numbers at the same pace, despite the fact that cell cycle lengths and differentiation times of the various cell types differ substantially. Since cell death plays no role in asexually growing polyps, survival and the production of new cells by proliferation and differentiation are the main players, and they require permanent cell communication throughout the entire body column. By experimentally manipulating the density of selected cell types, feedback mechanisms coming from beyond the ISC niche were uncovered. ISC self-renewal reacts to the ISC density in the surrounding gastric tissue. Low density causes an increase in the probability for self-renewal, and high density a decrease (Bode et al. 1976; David and MacWilliams 1978; Sproull and David 1979; Fujisawa 1992). This feedback mechanism seems to be strain-specific, since ISCs do not respond to host ISC densities in tests using donor and host cells from different *Hydra* strains (David et al. 1991). Transplantation studies introducing ISCs into host tissue with variable nerve cell densities demonstrated that the nerve cell density positively affects ISC proliferation (Heimfeld and Bode 1985; Bosch et al. 1991). Finally, Boehm and Bosch (2012) demonstrated that non-migratory ISCs are stimulated to migrate towards gastric tissue devoid of ISCs. They proposed two alternative models explaining the observed migration patterns. Either attractive signals from empty niches may activate and direct ISC migration over some distance, or gastric tissue holding normal ISC densities may constantly emit signals suppressing ISC migration. None of the proposed signals discussed above has been identified by now. In summary, our understanding of the molecular nature of the described feedback mechanisms is only at its very beginning.

3. Molecular Factors Acting in Somatic IC Decision Making in *Hydra*

3.1. ISC Maintenance/Self-Renewal Factors

According to the current paradigm, adult stem cell maintenance and the maintenance of pluri- or multi-potency is a result of the action of a distinct set of molecular factors, mostly stem cell-specific transcription factors. Among these factors, Oct4, Sox2, Kfl4, and c-Myc have become famous for inducing pluripotency in mammalian somatic cells (Takahashi and Yamanaka 2006). Based on this, they have been prime candidates in searches for stemness factors in other animals. However, there is little evidence that they play such a role across the animal kingdom. The *Hydra* genome does not encode homologs of *oct4* and *klf4* genes. Genes of related sub-families are encoded in the *Hydra* genome, but the closest relatives to *oct4* and *klf4* sub-families are not expressed in ISCs. Likewise, paralogs of the *sox* gene family are encoded. However, while several of them are expressed in the interstitial cell lineage, none of them is clearly and specifically activated in ISCs (Siebert et al. 2019). Taken together, these results suggest that different animal lineages have evolved different molecular signatures to maintain adult stem cells. A strong candidate factor for ISC maintenance in *Hydra* and some bilaterians including vertebrates is the transcription factor fork head box O, FoxO (Figure 3; Table 1). Overactivation of FoxO in normal polyps increased ISC proliferation (Boehm et al. 2012). It also activated expression of *vasa* and *piwi*, two known stem cell genes (see below), in ISCs and in differentiating nematocytes.

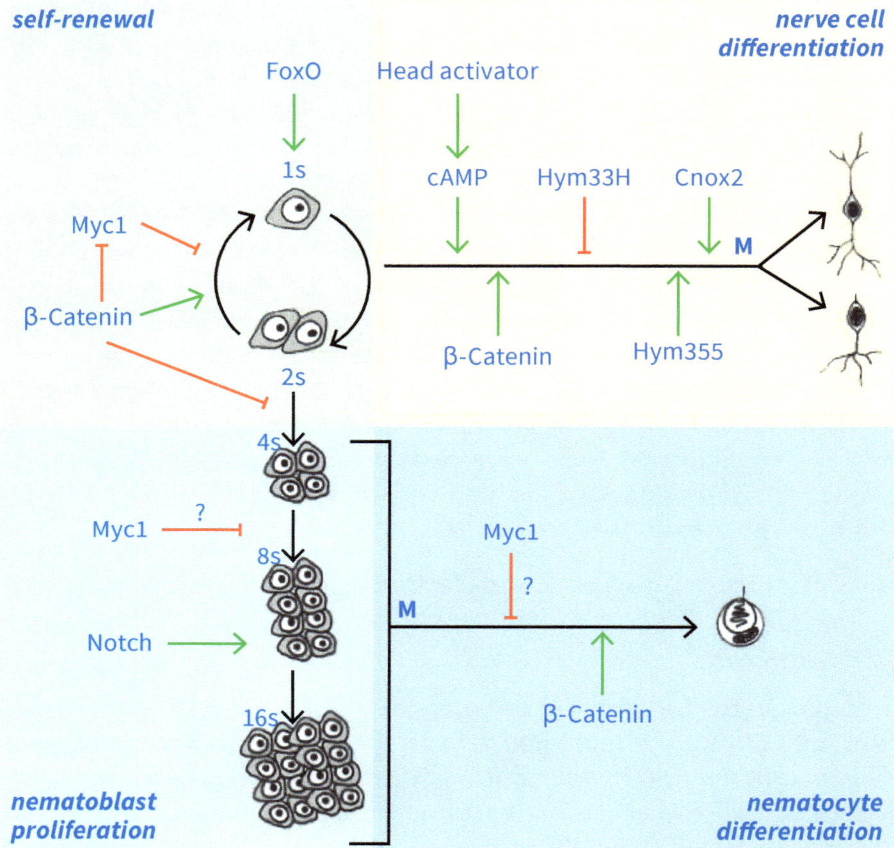

Figure 3. Schematic of the known molecular regulators of self-renewal and differentiation in the *Hydra* interstitial cell lineage. Positive and negative regulators are depicted in green and red, respectively. The precise time of action of these factors along a differentiation trajectory is in most cases unknown. Thus, the depicted position does not necessarily represent the precise sequence of events. Source: Graphic by authors.

Among the four *myc* gene homologs identified in *Hydra*, a structural and biochemical characterization showed HyMyc1 and HyMyc2 to share high similarities with c-Myc from vertebrates (Hartl et al. 2010, 2014). RNA interference suggested an initially unexpected role of *hymyc1* to decrease ISC self-renewal (Figure 3; Table 1;

Ambrosone et al. 2012). The *hymyc1* promoter turned out to be a target of repression by Wnt/β-Catenin signaling (Hartl et al. 2019). Furthermore, transgenic animals overexpressing nuclear β-Catenin show an increase in ISC density, which indicates an overall stimulating effect of Wnt/β-Catenin signaling on ISC self-renewal possibly by a double-negative cascade via HyMyc1 (Figure 3; Table 1; Hartl et al. 2019). The precise function of HyMyc2 in ISC maintenance is not clear at the present. The *hymyc2* gene is expressed in all proliferating cells in *Hydra*, including ectodermal and endodermal epithelial cells as well as gamete precursor cells, and its mRNA is a maternal contribution to early embryos (Hartl et al. 2014). Furthermore, the *hymyc2* exon/intron structure and the encoded amino acid sequence are slightly more similar to vertebrate *c-myc* than *hymyc1*. Thus, we proposed that *hymyc2* represents the functional *Hydra* homolog of vertebrate *c-myc* with a corresponding active role in cell cycle regulation and stem cell self-renewal (Hartl et al. 2014). Functional interference experiments to test this view are ongoing in our lab.

There are putative ISC stemness factors proposed primarily based on gene expression data. The so-called germline-specific genes *nanos*, *vasa* and *piwi* are strongly expressed in male- and female-restricted stem cells in *Hydra*, but also in multi-potent ISCs (Table 1; Mochizuki et al. 2000, 2001; Juliano et al. 2014; for review see Nishimiya-Fujisawa and Kobayashi 2012). Activation of these genes in somatic adult stem cells was observed in other sexually and asexually reproducing taxa such as Porifera, various other Cnidaria, Platyhelminthes, and Echinodermata, indicating that there is no clear soma-germline boundary in these species and that these genes contribute to maintaining adult stem cell multipotency (for review see Juliano and Wessel 2010). A recent in-depth *Hydra* single-cell transcriptome analysis identified only a single marker gene expressed specifically in the putative multipotent ISCs (Siebert et al. 2019). This new factor, Hy-icell1, has no homolog in the DNA data bases, and its function in ISCs is unknown at the present. Notably, single-cell transcriptomics failed to identify a specific set of ISC stemness factors (Siebert et al. 2019). Hence, it was argued that ISCs are largely defined by the absence of activity of cell type-specific differentiation genes. How this lack of differentiation activity is maintained is not known, but an understanding of its underlying molecular regulation will be essential to understand the potency and longevity of *Hydra* stem cells.

3.1.1. Nerve Cell Differentiation

Hydra exhibits a rather simple nervous system with distinct sub-clusters building three non-overlapping networks (Dupre and Yuste 2017; Siebert et al. 2019). ISC commitment for neuronal differentiation occurs in the late S-phase in the gastric region (Venugopal and David 1981). Committed nerve precursors then either migrate towards the head and foot areas or stay in the gastric region in order to support the growing neuronal network along the body column and the replacement of neurons lost at the terminal ends. Nerve precursors mostly undergo one terminal mitosis to yield two differentiated neurons; very few undergo one or two more divisions to yield four or eight neurons (Heimfeld and Bode 1985; Hager and David 1997; Technau and Holstein 1996).

Several studies revealed an unexpected action of small peptide signaling in nerve cell differentiation (Figure 3; Table 1). Intact polyps treated with either purified or synthetic Head Activator peptide (pEPPGGSKVILF) showed a significantly increased number of nerve cells throughout the body column (Holstein et al. 1986). This effect can be mimicked by cAMP, indicating that cAMP acts as second messenger in this cascade (Fenger et al. 1994). The gene encoding the Head Activator peptide sequence has yet not been found in the *Hydra* genome. Its origin thus remains elusive. Two other small peptides regulate neurogenesis in *Hydra*. Hym-355, a neuropeptide secreted along the entire body column (FPQSFLPRGa), enhances nerve cell differentiation in the early commitment phase, and treatment with this peptide leads to substantially higher numbers of nerve cells in the polyp (Takahashi et al. 2000). The epitheliopeptide Hym-33H (AALPW) counteracts nerve cell differentiation most likely also acting on early precursors (Takahashi et al. 1997).

Khalturin et al. (2007) showed that the differentiation of Hym-355-positive neurons is stimulated in *Hydra* treated with the β-Catenin-stabilizing small molecule Alsterpaullone. Wnt/β-Catenin signaling is also strongly elevated in transgenic polyps, in which a β-Catenin-GFP fusion protein is driven by the *actin1* promoter. The density of neurons in the body column of these transgenic polyps is more than twice as high as in controls (Hobmayer lab, unpublished data), clearly supporting the view that Wnt/β-Catenin signaling stimulates neurogenesis in *Hydra* (Figure 3; Table 1). Neurogenesis in the head of *Hydra* polyps is suppressed by the knock-down of the transcription factor Cnox-2, as shown by using RNA interference (Figure 3; Table 1). In addition, qPCR-data indicated that cnox-2 is an upstream regulator of the nerve cell marker genes *pradl-a*, *gsc*, *RFamide-B* and *hyCOUP-TF* (Miljkovic-Licina et al. 2007). Single-cell transcriptomics demonstrated that a few other transcription factors are specifically expressed in neuronal progenitor cells (HvSoxC) and in the population of precursors common to nerve and gland cell differentiation (Myb and Myc3; Table 1; Siebert et al. 2019).

3.1.2. Gland Cell Differentiation

Endodermal gland cells are differentiation products of ISCs, but they retain a capacity for proliferation in their differentiated state. Thus, very few gland cells are produced anew, while most of them reproduce by cell division (Schmidt and David 1986). Gland cells actually represent a set of different sub-populations distributed along the body column (Siebert et al. 2019), and they have been shown to change their phenotype by trans-differentiation when they change their axial position following global tissue movement (Siebert et al. 2008). Gland cell differentiation is not well studied in terms of regulatory molecular factors. As described above, gland cells seem to share a common precursor with differentiating neurons based on specific activation of *myb* and *myc3* genes (Table 1; Siebert et al. 2019).

3.1.3. Nematocyte Differentiation

ISCs committed for nematocyte differentiation undergo two to four steps of proliferation, resulting in cell nests of proliferating nematoblasts with a nest size of 4s to 16s. Nematoblast nests then undergo terminal mitosis and thereby form nests of differentiating nematocytes with a nest size of 8s to 32s (David and Challoner 1974). During the differentiation phase, every nest cell builds a fully functional nematocyte capsule. Upon completion of this process, nests break up, and individual and fully mature nematocytes start to migrate. Finally, nematocytes are taken up and mounted at the apical membrane in ectodermal epithelial cells, mostly in ectodermal battery cell complexes in the tentacles.

Wnt/β-catenin signaling may have two modes of action in this pathway (Figure 3; Table 1). First, the total number of nests of proliferating nematoblasts in the body column of a polyp is strongly reduced upon nuclear activation of β-catenin in Alsterpaullone-treated and in β-catenin transgenic polyps (Figure 3; Khalturin et al. 2007); Hobmayer lab, unpublished data). Second, post-mitotic differentiation of nematocytes seems to be strongly enhanced by Wnt/β-catenin signaling based on the observation that differentiating nests expressing the marker gene *nb035* disappear, whereas mature nematocytes expressing the marker gene *nb031* strongly increase in numbers upon Alsterpaullone treatment (Khalturin et al. 2007). In addition, Myc1 seems to be involved. The down-regulation of *myc1* mRNAs by RNA interference resulted in an increase in nests of proliferating nematoblasts and in an increased ratio of mature nematocytes/battery cells in the tentacles (Figure 3; Table 1; Ambrosone et al. 2012). Equivalent results were obtained after treatment of *Hydra* polyps with the c-Myc-specific small-molecule inhibitor 10058-F4 (Ambrosone et al. 2012).

Finally, Notch signaling has been reported to promote nematocyte differentiation in the early post-mitotic phase possibly by acting in nematoblast nests shortly before terminal mitosis (Figure 3; Table 1). This was shown using the small-molecule inhibitor DAPT, which inhibits gamma-secretase and therefore prevents downstream Notch signaling. Treating *Hydra* polyps with DAPT inhibited the expression of the nematocyte marker genes *nb031* and *nb035*, strongly reduced the numbers of nematocyte nests with small vacuoles, and it forced differentiating nematocytes to undergo programmed cell death (Käsbauer et al. 2007; Khalturin et al. 2007).

4. Conclusions

Deciphering the molecular regulation of decision making in the *Hydra* IC lineage is at its beginning. A detailed single-cell transcriptomic atlas and advanced methods for stable transgenesis and genetic knock-down join the available molecular tool kit, including a genome annotated at the chromosome level. A large set of sophisticated methods allows the analysis of *Hydra* ISCs and their lineage products at the cellular level. Due to its simple body plan, all this can be carried out in vivo in fully intact polyps. Furthermore, ISC behavior is also studied in related marine hydrozoans such as *Hydractinia* and *Clytia*. The action of the germline factors Nanos, Vasa and Piwi, as well as Myc function, may be conserved. However, there are unexpected differences among the hydrozoan polyp models. Polynem, a POU domain transcription factor more closely related to vertebrate Oct4 than any *Hydra* Pou transcription factor seems to keep cells undifferentiated in *Hydractinia* and is able to induce neoplasia when overactivated (Millane et al. 2011). While AP2 is a core activator for germ cell formation in *Hydractinia* and higher animals (DuBuc et al. 2020), single-cell transcriptome data do not support this role in *Hydra*. *Clytia* Sox proteins, in contrast to those in *Hydra*, seem to affect the balance between self-renewing stem cells and cells undergoing differentiation (Jager et al. 2011). Thus, there is obvious within-class diversity in the action of stem cell and differentiation factors among different hydrozoans, and each lineage may have evolved a stemness regulation adapted to its specific life cycle needs.

What are the imminent questions to be resolved? It is clear that we do not understand most of the key issues well enough. How is stemness and the non-differentiation state of ISCs in *Hydra* and other hydrozoan polyps defined at the molecular level? What are the molecular signals acting in direct niche interactions? How do long-range feedback mechanisms work? Finally, what roles do post-translational modifiers play, and which types of epigenetic mechanisms affect stem cell maintenance and differentiation? Isolating these regulatory factors will clearly contribute to a more general understanding of adult stem cell dynamics and decision making in the common ancestor of Bilateria, and more generally to the evolutionary ancestry of cellular plasticity, regeneration, and ageing.

Author Contributions: Conceptualization, B.H., M.L., M.A. and M.K.; formal analysis, M.L., M.A., M.K., W.S. and B.H.; writing—original draft preparation, M.L., M.A., M.K. and B.H.; writing—review and editing, M.L., M.A., M.K., W.S. and B.H.; funding acquisition, B.H.

Funding: This research was supported by the EC H2020 Marie Sklodowska-Curie COFUND research grant No 847681 "ARDRE—Ageing, Regeneration and Drug Research" and the EU-COST network 16203 "MARISTEM—Stem Cells of marine/aquatic invertebrates: From basic research to innovative applications".

Acknowledgments: The authors thank Lena Seppi, Natalie Kolb for technical support and animal culture, and Kevin Grüner for his help with electron microscopy.

Conflicts of Interest: The authors declare no conflict of interests.

Abbreviations

ISC	interstitial stem cell
scRNAseq	single-cell RNA Sequencing
smi	small molecular inhibitor
WISH	whole-mount in situ hybridization

References

Ambrosone, Alfredo, Valentina Marchesano, Angela Tino, Bert Hobmayer, and Claudia Tortiglione. 2012. *Hymyc1* downregulation promotes stem cell proliferation in *Hydra vulgaris*. *PLoS ONE* 7: e30660. [CrossRef] [PubMed]

Augustin, René, André Franke, Konstantin Khalturin, Rainer Kiko, Stefan Siebert, Georg Hemmrich, and Thomas C. G. Bosch. 2006. Dickkopf-related genes are components of the positional value gradient in *Hydra*. *Developmental Biology* 296: 62–70. [CrossRef] [PubMed]

Augustin, René, Katja Schröder, Andréa P. Murillo Rincón, Sébastian Fraune, Friederike Anton-Erxleben, Eva-Maria Herbst, Jörg Wittlieb, Martin Schwentner, Joachim Grötzinger, Trudy M. Wassenaar, and et al. 2017. A secreted antibacterial neuropeptide shapes the microbiome of *Hydra*. *Nature Communications* 8: 698. [CrossRef] [PubMed]

Bode, Hans R. 1996. The interstitial cell lineage of hydra: A stem cell system that arose early in evolution. *Journal of Cell Science* 109: 1155–64. [CrossRef] [PubMed]

Bode, Hans R., Kristine M. Flick, and G. Scott Smith. 1976. Regulation of interstitial cell differentiation in *Hydra attenuata*. I. Homeostatic control of interstitial cell population size. *Journal of Cell Science* 20: 29–46. [CrossRef] [PubMed]

Boehm, Anna-Marei, and Thomas C. G. Bosch. 2012. Migration of multipotent interstitial stem cells in *Hydra*. *Zoology* 115: 275–82. [CrossRef]

Boehm, Anna-Marei, Konstantin Khalturin, Friederike Anton-Erxleben, Georg Hemmrich, Ulrich C. Klostermeier, Javier A. Lopez-Quintero, Hans-Heinrich Oberg, Malte Puchert, Philip Rosenstiel, Jörg Wittlieb, and et al. 2012. FoxO is a critical regulator of stem cell maintenance in immortal *Hydra*. *Proceedings of the National Academy of Sciences of the United States of America* 109: 19697–702. [CrossRef]

Bosch, Thomas C. G. 2008. Stem cells in immortal Hydra. In *Stem Cells*. Edited by Bosch TCG. Heidelberg: Springer, pp. 37–58.

Bosch, Thomas C. G., and Charles N. David. 1987. Stem cells of *Hydra magnipapillata* can differentiate into somatic cells and germ line cells. *Developmental Biology* 121: 182–191. [CrossRef]

Bosch, Thomas C. G., and Charles N. David. 1990. Cloned interstitial stem cells grow as contiguous patches in *Hydra*. *Developmental Biology* 138: 513–15. [CrossRef]

Bosch, Thomas C. G., Friederike Anton-Erxleben, Georg Hemmrich, and Konstantin Khalturin. 2010. The *Hydra* polyp: Nothing but an active stem cell community. *Development, Growth & Differentiation* 52: 15–25.

Bosch, Thomas C. G., Rebecca Rollbühler, Birgit Scheider, and Charles N. David. 1991. Role of the cellular environment in interstitial stem cell proliferation in *Hydra*. *Roux's Archives of Developmental Biology* 200: 269–76. [CrossRef] [PubMed]

Broun, Mariya, Lydia Gee, Beate Reinhardt, and Hans R. Bode. 2005. Formation of the head organizer in hydra involves the canonical Wnt pathway. *Development* 132: 2907–16. [CrossRef] [PubMed]

Buzgariu, Wanda, Sarah Al Haddad, Szymon Tomczyk, Yvan Wenger, and Brigitte Galliot. 2015. Multi-functionality and plasticity characterize epithelial cells in *Hydra*. *Tissue Barriers* 3: e1068908. [CrossRef] [PubMed]

Campbell, Richard D., and Charles N. David. 1974. Cell cycle kinetics and development of *Hydra attenuata*. II. Interstitial cells. *Journal of Cell Science* 16: 349–58. [CrossRef] [PubMed]

Chapman, Jarrod A., Ewen F. Kirkness, Oleg Simakov, Steven E. Hampson, Therese Mitros, Thomas Weinmaier, Thomas Rattei, Prakash G. Balasubramanian, Jon Borman, and Dana Busam. 2010. The dynamic genome of *Hydra*. *Nature* 464: 592–96. [CrossRef] [PubMed]

David, Charles N. 2012. Interstitial stem cells in *Hydra*: Multipotency and decision-making. *International Journal of Developmental Biology* 56: 489–97. [CrossRef]

David, Charles N., and Alfred Gierer. 1974. Cell cycle kinetics and development of *Hydra attenuata*. III. Nerve and nematocyte differentiation. *Journal of Cell Science* 16: 359–75. [CrossRef]

David, Charles N., and Diane Challoner. 1974. Distribution of interstitial cells and differentiating nematocytes in nests in *Hydra attenuata*. *American Zoologist* 14: 537–42. [CrossRef]

David, Charles N., and Harry MacWilliams. 1978. Regulation of the self-renewal probability in *Hydra* stem cell clones. *Proceedings of the National Academy of Sciences of the United States of America* 75: 886–90. [CrossRef]

David, Charles N., and Ida Plotnick. 1980. Distribution of interstitial stem cells in hydra. *Developmental Biology* 76: 175–84. [CrossRef]

David, Charles N., and Susan Murphy. 1977. Characterization of interstitial stem cells in hydra by cloning. *Developmental Biology* 58: 372–383. [CrossRef]

David, Charles N., Thomas C. G. Bosch, Engelbert Hobmayer, Thomas W. Holstein, and Tobias Schmidt. 1987. Interstitial stem cells in Hydra. In *Genetic Regulation of Development*. Edited by William F. Loomis. New York: Alan R. Liss, pp. 189–208.

David, Charles N., Toshitaka Fujisawa, and Thomas C. G. Bosch. 1991. Interstitial stem cell proliferation in *Hydra*: Evidence for strain-specific regulatory signals. *Developmental Biology* 148: 501–7. [CrossRef]

DuBuc, Timothy Q., Christine E. Schnitzler, Eleni Chrysostomou, Emma T. McMahon, Febrimarsa, James M. Gahan, Tara Buggie, Sebastian G. Gornik, Shirley Hanley, Sofia N. Barreira, and et al. 2020. Transcription factor AP2 controls cnidarian germ cell induction. *Science* 367: 757–62. [CrossRef] [PubMed]

Dupre, Christophe, and Rafael Yuste. 2017. Non-overlapping neural networks in *Hydra vulgaris*. *Current Biology* 27: 1085–97. [CrossRef] [PubMed]

Fedders, Henning, René Augustin, and Thomas C. G. Bosch. 2004. A Dickkopf-3-related gene is expressed in differentiating nematocytes in the basal metazoan *Hydra*. *Development Genes and Evolution* 214: 72–80. [CrossRef] [PubMed]

Fenger, Ursula, Michael Hofmann, Brigitte Galliot, and H. Chica Schaller. 1994. The role of the cAMP pathway in mediating the effect of head activator on nerve-cell determination and differentiation in hydra. *Mechanisms of Development* 47: 115–25. [CrossRef]

Fujisawa, Toshitaka. 1992. Homeostatic recovery of interstitial cell populations in *Hydra*. *Developmental Biology* 150: 185–92. [CrossRef]

Gee, Lydia, Julia Hartig, Lee Law, Jörg Wittlieb, Konstantin Khalturin, Thomas C. G. Bosch, and Hans R. Bode. 2010. Beta-catenin plays a central role in setting up the head organizer in *Hydra*. *Developmental Biology* 340: 116–24. [CrossRef]

Genikhovich, Griogry, Ulrich Kürn, Georg Hemmrich, and Thomas C. G. Bosch. 2006. Discovery of genes expressed in *Hydra* embryogenesis. *Developmental Biology* 289: 466–81. [CrossRef]

Govindasamy, Niraimathi, Supriya Murthy, and Yashoda Ghanekar. 2014. Slow-cycling stem cells in hydra contribute to head regeneration. *Biology Open* 3: 1236–44. [CrossRef]

Grens, Ann, Elizabeth Mason, J. Lawrence Marsh, and Hans R. Bode. 1995. Evolutionary conservation of a cell fate specification gene: The *Hydra* achaete-scute homolog has proneural activity in *Drosophila*. *Development* 121: 4027–35. [CrossRef] [PubMed]

Guder, Corina, Sonia Pinho, Tanju G. Nacak, Heiko A. Schmidt, Bert Hobmayer, Christof Niehrs, and Thomas W. Holstein. 2006. An ancient Wnt-Dickkopf antagonism in *Hydra*. *Development* 133: 901–11. [CrossRef] [PubMed]

Hager, Gundel, and Charles N. David. 1997. Pattern of differentiated nerve cells in hydra is determined by precursor migration. *Development* 124: 569–76. [CrossRef] [PubMed]

Hartl, Markus, Anna-Maria Mitterstiller, Taras Valovka, Kathrin Breuker, Bert Hobmayer, and Klaus Bister. 2010. Stem cell-specific activation of an ancestral *myc* protooncogene with conserved basic functions in the early metazoan *Hydra*. *Proceedings of the National Academy of Sciences of the United States of America* 107: 4051–56. [CrossRef]

Hartl, Markus, Stella Glasauer, Sabine Gufler, Andrea Raffeiner, Kane Puglisi, Kathrin Breuker, Klaus Bister, and Bert Hobmayer. 2019. Differential regulation of myc homologs by Wnt/β-Catenin signaling in the early metazoan *Hydra*. *The FEBS Journal* 286: 2295–310. [CrossRef]

Hartl, Markus, Stella Glasauer, Taras Valovka, Kathrin Breuker, Bert Hobmayer, and Klaus Bister. 2014. Hydra myc2, a unique pre-bilaterian member of the myc gene family, is activated in cell proliferation and gametogenesis. *Biology Open* 3: 397–407. [CrossRef]

Heimfeld, Shelly, and Hans R. Bode. 1985. Growth regulation of the interstitial cell proliferation in hydra. I. Evidence for global control by nerve cells in the head. *Developmental Biology* 110: 297–307. [CrossRef]

Hemmrich, Georg, Konstantin Khalturin, Anna-Marei Boehm, Malte Puchert, Friederike Anton-Erxleben, Jörg Wittlieb, Ulrich C. Klostermeier, Philip Rosenstiel, Hans-Heinrich Oberg, Tomislav Domazet-Lošo, and et al. 2012. Molecular signatures of the three stem cell lineages in *Hydra* and the emergence of stem cell function at the base of multicellularity. *Molecular Biology and Evolution* 29: 3267–80. [CrossRef]

Hobmayer, Bert, Fabian Rentzsch, Kerstin Kuhn, Christoph M. Happel, Christoph Cramer von Laue, Petra Snyder, Ute Rothbächer, and Thomas W. Holstein. 2000. Wnt signalling molecules act in axis formation in the diploblastic metazoan *Hydra*. *Nature* 407: 186–89. [CrossRef]

Hobmayer, Bert, Marcell Jenewein, Dominik Eder, Stella Glasauer, Sabine Gufler, Markus Hartl, and Willi Salvenmoser. 2012. Stemness in *Hydra*—A current perspective. *International Journal of Developmental Biology* 56: 509–17. [CrossRef]

Holstein, Thomas W., Chica Schaller, and Charles N. David. 1986. Nerve cell differentiation in hydra requires two signals. *Developmental Biology* 115: 9–17. [CrossRef]

Jager, Muriel, Eric Quéinnec, Hervé Le Guyader, and Michael Manuel. 2011. Multiple *sox* genes are expressed in stem cells or in differentiating neuro-sensory cells in the hydrozoan *Clytia hemisphaerica*. *EvoDevo* 2: 1–17. [CrossRef] [PubMed]

Juliano, Celina E., Adrian Reich, Na Liu, Jessica Götzfried, Mei Zhong, Selen Uman, Robert A. Reenan, Gary M. Wessel, Robert E. Steele, and Haifan Lin. 2014. PIWI proteins and PIWI-interacting RNAs function in *Hydra* somatic stem cells. *Proceedings of the National Academy of Sciences of the United States of America* 111: 337–42. [CrossRef] [PubMed]

Juliano, Celina E., and Gary Wessel. 2010. Versatile germline genes. *Science* 329: 640–41. [CrossRef]

Käsbauer, Tina, Par Tow, Olga Alexandrova, Charles N. David, Ekaterina Dall'Armi, Andrea Staudigl, Beate Stiening, and Angelika Böttger. 2007. The Notch signaling pathway in the cnidarian *Hydra*. *Developmental Biology* 303: 376–90. [CrossRef]

Khalturin, Konstantin, Friederike Anton-Erxleben, Sabine Milde, Christine Plötz, Jörg Wittlieb, Georg Hemmrich, and Thomas C. G. Bosch. 2007. Transgenic stem cells in Hydra reveal an early evolutionary origin for key elements controlling self-renewal and differentiation. *Developmental Biology* 309: 32–44. [CrossRef]

Lindgens, Dirk, Thomas W. Holstein, and Ulrich Technau. 2004. Hyzic, the *Hydra* homolog of the zic/odd-paired gene, is involved in the early specification of the sensory nematocytes. *Development* 131: 191–201. [CrossRef]

Miljkovic-Licina, Marijana, Simona Chera, Luiza Ghila, and Brigitte Galliot. 2007. Head regeneration in wild-type hydra requires de novo neurogenesis. *Development* 134: 1191–201. [CrossRef]

Millane, R. Cathriona, Justyna Kanska, David J. Duffy, Cathal Seoighe, Stephen Cunningham, Günter Plickert, and Uri Frank. 2011. Induced stem cell neoplasia in a cnidarian by ectopic expression of a POU domain transcription factor. *Development* 138: 2429–39. [CrossRef]

Mochizuki, Kazufumi, Chiemi Nishimiya-Fujisawa, and Toshitaka Fujisawa. 2001. Universal occurrence of the *vasa-related* genes among metazoans and their germline expression in *Hydra*. *Development Genes and Evolution* 211: 299–308. [CrossRef]

Mochizuki, Kazufumi, Hiroko Sano, Satoru Kobayashi, Chiemi Nishimiya-Fujisawa, and Toshitaka Fujisawa. 2000. Expression and evolutionary conservation of *nanos-related* genes in *Hydra*. *Development Genes and Evolution* 210: 591–602. [CrossRef] [PubMed]

Müller, Werner A. 1967. Differenzierungspotenzen und Geschlechtsstabilität der I-Zellen von *Hydractinia echinata*. *Wilhelm Roux' Archiv für Entwicklungsmechanik der Organismen* 159: 412–32. [CrossRef] [PubMed]

Nishimiya-Fujisawa, Chiemi, and Satoru Kobayashi. 2012. Germline stem cells and sex determination in *Hydra*. *International Journal of Developmental Biology* 56: 499–508. [CrossRef] [PubMed]

Nishimiya-Fujisawa, Chiemi, and Tsutomu Sugiyama. 1993. Genetic analysis of developmental mechanisms in *Hydra*. XX. Cloning of interstitial stem cells restricted to the sperm differentiation pathway in *Hydra magnipapillata*. *Developmental Biology* 157: 1–9. [CrossRef]

Plickert, Günter, Uri Frank, and Werner A. Müller. 2012. *Hydractinia*, a pioneering model for stem cell biology and reprogramming somatic cells to pluripotency. *International Journal of Developmental Biology* 56: 519–34. [CrossRef]

Schmidt, Tobias, and Charles N. David. 1986. Gland cells in *Hydra*: Cell cycle kinetics and development. *Journal of Cell Science* 85: 197–215. [CrossRef]

Siebert, Stefan, Jeffrey A. Farrell, Jack F. Cazet, Yashodara Abeykoon, Abby S. Primack, Christine E. Schnitzler, and Celina E. Juliano. 2019. Stem cell differentiation trajectories in *Hydra* resolved at single-cell resolution. *Science* 365: eaav9314. [CrossRef]

Siebert, Stefan, Friederike Anton-Erxleben, and Thomas C. G. Bosch. 2008. Cell type complexity in the basal metazoan *Hydra* is maintained by both stem cell based mechanisms and transdifferentiation. *Developmental Biology* 313: 13–24. [CrossRef]

Sproull, Frederick, and Charles N. David. 1979. Stem cell growth and differentiation in *Hydra attenuata*. I. Regulation of the self-renewal probability in multiclone aggregates. *Journal of Cell Science* 38: 155–69. [CrossRef]

Sun, Shixiang, Ryan R. White, Kathleen E. Fischer, Zhengdong Zhang, Steven N. Austad, and Jan Vijg. 2020. Inducible aging in *Hydra oligactis* implicates sexual reproduction, loss of stem cells, and genome maintenance as major pathways. *GeroScience* 42: 1119–32. [CrossRef]

Takahashi, Kazutoshi, and Shinya Yamanaka. 2006. Induction of pluripotent stem cells from mouse embryonic and adult fibroblast cultures by defined factors. *Cell* 126: 663–76. [CrossRef] [PubMed]

Takahashi, Toshio, Osamu Koizumi, Yuki Ariura, Anna Romanovitch, Thomas C. Bosch, Yoshitaka Kobayakawa, Shirou Mohri, Hans R. Bode, Seungshic Yum, Masayuki Hatta, and et al. 2000. A novel neuropeptide, Hym-355, positively regulates neuron differentiation in *Hydra*. *Development* 127: 997–1005. [CrossRef] [PubMed]

Takahashi, Toshio, Yojiro Muneoka, Jan Lohmann, Maria S. Lopez de Haro, Gaby Solleder, Thomas C. G. Bosch, Charles N. David, Hans R. Bode, Osamu Koizumi, Hiroshi Shimizu, and et al. 1997. Systematic isolation of peptide signal molecules regulating development in hydra: LWamide and PW families. *Proceedings of the National Academy of Sciences of the United States of America* 94: 1241–46. [CrossRef] [PubMed]

Tardent, Pierre. 1954. Axiale Verteilungs-Gradienten der interstitiellen Zellen bei *Hydra* und *Tubularia* und ihre Bedeutung für die Regeneration. *Wilhelm Roux' Archiv für Entwicklungsmechanik der Organismen* 146: 593–643. [CrossRef] [PubMed]

Technau, Ulrich, and Thomas W. Holstein. 1996. Phenotypic maturation of neurons and continuous precursor migration in the formation of the peduncle nerve net in *Hydra*. *Developmental Biology* 177: 599–615. [CrossRef] [PubMed]

Teefy, Bryan B., Stefan Siebert, Jack F. Cazet, Haifan Lin, and Celina E. Juliano. 2020. PIWI-piRNA pathway-mediated transposable element repression in *Hydra* somatic stem cells. *RNA* 26: 550–63. [CrossRef] [PubMed]

Tomczyk, Szymon, Nenad Suknovic, Quentin Schenkelaars, Yvan Wenger, Kazadi Ekundayo, Wanda Buzgariu, Christoph Bauer, Kathleen Fischer, Steven Austad, and Brigitte Galliot. 2020. Deficient autophagy in epithelial stem cells drives aging in the freshwater cnidarian *Hydra*. *Development* 147: dev177840. [CrossRef]

Venugopal, Gopalan, and Charles N. David. 1981. Nerve commitment in *Hydra*: II. Localization of commitment in S Phase. *Developmental Biology* 83: 361–65. [CrossRef]

Watanabe, Hiroshi, Van Thanh Hoang, Robert Mättner, and Thomas W. Holstein. 2009. Immortality and the base of multicellular life: Lessons from cnidarian stem cells. *Sem Cell Developmental Biology* 20: 1114–25. [CrossRef]

Weismann, August. 1883. *Die Entstehung der Sexualzellen bei den Hydromedusen.* Jena: Gustav Fischer Verlag.

Wittlieb, Jörg, Konstantin Khalturin, Jan U. Lohmann, Friederike Anton-Erxleben, and Thomas C. G. Bosch. 2006. Transgenic *Hydra* allow in vivo tracking of individual stem cells during morphogenesis. *Proceedings of the National Academy of Sciences of the United States of America* 103: 6208–11. [CrossRef]

Planarian Stem Cells: Pluripotency Maintenance and Fate Determination

Gaetana Gambino, Leonardo Rossi and Alessandra Salvetti

Abstract: Basic molecular mechanisms that orchestrate stem cell maintenance and fate are widely conserved across kingdoms, allowing for cross-species studies from simple model systems to mammals. In this context, planarians offer extraordinary possibilities containing a reservoir of experimentally accessible adult pluripotent stem cells, "the neoblasts". Indeed, in vivo reverse genetic manipulation of crucial neoblast regulators allows a fine study of adult stem cell fate in their natural environment. Recent extensive transcriptomics analysis revealed that planarian neoblasts are a widely heterogeneous population including clonogenic and lineage-committed stem cells, constituting a dynamic compartment that talks with differentiated tissue for proper physiological homeostasis and tissue regeneration. In this chapter, we review, in a chronological perspective, the most recent findings in the comprehension of neoblast biology, including their embryonic origin, and compare the most accredited models of pluripotency maintenance and fate determination.

1. Introduction

You can hurt them, cut them, or even decapitate them, they will rapidly heal and regrow. This is not a mythological tale, nor is it a sentence of a fantasy book; this is the truth for regenerating organisms, especially planarians, flatworms of the phylum Platyhelminthes (Box 1). The ability to reconstitute missing body parts through the formation of a transient mass of undifferentiated cells, i.e., the epimorphic regeneration, relies on the coexistence of three fundamental factors: (i) a pluripotent reservoir of stem cells that will produce the "bricks" to form the blastema mass; (ii) a sophisticated molecular machinery to address undifferentiated cell fate and de novo tissue morphogenesis; (iii) a permissive inflammatory status that favors regeneration versus scarring. All these features enable planarians to rebuild an entire organism with perfect novel organs from almost any tiny piece of their body. The presence of a pluripotent reservoir of stem cells and active body patterning cues allows for continuous turn-over of specialized cells and tissue homeostasis also in intact organisms, thus making planarians virtually immortal.

"Planarian" is the generic name applied to free-living members of the order Tricladida of the phylum Platyhelminthes (the flatworms) (Sluys et al. 2009). A new higher classification of planarian flatworms (Platyhelminthes, Tricladida). Planarians are unsegmented acoelomates included in the Lophotrochozoan clade with bilateral symmetry and possess all three germ layers. They have a clear anteroposterior polarity with a head and a tail and are usually dorso-ventrally flattened. A mesenchyma intercalates among the various organs. The nervous system is composed of two cephalic ganglia connected to various sensory structures of the anterior part of the head and to two ventral longitudinal nerve cords, linked by commissural neurons and connected to a submuscular plexus that runs beneath the body wall musculature. Among sensory structures, the planarian eye is composed of two cell types, pigment cells, and photoreceptors. Pigment cells organize to form a cup-shaped structure, photoreceptors located at the opening of the pigment cup project their dendrites into the cup and their axons to the cephalic ganglia. Photoreceptor dendrites terminate with multiple microvilli-like structures called rhabdomeres and contain the photoreceptive molecule opsin.

The muscular system is organized into longitudinal, diagonal, and circular muscle fibers. In the midline of the animal, there is a muscular extensible organ, called the pharynx, connected to the digestive system, composed of three gut branches—one directed in the anterior part of the animal and two toward the tail region. The excretory system includes flame cells that remove unwanted liquids from the body by passing them through ducts, which lead to excretory pores on the dorsal surface of the body. The main nitrogenous waste product is soluble ammonia; thus, they are referred to as ammoniotelic. They lack circulatory, respiratory, and skeletal structures.

Freshwater planarians reproduce either asexually by transverse fission, generating two identical organisms (clones) or sexually as cross-fertilizing hermaphrodites.

If a planaria is cut, shortly after the amputation an unpigmented outgrowth, named the regenerative blastema, is observed near the site of injury, and cells within this structure will differentiate and spatially reorganize to restore the preexisting missing body part. Normal body proportions are attained after 3–4 weeks of regeneration. Freshwater planarians are easy and cheap to maintain in the laboratory and several species are used as model systems for cellular and molecular biology studies, in particular *Schmidtea Mediterranea* and *Dugesia japonica* species, belonging to the sister genus *Schmidtea and Dugesia*, respectively. Both species have excellent regenerative abilities, and clonal strains originating from single animals are used. Results from studies using either *S. mediterranea* or *D. japonica* are assumed comparable also in light of the preliminary *D. japonica* cell type atlas, which demonstrates that the two species share similar cell types in relatively comparable abundances (García Castro et al. 2021). Gene names in the planarian literature carry a prefix designating the species (i.e., Smed for *S. mediterranea* and Dj for *D. japonica*). Additional species might offer further features useful to understand complex patterning phenomena such as the *Dendrocoelum lacteum*, which is a regeneration-deficient planarian species in that its tailpieces are unable to regenerate a head and ultimately die. Indeed, downregulation of *Dlac-β-catenin-1*, the Wnt signal transducer, enables tailpieces to fully regenerate functional heads, rescuing *D. lacteum's* regeneration defect (Liu et al. 2013). An integrated web resource of data and tools to mine Planarian biology, PlanMine database: http://planmine.mpi-cbg.de/ (accessed on 18 July 2021)) has been created collecting all transcriptomics and genomic data and allowing for comparative analysis of flatworm biology (Rozanski et al. 2019).

From a research point of view, planarians represent a *"laboratory platform"* in which the most complicated cellular and developmental phenomena are continuously recapitulated in an in vivo context, thus offering the possibility to gain information about molecular regulatory mechanisms, cell-to-cell cross-talk, epigenetic phenomenon, ECM–cell interactions, and morphogenesis of tissues and organs (Figure 1).

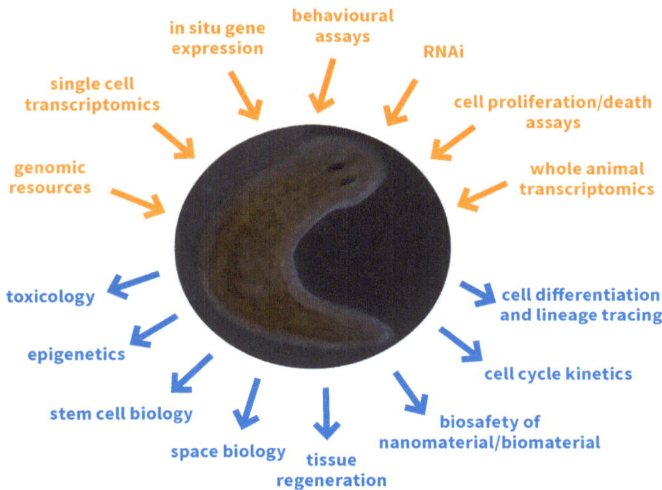

Figure 1. Scheme depicting the potential uses of the planarian model system. Orange arrows indicate methodological tools, while green arrows indicate some research fields. Source: Graphic by authors.

For this reason, this model system accompanies scientists since early 1900 up to nowadays despite, as perfectly described by Jaume Baguñà (2019) in his personal commentary, with "evident stumbling blocks due to hidden complexity and technical unfriendliness of planarians which explain why this model lagged, and still lags, behind other regeneration models and why and how they baffled and still baffle us" (ibid., p. 9). Today, most of the technical challenges have been overcome: interactive genomic/ transcriptomics databases are available (PlanMine database: http://planmine.mpi-cbg.de/ (accessed on 18 July 2021)) (Rozanski et al. 2019), even in the form of a single-cell atlas (Available online: https://digiworm.wi.mit.edu/ (accessed on 18 July 2021)) (Fincher et al. 2018; Zeng et al. 2018); RNAi is a widely used and validated technique (Sánchez and Newmark 1999); molecular markers for most of the differentiated tissues have been identified; protocols for several cellular assays have been successfully developed. Thus, in the last decade, molecular research in the planarian field jumped forward, revealing an extraordinary articulated cellular system in which multiple different specialized cell types, several postmitotic progenitors, and a complex population of stem cells, generally referred to as "the neoblasts", interact to orchestrate perfect physiological homeostasis and tissue regeneration program. Here, we review, in a chronological perspective, the most significant findings in the comprehension of neoblast biology, including their embryonic origin, and compare the most accredited models of pluripotency maintenance and fate determination.

2. The Clonogenic Neoblasts

All the neoblasts share a similar morphology and show the presence in their scanty undifferentiated cytoplasm (Figure 2A) of the so-called chromatoid bodies, electron-dense non-membrane-bound aggregates rich in RNA (Coward 1974). Requirements used nowadays to define a cell as a neoblast are widely described in Alessandra and Rossi (2019). Among them, the expression of PIWI-encoding genes (*smedwi-1* for *Schmidtea mediterranea* and *DjPiwiA* for *Dugesia japonica*) (Figure 2B), X-ray sensitivity (Figure 2C), and their proliferating activity (Figure 2D).

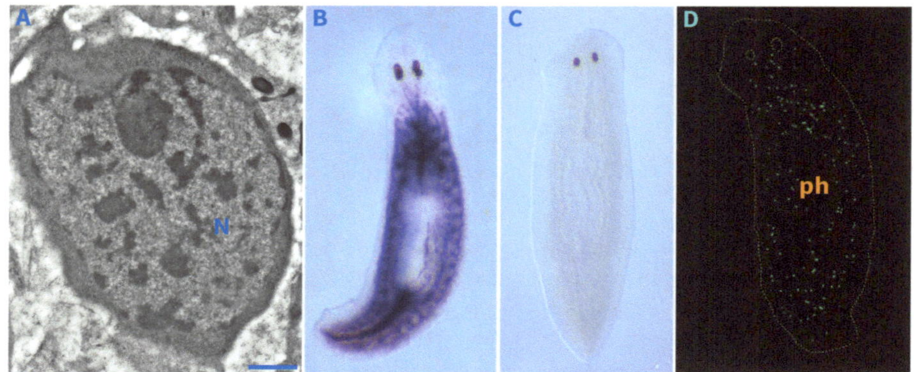

Figure 2. Neoblast features: (**A**) electron micrograph of a neoblast, mitochondria are highlighted in yellow. N—nucleus; (**B,C**) distribution of *DjPiwiA*-positive cells visualized by whole-mount in situ hybridization in wild-type (**B**) and in lethally irradiated (30 Gy) animals, 3 days after treatment; (**D**) phospho-H3 immunolabelling shows that proliferating cells are distributed throughout the entire planarian body with the exception of the pharynx and the anterior part of the head, especially behind the eyes. ph, pharynx. Scale bar corresponds to 800 nm in A and to 500 μm in (**B–D**). Source: Graphic by authors.

Despite these shared features, the neoblast population appears transcriptionally heterogeneous, as widely discussed in (Alessandra and Rossi 2019), and the discovery at the beginning of this century of the existence of some neoblasts able to resist low-dose X-ray treatment and repopulate the entire organism (Salvetti et al. 2009) opened the path toward the development of sophisticated assays owing to which some secrets of these extraordinary cells have been unveiled. A question that remained unsolved for several years was whether regeneration or tissue homeostasis was accomplished by pluripotent cells or by the cooperative activity of multiple lineage-committed cell types. In 2011, a breakthrough was achieved by an elegant paper of the Reddien's group (Wagner et al. 2011) in which by coupling ionizing radiation and single-cell transplantation, they demonstrated the existence of neoblasts that could give rise to progenies covering different germ layers and restore regeneration in lethally irradiated hosts. These pluripotent cells were defined as clonogenic neoblasts (the cNeoblasts).

2.1. From σ Neoblasts to Deep Clustering by Single-Cell Transcriptional Profiling

cNeoblasts, initially simply defined as a subpopulation of the *smedwi-1$^+$* cells, became the object of intense studies to try to characterize their molecular signature. Accordingly, the Reddien's group in 2014 identified, by a single-cell qRT-PCR assay, three prominent types of neoblasts—the ζ (zeta), the γ (gamma), and the σ (sigma) neoblast classes. σ-neoblasts proliferate in response to injury, possess broad lineage capacity, and can give rise to ζ-neoblasts, thus suggesting them as ideal candidates to include the cNeoblasts (van Wolfswinkel et al. 2014). The same authors also provide the observation that the conversion of the transcriptional profile from σ- to ζ-neoblasts begins directly upon entry into the S-phase. Indeed, the transcriptional profile of early S-phase ζ-neoblasts was more similar to that of G1-phase σ-neoblasts than to that of the G1-phase ζ-neoblasts, and that the ζ-neoblast identity became more resolved during progression through S-phase stages. Once produced, the majority of recently divided ζ-neoblasts are thought to exit the cell cycle permanently (van Wolfswinkel et al. 2014) and are not able of subsequent series of cell division and self-renewal (Lai et al. 2018).

Further evidence supported that σ-neoblasts might be the only neoblasts able to indefinitely proliferate (Lai et al. 2018) and, as a matter of fact, *Smed-soxP-1*, one of their molecular markers, is involved in stem cell self-renewal and is required in the rescue process after low-dose X-ray treatment for colony expansion (i.e., the ability of *smedwi-1$^+$* cell colonies, formed by radioresistant neoblasts after low-dose X-ray, to grow in size), strengthen the idea that σ-class neoblasts include cNeoblasts (Wagner et al. 2012). The discovery of σ-neoblasts has led scientists to imagine a well-defined population of stem cells with its own molecular signature endowed with pluripotency. However, later, the expression of some σ-neoblasts molecular markers was found to be dispersed across all the neoblast classes identified by van van Wolfswinkel et al. (2014), unlike the ζ marker *zfp-1* and γ marker *hnf4*, which are largely specific to their respective classes (Molinaro and Pearson 2016). For this and further additional reasons, Molinaro and Pearson (2016) wondered whether σ-neoblasts were a truly distinct neoblast class or simply a collection of non-ζ and non-γ cells. Accordingly, in a few years, advances in single-cell transcriptomic rapidly brought to light that σ-neoblasts are a heterogenous population themselves, not a single well-defined neoblast class. Indeed, both Fincher et al. (2018) and Zeng et al. (2018), focusing on the idea that *smedwi-1* differential expression levels might represent a discriminatory parameter for subclassifying neoblasts, identified several neoblast subclasses. Making the assumption that cNeoblasts might be included in the stem cell fraction with the highest level of *smedwi-1* transcript and its coded PIWI-1 protein, Zeng et al. (2018) identified a cluster (Nb2) that satisfies a series of selection criteria (expression of σ-neoblast markers and self-renewal regulators; negativity for fate specific transcription factors (FSTFs); increased expression of cluster markers within hours after amputation; decline in expression of cluster markers up to 6 days after sublethal irradiation with a markedly increased and sustained expression from 6 days after irradiation onward) was proposed to include the cNeoblasts. Nb2 cells show an enriched expression of *tspan-1* coding for the cell surface protein tetraspanin 1 (TSPAN-1).

Some concerns can be raised on the assumptions the author made delineating their strategy. First, the reason for which cells with the highest expression of *smedwi-1* should be considered as those that might contain cNeoblasts is a limiting assumption. For example, it is clearly a not sound strategy for *D. japonica* (the other principal planarian model system) in which both *DjpiwiA* and its coded protein show a very high expression in a dorsal midline population of neoblast-like cells that do not satisfy the previous selection criteria. Second, the "a priori" exclusion of clusters expressing FSTFs bias the analysis pre-assuming that a cell at the beginning of its commitments cannot revert its fate.

A common feature of adult pluripotent stem cells is that their self-renewal potential is proportional to their state of quiescence or deep dormancy (Post and Clevers 2019). This makes sense from an evolutionary point of view, as cell cycling exposes cells to propagate accidental DNA damage to daughter cells and future generations. Quiescent cells maintain DNA integrity and reenter the cell cycle only under appropriate stimuli.

A long debate on the existence of neoblasts with different cycling rates or on the length of the different cycle phases has characterized the 20th century. This question appeared definitely closed with the finding that up to 99% of neoblasts are labeled by BrdU in 3 days after treatment (Newmark and Sánchez 2000). However, very recently, the presence of a slow-cycling population of neoblasts, with low transcriptional activity (RNA[low] neoblasts), has been identified and proposed as a regeneration-reserved neoblast population (Molinaro et al. 2021). RNA[low] neoblasts show many characteristics reminiscent of quiescent stem cells, including very small size, slow division rate, and similarities in gene expression profile. RNA[low] neoblasts undergo morphological changes after injury or low-dose X-ray and enter the cell cycle during regeneration by a TORC1-dependent mechanism (Molinaro et al. 2021). A small fraction of RNA^{low} neoblasts expresses the tspan-1 marker, suggesting that some of them are part of the N2b cluster. Diverging lineage markers were often detected within individual RNA^{low} neoblasts, suggesting that some of these cells may not be specified to any one lineage (Molinaro et al. 2021). Further studies are necessary to characterize this novel subpopulation and its relationship with cNeoblasts and/or other neoblast subpopulations.

2.2. The Neoblast Fate Restriction Model

The last 15 years of scientific research in the planarian stem cell field was dominated by the line of reasoning that a clear hierarchical organization exists between neoblast subpopulations, with pluripotent stem cells (the cNeoblasts) giving rise to neoblasts with a restricted potency, the so-called lineage-committed neoblasts (Figure 3, left side). The first evidence of the existence of lineage-committed neoblasts was provided in 2006 owing to the work of Sato et al. (2006) that identified, in asexual *D. japonica*, germline stem cells that specifically express a *nanos*-related gene (*Djnos*), localized in the presumptive ovary or testis-forming regions, and morphologically indistinguishable from neoblasts. Although these *Djnos$^+$* cells highly express the PCNA protein, they are blocked in the cell cycle and incapable to incorporate BrdU. Following the discovery of epidermal-committed ζ-neoblast and gut-committed γ-neoblasts (van Wolfswinkel et al. 2014), intense research was focused on identifying FSTFs that were also expressed in *smedwi-1$^+$* cells and, thus, probably involved in neoblast commitment versus a specific lineage. In this way, putative neoblast precursors for cells of the eye, protonephridia, nervous system, pharynx, anterior pole, and gut were identified (Scimone et al. 2011, 2014a, 2014b; Lapan and Reddien 2012; Currie and Pearson 2013; Cowles et al. 2013; Adler et al. 2014; Vásquez-Doorman and Petersen 2014; Flores et al. 2016). Recently, Plass et al. (2018) performed highly parallel droplet-based single-cell transcriptomics and by applying a partition-based graph abstraction algorithm, combined with independent computational and experimental approaches, derived a consolidated lineage tree that includes all identified cell types rooted to a single stem cell cluster. In this tree, they identified gene sets that are co-regulated during the differentiation of specific cell types, thus providing a single tree that models stem cell differentiation trajectories into all identified cell types of adult planarians. According to the consolidated lineage tree, neoblasts differentiate into at least 23 independent cell lineages and several progenitors have been identified. In addition to all these putative subpopulations, in the planarian species, *D. japonica* a spatially well-defined abundant group of cells that show morphological features of neoblasts, are sensitive to irradiation, express *DjpiwiA* transcripts and genes involved in cell cycle progression, and is localized in the dorsal midline. This population is specifically identifiable by the expression of *DjPiwi-1* (Rossi et al. 2006, 2008), a *piwi* homolog gene that has been found in planarians from the *Dugesia* genus and not in *S. mediterranea* and *Girardia dorotocephala* (Kashima et al. 2020). The function/fate of *DjPiwi-1*$^+$ cells is still unknown; however, we recently demonstrated that they are part of a population of *soxP-1*-negative lineage-committed neoblasts that, as a consequence of their very slow-cycling rate, are transiently resistant to continuous high-dose 5-fluorouracil (5FU) treatment (Gambino et al. 2020). In case of short low-dose 5FU treatment, cells of this dorsal midline subpopulation never disappear, activate proliferation after cutting but never change their expression pattern, remain negative for *soxP-1*, and do not seem to contribute to the repopulation process (Gambino et al. 2021). On the contrary, *DjPiwi-1*$^+$ expression appears to be associated with cells reentering the cell cycle at the ventral surface of the animal in challenging conditions, as demonstrated after a short 5FU low-dose and sublethal X-ray treatment (Gambino et al. 2021; Salvetti et al. 2009). The recent advances in single-cell transcriptomics in *D. japonica* species (García Castro et al. 2021) will allow more information to be obtained on *DjPiwi-1*$^+$cells of the dorsal midline, which may represent a valuable resource for the understanding of planarian stem cell biology.

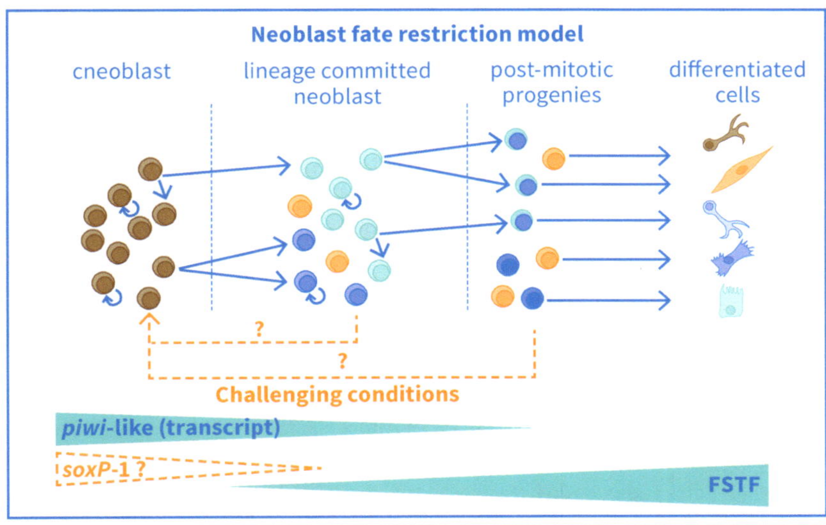

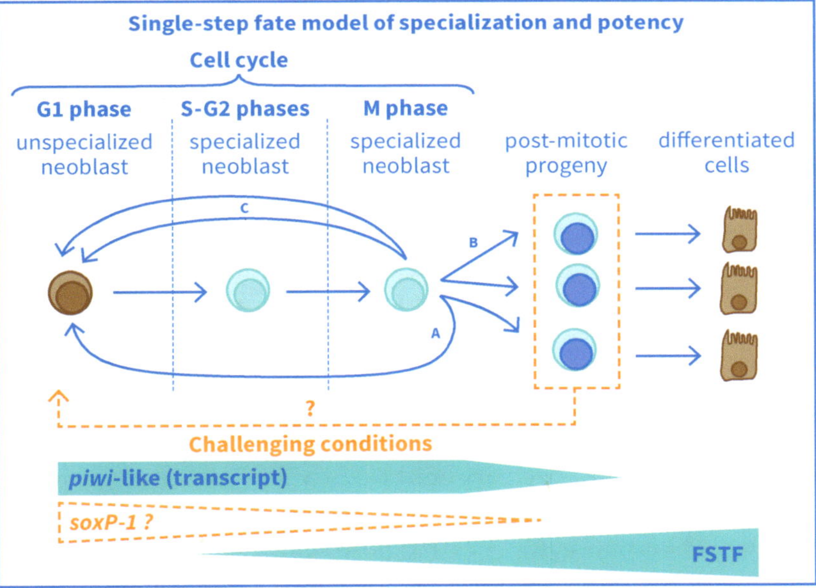

Figure 3. Scheme depicting the comparison between the neoblast fate restriction model and the single-step model of specialization and potency. In the first, the cNeoblasts can symmetrically self-maintain or asymmetrically divide to give rise to a daughter cNeoblast and a daughter specialized neoblast or even symmetrically divide to give rise to two specialized daughters. It is not clear how many times specialized neoblasts can divide, but in any case, they shortly produce postmitotic progenies, which gradually differentiate into a specialized cell. *piwi-like* gene transcripts are highly expressed in cNeoblasts, and their expression gradually declines in parallel

to fate restriction being very low or undetectable in postmitotic progenies and differentiated cells. Although *soxP-1* expression has been found to span several *piwi*-positive subclasses of neoblasts, several lines of evidence suggest that its expression is limited to pluripotent stem cells and declines in specialized neoblasts. FSTFs are specifically expressed in specialized neoblasts. In challenging conditions, including sublethal X-ray doses, short (5FU) treatment, and regeneration for some sexual planarians, bodies of evidence suggest that postmitotic cells or at least specialized neoblasts can revert their fate and acquire a wider differentiation potency. In the single-step model of specialization and potency, an unspecialized G1 neoblast become specialized, progressing through the cell cycle and starting to express FSTF from phase S. Concomitantly, a reduction in *soxP-1* expression could be hypothesized. Following the G2 phase, in many cases, an asymmetric division (A) generates an unspecialized neoblast, which will progress through a novel cell cycle, acquiring the same or a different specialization, and a postmitotic *piwi-soxP-1*-negative progeny that will differentiate. In other cases, a symmetric division can generate two unspecialized neoblasts (C) or even two postmitotic cells (B). It cannot be excluded that in challenging conditions, early postmitotic progenies might revert their fate and reenter the cell cycle. Source: Graphic by authors.

2.3. The Single-Step Fate Model of Specialization and Potency

An emerging viewpoint that opposes the historical idea that cNeoblasts are a subpopulation with a specific molecular signature refutes the existence of exclusive transcripts for pluripotent stem cells and accepts the concept of a modulation in the expression levels of neoblast specific transcripts. In this view, a recent paper by Gambino et al. (2021) demonstrated that following a short 5FU treatment, *soxP-1* expression is extensively downregulated below the detection limit of in situ hybridization. However, *soxP-1*-positive cells remain in the animal body, and only after a few weeks, some of these cells upregulate again *soxP-1* expression, restart proliferation and repopulate the entire planarian body. The idea that cell behavior is dependent on the transitory enrichment of specific transcripts invalidates the existence of clearly defined subpopulations organized in a strict hierarchy and opens to the concept of blurred borders between neoblast populations, with cells that possibly fluctuate from wider to restricted differentiative potential and "vice versa".

In this line of thought, a cutting-edge interpretation of the neoblast fate specification mechanism questions the idea of the existence of a limited population of cNeoblasts, and a strictly organized hierarchy of neoblasts with progressive restriction in differentiation potency has been recently published by Raz et al. (2021). They demonstrated, by single-cell transplantation in irradiated animals and colony assays, that no known neoblast subpopulation is exclusively pluripotent and neoblasts from different subpopulations can be clonogenic. They proposed a single-step fate model of specialization and potency: newly produced G1 neoblasts are pristine but become specified by progressive enrichment in FSTFs during the progression through S/G2/M phases of cell cycle; a G2 specialized neoblast then asymmetrically divide to give rise to a non-neoblast daughter cell that will differentiate and a daughter cell that remains a neoblast and can again specialize to a different fate during the next cell cycle progression without progenies with intermediary potency. In other terms, this means that specialized neoblasts can return to pluripotency after cell division. This model fits well with the proposed switch from σ- to ζ-neoblasts in the S phase and is in line with some recent publications: (i) Gambino et al. (2021) demonstrated that in challenging conditions after 5FU treatment, neoblasts early postmitotic cells could modify their expression profile reacquiring a broader differentiative potential; (ii) Davidian et al. (2021) showed that subthreshold direct current stimulation rapidly restores pluripotent stem cell populations previously eliminated by lethal irradiation promoting cell cycle entry of postmitotic cells. However, even more remarkably, this new line of thought brings us back to findings obtained at the beginning of the 1980s from Gremigni's group (Gremigni and Miceli 1980; Gremigni et al. 1980a, 1980b, 1982) by using a triplo-hexaploid biotype of *D. polychroa* that provided a useful karyological marker because embryonic and somatic cells are triploid (3n = 12 chromosomes) and could be easily distinguished from male diploid (2n = 8 chromosomes) and female hexaploid (6n = 24 chromosomes) germ cells by their chromosome number. Gremigni et al. (1980a, 1980b, 1982) showed that a small percentage of male and (to a much lesser extent) female germ cells are involved in blastema formation and somatic tissue reconstruction, along with a large number of neoblasts, suggesting that germ cells, at the very beginning of their differentiation process, can interrupt their pathway toward specialization and return to the pluripotent state. This interpretation fits perfectly with the single-step fate model of specialization and potency proposed by Raz et al. (2021). In this case, germ cells can be interpreted as specialized neoblasts, which asymmetrically divide to give rise on the one hand, to a gamete precursor and, on the other hand, to a cell that specialized to a different fate. Unfortunately, the triplo-hexaploid biotype did not survive up to the molecular age, and thus, it was not possible to provide additional demonstrations (Salvetti and Rossi 2012).

A key event for the single-step fate model of specialization is the asymmetric division, which is still an unexplored field in planarian mainly due to technical difficulties. The first molecular evidence of asymmetric stem cell division has been provided by a comprehensive paper of the Sanchez Alvarado's group (Lei et al. 2016) in which, by applying a combined approach of RNAi and colony expansion assay after low-dose X-ray, they demonstrated that the epidermal growth factor pathway and its receptor *egfr-3* are involved in the expansion of neoblasts when their number is diminished by sublethal radiation. *egfr-3* is also fundamental in physiological conditions for the second peak of hyperproliferation at 48 h postamputation. Strikingly, *egfr-3* protein frequently shows an asymmetric distribution on the neoblast membrane, and *egfr-3* distribution during mitoses was associated with symmetric/asymmetric distribution of *smedwi-1* transcripts and the chromatoid bodies. Thus, the authors hypothesize that *egfr-3* controls the repopulation of neoblast by regulating asymmetric versus symmetric cell division. Additional lines of evidence emerged from the analysis of the function of the planarian homolog of *mex*3 RNA-binding protein (*Smed-mex3-1*) that is expressed in both stem cell and immediate postmitotic progeny populations (Zhu et al. 2015). Knockdown of *mex*3-1 leads to a rapid decline of progenitor markers for multiple lineages but not of stem cells, suggesting its specific role in specifying committed progeny. Despite *Smed-mex3-1* mRNA showing no asymmetric distribution into stem cells, on the basis of its proven function in other model systems, the authors speculate that it may function to maintain asymmetry in stem cell lineage progression by promoting postmitotic fates and suppressing self-renewal (Zhu et al. 2015). Despite these pioneering papers, much research needs to be performed to demonstrate and understand asymmetric cell division in neoblasts.

In conclusion, the molecular classification of planarian neoblasts is still a work in progress, and although many efforts to link the molecular and functional definitions of cNeoblasts have been made, unambiguous cNeoblast markers have not been yet identified. Thus, we cannot picture cNeoblasts as a special subpopulation; on the contrary, we had to assume, according to the single-step fate model, that pluripotency is the consequence of transitory and cell-cycle related fluctuations in the quantitative transcriptional profile, rather than expression of specific genes; asymmetric distribution of cell fate determinants such as chromatin remodeling factors might be at the basis of self-maintenance mechanisms. In this view, all neoblasts are potentially clonogenic, and the expression of FSTFs is correlated with cell-cycle progression rather than limited to lineage-committed neoblasts with intermediate potency. This hypothesis is groundbreaking and relates to a previous probabilistic model that considers the possibility that pluripotency may be a transient, probabilistic state exhibited by stem cells. In this view, self-renewal becomes a feature not possessed by a discrete population of cells but transiently held by a small number of cells and arising depending on the demands of the animal (Adler and Sánchez 2015). However, many open questions still remain. For example, which precise changes in transcriptional profile drives the switch from G1 pluripotent state to S-G2 m lineage-committed state? Is the increase in the expression of FSTFs necessary and sufficient for the downregulation of self-renewal regulators such as members of the polycomb complex PRC2, the transcription factor *soxP-1* (Wagner et al. 2012), the RNA-binding proteins PIWI-like (*smedwi-2* and *3*), Bruno-like (*Bruli*), *Pumilio*, and *CIP29* (Reddien et al. 2005; Wagner et al. 2012; Guo et al. 2006; Salvetti et al. 2005; Rossi et al. 2007), histone-2B (Solana et al. 2012), the Retinoblastoma homolog (Zhu and Pearson 2013), and the epidermal growth factor receptor *egfr-3* (Lei et al.

2016)? What is the role of *p53* known to inhibit proliferation and stem cell identity and induce differentiation in the early progeny (Pearson and Sánchez 2010)? Can all the S-G2 m lineage-committed neoblasts reverse their differentiation fate with the same effectiveness? What is the role played by epigenetic inheritance? For example, the asymmetric inheritance of chromatoid bodies and *piwi* transcripts might bring into one daughter cell transcripts that, once translated, produce a specific chromatin condensation pattern that promotes self-renewal. Indeed, evidence that links PIWI proteins and chromatoid bodies to histone mRNA regulation in planarian stem cells has been provided (Rouhana et al. 2014). Finally, which positional information signals drive the decision of a G1 daughter cell to specialize toward a specific fate? Indeed, a classical niche, meant as a specific anatomical structure in which stemness is maintained, has not yet been proven in planarians. However, neoblast dynamics appear to be under the control of signaling from multiple tissues, suggesting that a global niche, a macroenvironment, comprehensive to the entire planarian body might exist (Rossi and Salvetti 2019).

3. From Lineage-Committed Neoblasts to Differentiated Cells: The Case of Epidermal Cell Differentiation

Independently from which specification model is valid, a committed neoblast should first become a postmitotic cell and then progress toward a fully differentiated fate. Several examples exist in the literature describing the role of molecular regulators in the differentiation of multiple planarian tissues including the eye (Lapan and Reddien 2012) and excretory system (Scimone et al. 2011). However, owing to the prolific production of some research groups in the last decade, the most comprehensive overview is available for the differentiation of epidermis. The Planarian epidermis is a monostratified tissue of multiple multiciliated and nonciliated cell types (Rompolas et al. 2010). However, despite similar morphological appearance, Wurtzel et al. (2017) identified eight different spatial transcriptional identities by the analysis of epidermis-enriched RNAseq libraries, demonstrating that planarian epidermis is a complex tissue with distinct cell types, all originating from the single lineage-committed class of ζ-neoblasts (van Wolfswinkel et al. 2014). ζ-neoblasts, characterized by the expression of a group of molecular markers such as *zfp-1*, divide and produce postmitotic progenitors that express the marker *prog-1* (*NB.21.11e*)—the so-called early epidermal progeny. Early progeny cell identity is maintained for a short period of time; indeed, *prog-1*$^+$ cells disappear 2 days after lethal X-ray treatment and rapidly differentiate in the late epidermal progeny, characterized by the expression of *AGAT-1*, *AGAT-2*, and *AGAT-3* transcripts. The early growth response family transcription factor, *egr-5*, seems implicated in switching off the expression of *prog-1* and turning on the expression of genes necessary for the *AGAT-1+* transition stage. *AGAT-1*$^+$ cells localize more distal with respect to *prog-1*$^+$ cells and disappear 7 days after lethal irradiation (Tu et al. 2015; Eisenhoffer et al. 2008). The spatial expression pattern of *prog-1* and *AGAT-1* indicates that these progeny cells migrate to the outer surface of the animal during epidermal differentiation. Zhu and Pearson (2018) identified *myb-1* as a key regulator of the temporal phase of early progenitor specification during epidermal lineage differentiation. Indeed, *myb-1* (RNAi) resulted in a selective loss of the early progeny fate, causing *prog-1*$^+$ cells prematurely to adopt the late progeny transcriptional profile. Despite this heterochronic temporal shift, late progenitors resumed differentiation. The early transition state into the planarian epidermis is marked by the expression of *zpuf-6* transcript, which labels all the *AGAT-1*$^+$ cells, but also some *AGAT-1*$^-$ cells and some cells located into the epidermis monolayer. Final differentiation steps are then marked by *vimentin 3* and *rootletin* expression (Tu et al. 2015). Recent findings demonstrate that the terminal identity of epidermal cells is acquired early during the differentiation process, indicating that epidermal progenitors recognize their position and modulate their gene expression in a way to reflect the array of transcripts in the mature epidermis (Wurtzel et al. 2017). Strikingly, also the expression of cilia specific genes starts in early progenitors despite the fact that the formation of cilia is restricted to the mature epidermis, and accordingly, it has been demonstrated that genes involved in ciliogenesis that are inactive loci in the stem cell population are methylated in order to poise them for activation later in development (Duncan et al. 2015). This also demonstrates that the identity as a ciliated or not-ciliated epidermal cell is acquired in migratory progenitors before their terminal differentiation (Wurtzel et al. 2017). Interestingly, it has been demonstrated that terminal epidermal cell differentiation is finely regulated by the two specific components of the nucleosome remodeling deacetylase (NuRD) complex: the methyl-CpG-binding domain 2/3 (mbd2/3) gene (Jaber-Hijazi et al. 2013) and the GATA-type zinc-finger-domain-containing gene p66 (Vásquez-Doorman and Petersen 2016) revealing opening future avenues of research on how neoblast processes are coordinated at the epigenetic level (Dattani et al. 2019).

4. Embryonic Origin of Neoblasts

Where do cNeoblasts originate from? Are they the heritage of naïve embryonic stem cells? Or are they formed as a specific need for adult tissue maintenance? Yes and no! Planarians show an ectolecitic embryonic development in which blastomeres undergo dispersed cleavage among yolk cells, do not contact with one another, and divide asynchronously. During sphere formation, temporary embryonic tissues are formed and then degenerate as adult organs are shaped, owing to undifferentiated blastomeres remaining after sphere formation (Martín-Durán et al. 2012). Recently, a very comprehensive study on *S. mediterranea* embryonic development demonstrates that pluripotent neoblasts and lineage-dedicated progenitors arise when the morphogenesis of definitive organs begins (Davies et al. 2017). The authors demonstrate that *smedwi-1* transcripts are expressed in the zygote and *smedwi-1*[+] cells, endowed with proliferative capability, are detectable throughout all embryogenesis. However, large-scale changes in gene expression occur in *smedwi-1*[+] cells, as definitive organogenesis begins (developmental stage S5). At this time, early embryo-enriched (EEE) transcripts, specifically expressed by blastomeres, dramatically decline and *smedwi-1*[+] cells start to be enriched in FSTF. Transplantation experiments of blastomeres collected from different developmental stages into lethally irradiated hosts demonstrate that the change in transcriptional profile reflects important functional differences. Indeed, *smedwi-1*[+] cells from S4 and S5 embryonic donor cells did not rescue lethally irradiated animals, while cells from S6–S8 embryos acted similarly to adult neoblasts and rescued the lethal phenotype. These findings suggest that cNeoblast specification occurs during S5. Considering that transcripts of genes previously implicated in neoblast maintenance such as *SoxP-1* or *bruli-1* show an expression profile similar to that of *smedwi-1* during embryogenesis, the authors suggest that the expression of pluripotency factors is probably necessary but not sufficient for the assumption of neoblast fate. Indeed, EEE transcripts downregulation in blastomeres is necessary for neoblast specification, suggesting that they might represent repressors of neoblast fate. This hypothesis is very intriguing; however, no direct proof has been provided such as analyzing the ability to rescue irradiated hosts after blocking the downregulation of EEE transcripts. Despite the significant advances in the comprehension of cNeoblast origin, several questions are still open—are EEE transcripts maternally deposited? If this is the case, which mechanisms affect maternal transcript degradation? When does zygotic genome activation occur? Further, how does it influence neoblast specification?

Author Contributions: Conceptualization, writing, review and editing, G.G., L.R. and A.S.

Funding: This research received no external funding.

Conflicts of Interest: The authors declare no conflict of interest.

References

Adler, Carolyn E., and Alvarado A. Sánchez. 2015. Types or States? Cellular Dynamics and Regenerative Potential. *Trends in Cell Biology* 25: 687–96. [CrossRef] [PubMed]

Adler, Carolyn E., Chris W. Seidel, Sean A. McKinney, and Alvarado A. Sánchez. 2014. Selective amputation of the pharynx identifies a FoxA-dependent regeneration program in planaria. *Elife* 3: e02238. [CrossRef] [PubMed]

Alessandra, Salvetti, and Leonardo Rossi. 2019. Planarian stem cell heterogeneity. *Stem Cells Heterogeneity-Novel Concepts* 1123: 39–54. [CrossRef]

Baguñà, Jaume. 2019. Planarian regeneration between 1960s and 1990s: From skillful baffled ancestors to bold integrative descendants. A personal account. *Seminars in Cell & Developmental Biology* 87: 3–12. [CrossRef]

Coward, Stuart J. 1974. Chromatoid bodies in somatic cells of the planarian: Observations on their behavior during mitosis. *Anatomical Record* 180: 533–45. [CrossRef]

Cowles, Martis W., David D. Brown, Sean V. Nisperos, Brianna N. Stanley, Bret J. Pearson, and Ricardo M. Zayas. 2013. Genome-wide analysis of the bHLH gene family in planarians identifies factors required for adult neurogenesis and neuronal regeneration. *Development* 140: 4691–702. [CrossRef]

Currie, Ko W., and Bret J. Pearson. 2013. Transcription factors lhx1/5-1 and pitx are required for the maintenance and regeneration of serotonergic neurons in planarians. *Development* 140: 3577–88. [CrossRef]

Dattani, Anish, Divya Sridhar, and Aziz A. Aboobaker. 2019. Planarian flatworms as a new model system for understanding the epigenetic regulation of stem cell pluripotency and differentiation. *Seminars in Cell & Developmental Biology* 87: 79–94. [CrossRef]

Davidian, Devon, Melanie LeGro, Paul G. Barghouth, Salvador Rojas, Benjamin Ziman, Eli I. Maciel, David Ardell, Ariel L. Escobar, and Néstor J. Oviedo. 2021. Restoration of DNA Integrity and Cell Cycle by Electric Stimulation 6 in Planarian Tissues Damaged by Ionizing Radiation. *bioRxiv Preprint*. [CrossRef]

Davies, Erin L., Kai Lei, Christopher W. Seidel, Amanda E. Kroesen, Sean A. McKinney, Longhua Guo, Sofia M. Robb, Eric J. Ross, Kirsten Gotting, and Alvarado A. Sánchez. 2017. Embryonic origin of adult stem cells required for tissue homeostasis and regeneration. *Elife* 6: e21052. [CrossRef] [PubMed]

Duncan, Elizabeth M., Alex D. Chitsazan, Chris W. Seidel, and Alvarado A. Sánchez. 2015. Set1 and MLL1/2 Target Distinct Sets of Functionally Different Genomic Loci In Vivo. *Cell Reports* 13: 2741–55. [CrossRef] [PubMed]

Eisenhoffer, George T., Hara Kang, and Alvarado A. Sánchez. 2008. Molecular analysis of stem cells and their descendants during cell turnover and regeneration in the planarian Schmidtea mediterranea. *Cell Stem Cell* 3: 327–39. [CrossRef] [PubMed]

Fincher, Christopher T., Omri Wurtzel, Thom de Hoog, Kellie M. Kravarik, and Peter W. Reddien. 2018. Cell type transcriptome atlas for the planarian *Schmidtea mediterranea*. *Science*. [CrossRef] [PubMed]

Flores, Natasha M., Néstor J. Oviedo, and Julien Sage. 2016. Essential role for the planarian intestinal GATA transcription factor in stem cells and regeneration. *Developmental Biology* 418: 179–88. [CrossRef] [PubMed]

Gambino, Gaetana, Chiara Ippolito, Letizia Modeo, Alessandra Salvetti, and Leonardo Rossi. 2020. 5-Fluorouracil-treated planarians, a versatile model system for studying stem cell heterogeneity and tissue aging. *Biology of the Cell* 112: 335–48. [CrossRef] [PubMed]

Gambino, Gaetana, Chiara Ippolito, Monica Evangelista, Alessandra Salvetti, and Leonardo Rossi. 2021. Sub-Lethal 5-Fluorouracil Dose Challenges Planarian Stem Cells Promoting Transcriptional Profile Changes in the Pluripotent Sigma-Class Neoblasts. *Biomolecules* 11: 949. [CrossRef]

García Castro, Helena, Nathan J. Kenny, Marta Iglesias, Patricia Álvarez Campos, Vincent Mason, Anamaria Elek, Anna Schönauer, Victoria A. Sleight, Jakke Neiro, Aziz Aboobaker, and et al. 2021. ACME dissociation: A versatile cell fixation-dissociation method for single-cell transcriptomics. *Genome Biology* 22: 89. [CrossRef]

Gremigni, Vittorio, and Cristina Miceli. 1980. Cytophotometric evidence for cell 'transdifferentiation' in planarian regeneration. *Wilhelm Roux's Archives of Developmental Biology* 188: 107–13. [CrossRef]

Gremigni, Vittorio, Cristina Miceli, and Ileana Puccinelli. 1980a. On the role of germ cells in planarian regeneration: I. A karyological investigation. *Journal of Embryology and Experimental Morphology* 55: 53–63.

Gremigni, Vittorio, Cristina Miceli, and Eugenio Picano. 1980b. On the role of germ cells in planarian regeneration: II. Cytopliotometric analysis of the nuclear Feulgen-DNA content in cells of regenerated somatic tissues. *Journal of Embryology and Experimental Morphology* 55: 65–76.

Gremigni, Vittorio, Marco Nigro, and Ileana Puccinelli. 1982. Evidence of male germ cell redifferentiation into female germ cells in planarian regeneration. *Journal of Embryology and Experimental Morphology* 70: 29–36. [CrossRef] [PubMed]

Guo, Tingxia, Antoine H. F. M. Peters, and Phillip A. Newmark. 2006. A Bruno-like gene is required for stem cell maintenance in planarians. *Developmental Cell* 11: 159–69. [CrossRef] [PubMed]

Jaber-Hijazi, Farah, Priscilla J. Lo, Yuliana Mihaylova, Jeremy M. Foster, Jack S. Benner, Belen Tejada Romero, Chen Chen, Sunir Malla, Jordi Solana, Alexey Ruzov, and et al. 2013. Planarian MBD2/3 is required for adult stem cell pluripotency independently of DNA methylation. *Developmental Biology* 384: 141–53. [CrossRef] [PubMed]

Kashima, Makoto, Kiyokazu Agata, and Norito Shibata. 2020. What is the role of PIWI family proteins in adult pluripotent stem cells? Insights from asexually reproducing animals, planarians. *Development, Growth & Differentiation* 62: 407–22. [CrossRef]

Lai, Alvina G., Nobuyoshi Kosaka, Prasad Abnave, Sounak Sahu, and Aziz A. Aboobaker. 2018. The abrogation of condensin function provides independent evidence for defining the self-renewing population of pluripotent stem cells. *Developmental Biology* 433: 218–26. [CrossRef] [PubMed]

Lapan, Sylvain W., and Peter W. Reddien. 2012. Transcriptome analysis of the planarian eye identifies ovo as a specific regulator of eye regeneration. *Cell Reports* 2: 294–307. [CrossRef]

Lei, Kai, Hanh Vu hi-Kim, Ryan D. Mohan, Sean A. McKinney, Chris W. Seidel, Richard Alexander, Kirsten Gotting, Jerry L. Workman, and Alvarado A. Sánchez. 2016. EGF signaling directs neoblast repopulation by regulating asymmetric cell division in planarians. *Developmental Cell* 38: 413–29. [CrossRef]

Liu, Shang-Yun, Claudia Selck, Bärbel Friedrich, Rainer Lutz, Miquel Vila-Farré, Andreas Dahl, Holger Brandl, Naharajan Lakshmanaperumal, Ian Henry, and Jochen C. Rink. 2013. Reactivating head regrowth in a regeneration-deficient planarian species. *Nature* 500: 81–4. [CrossRef]

Martín-Durán, José M., Francisco Monjo, and Rafael Romero. 2012. Planarian embryology in the era of comparative developmental biology. *International Journal of Developmental Biology* 56: 39–48. [CrossRef]

Molinaro, Alyssa M., and Bret J. Pearson. 2016. In silico lineage tracing through single cell transcriptomics identifies a neural stem cell population in planarians. *Genome Biology* 17: 87. [CrossRef]

Molinaro, Alyssa M., Nicole Lindsay-Mosher, and Bret J. Pearson. 2021. Identification of TOR-responsive slow-cycling neoblasts in planarians. *EMBO Reports* 22: e50292. [CrossRef] [PubMed]

Newmark, Philip A., and Alvarado A. Sánchez. 2000. Bromodeoxyuridine specifically labels the regenerative stem cells of planarians. *Developmental Biology* 220: 142–53. [CrossRef] [PubMed]

Pearson, Bret J., and Alvarado A. Sánchez. 2010. A planarian p53 homolog regulates proliferation and self-renewal in adult stem cell lineages. *Development* 137: 213–21. [CrossRef] [PubMed]

Plass, Mireya, Jordi Solana, Alexander F. Wolf, SalahAyoub, Aristotelis Misios, Petar Glažar, Benedikt Obermayer, Fabian J. Theis, Christine Kocks, and Nikolaus Rajewsky. 2018. Cell type atlas and lineage tree of a whole complex animal by single-cell transcriptomics. *Science*. [CrossRef]

Post, Yorick, and Hans Clevers. 2019. Defining adult stem cell function at its simplest: The ability to replace lost cells through mitosis. *Cell Stem Cell* 25: 174–83. [CrossRef]

Raz, Amelie A., Omri Wurtzel, and Peter W. Reddien. 2021. Planarian stem cells specify fate yet retain potency during the cell cycle. *Cell Stem Cell* 28: 1307–22.e5. [CrossRef]

Reddien, Peter W., Nestor J. Oviedo, Joya R. Jennings, James C. Jenkin, and Alvarado A. Sánchez. 2005. SMEDWI-2 is a PIWI-like protein that regulates planarian stem cells. *Science* 310: 1327–30. [CrossRef]

Rompolas, Panteleimon, Ramila S. Patel-King, and Stephen M. King. 2010. An outer arm Dynein conformational switch is required for metachronal synchrony of motile cilia in planaria. *Molecular Biology of the Cell* 21: 3669–79. [CrossRef]

Rossi, Leonardo, and Alessandra Salvetti. 2019. Planarian stem cell niche, the challenge for understanding tissue regeneration. *Seminars in Cell & Developmental Biology* 87: 30–36. [CrossRef]

Rossi, Leonardo, Alessandra Salvetti, Annalisa Lena, Renata Batistoni, Paolo Deri, Claudio Pugliesi, Elena Loreti, and Vittorio Gremigni. 2006. DjPiwi-1, a member of the PAZ-Piwi gene family, defines a subpopulation of planarian stem cells. *Development Genes and Evolution* 216: 335–46. [CrossRef]

Rossi, Leonardo, Alessandra Salvetti, Francesco M. Marincola, Annalisa Lena, Paolo Deri, Linda Mannini, Renata Batistoni, Ena Wang, and Vittorio Gremigni. 2007. Deciphering the molecular machinery of stem cells: A look at the neoblast gene expression profile. *Genome Biology* 8: 1–17. [CrossRef] [PubMed]

Rossi, Leonardo, Alessandra Salvetti, Renata Batistoni, Paolo Deri, and Vittorio Gremigni. 2008. Planarians, a tale of stem cells. *Cellular and Molecular Life Sciences* 65: 16–23. [CrossRef] [PubMed]

Rouhana, Labib, Jennifer A. Weiss, Ryan S. King, and Philip A. Newmark. 2014. PIWI homologs mediate histone H4 mRNA localization to planarian chromatoid bodies. *Development* 141: 2592–601. [CrossRef] [PubMed]

Rozanski, Andrei, HongKee Moon, Holger Brandl, José M. Martín-Durán, Markus A. Grohme, Katja Hüttner, Kerstin Bartscherer, Ian Henry, and Jochen C. Rink. 2019. PlanMine 3.0-improvements to a mineable resource of flatworm biology and biodiversity. *Nucleic Acids Research* 47: D812–D820. [CrossRef] [PubMed]

Salvetti, Alessandra, and Leonardo Rossi. 2012. The past and present of planarians-An interview with Vittorio Gremigni. *International Journal of Developmental Biology* 56: 49–52. [CrossRef]

Salvetti, Alessandra, Leonardo Rossi, Annalisa Lena, Renata Batistoni, Paolo Deri, Giuseppe Rainaldi, MariaT. Locci, Monica Evangelista, and Vittorio Gremigni. 2005. DjPum, a homologue of Drosophila Pumilio, is essential to planarian stem cell maintenance. *Development* 132: 1863–74. [CrossRef]

Salvetti, Alessandra, Leonardo Rossi, Lucia Bonuccelli, Annalisa Lena, Claudio Pugliesi, Giuseppe Rainaldi, Monica Evangelista, and Vittorio Gremigni. 2009. Adult stem cell plasticity: Neoblast repopulation in non-lethally irradiated planarians. *Developmental Biology* 15: 305–14. [CrossRef]

Sánchez, Alvarado A., and Philip A. Newmark. 1999. Double-stranded RNA specifically disrupts gene expression during planarian regeneration. *Proceedings of the National Academy of Sciences of the United States of America* 96: 5049–54. [CrossRef]

Sato, Kimihiro, Norito Shibata, Hidefumi Orii, Reiko Amikura, Takashige Sakurai, Kiyokazu Agata, Satoru Kobayashi, and Kenji Watanabe. 2006. Identification and origin of the germline stem cells as revealed by the expression of nanos-related gene in planarians. *Development, Growth & Differentiation* 48: 615–28. [CrossRef]

Scimone, M. Lucila, Mansi Srivastava, George W. Bell, and Peter W. Reddien. 2011. A regulatory program for excretory system regeneration in planarians. *Development* 138: 4387–98. [CrossRef]

Scimone, M. Lucila, Sylvain W. Lapan, and Peter W. Reddien. 2014a. A forkhead transcription factor is wound-induced at the planarian midline and required for anterior pole regeneration. *PLoS Genetics* 10: e1003999. [CrossRef] [PubMed]

Scimone, M. Lucila, Kellie M. Kravarik, Sylvain W. Lapan, and Peter W. Reddien. 2014b. Neoblast specialization in regeneration of the planarian *Schmidtea mediterranea*. *Stem Cell Reports* 3: 339–52. [CrossRef] [PubMed]

Sluys, Ronald, Masaharu Kawakatsu, Marta Riutort, and Jaume Baguñà. 2009. A new higher classification of planarian flatworms (Platyhelminthes, Tricladida). *Journal of Natural History* 43: 1763–77. [CrossRef]

Solana, Jordi, Damian Kao, Yuliana Mihaylova, Farah Jaber-Hijazi, Sunir Malla, Ray Wilson, and Aziz A. Aboobaker. 2012. Defining the molecular profile of planarian pluripotent stem cells using a combinatorial RNAseq, RNA interference and irradiation approach. *Genome Biology* 13: R19. [CrossRef]

Tu, Kimberly C., Li C. Cheng, Hanh T. K Vu, Jeffrey J. Lange, Sean A. McKinney, Chris W. Seidel, and Alvarado A. Sánchez. 2015. Egr-5 is a post-mitotic regulator of planarian epidermal differentiation. *Elife* 4: e10501. [CrossRef]

van Wolfswinkel, C. Josien, Daniel E. Wagner, and Peter W. Reddien. 2014. Single-cell analysis reveals functionally distinct classes within the planarian stem cell compartment. *Cell Stem Cell* 15: 326–39. [CrossRef]

Vásquez-Doorman, Constanza, and Christian P. Petersen. 2014. zic-1 Expression in Planarian neoblasts after injury controls anterior pole regeneration. *PLoS Genetics* 10: e1004452. [CrossRef]

Vásquez-Doorman, Constanza, and Christian P. Petersen. 2016. The NuRD complex component p66 suppresses photoreceptor neuron regeneration in planarians. *Regeneration* 3: 168–78. [CrossRef]

Wagner, Daniel E., Irving E. Wang, and Peter W. Reddien. 2011. Clonogenic neoblasts are pluripotent adult stem cells that underlie planarian regeneration. *Science* 332: 811–16. [CrossRef]

Wagner, Daniel E., Jaclyn J. Ho, and Peter W. Reddien. 2012. Genetic regulators of a pluripotent adult stem cell system in planarians identified by RNAi and clonal analysis. *Cell Stem Cell* 10: 299–311. [CrossRef]

Wurtzel, Omri, Isaac M. Oderberg, and Peter W. Reddien. 2017. Planarian epidermal stem cells respond to positional cues to promote cell-type diversity. *Developmental Cell* 40: 491–504.e5. [CrossRef] [PubMed]

Zeng, An, Hua Li, L.onghua Guo, Xin Gao, Sean McKinney, Yongfu Wang, Zulin Yu, Jungeun Park, Craig Semerad, Eric Ross, and et al. 2018. Prospectively isolated tetraspanin+ neoblasts are adult pluripotent stem cells underlying planaria regeneration. *Cell* 173: 1593–608.e20. [CrossRef] [PubMed]

Zhu, Shu Jun, and Bret J. Pearson. 2013. The Retinoblastoma pathway regulates stem cell proliferation in freshwater planarians. *Developmental Biology* 373: 442–52. [CrossRef] [PubMed]

Zhu, Shu Jun, and Bret J. Pearson. 2018. Smed-myb-1 specifies early temporal identity during planarian epidermal differentiation. *Cell Reports* 25: 38–46. [CrossRef] [PubMed]

Zhu, Shu Jun, Stephanie E. Hallows, Ko W. Currie, ChangJiang Xu, and Bret J. Pearson. 2015. A *mex3* homolog is required for differentiation during planarian stem cell lineage development. *Elife* 4: e07025. [CrossRef] [PubMed]

Pigment Cell-Specific Genes throughout Development and in Cell Cultures of Embryonic Stem Cells of *Scaphechinus mirabilis*, a Sand Dollar

Natalya V. Ageenko, Konstantin V. Kiselev and Nelly A. Odintsova

Abstract: Pigmentation, a natural mechanism, plays an important role in photo-protecting larvae and embryos of sea urchins from harmful impacts of solar radiation, hypoxia, pathogens, metals and toxicants and might be useful as a marker of environmental stresses. The use of sea urchin embryos and gametes in testing developmental and production effects has been successfully developed by a number of laboratories worldwide. The objective of this study was to find the maximal expression level of the genes encoding enzymes expressed in pigment cells throughout the development of *Scaphechinus mirabilis* and in cell cultures of this sand dollar. Two genes related to different gene families (*pks* and *sult*) were selected for analysis in pigmentation, and their expression level was evaluated by quantitative real-time PCR. The naphthoquinoid pigments of echinoderms and related compounds form a new class of highly effective antioxidants of the phenol type, exhibiting high bactericidal, algicidal, hypotonic and psychotropic activity. Studying marine invertebrate stem cells and primarily differentiation processes and growth regulation may open novel biotechnological avenues such as new applications including basic research in translational medicine.

1. Introduction

Marine organisms are known to possess various compounds with significant and valuable biotechnological potential for the pharmaceutical industry (Martins et al. 2014). Protostomes (Porifera, Cnidaria) and some deuterostomes (Echinodermata) have been reported to contain very high concentrations of bioactive compounds, many of which are not found in terrestrial organisms. In particular, sand dollar pigment cells are the source of organic naphthoquinone pigments, known as potent antioxidant substances (Koltsova et al. 1981). It has previously been reported that naphthoquinone pigments from purple sea urchins (marked with various phenolic hydroxyl groups) demonstrated the antioxidant ability to depress lipid peroxidation, driving purple sea urchins as an original and natural source of antioxidants. Moreover, the sea urchin naphthoquinone pigment response may manifest via an increased antioxidant activity (Vasileva et al. 2020) due to carotenoids. The pigment–protein complex seems

"

to have appeared 2–3 billion years ago in primitive purple photosynthetic bacteria. Later, the resulting strong electronic donor became the naphthoquinoid pigment (Sakuragi et al. 2005).

Pigmentation, a natural mechanism, plays an important role in photo-protecting the larvae and embryos of sea urchins from harmful impacts of solar radiation, hypoxia, pathogens, metals and toxicants (Pinsino and Matranga 2015; Calestani and Wessel 2018). Pigmentation might be useful as a marker of environmental stresses in adult and larval individuals. In fungi, the biosynthesis of naphthoquinone pigments is an important response to stress exposure, as shown by Medentsev et al. (2005). In sea urchins, the main pigments produced in normal conditions and in response to different types of stress are also naphthoquinone pigments. For example, there are echinochrome A and some spinochromes in the cytoplasm of sand dollar pigment cells. One of the cellular defense mechanisms in sea urchins is the activation of specific coelomocytes—red spherule cells. These specific coelomocytes participate in recognizing and neutralizing pathogens. Only polyketide synthase (*pks*) and some flavin-containing monooxygenases (*fmos*) have been previously reported to have a role in the biosynthesis of naphthoquinone pigments of *Strongylocentrotus purpuratus*, a sea urchin (Perillo et al. 2020; Wessel et al. 2020). The sea urchin *pks* gene encodes an enzyme important in echinochrome synthesis. Other functions of pigment cells can be connected with their immune system (Smith et al. 2018; McClay et al. 2020). It is possible that a sulfotransferase gene (*sult*) is necessary for some enzymes also participating in naphthohynoid synthesis (Ageenko et al. 2011; Ageenko et al. 2014).

The chemical synthesis of naphthoquinones was reported in 1985, but commercial applications have been hampered by the toxicity of some synthesized substances, as demonstrated Klotz et al. (2014). Echinochrome A has been reported to usually be produced in sand dollar pigment cells, while the spinochromes are synthesized in cells of several sea urchin species (Koltsova et al. 1981; Ageenko et al. 2014). To protect sea urchins and their habitat from over-exploitation, some authors have developed in vitro approaches for the induction of pigment differentiation through gene transfection in embryonic cell cultures of two echinoderms, the sand dollar *Scaphechinus mirabilis* and *Strongylocentrotus intermedius*. After two-month cultivation, the cells of sand dollar embryos transfected with plasmid DNA containing the yeast *gal4* produced naphthoquinone pigments with an absorbance spectrum similar to the echinochrome spectrum in vivo. A new in vitro technology that does not consider gene transfection into embryos of sea urchins was developed, supported by sea urchins' coelomic fluid components (Ageenko et al. 2011).

The *pks* and *sult* expressions were evaluated by quantitative real-time PCR (qRT-PCR) in order to identify an association with the biosynthesis of naphthoquinone pigments in the sand dollar *S. mirabilis*. Peak expression levels of *pks* and *sult* in sand dollar embryos were detected at the blastula and gastrula stages. In vitro, sand dollar pigment cell numbers were higher when cultured in sea urchin coelomic fluids than in seawater.

Currently, the knowledge on the growth actor genes expressed in the tissues of marine invertebrates is meager. For vertebrates, the key genes regulating the stem state of cells and ensuring a high level of proliferation of embryonic stem cells in culture are, mainly, nanog and oct-4. The mechanism of the realization of stem cell programming for the toti- and pluripotential states is determined by the key genes regulating "stemness". Some authors previously discovered one of the conserved genes, *SpOct*, in the sea urchin *S. purpuratus*. In addition, a homologue of the pluripotent gene nanog was found in the genome of sea urchins, with clear expression at the mesenchymal blastula stage (the beginning of gastrulation). This nanog gene exhibited 64% homology and 44.7% identity in amino acid residues, further revealing high similarities with the mouse brain-specific homeobox gene *bsx* (80.8% homology, 61.7% identity; Odintsova 2009). The aim of this article was to describe naphthoquinoid pigments of sand dollars, obtained in vivo and in vitro for practical application.

2. Methods

2.1. Collection of Biological Material

Adult sand dollars (*S. mirabilis*) were collected from Vostok Bay (Sea of Japan, Russia) throughout the breeding season (at the beginning–middle of August) and were maintained in running aerated seawater aquaria at 17 °C for 1–3 days. There are different groups of Echinoderms: sea urchins, sea stars, holothurians and sea lilies. In terms of body shape, sea urchins (*Echinoidea*) are divided into two types: regular (spherical) and irregular (flat and heart-shaped) sea urchins. The irregular, flat sea urchin *S. mirabilis* (sand dollar)) is one of the widespread representatives of shallow-water benthos. The larvae at 48 hpf (hours post-fertilization) were fed daily with *Isochrysis galbana* (100,000 cells mL^{-1}). The larvae of the mesenchymal blastula (14 hpf) were collected on a 30 μm nylon mesh and cell cultures were obtained, as described (Ageenko et al. 2014). The coelomic fluids from intact or injured (after needle pricks around Aristotle's lantern) adult sea urchins were named normal coelomic fluid (CF$_n$) or wounded coelomic fluid (CF$_w$), respectively. SW was used as a control medium. All culture media (SW and the coelomic fluids) were supplemented with 2% fetal calf serum and gentamicin (40 mg·L^{-1}). All reagents were purchased from Sigma-Aldrich Co. LLC (USA). We used CF of the wounded sea urchins and CF of normal sea urchins, but not CF of sand dollars because it is very difficult to collect the required amount of CF. In our first experiments, we tested all sand dollar CFs: the effect was the same (data not shown).

2.2. RNA Isolation

Total RNAs from sand dollar gametes, embryos and larvae, as well as from their cultivated embryonic cells, were extracted using Yellow Solve reagent (Clonogen, Russia) followed by DNase I treatment (Sileks, Russia) to remove genomic DNA. The first strand of cDNA was synthesized using 1.5 μg of total RNA as a template with the Reverse Transcription System (Sileks) in a 50 μL reaction volume. The PCR reactions were conducted using an iCycler thermocycler (Bio-Rad Laboratories, Minneapolis/Saint Paul, Minnesota, USA) under the following conditions: one cycle of 2 min at 95 °C followed by 40 cycles of 15 s at 95 °C, 15 s at 50 °C and 35 s at 72 °C, with a final extension cycle of 10 min at 72 °C.

2.3. Quantitative Real-Time PCR (q-RT-PCR)

Q-RT-PCR (iCycler thermocycler equipped with the iQ5 Multicolor q-RT-PCR detection system; Bio-Rad Laboratories) was performed using an established protocol (Ageenko et al. 2011). cDNAs were amplified in 20 μL of the reaction mixture containing 1× TaqMan Buffer, 2.5 mM $MgCl_2$, 250 μM of each deoxynucleotide, 1U Taq DNA polymerase, 0.5–2 μL cDNA samples and 0.25 μM of each primer and probe (Real-Time PCR Kit, Syntol, Russia). Amplification conditions: one 2 min cycle at 95 °C followed by 50 cycles of 10 s at 95 °C and 25 s at 62 °C. Results were analyzed with the iQ5 Optical System Software v.2.0 and presented in relative units. The *S. mirabilis* actin gene (GenBank accession number DQ222227) and ubiquitin gene (PRJEB33560) were used as endogenous controls. Results were summarized from five independent experiments, each with three technical replicates in relative units. The primer and TaqMan probe used in q-RT-PCR, namely, 5'CTT CGC CAG CCC ATG ATC AAC3' and 5'ACT CGC CCA CGT CAC CAT CT3', were developed for expression analysis of the *pks* gene. The primer 5'GAT CTT CGC TGG CAA GCA GCT3' and TaqMan probe 5'CCT TCT GGA TGT TGT AGT CGG ACA3' were used for expression analysis of the *sult* gene (Ageenko et al. 2011; Kiselev et al. 2013).

3. Results

3.1. Pks and Sult Expression Profiles in Sand Dollar Gametes and throughout Development

We chose the sand dollar *S. mirabilis* from available species of regular and irregular Echinoids because the sand dollar embryos and larvae contain many pigment cells in their body (Figure 1). The gene expression profiles for the two genes tested, *pks* and *sult*, were different in gametes and throughout development.

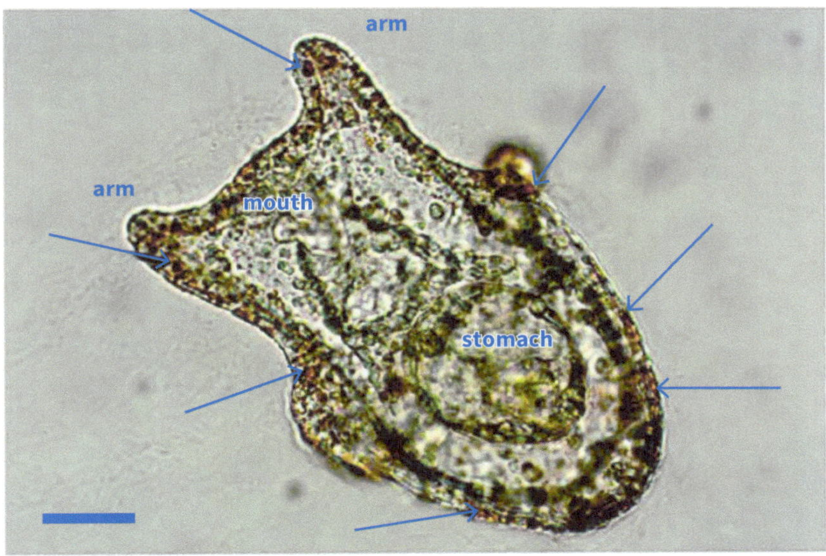

Figure 1. Sand dollar *Scaphechinus mirabilis* pluteus larva. Arrows show the pigment cells. Bar 20 μm. Source: Graphic by author Natalya V. Ageenko.

We detected trace levels of *pks* and *sult* transcripts in spermatozoids and unfertilized eggs. The highest *pks* level of expression was observed at the gastrula stage (Figure 2A), while the maximum *sult* level was found at the earlier blastula stage (Figure 2B). Then, at the prism stage, the level of *pks* expression fell by more than three times but increased in the pluteus larvae, without reaching the previous peak gastrula levels. In contrast, the level of *sult* expression fell after gastrulation and equally increased in abundance at the prism and pluteus stages but also did not reach the level of *sult* expression at the blastula stage.

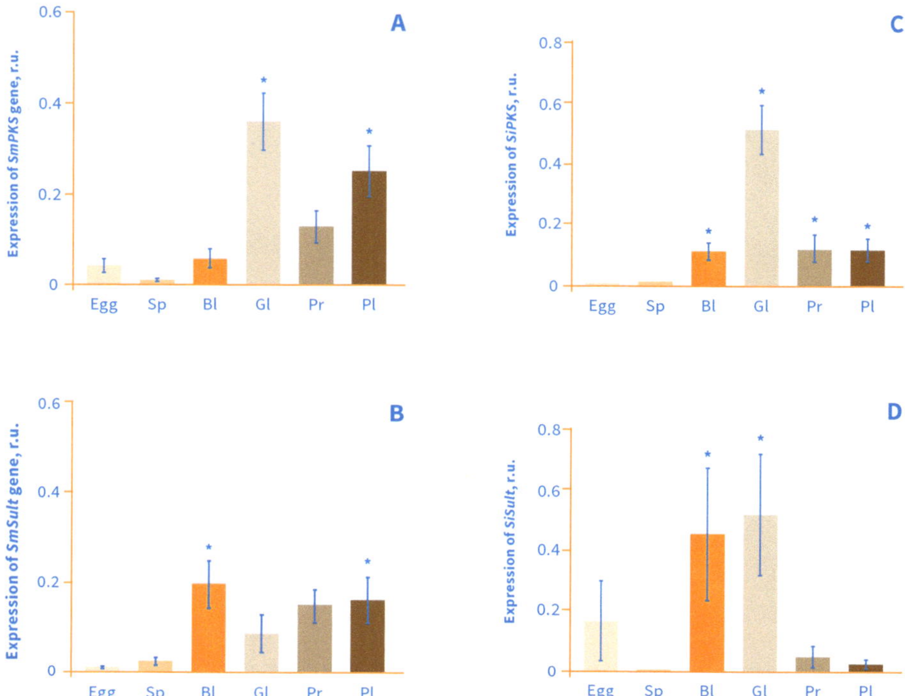

Figure 2. Expression of *pks* and *sult* in vivo: in unfertilized eggs (Egg), spermatozoids (Sp), embryos and larvae of the sand dollar *Scaphechinus mirabilis* (**A**,**B**) and *Strongylocentrotus intermedius* (**C**,**D**) at various stages of development: blastula, 12 h post-fertilization (Bl); gastrula, 24 hpf (Gl); prism, 34 hpf (Pr); and pluteus, 72 hpf (Pl). * $p < 0.05$; ** $p < 0.01$. Mean ± SD, five biological replicates, each with three technical replicates in relative units. One-way analysis of variance (ANOVA) followed by Tukey's pairwise comparison test Significant at $p < 0.05$. *Y*-axis—relative units (r.u.). Source: Graphics 2A,B by author Natalya V. Ageenko; graphics 2C,D from Ageenko et al. (2011).

3.2. Pigment Differentiation in Cell Culture

We found significant differences in the pigment gene expression profiling for the embryonic sand dollar cells cultivated for four days under various culture media (Figure 3). Three media were employed to test the effects of various culture conditions on the sand dollar pigment differentiation under in vitro conditions (Figure 3): SW, and the coelomic fluids of normal and wounded sea urchins. The appearance of sand dollar cells cultivated for 4 days is presented in Figure 3A–C. In the blastula-derived cultures, pigment cells were detected in all media tested at all cultivations. Pigment cell numbers (Figure 3D) were associated with the coelomic fluids tested, and we revealed a >2-fold increase in pigment cell numbers in the CF_w as compared to the CF_n after 4 days of cultivation. After 10 days of cultivation, the pigment cell numbers drastically reduced in all tested media.

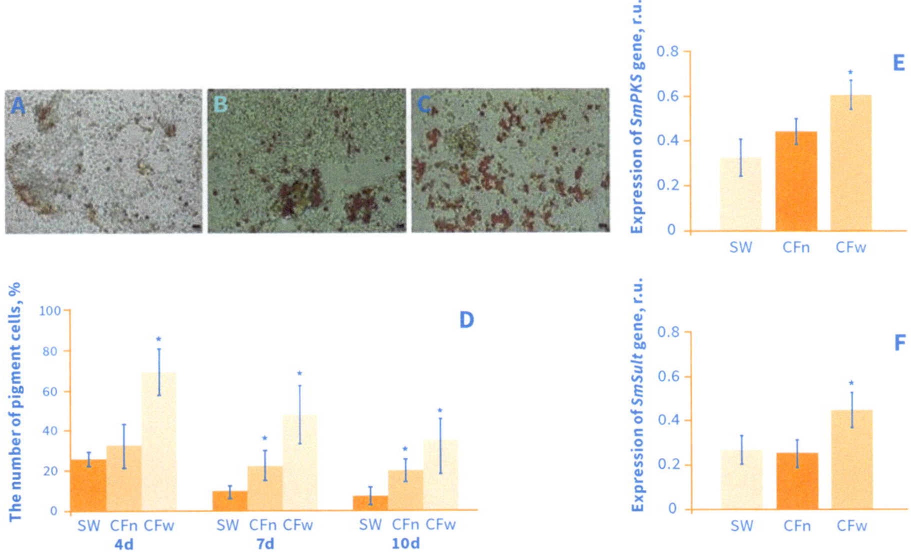

Figure 3. Pigment cells of the sand dollar *Scaphechinus mirabilis*. (**A–C**) Embryonic pigment cells in a blastula-derived cell culture of the sea urchin *S. mirabilis* cultivated for 4 days (bar 10 μm). The cells were cultivated in seawater (**A**); the coelomic fluid of intact sea urchins (**B**); and the coelomic fluid of injured sea urchins (**C**). (**D**) Cellular dynamics of the sand dollar pigment cells cultivated in the three culture media (SW, CF_n and CF_w) over 4–10 days of incubation. In total, >500 cells were counted for each studied culture medium. (**E,F**) Expression of two genes associated with the biosynthesis of naphthoquinone pigments in sand dollar cells maintained under various culture media (SW, CF_n and CF_w) over four days: E. The *pks* expression level; F. the *sult* expression level. Bars represent the mean ± SD, five biological replicates, each with three technical ones. One-way ANOVA followed by Tukey's pairwise comparison test. Significant at $p < 0.05$. Source: Graphics by author Natalya V. Ageenko.

The expression of *pks* in sand dollar cells cultivated in the coelomic fluids was increased when compared with cells cultivated in SW, and the expression levels in CF_w-cultivated cells were significantly (>2-fold) higher than those in cells cultivated in SW (Figure 3E). These results coincide with our data about the number of pigment cells cultivated in the various culture media. The expression profile of *sult* (Figure 3F) had a similar trend to that of the *pks* expression profile.

4. Discussion

Kominami and colleagues reported that sand dollar larvae contain one of the largest amounts of pigment cells compared with larvae of all other tested species of Echinoids (Kominami and Takata 2002). In contrast to our in vivo data obtained previously for the sea urchin *S. intermedius* (Ageenko et al. 2011), in the sand dollar *S. mirabilus*, the timing of *pks* and *sult* expressions differed between the blastula and gastrula stages, indicating the specificity of pigment cell appearance in these two echinoid species. Yet, we confirm previous results on the effects of the culture medium composition on the appearance rates of echinoid pigment cell precursors (Ageenko et al. 2014). As shown in this study, pigment differentiation in cultured sand dollar cells grown in coelomic fluids intensified when compared with cells grown in SW. The distinct changes in the proportion of pigment cells under the CF_w versus the CF_n conditions could be explained by the specific components of coelomic fluids: alternations in protein compositions of coelomic fluids after sea urchin injury compared to control (uninjured) animals and the considerable shift in the absorption maxima for some proteins were previously reported (Ageenko et al. 2014).

However, there is an alternative explanation for these effects, which is connected with carbonic anhydrases (CAs). CA is a participant in the calcification process in numerous invertebrates. Recently, very low concentrations of a specific inhibitor of biomineralization and a potent inhibitor of CAs, acetazolamide, have been found to inhibit pigment cell precursor differentiation, as well as the production of echinochrome in echinoid larvae (Zito et al. 2015). The authors suggest that some isoforms of CAs might be implicated in the production of echinochrome, providing plausible support for the impact of acetazolamide on the pigment cell number in the sea urchin larvae. Currently, the roles of CAs in echinoid larval pigment cell formation are still elusive, requiring further study and analysis.

Numerous endeavors focusing on the development of immortal cell lines from a wide range of marine invertebrate species have been reported, but all have been unsuccessful (Cai and Zhang 2014). Thus, we could not use any echinoid cell line. In this study, q-RT-PCR and cell culture applications were used for the quantitative assessment of pigment cell precursor differentiation in sand dollar primary cell cultures. We revealed that the maximum level of pigment differentiation was reached when the cells of the sand dollar *S. mirabilus* were cultivated in CF_w. The same has previously been reported for *S. intermedius* cultivated cells, and pigment cell numbers were higher when cultured in sea urchin coelomic fluids than in seawater (Ageenko et al. 2014).

In vivo, the highest level of pigment expression in sand dollar embryos (the Sea of Japan, Russia) was observed at the blastula and gastrula stages. In vitro, genes of interest are also significantly expressed in blastula-derived cell cultures, confirming that primary embryonic cell cultures are suitable models for in vitro investigation of pigment differentiation. Further, the employed assay has emerged as a valuable tool for naphthoquinone pigment assessment throughout development and in cell cultures of these sand dollars. The findings contribute to the understanding of the pigment biology of Echinoid cells (Calestani and Wessel 2018; Perillo et al. 2020; Wessel et al. 2020) and create opportunities for the commercial production of natural antioxidants of marine origin. The naphthoquinoid pigments of sea urchins are a promising source for the production of drugs with various pharmacological activities (Lebedev et al. 2005). The use of aqueous solutions of sodium salts of naphthoquinone—echinochrome A in experiments to study the level of emission of the cardiac isoenzyme creatine phosphokinase in the coronary effluent showed a decrease in the size of the necrosis zone (Elyakov et al. 1999b). Thus, echinochrome A was found to have some ophthalmological and cardioprotective properties (Elyakov et al. 1999a, 1999b). Based on the data obtained, new effective drugs with unique therapeutic properties, such as "Histochrome for cardiology" and "Histochrome for ophthalmology", have been developed.

Author Contributions: A.N.V. and O.N.A.: conceptualization, methodology, visualization, investigation, original draft preparation, writing—reviewing and editing. In addition, they took part in the experiments with sand dollar embryos and larvae. A.N.V. and K.K.V.: software, validation. They performed q-RT-PCR experiments and analyzed the obtained data.

Funding: This research received no external funding.

Acknowledgments: This study was partially performed at the Federal Scientific Center of the East Asia Terrestrial Biodiversity, FEB RAS, Vladivostok, Russia.

Conflicts of Interest: The authors declare no conflict of interest. The founding sponsors had no role in the design of the study; in the collection, analyses or interpretation of data; in the writing of the manuscript; or in the decision to publish the results.

References

Ageenko, Natalya V., Konstantin V. Kiselev, and Nelly A. Odintsova. 2011. Expression of pigment cell-specific genes in the ontogenesis of the sea urchin *Strongylocentrotus intermedius*. *Evidence-Based Complementary and Alternative Medicine* 2011: 730356. [CrossRef] [PubMed]

Ageenko, Natalya V., Konstantin V. Kiselev, Pavel S. Dmitrenok, and Nelly A. Odintsova. 2014. Pigment cell differentiation in blastula-derived primary cell cultures of sea urchins. *Marine Drugs* 12: 3874–91. [CrossRef] [PubMed]

Cai, Xiaoqing, and Yan Zhang. 2014. Marine invertebrate cell culture: A decade of development. *Journal Oceanography* 70: 405–14. [CrossRef]

Calestani, Cristina, and Gary M. Wessel. 2018. These Colors Don't Run: Regulation of Pigment Biosynthesis in Echinoderms. *Results and Problems in Cell Differentiation* 65: 515–25. [CrossRef] [PubMed]

Elyakov, George B., Oleg B. Maksimov, Natalia P. Mischenko, Evgeniya A. Koltsova, Sergei A. Fedoreev, Lucia I. Glebko, Natalia P. Krasovskaya, and Alexander A. Artyukov. 1999a. Medicinal Drug "Histochrome" for Treatment of Inflammatory Diseases of Eye Retina and Cardiac Ischemia. Patent RUS No 2134107, August 10, Byull. Izobret. 22, p. 5. (In Russian)

Elyakov, George B., Oleg B. Maksimov, Natalia P. Mischenko, Evgeniya A. Koltsova, Sergei A. Fedoreev, Lucia I. Glebko, Natalia P. Krasovskaya, and Alexander A. Artyukov. 1999b. Medicinal Drug "Histochrome" for Treatment of Acute Cardiac Infarction and Cardiac. Patent RUS No 2137472, September 20, Byull. Izobret. 26, p. 9. (In Russian)

Kiselev, Konstantin V., Natalya V. Ageenko, and Valeria V. Kurilenko. 2013. Involvement of the cell-specific pigment genes pks and sult in the bacteria defense response of the sea urchin Strongylocentrotus intermedius. *Diseases of Aquatic Organisms* 103: 121–132. [CrossRef] [PubMed]

Klotz, Lars-Oliver, Xiaoqing Hou, and Claus Jacob. 2014. 1,4-Naphthoquinones: From oxidative damage to cellular and inter-cellular signaling. *Molecules* 19: 14902–18. [CrossRef] [PubMed]

Koltsova, Evgeniya A., Larisa V. Boguslavskaya, and Oleg B. Maximov. 1981. On the functions of quinonoid pigments in sea-urchin embryos. *International Journal of Invertebrate Reproduction* 4: 17–23. [CrossRef]

Kominami, Tetsuya, and Hiromi Takata. 2002. Process of pigment cell specification in the sand dollar, *Scaphechinus mirabilis*. *Development, Growth and Regeneration* 44: 113–25. [CrossRef] [PubMed]

Lebedev, Alexander V., Marina V. Ivanova, and Dmitri O. Levitsky. 2005. Echinochrome, a naturally occurring iron chelator and free radical scavenger in artificial and natural membrane systems. *Life Sciences* 7: 863–75. [CrossRef] [PubMed]

Martins, Ana, Helena Vieira, Helena Gaspar, and Susana Santos. 2014. Marketed marine natural products in the pharmaceutical and cosmeceutical industries: Tips for success. *Marine Drugs* 12: 1066–101. [CrossRef] [PubMed]

McClay, David R., Jacob Warner, Megan Martik, Esther Miranda, and Leslie Slota. 2020. Gastrulation in the sea urchin. *Current Topics in Developmental Biology* 136: 195–218. [CrossRef] [PubMed]

Medentsev, Alexander G., Anna Yu Arinbasarova, and Vasilii K. Akimenko. 2005. Biosynthesis of naphthoquinone pigments by fungi of the genus Fusarium. *Prikladnaia Biokhimiia i Mikrobiologiia.* 41: 573–77. [CrossRef] [PubMed]

Odintsova, Nelly A. 2009. Stem Cells of Marine Invertebrates: Regulation of Proliferation and Induction of Differentiation in vitro. *Cell and Tissue Biology* 3: 403–8. [CrossRef]

Perillo, Margherita, Nathalie Oulhen, Stephany Foster, Maxwell Spurrell, Cristina Calestani, and Gary Wessel. 2020. Regulation of dynamic pigment cell states at single-cell resolution. *eLife* 9: e60388. [CrossRef] [PubMed]

Pinsino, Annalisa, and Valeria Matranga. 2015. Sea urchin immune cells as sentinels of environmental stress. *D Developmental & Comparative Immunology* 49: 198–205. [CrossRef]

Sakuragi, Yumiko, Boris Zybailov, Gaozhong Shen, Donald A. Bryant, John H. Golbeck, Bruce A. Diner, Irina Karygina, Yulia Pushkar, and Dietmar Stehlik. 2005. Characterization of a menB rubA double deletion mutant in *Synechococcus* sp. PCC 7002 devoid of FX, FA, and FB and containing plastoquinone or exchanged 9,10-anthraquinone. *Journal of Biological Chemistry* 280: 12371–81. [CrossRef] [PubMed]

Smith, L. Courtney, Vincenzo Arizza, Megan A. Barela Hudgell, Gianpaolo Barone, Andrea G. Bodnar, Katherine M. Buckley, Vincenzo Cunsolo, Nolwenn M. Dheilly, Nicola Franchi, Sebastian D. Fugmann, and et al. 2018. Echinodermata: The Complex Immune System in Echinoderms. In *Advances in Comparative Immunology*. Edited by Edwin L. Cooper. Cham: Springer International Publishing AG, pp. 409–501. [CrossRef]

Vasileva, Elena A., Natalia P. Mishchenko, and Sergey A. Fedoreyev. 2020. Spinochromes of Pacific Sea Urchins: Distribution and Bioactivity. *Marine Drugs* 18: 77. [CrossRef]

Wessel, Gary M., Masato Kiyomoto, Tun-Li Shen, and Mamiko Yajima. 2020. Genetic manipulation of the pigment pathway in a sea urchin reveals distinct lineage commitment prior to metamorphosis in the bilateral to radial body plan transition. *Scientific Reports* 10: 1973. [CrossRef] [PubMed]

Zito, Francesca, Demian Koop, Maria Byrne, and Valeria Matranga. 2015. Carbonic anhydrase inhibition blocks skeletogenesis and echinochrome production in *Paracentrotus lividus* and *Heliocidaris tuberculata* embryos and larvae. *Development, Growth & Differentiation* 57: 507–14. [CrossRef]

The Separation of Cell Suspensions Isolated from Coelomic Fluid and Coelomic Epithelium of the Starfish *Asterias rubens* in Percoll Density Gradients

Natalia Sharlaimova, Sergey Shabelnikov, Dan Bobkov and Olga Petukhova

Abstract: The regeneration process assumes the presence in the body of cells capable of self-renewal and subsequent differentiation into specialized cells. Whether these cells are stem cells or are present in circulating fluids or tissues as a pool of reserve progenitor cells, or whether they appear following dedifferentiation/transdifferentiation of specialized cells of individual tissues, are the main questions that scientists are focusing on. Understanding the origin and pathways of differentiation in coelomic fluid cells and coelomocytes of the starfish *Asterias rubens* was the aim of this research. The coelomic epithelium is considered as a possible source of coelomocytes. Further effective studies of coelomocyte replenishment are difficult due to the lack of protein markers characterizing various cell morphotypes. Additional difficulties lie in the heterogeneity of analyzed cell populations. In the present study, we separated cells of the coelomic fluid and the coelomic epithelium, and a subpopulation of the coelomic epithelium enriched with poorly differentiated cells, which are proposed precursors of some types of coelomocytes, in a Percoll density gradient. Characterization of the cell morphology of different fractions and their behavior in vitro (functional characteristics) revealed an enrichment of the gradient fractions in two of eight types of coelomocytes and three of eight morphotypes of cells of the coelomic epithelium.

1. Introduction

The origin of cells contributing to tissue homeostasis and regeneration is one of the fundamental questions of biology (Rinkevich et al. 2022). Typical examples of adult invertebrate stem cells include sponge archaeocytes and choanocytes (Simpson 1984; Funayama 2018), cnidarian interstitial cells (Bosch 2009), flatworm neoblasts (De Mulder et al. 2009; Rossi and Salvetti 2019), and annelid teloblasts (Sugio et al. 2012).

Deuterostome invertebrates provide a significant pool of results supporting the hypothesis of the dominant role of dedifferentiation or transdifferentiation of body cells as the main mechanisms of regeneration, while the participation of stem cells has been proven to tunicate hemoblasts (Ferrario et al. 2020; Kassmer et al. 2020). Studying the origin and fate of individual cells will answer many questions related to elucidating the mechanisms of tissue renewal. A prerequisite for this is the characterization of molecular markers of specialized and undifferentiated cells.

The coelomocyte replenishment of the starfish *Asterias rubens* is an example of maintenance of tissue homeostasis (Pellettieri and Alvarado 2007; Blanpain and Fuchs 2014). Coelomocytes are a heterogeneous cell population of the main body cavity of a starfish, or the coelom. Coelomocytes are responsible for various functions including immune defense, nutrient transport, and formation of aggregates in the zones of body damage (clotting reaction) (Smith 1981; Dogel 1981; Chia and Xing 1996). The coelomic cavity and organs located in it are lined with a ciliated epithelium, called the coelomic epithelium (CE) (Dogel 1981; Blowes et al. 2017).

The concept of the origin of mature coelomocytes from the CE of the starfish *A. rubens* is based on the work of French authors (Bossche and Jangoux 1976). Our own data showed the presence of a significant pool of cells on the surface of the CE, including small poorly differentiated cells named small epithelial cells of type 1 (SECs-1) (Sharlaimova et al. 2014, 2020). SECs-1 comprise about 50% of an individual subpopulation of weakly attached CE cells (CE-W), which can be collected and analyzed separately (Sharlaimova et al. 2014, 2020). Indirect data (experiments with a conditionally intact epithelium) of the same work suggested cell migration from the epithelium. Moreover, morphological analysis suggests that these cells may be precursors of coelomocytes (Sharlaimova and Petukhova 2012).

The question of the origin of SECs-1 on the surface of the CE remains unclear. It could be due to the activity of stem cells that serve as a pool of reserve cells, or it could result from dedifferentiation of specialized cells, for example, ciliated cells of the CE (Bossche and Jangoux 1976) or myoepithelial cells (García-Arrarás and Dolmatov 2010). Further effective studies answering this question are difficult due to the lack of protein markers characterizing various cell populations. To track the fate of cells, it is necessary to identify molecular markers for different types of cells, both undifferentiated and specialized.

The heterogeneity of cell populations is an additional problem for the search for markers of certain types of cells, since the concentration of marker proteins specific for a certain type of cell decreases in the mixture of cells, which leads to a decrease in the efficiency of mass spectrum analysis.

The dominant types of cells in the coelomic fluid (CF) of *A. rubens* identified after azure-eosin staining were small and large petaloid agranulocytes. A homogeneous substance characterized the cytoplasmic matrix of these cells. Eosinophilic granulocytes, roundish agranulocytes, fusiform cells, two types of small cells with a high nuclear–cytoplasmic ratio, and bi- or trinucleated cells were less represented among coelomocytes (Sharlaimova et al. 2020).

Mass spectrometric analysis, performed in our previous work (Sharlaimova et al. 2020), identified only two proteins (integrin alpha 8 and integrin beta 1) that could serve as markers in a cell differentiation study among the proteins in the total coelomocyte suspensions.

The dominant cell type of the CE (30%) is represented by small agranulocytes. A significant part of small agranulocytes, possessing irregularly shaped nuclei and tending to form aggregates, was identified as ciliated epithelial cells after alpha-tubulin staining (Sharlaimova et al. 2014, 2020). Other cell types included large agranulocytes, small azurophilic granulocytes, large eosinophilic granulocytes, morula cells, myoepithelial cells, and two types of small cells with a high nuclear–cytoplasmic ratio (Sharlaimova et al. 2014). Unique CE proteins identified in the total cell suspensions were represented (according to Gene Ontology analysis) by oxidoreductase activities only, although visual inspection revealed one regulatory protein unique to this population (ninjurin) (Sharlaimova et al. 2020). However, several proteins involved in the regulation of proliferation and differentiation processes were identified in a subpopulation of weakly attached CE cells (CE-W), which can be collected and analyzed separately (Sharlaimova et al. 2014). The CE-W cell subpopulation is 50% enriched with small cells with a high nuclear–cytoplasmic ratio (SECs-1). Importantly, SECs-1 demonstrate proliferative activity in vivo and in vitro. They were proposed as precursors of some coelomocyte types (Sharlaimova and Petukhova 2012).

One of the approaches to solving the problem of heterogeneous cell suspension analysis is the preliminary fractionation of cells in density gradients to obtain fractions enriched in specialized morphotypes. Examples of successful separation of invertebrate cells have been reported in the literature (Kudryavtsev et al. 2016; Lin et al. 2001; Kauschke et al. 2001; Hamed et al. 2005).

The aim of this study was the separation of CF and CE cell suspensions by Percoll density gradient centrifugation, and the characterization of cell fractions by histological and immunofluorescent staining and by functional tests in vitro. Morphological and functional analysis of cells of different CF and CE fractions showed that cell separation in the Percoll density gradients allows obtaining fractions of CF and CE cells enriched with certain morphotypes.

Morphological and functional analysis of CF cells showed the enrichment of fraction 1 with roundish coelomocytes unable to form networks (a characteristic property of coelomocytes) and fraction 4 circulatory coelomocytes with petaloid agranulocytes that form networks in vitro. For CE cells, enrichment with ciliated cells in fractions 3 and 4 and enrichment with small epithelial cells of the second type (SEC-2), another type of proliferating CE cell (Sharlaimova et al. 2014), in fractions 1 and 2 were found. For CE-W cells, additional enrichment with SECs-1 in fractions 1–3 was obtained.

2. Materials and Methods

2.1. Animal Manipulation

Experiments were performed at the Biological Station of the Zoological Institute, Russian Academy of Sciences, on Cape Kartesh (Kandalaksha Bay, the White Sea), in September 2018–2020. Intact *A. rubens* L. (Asteroidea, Echinodermata) specimens, 10–15 cm in diameter, were collected off Fettakh Island and kept in cages at a depth of 3–5 m throughout the experimental period. They were fed ad libitum with a diet of mussels. Some experiments were also carried out on animals deprived of food for 4 days.

2.2. Isolation of Circulatory Coelomocytes

The CF was collected after cutting off an arm tip and filtering the fluid through a nylon gauze (70 mesh) into a test tube with saline solution free of Ca^{2+} and Mg^{2+} (CMFSS, Kanungo 1982) and then supplemented with 15 mM EDTA (anticoagulant buffer) (Sharlaimova et al. 2020). The cells were pelleted by centrifugation at $550 \times g$ for 10 min and washed twice in CMFSS. About 200×10^6 circulatory coelomocytes could be isolated from 4 freshly caught starfish with a diameter of 10–15 cm.

2.3. Isolation of Coelomic Epithelium Cells (Epitheliocytes)

Fragments of the CE were detached with forceps from the inner surface of the aboral body wall of the arm and washed with CMFSS. The washing solution obtained at this step contained a considerable number of cells, which were classified as an individual subpopulation of weakly attached CE cells (CE-W) (Sharlaimova et al. 2014, 2020). They were collected and analyzed separately. Remaining CE fragments were treated with 0.05%–0.1% crab hepatopancreas collagenase (Biolot, Russia) in CMFSS for 15 min with periodic pipetting to obtain the dissociated cells. The CE-W cell preparation and dissociated CE cells were filtered through a nylon gauze, pelleted from the suspension by centrifugation at $550 \times g$ for 10 min, and washed twice with CMFSS. About 500×10^6 CE cells and 100×10^6 CE-W cells could be isolated from 4 freshly caught starfish with a diameter of 10–15 cm.

2.4. Cell Separation in Discontinuous (Step) Percoll Density Gradients

CF and CE cell separation was performed in discontinuous (steps 50%–45%–40%–35%–30%–25%, 1 mL each) Percoll density gradients using the

Amersham protocol with modification (Percoll Methodology and Applications Amersham Biosciences[1]).

CF and CE cell separation was performed in discontinuous (steps 50%–45%–40%–35%–30%–25%, 1 mL each) Percoll density gradients using the manufactory protocol (GE Healthcare, Sweden, Uppsala) with modification.

To prepare a stock isotonic Percoll (SIP) solution, 9 parts (v/v) of Percoll were added to 1 part (v/v) of $10\times$ CMFSS solution. To form Percoll density gradients, SIP was diluted to lower densities by adding CMFSS and then layered in 15 mL polycarbonate centrifuge tubes (Sarstedt, Germany), starting with the densest at the bottom of the tube using a 3 mL syringe fitted with a 21 G needle.

The coelomocyte suspension was layered on top in 0.5 mL of CMFSS/5 mM EDTA (about 33×10^6 per gradient), and tubes were centrifuged at $400\times g$ for 20 min at 8 °C using a swing-out bucket. CE and CE-W cell suspensions were layered in 0.5 mL of CMFSS (about 70×10^6 CE cells per gradient and 90–110×10^6 CE-W cells per gradient), and tubes were centrifuged at $400\times g$ for 20 min at 8 °C using a swing-out bucket.

The visible layers of cells at phase boundaries were collected with a Pasteur pipette, transferred into the tubes with 7 mL CMFSS, and then centrifugated at $550\times g$ for 10 min at 8 °C. Cells were resuspended in 1 mL of CMFSS, the number of cells in each fraction was counted with a hemocytometer, and cell suspensions were subdivided for fixation and functional tests. The number of cells in all fractions was summed up, and the proportion of cells in each fraction was calculated (%).

2.5. Histological and Immunofluorescent Staining of CF and CE Cell Suspensions

The circulatory cells of the CF, and CE and CE-W cells were fixed with 4% paraformaldehyde (PFA), placed onto coverslips coated with poly-L-lysine (Sigma) (0.5–1.0×10^6 cells), and stained with azure-eosin or DAPI. Preparations were examined in transmitted light or in fluorescence light at a $\times 100$ objective lens magnification under an Axiovert 200M microscope (Carl Zeiss, Jena, Germany) fitted with a Leica DFC420 digital camera.

The efficiency of cell fractionation in Percoll density gradients was estimated by comparing the proportion of distinct cell morphotypes, identified after staining the nuclei with DAPI under an Axiovert 200M microscope (Carl Zeiss, Germany), in each gradient fraction. In order to use the previously proposed terminology to characterize cell morphotypes (Sharlaimova et al. 2014, 2020), we compared the morphotypes of cells after staining with azure-eosin and DAPI. The criteria were cell and nucleus size and shape, presence or absence of granules, and the pattern

---

[1]   Available online: www.amershambiosciences.com (accessed on 23 July 2021).

and intensity of staining. Cell sizes were determined from images obtained with a camera with a known resolution. Cell counts in three parallel samples were taken in several randomly selected microscopic fields (in total, no less than 400 cells were analyzed in each sample).

2.6. Functional Test of Cell Fractions In Vitro

For the functional test, cells of individual fractions were washed in CMFSS, precipitated by centrifugation at $550\times g$ for 10 min, resuspended in sterile seawater, and plated onto 96-well plates in a volume of 100 µL (0.5×10^6 of CF cells, and 0.8×10^6 of CE and CE-W cells per well). The samples were photographed in transmitted light with an inverted Biolam microscope at different intervals of incubation in seawater: 1.5 and 18 h for coelomocytes, and 18 h for CE cells. For detecting ciliated CE cells, cells from Percoll density gradient fractions were placed onto the coverslips for 18 h, fixed with 4% PFA, permeabilized with 0.1% Triton X-100, and stained with anti-tubulin antibodies, 1:1500 (Sigma, New York, NY, USA), and DAPI (Sigma, USA). Preparations were examined at a $\times 63$ objective lens magnification under a Leica TSC SP5 confocal microscope (Leica Microsystems, Wetzlar, Germany).

2.7. Statistics

The data of CF and CE cell population compositions are expressed as mean $\pm$ SEM ($p < 0.05$); the data concerning the proportion of cells in different fractions are expressed as box and whisker plots with the designation of means, SE, SD, and outliers. All data were processed statistically by ANOVA with Tukey's HSD multiple comparison test to determine significant differences ($p < 0.05$), using the STATISTICA 7.0 Software.

3. Results

3.1. Separation of Cell Suspensions in Percoll Density Gradients

Separation of CF cells and CE cells in Percoll density gradients was carried out for cell suspensions isolated from animals kept in cages.

Separation of both coelomocytes and CE cell suspensions in a six-step gradient led to the concentration of cells in six zones: at the border of 25%—fraction 1; at the border of 25%–30%—fraction 2; at the border of 30%–35%—fraction 3; at the border of 35%–40%—fraction 4; at the border of 40%–45%—fraction 5; and at the border of 45%–50%—fraction 6 (Figure 1A). Preliminary experiments showed that fraction 6 of coelomocytes and the CE contained an insignificant proportion of cells (<0.5%), and they were represented, to a large extent, by cells with fragmented nuclei. Therefore, no further analysis of this fraction was carried out.

3.2. Coelomocyte Separation in Percoll Density Gradients

3.2.1. Composition of Gradient Fractions

The relative proportion of the total cells across the different fractions varied in different experiments. The fact of significant variability in the number of cells in fractions 1 and 4–5 was confirmed by statistical analysis ($n = 12$), showing the maximal variation in cell number in these fractions (Figure 1B).

The efficiency of cell separation in gradients was assessed by comparing the composition of cell fractions with the composition of the total cell suspension. Images were obtained, using a combination of immunofluorescence and bright-field light microscopy, after staining the nuclei with DAPI (Figure 1C). The cell size was determined at the light optical level after the attachment of the fixed cells to the coverslips coated with poly-L-lysine. The cells corresponding to them in morphology, revealed in CF previously after staining of cell suspensions with azure-eosin, are shown next to each type (Figure 1C).

These cells comprise eight morphotypes: small coelomocytes (4.4 ± 0.25 µm) with a high nuclear–cytoplasmic ratio, having discretely (Figure 1C(a)) or densely (Figure 1C(b)) stained nuclei and invisible cytoplasm; petaloid agranulocytes of small (7.33 ± 0.2 µm) and large (11.7 ± 0.7 µm) sizes with densely stained nuclei (Figure 1C(c,d), respectively); roundish agranulocytes with densely stained nuclei (8.2 ± 0.46 µm) (Figure 1C(e)); eosinophilic granulocytes with weakly stained nuclei (10.84 ± 0.28 µm) (Figure 1C(f)); bi- or trinucleated cells with densely stained nuclei (12.4 ± 0.5 µm) (Figure 1C(g)); and fusiform cells (15 ± 0.5 µm) (Figure 1C(h)). In the calculations presented in Figure 1C(a,b), small coelomocytes with a high nuclear–cytoplasmic ratio or discretely or densely stained nuclei were combined into one type due to their insignificant proportion. Enrichment for these types of cells was found in fraction 3. Moreover, fusiform cells were combined with the class of large agranulocytes due to their insignificant proportion (less than 1% in each field of view).

The data analysis showed that fractions 1 and 2 were enriched (58% and 41%, respectively) with roundish (not petaloid) agranulocytes mainly with fine-grained or smooth cytoplasm (Figure 1C(e)). These cells in the total suspension make up an insignificant fraction. Azure-eosin staining showed that roundish cells possessed densely stained nuclei, unlike weakly stained nuclei of eosinophilic granulocytes, and there were no granules detected (Figure 1C(e)). Therefore, it is unlikely that they are granulocytes. Most of the large petaloid agranulocytes are distributed between fractions 3 and 5, while in fractions 1–2, their share is significantly reduced (up to 15% in fraction 1).

Granulocytes and two nuclear cells were more abundant in fraction 5 (35% and 5%, respectively) (Figure 1C(f,g)).

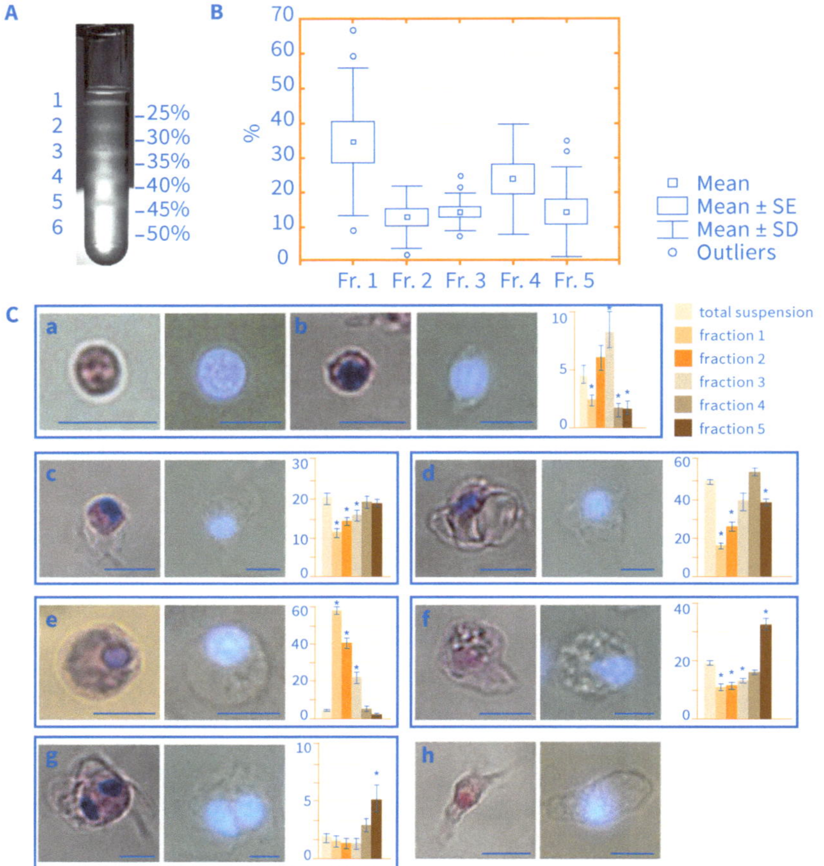

Figure 1. Coelomocyte separation in discontinuous Percoll density gradients. (**A**) An example of *Asterias rubens* coelomocyte separation using six-step Percoll density gradients, (**B**) the portion of cells in different fractions, and (**C**) the coelomocyte types revealed after azure-eosin staining and staining of cell suspensions with DAPI, and the proportion of this type in each gradient fraction: (**a**) small coelomocyte with discretely stained nuclei and invisible cytoplasm; (**b**) small coelomocyte with densely stained nuclei and invisible cytoplasm (data for types (**a**,**b**) were combined); (**c**) small petaloid agranulocyte with densely stained nuclei; (**d**) large petaloid agranulocyte with densely stained nuclei; (**e**) roundish agranulocyte with densely stained nuclei; (**f**) granulocyte with weakly stained nuclei; (**g**) binucleated cell; and (**h**) fusiform cell. Values were obtained from three independent separations. Bar 5 μm. Source: Graphic by authors.

3.2.2. Functional Analysis

Functional analysis was performed for each gradient fraction in each of the 12 cell separation experiments.

The test revealed different behaviors of coelomocytes from Percoll density gradient fractions in vitro and from incubation in seawater. We used the term "behavior" for adhesive characteristics of cells, and both cell–cell and cell–substrate interactions. A typical picture of total coelomocyte suspension behavior is presented in Figure 2A(a): cells after 1.5 h of incubation in seawater formed the nets of cells. However, cells of fractions 1 and 2 did not form networks after 1.5 h (Figure 2A(b,c)) or after 18 h (Figure 2B(b,c)), while cells of fractions 4–5 formed networks after 1.5 h (Figure 2A(e,f)), and a clotting reaction occurred after 18 h (Figure 2B(e,f)). Cells of fraction 3 demonstrated an intermediate behavior (Figure 2A(d),B(d)).

These patterns of cell behavior in vitro were typical for cells of different fractions in about 90% of the experiments. In 1 experiment of the 12 carried out, the coelomocytes did not form networks in the total suspension or in the fractions. Circulatory coelomocytes were isolated from the starfish after 4 days of starvation in this case. The maximum number of cells was found in fraction 1 (Figure 2C). The composition of the total suspension of these coelomocytes significantly differed from that usually observed—the suspension was dominated by roundish (not petaloid) cells with fine-grained cytoplasm (Figure 2D), which in other experiments were observed mainly in fractions 1 and 2.

3.2.3. Separation of CE Cells in Percoll Density Gradients

A comparison of the proportion of cells in different CE fractions revealed the maximum proportion of cells in fractions 3 and 4 and the minimum in fraction 5 (Figure 3A), while for CE-W cells, these fluctuations were not so significant. The maximum variability for cell number was found for fractions 3 and 4 of the CE and fraction 4 of CE-W cells (Figure 3B).

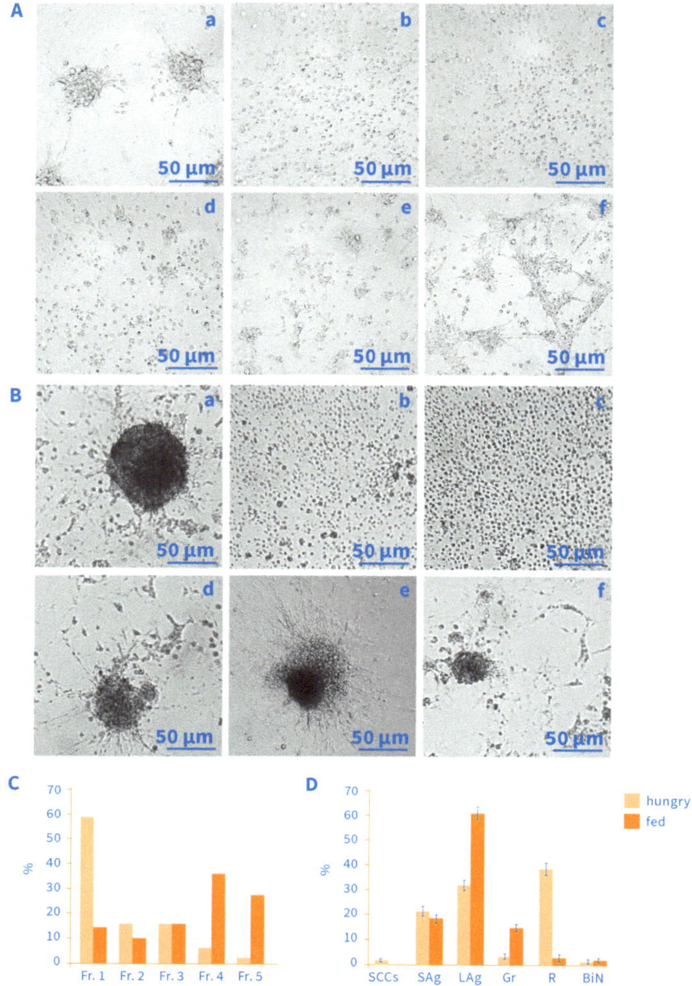

Figure 2. Functional test for Percoll-separated coelomocyte suspensions. (**A**) Behavior of coelomocytes of different fractions after 1.5 h cell incubation in seawater: (**a**) a typical picture of total cell suspension behavior, net formation; (**b–f**) behavior of coelomocytes from fractions 1–5. (**B**) Behavior of coelomocytes of different fractions after 18 h cell incubation in seawater: (**a**) clotting reaction of total cell suspensions; (**b–f**) behavior of coelomocytes from fractions 1–5. (**C**) Comparison of cell number in each fraction in hungry (white columns) and fed (gray columns) animals. (**D**) Comparison of the cell composition of total coelomocyte suspensions for hungry (white columns) and fed (gray columns) animals. SCCs—small coelomocytes; Sag—small agranulocytes; Lag—large agranulocytes; Gr—granulocytes; R—roundish coelomocytes; BiN—binucleated cells. Source: Graphic by authors.

3.2.4. The Composition of CE Cell Gradient Fractions

The cell types of the CE and of CE-W cells (epitheliocytes) are presented in Figure 3C. Analysis of images after staining with DAPI made it possible to subdivide small agranulocytes of the CE into two subtypes. Dominant cell types in the total CE cell suspension (Figure 3C(a)) were small agranulocytes with irregularly shaped nuclei (4.8 ± 0.17 µm), which tended to form aggregates. Only this type of agranulocyte was attributed to ciliated epithelial cells. Small agranulocytes with roundish nuclei (6 ± 0.33 µm) were assigned to a different type (Figure 3C(b)). Other cell types were as follows: two types of small cells with a high nuclear–cytoplasmic ratio: small epitheliocytes with discretely stained roundish (4.06 ± 0.2 µm) or oval ($4 \times 4.06 \pm 0.5$ µm) nuclei (Figure 3C(c)) and invisible cytoplasm (SECs-1) and small cells (3.3 ± 0.2 µm) with densely stained nuclei (SECs-2) (Figure 3C(d)); large agranulocytes with densely stained roundish nuclei (9.5 ± 0.33 µm) (Figure 3C(e)); large eosinophilic granulocytes with weakly stained nuclei and two or more eosinophilic granules in the cytoplasm (9.3 ± 0.5 µm) (Figure 3C(f)). Other cells identified in the CE previously included: small azurophilic granulocytes with densely stained nuclei, morula cells with weakly stained bean-shaped acentric nuclei, and enucleated cells varying in size (2–12 µm) and shape. These were not identified after DAPI staining. Therefore, they were not evaluated in these experiments.

Analysis of the composition of CE gradient fractions revealed the presence of ciliated cells in all fractions of the gradient (Figure 3C(a)). This is the dominant type of cell. In addition, small agranulocytes with roundish nuclei were found in the same fractions in a significant amount. Enrichment of fraction 1 with small cells of type 2 (SECs-2) was found. In addition, the smallest number of large eosinophilic granulocytes was found in fraction 1; their share increased in heavier fractions.

Separation of the subpopulation of CE-W cells, enriched with SECs-1, showed other traits (Figure 3C). SECs-1 were abundantly revealed in fractions 1–3 (Figure 3C(c)). Ciliated cells occupy a much smaller proportion in the total population of CE-W cells compared to the CE. Their share in fractions 1–3 is even lower. Enrichment of fractions 4–5 with small agranulocytes (Figure 3C(b)) and fraction 4 with large agranulocytes (Figure 3C,E) was revealed.

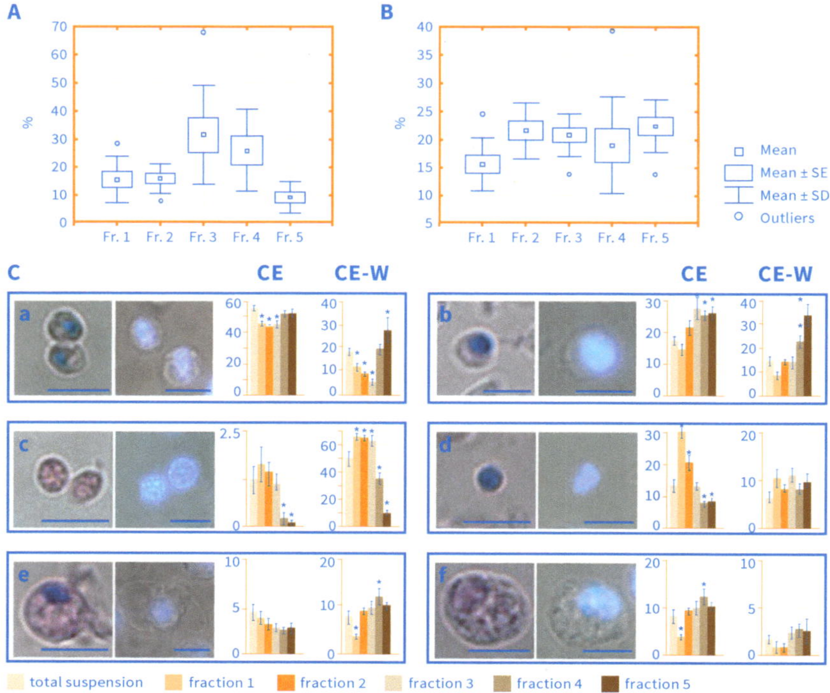

Figure 3. Coelomic epithelium (CE) and CE weakly bound (CE-W) cell separation in discontinuous Percoll density gradients. (**A**) The portion of cells in different fractions of the CE. (**B**) The portion of cells in different fractions of CE-W cells. (**C**) The epitheliocyte types revealed after azure-eosin staining and staining of cell suspensions with DAPI and the portion of this type in each gradient fraction of CE and CE-W cells: (**a**) small agranulocyte with irregularly shaped nuclei (ciliated cell); (**b**) small agranulocyte with roundish nuclei; (**c**) small epitheliocyte with discretely stained nuclei and invisible cytoplasm; (**d**) small epitheliocyte with densely stained nuclei and invisible cytoplasm; (**e**) large agranulocytes with densely stained roundish nuclei; and (**f**) large granulocytes with weakly stained nuclei and two or more granules in the cytoplasm. Bar 5 µm. Source: Graphic by authors.

3.2.5. Functional Test

Cells of different CE fractions exhibited different behavior when incubated in seawater (Figure 4A–F).

The peculiarities were revealed only after 18 h of incubation. Cells of fractions 1 and 2 remained solitary during the entire period of time, while cells of fractions 3–5 formed aggregates. The staining of the cells with anti-tubulin antibody and DAPI showed that these aggregates are composed of ciliated cells (Figure 4G).

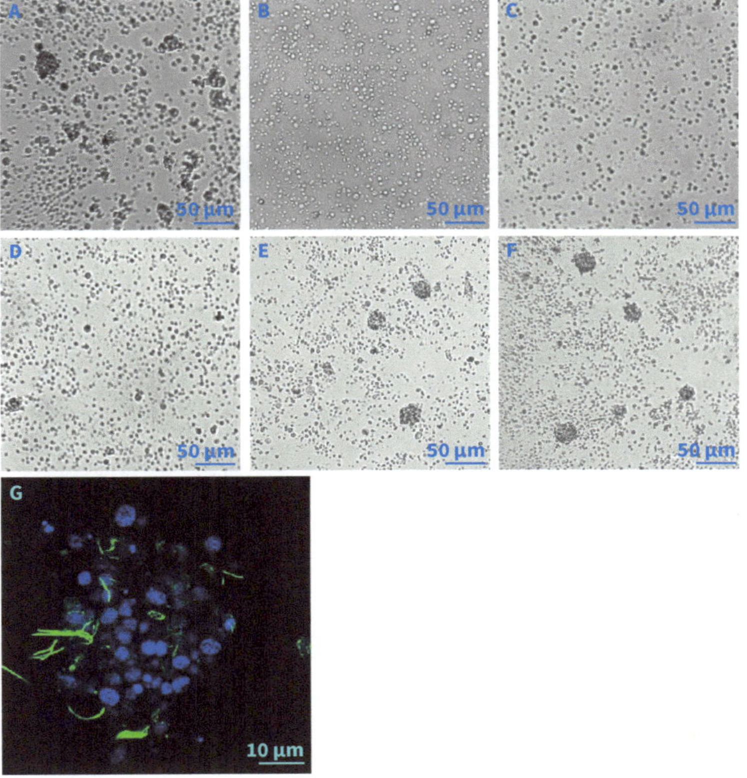

Figure 4. Functional test for Percoll-separated epitheliocyte suspensions after 18 h of cell incubation in seawater. (**A**) A typical picture of total cell suspension behavior, aggregate formation. (**B–F**) Behavior of epitheliocytes from fractions 1–5. (**G**) Aggregate of ciliated cells after the staining of the CE cells with anti-tubulin antibody (green) and DAPI (blue). Source: Graphic by authors.

4. Discussion

In this study, circulatory cells of the CF and cells of two subpopulations of the *A. rubens* CE were separated in Percoll density gradients, and the morphology and behavior of cells in each fraction of the gradient in vitro were characterized.

This study was undertaken with the aim of isolating subpopulations of cells enriched with certain morphotypes for subsequent proteomic analysis. The need for this is associated with the difficulties of assessing the origin of cells in heterogeneous populations based only on morphological data. The general problem can be formulated as follows: to find out whether coelomocytes are a single line of cells at different stages of differentiation, or whether there are distinct sources for different types of cells.

The literature provides examples of the separation of invertebrate cells in density gradients, followed by the characterization of the cytotoxicity and immune characteristics of the cells.

Separation of earthworm (*Eisenia foetida*) coelomocytes in Percoll density gradients revealed four fractions of cells. Four cell types that differ significantly morphologically and functionally (Kauschke et al. 2001; Hamed et al. 2005) were classified in them: acidophils, basophils, chloragocytes, and neutrophils. The second fraction was composed mainly of basophils (40%), and the fourth fraction was enriched with basophils (60%) and neutrophils (35%). Basophils and neutrophils showed the greatest cytotoxic activity against the human immortalized myelogenous leukemia cell line K562 (Kauschke et al. 2001).

Separation of coelomocytes of purple sea urchins, *Arbacia punctulata*, in Percoll density gradients led to isolation of four cell types: white cells, which are called phagocytic amebocytes (>99.5% pure), vibratile cells (>93% pure), granular white spherule cells (white morula cells), and red spherule cells (red morula cells, >99% pure) (Lin et al. 2001). White phagocytic amebocytes showed the greatest cytotoxic activity against human K562 target cells compared to total coelomocytes.

Separation of the blood cells of *Styela rustica* (Styelidae, Stolidobranchiae) in discontinuous Percoll gradients led to more than 90% enrichment with morula cells of the bottom fraction (Podgornaya and Shaposhnikova 1998).

Coelomocytes of *Asterias rubens* were firstly separated in a four-step discontinuous density gradient of sodium diatrizoate (Kudryavtsev et al. 2016). Three cell fractions were obtained after centrifugation of circulatory coelomocytes. Small cells (lymphocyte-like agranulocytes) with a high nuclear–cytoplasmic ratio predominated in the upper fraction (≥95%). These cells expressed a homolog of the C3 gene, a component of the complement system, in response to stimulation with bacterial lipopolysaccharides. Cells with small granules evenly distributed in the cytoplasm were typical for the middle fraction (73%–80%). They demonstrated an ability to produce reactive oxygen species and phagocytosis. The cells of the lower fraction, large coelomocytes with a high content of large granules and vesicles in the perinuclear space (75%–85%), had a high level of hemolytic activity and neutral red uptake.

Centrifugation of coelomocytes and CE cells in a six-step Percoll density gradient, undertaken in the present study, resulted in the appearance of six zones of cells formed at the boundaries of the steps. Cells were mainly redistributed among fractions 1–5. The sixth fraction (55% Percoll) contained an insignificant proportion of cells, mostly destroyed, and was excluded from consideration.

In this study, we identified eight conditional morphotypes of coelomocytes after staining with azure-eosin and DAPI, in contrast to the three types proposed by Kudryavtsev et al. (2016). This led to a more complex picture when characterizing the composition of the fractions. Compared to our previous classification, we identified and isolated another type of CF cells, roundish agranulocytes, which were rarely represented in the total population and previously referred to as agranulocytes. Moreover, DAPI staining revealed a new cell type among the CE cell population, small agranulocytes with roundish nuclei. Roundish coelomocytes and small agranulocytes of the CE with a roundish nucleus are probably independent types of cells of the CF and CE, respectively.

Analysis of the composition of fractions revealed significant enrichment of fractions 1 and 2 with roundish (not petaloid) cells. Their share in the total suspension of coelomocytes was very small. Granules were not detected in the cytoplasm of these cells after staining with azure-eosin. They did not form networks and did not show a clotting reaction in the functional test. Therefore, these cells can be considered separate from petaloid agranulocytes. They could be isolated with an enrichment of 58% in fraction 1 and 40% in fraction 2. Fractions 4–5 were enriched (55% and 40%, respectively) in petaloid agranulocytes, which showed network formation and clotting response in the functional test.

The stability of the results of the functional test was disturbed in only 1 case out of 12, when the cells did not even form networks in the total coelomocyte suspension or in fraction 4. In this case, the total cell population contained 40% of roundish cells, the maximum number of cells was detected in fraction 1, and the proportion of petaloid cells was 30% compared to 50% in other cases. This fact correlated with the 4-day starvation of the animals. A previous study showed the importance of such a physiological parameter as "fed-hungry" for the coelomocyte concentration: starvation leads to a decrease in the concentration of circulating coelomocytes, while feeding leads to an increase in this value (Sharlaimova et al. 2020). The fact of the relationship between starvation and the composition of circulating coelomocytes needs to be confirmed by further research.

The less represented types of cells, small cells with a high nuclear–cytoplasmic ratio, granulocytes, and binucleated cells, also demonstrated enrichment in distinct gradient fractions. However, their number was insufficient for subsequent proteomic analysis. Other approaches are required to isolate these cell morphotypes.

Separation of CE cells in a Percoll density gradient revealed ciliated cells in all fractions of the gradient. However, the functional test exhibited unequal behavior of ciliated cells of different fractions: cells of fractions 1 and 2 remained singular during the entire observation period, and cells of fractions 4 and 5 formed cell aggregates. This indicates the heterogeneity of the buoyant density of ciliated cells, which can be explained by the presence of several types of epithelia in the CE: a flat epithelium on the mesentery, and a cuboid or cylindrical epithelium in other regions of the CE (Sharlaimova et al. 2020).

The more interesting result in the case of CE cell separation is the enrichment in fraction 1 with SECs-2. These cells are the second type of small proliferating cells of the CE with unknown functions, identified both in the cellular suspension and with electron microscopy (Sharlaimova et al. 2014).

The total suspension of the subpopulation of CE-W cells was enriched in SECs-1 cells, proposed to be the progenitor for some coelomocytes. This fact made it possible to identify proteins that can serve as markers of these cells (Sharlaimova et al. 2020). Gradient fractionation further increases the proportion of SECs-1 in the suspension. SECs-1 were distributed among fractions 1–3, that is, in three Percoll densities, which confirms the heterogeneity of the population of these cells. Earlier, the assumption about their heterogeneity was made on the basis of electron microscopic studies (Sharlaimova et al. 2020). Separation of CE-W cells also makes it possible to obtain enrichment with small agranulocytes with roundish nuclei and ciliated epithelial cells in fraction 5, in which the share of SECs-1 and SECs-2 is insignificant. The position and function of small agranulocytes with roundish nuclei, cells that we separated into an independent morphotype based on image analysis after cell staining with DAPI in this study, are unclear.

5. Conclusions

Separation of the coelomic fluid and coelomic epithelial cells in Percoll density gradients made it possible to isolate several enriched morphotypes of cells from heterogeneous populations. For cells of the coelomic fluid, these were roundish agranulocytes and petaloid agranulocytes, presumably two stages of coelomocyte differentiation. Separation of the coelomic epithelium allowed the isolation of small epithelial type 2 cells and ciliated cells, characterized by the ability to form aggregates in vitro. Separation of CE-W cells primarily permitted the isolation of small epithelial type 1 cells, which are proposed progenitors for some types of coelomocytes, and small agranulocytes with roundish nuclei, which are cells with unclear functions. This study creates the basis for proteomic analysis of cell fractions enriched with a certain morphotype.

Identification of surface and membrane protein markers of poorly differentiated cells of the coelomic epithelium, as well as protein markers of specialized cells of the coelomic fluid and coelomic epithelium, allows tracking the differentiation or dedifferentiation of cells. The results of this study contribute to the elucidation of the mechanisms of coelomocyte replenishment.

Author Contributions: Designed the study: O.P.; conducted the experiments and processed the data: O.P., S.S., N.S., and D.B.; prepared the figures: N.S.; writing—original draft preparation, O.P.; discussion and editing: N.S. and O.P. All authors have read and agreed to the published version of the manuscript.

Funding: This research was carried out within the state assignment of the Ministry of Science and Higher Education of the Russian Federation (No. AAAA-A19-119020190093-9) and was funded by the Director's Fund of the Institute of Cytology of the Russian Academy of Sciences.

Acknowledgments: We are grateful to the administration and staff of the White Sea Biological Station "Kartesh" of the Zoological Institute, Russian Academy of Sciences, for providing the conditions for the work and very valuable assistance.

Conflicts of Interest: The authors declare no conflict of interest.

Abbreviations

CE   Coelomic epithelium
CE-W   Subpopulation of weakly attached CE cells
CF   Coelomic fluid
SECs-1   Small epithelial cells of type 1
SECs-2   Small epithelial cells of type 2

References

Blanpain, Cédric, and Elaine Fuchs. 2014. Plasticity of epithelial stem cells in tissue regeneration. *Science* 344: 1242281. [CrossRef]

Blowes, Liisa M., Michaela Egertová, Yankai Liu, Graham R. Davis, Nick J. Terrill, Himadri S. Gupta, and Maurice R. Elphick. 2017. Body wall structure in the starfish *Asterias rubens*. *Journal of anatomy* 231: 325–41. [CrossRef] [PubMed]

Bosch, Thomas C.G. 2009. Hydra and the evolution of stem cells. *BioEssays* 31: 478–86. [CrossRef] [PubMed]

Bossche, J.-P. Vanden, and Michel Jangoux. 1976. Epithelial origin of starfish coelomocytes. *Nature* 261: 227–28. [CrossRef] [PubMed]

Chia, Fu-Sh, and Jun Xing. 1996. Echinoderm Coelomocytes. *Zoological Studies-Taipei-* 35: 231–54.

De Mulder, Katrien, Georg Kuales, Daniela Pfister, Maxime Willems, Bernhard Egger, Willi Salvenmoser, Marlene Thaler, Anne-Kathrin Gorny, Martina Hrouda, Gaëtan Borgonie, and et al. 2009. Characterization of the stem cell system of the acoel *Isodiametra pulchra*. *BMC Developmental Biology* 9: 69. [CrossRef] [PubMed]

Dogel, Valentine A. 1981. *Zoology of Invertebrates*, 7th ed. Moscow: Higher School, p. 606. (In Russian)

Ferrario, Cinzia, Michela Sugni, Ildiko M. L. Somorjai, and Loriano Ballarin. 2020. Beyond Adult Stem Cells: Dedifferentiation as a Unifying Mechanism Underlying Regeneration in Invertebrate Deuterostomes. *Frontiers in Cell and Developmental Biology* 8: 587320. [CrossRef]

Funayama, Noriko. 2018. The cellular and molecular bases of the sponge stem cell systems underlying reproduction, homeostasis and regeneration. *The International Journal of Developmental Biology* 62: 513–25. [CrossRef]

García-Arrarás, José E., and Igor Yu Dolmatov. 2010. Echinoderms: Potential Model Systems for Studies on Muscle Regeneration. *Current Pharmaceutical Design* 16: 942–55. [CrossRef]

Hamed, Sherifa S., E. Kauschke, and E. L. Cooper. 2005. Cytochemical properties of earthworm coelomocytes enriched by percoll. *International Journal of Zoological Research* 1: 74–83. [CrossRef]

Kanungo, Kalpataru T. 1982. In vitro studies on the effect of cell–free coelomic fluid, calcium, and/or magnesium on clumping of coelomocytes of the sea star *Asterias forbesi* (Echinodermata: Asteroidea). *The Biological Bulletin* 163: 438–52. [CrossRef]

Kassmer, Susannah H., Adam D. Langenbacher, and Anthony W. De Tomaso. 2020. Integrin alpha-6+ candidate stem cells are responsible for whole body regeneration in the invertebrate chordate *Botrylloides diegensis*. *Nature communications* 11: e4435. [CrossRef] [PubMed]

Kauschke, Ellen, Kazuo Komiyama, Itaru Moro, Ines Eue, Simone König, and Edwin L. Cooper. 2001. Evidence for perforin-like activity associated with earthworm leukocytes. *Zoology* 104: 13–24. [CrossRef] [PubMed]

Kudryavtsev, Igor V., Ivan S. D'yachkov, Denis A. Mogilenko, Alexander N. Sukhachev, and Alexander V. Polevshchikov. 2016. The Functional Activity of Fractions of Coelomocytes of the Starfish *Asterias rubens* Linnaeus, 1758. *Russian Journal of Marine Biology* 42: 158–65. [CrossRef]

Lin, Wenyu, Haiyan Zhang, and Gregory Beck. 2001. Phylogeny of Natural Cytotoxicity: Cytotoxic Activity of Coelomocytes of the Purple Sea Urchin, *Arbacia punctulata*. *Journal of Experimental Zoology* 290: 741–50. [CrossRef] [PubMed]

Pellettieri, Jason, and Alejandro Sánchez Alvarado. 2007. Cell Turnover and Adult Tissue Homeostasis: From Humans to Planarians. *Annual Review of Genetics* 41: 83–105. [CrossRef]

Podgornaya, Olga I., and Tatjana G. Shaposhnikova. 1998. Antibodies with the Cell-type Specificity to the Morula Cells of the Solitary Ascidians *Styela Rustica* and *Boltenia Echinata*. *Cell structure and Function* 23: 349–55. [CrossRef]

Rinkevich, Baruch, Loriano Ballarin, Pedro Martinez, Ildiko Somorjai, Oshrat Ben-Hamo, Ilya Borisenko, Eugene Berezikov, Alexander Ereskovsky, Eve Gazave, Denis Khnykin, and et al. 2022. A pan-metazoan concept for adult stem cells: The wobbling Penrose landscape. *Biological Reviews* 97: 299–325. [CrossRef]

Rossi, Leonardo, and Alessandra Salvetti. 2019. Planarian stem cell niche, the challenge for understanding tissue regeneration. *Seminars in Cell & Developmental Biology* 87: 30–36. [CrossRef]

Sharlaimova, Natalia, Sergey Shabelnikov, and Olga Petukhova. 2014. Small coelomic epithelial cells of the starfish *Asterias rubens* L. that are able to proliferate in vivo and in vitro. *Cell and Tissue Research* 356: 83–95. [CrossRef]

Sharlaimova, Natalia, Sergey Shabelnikov, Dan Bobkov, Marina Martynova, Olga Bystrova, and Olga Petukhova. 2020. Coelomocyte replenishment in adult *Asterias rubens*: The possible ways. *Cell and Tissue Research* 383: 1043–60. [CrossRef] [PubMed]

Sharlaimova, N.S., and O.A. Petukhova. 2012. Characteristics of populations of the coelomic fluid and coelomic epithelium cells from the starfish *Asterias rubens* L. able attach to and spread on various substrates. *Cell and Tissue Biology* 6: 176–88. [CrossRef]

Simpson, Tracy L. 1984. *The Cell Biology of Sponges*. New York: Springer, p. 662. [CrossRef]

Smith, Valerie J. 1981. The echinoderms. In *Invertebrate Blood Cells*. Edited by Norman A. Ratcliffe and Andrew F. Rowley. London: Academic Press, pp. 513–62.

Sugio, Mutsumi, Chikako Yoshida-Noro, Kaname Ozawa, and Shin Tochinai. 2012. Stem cells in asexual reproduction of Enchytraeus japonensis (Oligochaeta, Annelid): Proliferation and migration of neoblasts. *Development, Growth & Differentiation* 54: 439–50. [CrossRef]

Current Knowledge on Stem Cells in Ascidians

Virginia Vanni, Chiara Anselmi, Loriano Ballarin, Laura Drago, Fabio Gasparini, Tal Gordon, Anna Peronato, Benyamin Rosental, Amalia Rosner, Baruch Rinkevich, Antonietta Spagnuolo, Lucia Manni and Ayelet Voskoboynik

Abstract: Ascidians belong to tunicates, the sister group of vertebrates. Ascidians are cosmopolitan marine filter-feeding organisms that, along with other members of the chordate subphylum, maintain remarkable regenerative abilities throughout their life. Ascidians' high stem-cell-mediated regenerative capacity, which allows colonial species to continuously generate new individuals, has fascinated researchers and scientists. In this chapter, we emphasize what is currently known about the biology and level of involvement of stem cells in ascidian development and regeneration for both solitary and colonial species. The chapter focuses on the methods used to identify stem cells and stem cell niches and discusses hypotheses regarding their role in biological phenomena such as budding, torpor, regeneration, aging, and chimerism. Future areas of study on stem cells using regenerative ascidians are discussed.

1. Introduction

Tunicates, the sister group of vertebrates (Figure 1) (Delsuc et al. 2006, 2018), are filter-feeding marine invertebrates found in harbors, estuaries, and oceans around the world (Burighel and Cloney 1997; Holland 2016).

As members of the phylum Chordata, tunicates develop from swimming larvae that contain all the primary chordate features such as a notochord, dorsal neural tube, segmented musculature, and gill slits (Brusca et al. 2016). After a swimming phase, the larva loses many of its chordate characteristics, metamorphosing into a sessile or pelagic individual (Brusca et al. 2016). Ascidians are the group most studied among tunicates and include both solitary (Figure 1b–e) and colonial (Figure 1f–l) species. Solitary ascidians reproduce via embryogenesis, with individuals developing from a single fertilized egg, while the colonial species produce an adult body through both embryogenesis and diverse types of asexual reproduction (Brien and Brien-Gavage 1928; Oka and Watanabe 1957a; Freeman 1964; Sköld et al. 2011; Lemaire 2011; Manni et al. 2019; Kowarsky et al. 2021). These two disparate reproductive methods ultimately give rise to a similar adult body plan consisting of a simple central nervous system, digestive system, respiratory system, circulatory system, and reproductive system (Manni et al. 2019; Kowarsky et al. 2021) (Figure 1a–d).

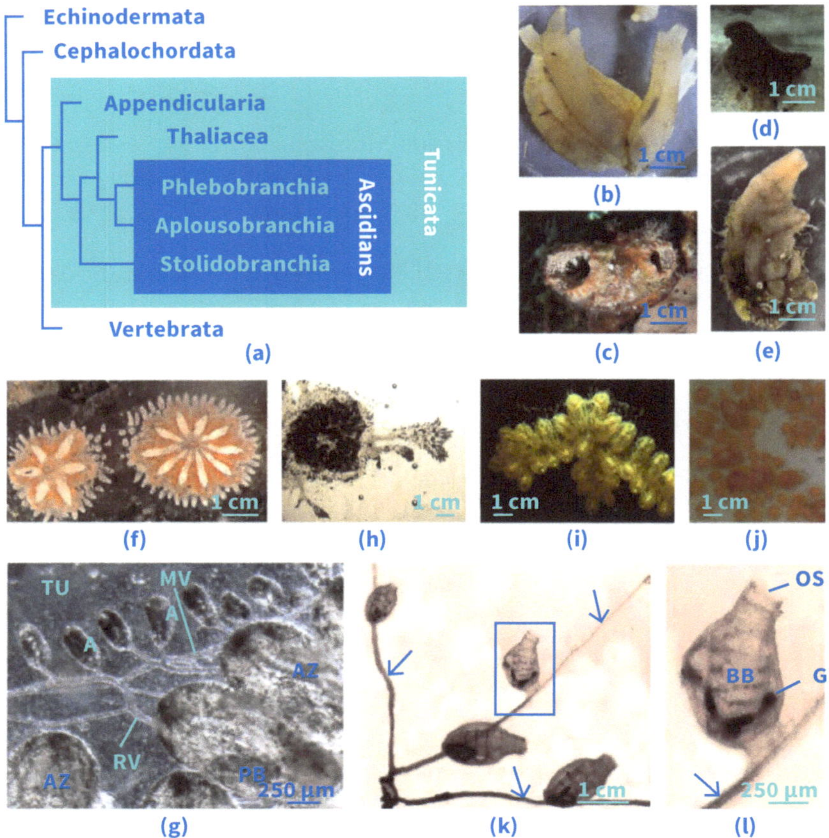

Figure 1. (**a**) Phylogenetic tree of deuterostomes (modified from (Delsuc et al. 2018). Ascidians are considered a paraphyletic group. The three taxa—namely, Phlebobranchia, Aplousobranchia, and Stolidobranchia, also include colonial species able of budding. (**b–e**) Solitary ascidians: *Ciona robusta*, lateral view (**b**); *Polycarpa mytiligera*, upper view (**c**); *Microcosmus exasperatus*, upper-lateral view (**d**); *Styela plicata*, lateral view (**e**). (**f–l**) Colonial ascidians (dorsal view): *Botryllus schlosseri* (**f**) and detail of the colonial circulatory system (**g**, ventral view); *Botryllus primigenus* (**h**); *Botrylloides leachii* (**i**); *Polyandrocarpa zorritensis* (**j**); *Perophora viridis* (**k,l**); arrowheads in (**k,l**): stolon. The Square area in (**k**) is enlarged in (**l**). A: ampulla; AZ: adult zooid; BB: branchial basket; G: gut; MV: marginal vessel; OS: oral siphon; PB: primary bud; RV: radial vessel; TU: tunic. Source: Graphic by authors.

Both solitary and colonial species have high regenerative capacities, with colonial species regenerating an entire body plan from a small fragment of its vasculature (Oka and Watanabe 1957a; Sabbadin et al. 1975; Rinkevich et al. 2007a, 2007b, 2008; Voskoboynik et al. 2007; Manni et al. 2014, 2019; Alié et al. 2021). When colonies come in contact with each other, colonies may form natural chimeras with adjacent colonies by vascular fusion if they share one or two alleles in their highly polymorphic histocompatibility gene, the *Botryllus* histocompatibility factor (BHF); if the colonies are incompatible, a barrier forms between them, and they reject (Oka and Watanabe 1957b; Sabbadin 1962; Scofield et al. 1982; Voskoboynik et al. 2013b).

Following fusion, the circulating stem cells of the chimeric partners compete to replace the germline and/or the soma of the other partner in a process similar to allogeneic transplantation (Oka and Watanabe 1957b, 1959; Sabbadin and Zaniolo 1979; Pancer et al. 1995; Stoner and Weissman 1996; Stoner et al. 1999; Laird et al. 2005; Voskoboynik et al. 2008; Rinkevich et al. 2013).

The colonial ascidians' abilities to reproduce sexually and asexually, regenerate whole body plans, and replace the genotypes of germline and somatic tissues in chimeras has prompted studies aiming to identify and prospectively isolate the stem cells involved in these events.

In this chapter, we review the current knowledge on the stem and progenitor cells in solitary and colonial ascidians, and their involvement in developmental/regeneration processes. Special emphasis is given to the methods used to identify/isolate candidate stem cells and their niches.

2. Ascidians as Model Organisms for Developmental Studies

At the end of the 18th century, studies performed on ascidians established them as key models of chordate development (Corbo et al. 2001; Satoh 2001; Lemaire 2011; Stolfi and Christiaen 2012), sexual and asexual reproduction (Manni et al. 2019; Kowarsky et al. 2021), and the evolution of the immune system (Scofield et al. 1982; Cooper et al. 1992; Oren et al. 2013; Voskoboynik et al. 2013b; Ballarin et al. 2015, 2021a; Franchi et al. 2017; Rosental et al. 2018; Mueller and Rinkevich 2020).

Through a classic chordate embryogenesis process, ascidians produce swimming tadpole-like larvae that, following metamorphosis, lose their chordate characteristics (Lemaire et al. 2008). Taking advantage of the transparent embryos of solitary ascidians, Conklin (1905) performed the first cell lineage experiment in *Styela partita* embryos and discovered that, at the cleavage stage, cells (blastomeres) are committed to the three germ layers: ectoderm, mesoderm, and endoderm. Conklin's studies established ascidians as a key model for embryogenesis. Today's advanced transgenic lineage tracing techniques and single-cell transcriptome trajectories are used on ascidian *Ciona* species to build comprehensive embryonic cell fate maps (Dehal et al. 2002; Lemaire 2011; Oonuma et al. 2016; Tolkin and Christiaen 2016; Cao et al. 2019; Lemaire et al. 2021).

The most studied solitary ascidians for developmental research are the widely distributed *Ciona robusta* (Figure 1b) and *Ciona intestinalis*. The genome of *C. robusta* was one of the first genomes assembled (Dehal et al. 2002), allowing molecular studies on the origin of chordates. Embryos are obtained by in vitro fertilization, and gene reporter assays are used to monitor and manipulate gene expression in vivo as the embryo develops (Squarzoni et al. 2011; Stolfi and Christiaen 2012; Racioppi et al. 2014; Farley et al. 2015; Fujiwara and Cañestro 2018).

Among colonial ascidians, *Botryllus schlosseri* is one of the reference colonial species. Several features make *B. schlosseri* an excellent model organism (Figure 1f,g)—namely, (i) it is abundant in shallow waters and easily cultured in the laboratory; (ii) its genome and transcriptome are available (Voskoboynik et al. 2013a, 2013b; Corey et al. 2016; Campagna et al. 2016; Rosental et al. 2018; Kowarsky et al. 2021; Voskoboynik et al. 2020; Anselmi et al. 2021); (iii) asexual reproduction results in identical individuals, facilitating the ability to separate one colony (genotype) into several clonal replicates (Manni et al. 2007, 2014; Kowarsky et al. 2021); (iv) it naturally forms chimeras, which allow lineage tracing by DNA fingerprints (Stoner and Weissman 1996; Laird et al. 2005); (v) its transparent tissue allows in vivo tracing of labeled cells (Voskoboynik et al. 2008; Rinkevich et al. 2013; Rosental et al. 2018).

3. Stem Cells and Their Identification

The term stem cell derives in part from the word Stammzelle, first used by Ernst Haeckel in the mid-1800s to describe both the single-celled organism precursors to multicellular life and the single-celled embryo that develops into a multicellular organism. The term and its concept were later used by August Weissman to describe cells that he hypothesized to be the common precursor of a specific tissue.

Stem cells must satisfy the following criteria to be classified as stem cells: (i) they can divide and create an identical copy of themselves (self-renewal), and (ii) they can divide to produce other cell types (e.g., hematopoietic stem cells (HSCs) produce all blood cells). They can also express a specific gene signature (e.g., piwi, vasa) and demonstrate a high nucleus:cytoplasm ratio.

Studies on the proliferation state of mammalian HSCs demonstrate that HSCs are quiescent most of the time (remaining in G0) and only on rare occasions enter the cell cycle (Passegué et al. 2005; Forsberg et al. 2010); therefore, proliferation markers, including EdU and PCNA, that detect proliferating progenitor cell populations in many cases do not identify stem cells.

To isolate a pure population of self-renewing HSCs, the Weissman group (Spangrude et al. 1988; Uchida and Weissman 1992; Morrison and Weissman 1994) developed methods that used (i) FACS-based monoclonal antibody cell separation technologies to isolate specific cell populations; (ii) transplantation of limited dilutions of these cell populations to irradiated mice and long term tracing of transplanted cells to assay multipotentiality; (iii) reisolation and transplantation of candidate stem cells from primary recipients to secondary hosts and long term tracing of transplanted cells to assay self-renewal. These became the standard methods to isolate adult tissue-specific stem cells and were used to isolate various tissue-specific stem cells including neural (Uchida et al. 2000) and skeletal stem cells (Chan et al. 2018). A genetic approach that uses fluorescent reporter genes to trace differentiation of single cells was developed to isolate the gut stem cells (Barker et al. 2007). This genetic tracing method also reveals self-renewal and multipotency characteristics.

In order to confirm the involvement of candidate stem cells in regeneration in ascidians, cellular and transgenic methods such as the development of ascidian specific monoclonal antibodies, FACS protocols, transplantation protocols, transgenic animals, lineage tracing, and in vivo cell tracking are required.

4. Stem Cells in Ascidians

Observing *C. intestinalis* hemolymph, Rowley (1982) used the term to indicate cells with a high nucleus:cytoplasm ratio typical of undifferentiated cells. Kawamura et al. (1991) used the same term for cells with similar morphology that migrate and aggregate in the developing buds of *Polyandrocarpa misakiensis*.

The vast majority of data on stem cells in ascidians emerged from studies on *Botryllus schlosseri*. Observing genotype replacement of germline and somatic tissues in *B. schlosseri* chimeras led Pancer et al. (1995) and Stoner and Weissman (1996) to hypothesize that *B. schlosseri* chimerism, cell parasitism, and budding are mediated by stem cells (Pancer et al. 1995; Stoner and Weissman 1996; Stoner et al. 1999; Rinkevich and Yankelevich 2004).

The ability of allogeneic *B. schlosseri* colonies to form chimeras if they share one or two alleles in their histocompatibility gene BHF (Voskoboynik et al. 2013b) allows lineage tracing of transplanted cells using allele-specific markers of host and donor as genotype barcodes (Figure 2) (Pancer et al. 1995; Stoner and Weissman 1996; Stoner et al. 1999; Rinkevich and Yankelevich 2004; Laird et al. 2005; Voskoboynik et al. 2008; Rinkevich et al. 2013). By microinjecting $2.5–5 \times 10^4$ hemocytes into allogeneic partners, Pancer et al. (1995) documented co-sharing and even replacement of the gonads in the recipient partners by the donor cells, as well as continuous somatic chimerism. By transplanting a small number of cells that expressed high aldehyde dehydrogenase activity (ALDH, a stem cell marker), and a set of serial engraftment assays (Figure 2a,b), Laird et al. (2005) further proved that in *B. schlosseri* stem cells are mediating both chimerism and budding. Using transplantation experiments, in vivo cell labeling, and tracing, the anterior ventral side of the endostyle and the cell islands were identified as niches for somatic and germline stem cells (Voskoboynik et al. 2008; Rinkevich et al. 2013). Cells from the endostyle niche migrated via the branchial sac sinuses to buds and contributed to their development (Voskoboynik et al. 2008). Cells from the cell islands migrated to the developing gonads and contributed to their development (Rinkevich et al. 2013). Rosental et al. (2018) adapted FACS to characterize *B. schlosseri* circulating cells and isolated 24 populations. Transcriptome analysis of these populations revealed a cluster of 3 cell populations that differentially upregulated 235 genes homologous (based on sequence) to mammalian genes, known to be expressed in the mammalian hematopoietic stem, progenitor, and myeloid lineage cells. It also revealed three cell populations that highly expressed genes homologous to mammalian genes expressed in cells and tissues of the human reproductive system (testes, ovary, placenta, sperm, and germline). Transplantation experiments and lineage tracing further demonstrated the multipotent potential of the cHSC populations (Rosental et al. 2018) (Figure 2d). Transplantation of labeled cells and in vivo tracing experiments demonstrated migration of cHSC to the endostyle niche, while the cGSC identified based on sequence migrated to the cell island niche (Rosental et al. 2018).

In both colonial and solitary ascidians, it was suggested that wound response and tissue regeneration are mediated by stem cells (Voskoboynik et al. 2007; Voskoboynik and Weissman 2015; Blanchoud et al. 2018; Jeffery 2019; Kassmer et al. 2020; Qarri et al. 2020). A population of circulatory cells was proposed to have stem cell potency (Kawamura et al. 1991; Stoner and Weissman 1996; Voskoboynik et al. 2007; Tiozzo et al. 2008a; Brown et al. 2009; Jeffery 2015a; Kassmer et al. 2020). Genotyping of somatic and germline tissues several months after transplantation of a few cells isolated from specific niches (e.g., endostyle niche, cell islands) or expression specific markers (ALDH) (Figure 2a; Table 1) demonstrated the ability of these cells to contribute to somatic or germline organs (Laird et al. 2005; Voskoboynik et al. 2008; Rinkevich et al. 2013). Transplantation of single cells with high ALDH expression and lineage tracing of their contribution to germline or somatic *B. schlosseri* tissues revealed contribution to either soma (buds) or germline (tests) but not both (Laird et al. 2005) (Figure 2a,b). These results strongly suggest that *B. schlosseri* stem cells are not pluripotent, i.e., they do not produce both germline and soma. The identification of candidate hematopoietic stem cell and germline cell populations in the colonial ascidian *B. schlosseri* (Rosental et al. 2018) (Figure 2c,d) suggests that tissue-specific stem cells mediate bud formation in colonial ascidians. Tissue-specific transcriptional signature and organogenesis timeline during embryogenesis and blastogenesis also strongly support this notion (Kowarsky et al. 2021).

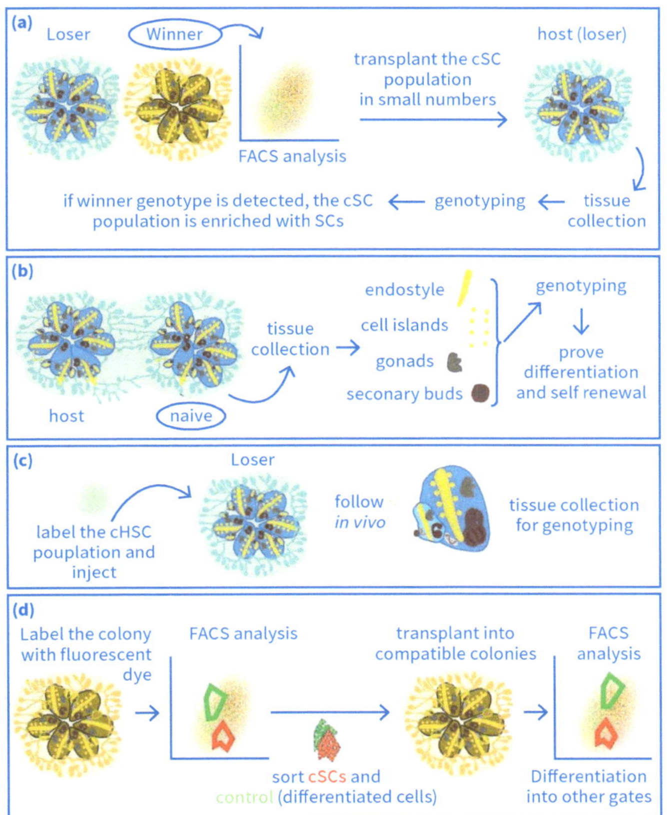

Figure 2. Assaying multilineage contribution, self-renewal capacities, and homing sites of *B. schlosseri* prospective isolated stem cells: (**a**) transplantation of candidate stem cell populations between genetically distinct but compatible colonies and use of tissue genotyping to determine the full developmental potential of transplanted cells; (**b**) primary recipients are fused with secondary naive hosts several months following initial transplantation, tissue genotyping of secondary hosts is used to assay self-renewal capacities; (**c**) candidate stem cells are isolated by FACS, labeled with fluorescent dyes, and transplanted to *Botryllus* blood vessels. Cells are traced in vivo using confocal microscopy via the transparent body of the colonies to identify the location of stem cell niches; (**d**) FACS-based analysis used to demonstrate candidate stem cells differentiation ability. Candidate stem cell and control populations are labeled with fluorescent dyes and transplanted into compatible hosts, a few weeks following transplantation the fluorescent cells from the recipient colonies are analyzed by FACS. While the majority of the transplanted control cell populations are expected to remain in their original gate, the majority of the transplanted candidate stem cell populations are expected to be detected in gates different from their original gates (suggesting they were differentiated). Source: Graphic by authors.

Table 1. Genes expressed in ascidian candidate stem and progenitor cells.

Gene	Species	Expressing Cell(S)	Methods	References
RNA-binding proteins				
Argonaute family silencing genes				
piwi	*Botryllus schlosseri* *Botrylloides leachii* *Botrylloides violaceus* *Botrylloides diegensis* *Ciona intestinalis* *Styela plicata*	hemoblasts, phagocytes, tunic cells, stomach cells, cell islands, endostyle, epithelial cells of the vasculature	ISH, IHC; iRNA	(Brown et al. 2009; Rosner et al. 2009; Rinkevich et al. 2010, 2013; Jeffery 2015d; Jiménez-Merino et al. 2019; Kassmer et al. 2020)
DEAD and DEAH-box-containing helicases				
vasa	*Botryllus schlosseri* *Botrylloides violaceus* *Botrylloides diegensis*	hemoblasts, epithelial cells, phagocytes, stomach cells, cell islands	ISH, IHC	(Brown and Swalla 2007; Rosner et al. 2009; Rinkevich et al. 2013; Kassmer et al. 2020)
pl10	*Botryllus schlosseri*	epithelial cells, some blood cells, phagocytes in cell islands, stomach cells	ISH, IHC, ICC	(Rosner et al. 2006, 2009)
ddx1	*Botryllus schlosseri*	cell islands	IHC, ISH	(Rosner et al. 2013)
Nanos family proteins				
nanos	*Botryllus primigenus*	pharyngeal epithelia of developing budlets	ISH, IHC	(Sunanaga et al. 2008)
RNA recognition motif (RRM)-containing proteins				
dazap1	*Botryllus schlosseri*	buds, during blastogenesis, cHSC	ISH; RNAseq	(Gasparini et al. 2011; Rosental et al. 2018)
Signal transduction pathways				
Wnt				
wnt2B, *wnt5a,* *wnt7a,* *nt5A,* *wnt9Aβ-cat*	*Botryllus schlosseri*	secondary buds (stages 1-3), developing gonads, primary buds, cHSC, endostyle	ISH, RNAseq	(Rinkevich et al. 2013; di Maio et al. 2015; Rosental et al. 2018)
fzd5/8, *β-cat, dsh*	*Botrylloides diegensis*	cycling hemoblasts	ISH	(Kassmer et al. 2020)
TGF-β/BMP				
smad1/2/5/8	*Botryllus schlosseri*	phagocytes, endostyle, cHSC	IHC, ISH, RNAseq	(Rosner et al. 2013; Rosental et al. 2018)

Table 1. *Cont.*

Gene	Species	Expressing Cell(S)	Methods	References
		Notch		
notch1, notch2, notch3, hes1	*Botrylloides diegensis, Botryllus schlosseri*	cycling hemoblasts during WBR, endostyle, cHSC	ISH, RNAseq	(Rosental et al. 2018; Kassmer et al. 2020)
		Kinases		
pm-rack1	*Polyandrocarpa misakiensis*	atrial epithelium of developing buds, undifferentiated mesenchymal cells, pharynx epithelium	ISH, IHC	(Tatzuke et al. 2012)
		Homeobox-containing proteins		
pitx, oct-4, pou-3	*Botryllus schlosseri*	budlets at stage 1-3, oral siphon and tentacles; forming cerebral ganglion, endostyle, developing gut, epithelial cells	ISH, IHC, qRT-PCR	(Tiozzo et al. 2005, 2009; Rosner et al. 2009; Tatzuke et al. 2012; Rinkevich et al. 2013; Ricci et al. 2016)
pou3	*Botrylloides diegensis*	hemoblasts	ISH	(Kassmer et al. 2020)
		Zinc-finger proteins		
GATA4/5/6	*Botryllus schlosseri*	atrial epithelium of the bud at stage 3	ISH	(Ricci et al. 2016)
myc	*Botryllus primigenus, Polyandrocarpa misakiensis*	branchial epithelia, circulating hemocytes of growing palleal and vascular buds (Bs); cells of the atrial epithelium and fibroblast-like cells involved in organogenesis (Pm)	ISH	(Sunanaga et al. 2008; Fujiwara et al. 2011; Kawamura and Sunanaga 2011)
		Chromatin modification/cell cycle/differentiation		
		Histones		
p-h3	*Botryllus schlosseri, Botrylloides diegensis, Styela plicata*	primary and secondary buds, zooidal stomach (Bs, Bd) hemoblasts, intestine submucosa, adults (Sp)	IHC, ISH	(Rosner et al. 2014; Jiménez-Merino et al. 2019; Kassmer et al. 2020)

Gene	Species	Expressing Cell(S)	Methods	References
		Proliferation markers		
pcna	*Polyandrocarpa misakiensis, Botrylloides violaceus, Botryllus schlosseri*	cells of the atrial epithelium of the developing buds, during WBR, hemocytes, cHSC	IHC, RNAseq	(Brown et al. 2009; Kawamura et al. 2012; Rosental et al. 2018)
cyclin b	*Botrylloides diegensis*	hemoblasts during WBR	ISH	(Kassmer et al. 2020)
		Cytostatic proteins		
tc14-1, tc14-3	*Polyandrocarpa misakiensis*	atrial epithelial cells of growing buds, hemoblasts	IHC	(Kawamura et al. 1991; Matsumoto et al. 2001)
		Telomere protection		
pot1	*Botryllus schlosseri*	multipotent epithelia of budlets	ISH	(Ricci et al. 2016)
		Proteins involved in autophagy		
Pm-atg7	*Polyandrocarpa misakiensis*	atrial epithelium of developing buds	ISH	(Kawamura et al. 2018)
		Control of differentiation		
raldh	*Botrylloides leachii, Botryllus schlosseri*	circulating phagocytes, inner epithelium of the bud, endostyle	ISH	(Rinkevich et al. 2007b, 2013; Ricci et al. 2016)
if-b	*Botrylloides leachii*	atrial epithelium of buds	ISH	(Ricci et al. 2016)
		Niche interaction		
cadherin	*Botryllus schlosseri*	aggregates of hemoblasts, aggregates of phagocytes near the endostyle, bud epithelia	ISH, IHC	(Rosner et al. 2007)
cd133	*Botryllus schlosseri*	ampullae epithelium during vasculature regeneration, some hemocytes	ISH, FACS	(Braden et al. 2014)
Ia-6	*Botrylloides diegensis*	hemoblasts	ISH	(Kassmer et al. 2020)
		Others		
Pm-pumpA	*Polyandrocarpa misakiensis*	atrial epithelium of developing buds	ISH	(Kawamura et al. 2018)

Table 1. *Cont.*

Gene	Species	Expressing Cell(S)	Methods	References
Pathways associated with stem cell activity and stem cell niches				
Wnt signaling, Signaling by *Notch*, *SMAD* signaling (See Figure 4g for more)	*Ciona*	endostyle	ISH	(Ogasawara et al. 2002)
Wnt/RET/ HIF-1/ VEGF/ signaling, *Nanog* ESC pluripotency (See Figure 6e for more)	*Botryllus schlosseri*	endostyle cHSC	RNAseq	(Rosental et al. 2018)

5. Development and Regeneration in Ascidians

As stated above, ascidians reproduce through two different pathways—either classic embryogenesis or blastogenesis where an adult organism develops via budding. Solitary ascidian species are restricted to the sexual mode of reproduction, while colonial species reproduce both ways.

The ability to replace or restore cells, tissues, and organs in response to either damage or loss is a remarkable regenerative function shared by many organisms. Ascidians vary in their regeneration capacities from those that have a limited regeneration to those that can replace any missing body part or even regenerate a complete organism.

5.1. Development, Regeneration, and Stem Cells in Solitary Ascidians

5.1.1. Embryonic Development

In solitary ascidians, fertilization occurs externally, and embryos develop on the water's column; a vacuolated layer of internal follicular cells can keep the embryos floating. During embryogenesis, pluripotent embryonic cells gradually restrict their developmental potential as they become committed toward particular tissues or cells. By correlating single-cell transcriptomic data with knowledge regarding cell lineages, recent works systematically examine lineage specification during development in solitary species (Kobayashi et al. 2013; Cao et al. 2019; Ilsley et al. 2020; Sladitschek et al. 2020; Zhang et al. 2020). These studies reveal asymmetric cell divisions and conserved expression of transcription factors involved in cell differentiation trajectories between ascidians and mice and may lead to the identification of the precursors of somatic or germ stem cells in adults. Additional information on germ stem cells is shown below.

5.1.2. Regeneration in Solitary Ascidians

Solitary ascidian reproduction is strictly sexual, and their regeneration capacity has been investigated in a few chosen species since the 19th century (reviewed in Jeffery 2015b). The majority of these species can only regenerate specific body parts, such as the siphons and neural complex (i.e., the brain and the associated neural gland) (Table 1).

The first report on regeneration in *Ciona* dates back to 1891, when Mingazzini, at the Stazione Zoologica in Naples (Italy), demonstrated that the oral and atrial siphons, as well as the brain, could regenerate following ablation (Mingazzini 1891). Later, Hirschler (1914) discovered *Ciona*'s basal body portion can regenerate distal organs, even though the distal portion cannot similarly regenerate basal organs such as the digestive system and heart. The basal body part is able to regenerate distal organs within one month as long as a fragment of branchial sac remains in the basal portion of the body (Jeffery 2015c). During the last century, new studies have confirmed these results and further described the cellular and molecular processes underlying tissue regeneration in this model system (Jeffery 2015a, 2015b, 2015c, 2019).

Partial body regeneration was also studied in the solitary ascidian *Styela plicata* (Stolidobranchia) (Table 2). This species can regenerate both the oral and atrial siphon following ablation (Gordon et al. 2019). By using the niacinamide antagonist 3-acetylpyridine (3AP) that causes lesions in the brain and reduction of glial and neuronal cells, Medina et al. (2015) demonstrated neuron regeneration in this species and the recruitment of circulating candidate stem cells to the lesion site.

A comparative study on the regenerative abilities of four solitary stolidobranch ascidians, *Polycarpa mytiligera*, *Herdmania momus*, *Microcosmus exasperatus*, and *S. plicata* (Figure 3) reported variation in regeneration potential among these species (Gordon et al. 2019). While all species survived and initiated regeneration following ablation of their siphons, only *P. mytiligera* survived the ablation of a larger portion of its body, including both siphons and the brain. A recent study further examined *P. mytiligera*'s regenerative capacity (Gordon et al. 2021). In this study, individuals were cut in two or three fragments along the longitudinal and transverse body axis. After a month, each fragment had reconstituted the whole body and was physiologically active, able to filter feed and respond to stimuli (Figure 3a–c). *P. mytiligera*'s ability to regenerate all tissue and organs distinguishes it from the other solitary species studied so far (Gordon et al. 2021), emphasizing the wide range of regenerative abilities among closely related species. Comparative studies of these species will shed light on the mechanisms underlying regeneration and the evolution of this complex process.

Table 2. Solitary ascidians regeneration capacities and the source of candidate stem cells that mediate it. N/A: not available.

Species	Regenerative Body Structures	Candidate Stem Cell Source	Methods Used to Identify Candidate Stem Cells	References
Ciona intestinalis *Ciona robusta*	Siphons, neural complex, and branchial basket	Branchial basket	Proliferation (EdU, Notch signaling) and stemness markers (PIWI, alkaline phosphatase), Transplantation experiments	(Dahlberg et al. 2009; Auger et al. 2010; Jeffery 2015a, 2015b, 2015c, 2019; Hamada et al. 2015)
Polycarpa mytiligera	Siphons, neural complex, branchial basket, digestive system, and heart	N/A	N/A	(Shenkar and Gordon 2015; Gordon et al. 2019, 2021)
Styela plicata	Siphons and neural complex	Intestinal submucosa, Branchial basket	Morphological characterization proliferation (pHH3) and stemness markers (Aldehyde dehydrogenase activity, PIWI, CD34)	(Medina et al. 2015; Gordon et al. 2019; Jiménez-Merino et al. 2019)
Microcosmus exasperatus	Siphons	N/A	N/A	(Gordon et al. 2019)
Herdmania momus	Siphons	N/A	N/A	(Gordon et al. 2019)

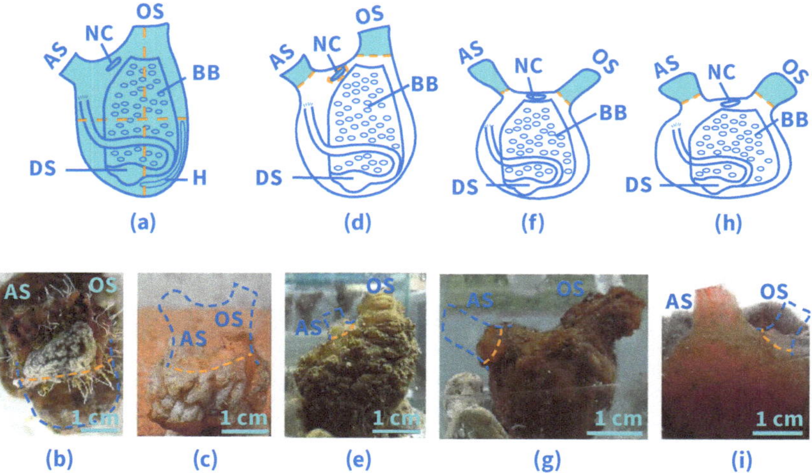

Figure 3. Solitary ascidians regeneration capacity: (**a**–**c**) *Polycarpa mytiligera*: (**a**) illustration summarizing body structures and regeneration capabilities. Dashed red lines indicate ablation lines. Regenerative body parts are highlighted in purple. Note that all organs can be regenerated following amputations; (**b**,**c**) in vivo images of regenerated animals 30 days following ablation along the anterior–posterior body axes: (**b**) anterior body part following ablation of posterior structures (indicated by a black dashed line). Note the open oral (OS) and atrial (AS) siphons; (**c**) posterior body part following ablation of anterior structures. Note the regenerated oral and atrial siphons. (**d**,**e**) *Styela plicata*: (**d**) illustration summarizing body structures and regeneration capabilities; (**e**) in vivo image of atrial siphon regeneration 30 days following ablation. Note the regenerated atrial siphon. (**f**,**g**) *Microcosmus exasperates*: (**f**) illustration summarizing body structures and regeneration capabilities; (**g**) in vivo image of atrial siphon regeneration 30 days following ablation. Note the regenerated atrial siphon. (**h**,**i**) *Herdmania momus*: (**h**) illustration summarizing body structures and regeneration capabilities; (**i**) in vivo image of atrial siphon regeneration 30 days following ablation. Note the regenerated oral siphon. BB: branchial basket; DG: digestive system; H: heart; NC: neural complex. Source: Graphic by authors.

5.1.3. Regeneration in *Ciona*

The involvement of candidate stem cells in solitary ascidian regeneration in response to injury is suggested by morphological, proliferation, and cell migration studies, as well as stem-cell-associated gene expression (Ermak 1975; Jeffery 2019; Jiménez-Merino et al. 2019; Kassmer et al. 2019). However, it is still unclear whether regeneration is accomplished by the proliferation of differentiated cells, activation of quiescent stem cells, recruitment of progenitor cells, or a combination of these strategies.

As stated above, *C. robusta* and *C. intestinalis* are the main solitary species used to study regeneration (Bollner et al. 1992; Dahlberg et al. 2009; Auger et al. 2010; Jeffery 2015a, 2015b, 2015c, 2019). Members of the genus *Ciona* are among the most abundant invasive marine species with a wide geographic distribution (Lambert and Lambert 1998; Lambert 2001; Madariaga et al. 2014). The high accessibility of the species, combined with a simple, transparent body structure and a relatively short life span, associated with the availability of a sequenced genome and various transcriptomes, render these organisms useful models for experimental studies on regeneration (Millar 1952; Jeffery 2015c; Satoh 2019).

Early studies suggested a possible role for circulatory hemoblasts in regeneration (Hirschler 1914; Sutton 1953) (Figure 4). Using histological and light microscopy, these pioneering researchers described tissue regeneration in detail. Following these reports, *Ciona* regeneration research remained relatively silent, until a renewed interest in the field arose with the emergence of stem cell research in the 1990s. Recent studies introduced advanced molecular tools, such as in situ hybridization, immunofluorescent staining, and gene expression to further analyze the cellular and molecular process underlying *Ciona* regeneration. They focused, in particular, on the possible role of adult stem cells in wound response and regeneration (Hamada et al. 2015; Spina et al. 2017; Jeffery 2019; Kassmer et al. 2019; Jeffery and Gorički 2021).

Most regeneration studies on *Ciona* focused on the ability of this species to regrow its oral siphon (OS) (Figure 4c). The OS is composed of longitudinal and circular muscle fibers entrapped within a dense extracellular matrix where vascular sinuses and nerve fibers are present. Externally, the epidermis and the tunic layer cover it. At the base of the OS, a ring of tentacles embeds the coronal organ, a mechanosensory structure (Manni et al. 2006). Ciliated receptor cells at the center of a cup-like structure of orange pigmented cells form the eight pigmented oral siphon sensory organs (OPOs) located along the rim of the OS. The brain, from which several nerves originate, lies at the OS base (Dilly and Wolken 1972; Auger et al. 2010).

Following amputation, the OS regeneration proceeds through the following three phases: (i) formation of wound epidermis, (ii) OPO replacement, and (iii) OS regrowth (Auger et al. 2010). Full siphon regeneration requires a blastema formation supported by the migration of proliferating cells from the branchial sac (long-distance regeneration). However, the regeneration of siphon tip (including OPOs) most likely involves the differentiation of candidate quiescent stem cells already present in the siphonal tissues (short-distance regeneration) (Auger et al. 2010; Jeffery 2015b). The position of the amputation line controls the extent to which short- and long-distance regeneration processes are used (Figure 4c): removal of the entire OS leads to a complex regeneration process that involves both local cells and cells migrating from the lymph nodules of the branchial sac. Conversely, when the amputation line is close to the siphon tip, it results in a faster regeneration process, relying only on local cell reservoirs (Jeffery 2015c). The latter assumption results from the observation that when the OS is fully removed from its base, the patterning of OPO regeneration is not fully conserved, showing duplications in the OPO number. Conversely, the removal of the distal part of the OS results in the complete replacement of the OPOs, both in numbers and structure, even after repeated amputations (Auger et al. 2010). In addition, UV irradiation of the siphon blocks OPO replacement following the removal of the siphon tip, supporting the idea that regeneration is mediated by local events (Auger et al. 2010). The regenerative capacities of *Ciona* are related to age and are compromised in older animals; when siphons are amputated in old animals at any position, the regeneration is often delayed or absent (Jeffery 2015a, 2015d).

Gene expression analysis of regenerating structures in *C. intestinalis* shows upregulation of conserved regulatory signaling pathways such as Notch and TGF-beta, as well as apoptosis-related genes (Hamada et al. 2015; Spina et al. 2017; Jeffery and Gorički 2021) (Figure 4g). Consequently, when the Notch pathway is inhibited, the levels of cell proliferation in the *Ciona* branchial sac and OS blastema is reduced (Hamada et al. 2015), and differentiation of OS muscle cells in the regenerating area is also affected. In particular, recent data indicate that apoptosis is required for OS regeneration and branchial sac homeostasis through activation of Wnt signaling. Notably, after mid-body amputation, these processes are unilateral, since they involve only the basal fragments and not the distal ones (Jeffery and Gorički 2021).

Brain regeneration was first described in 1964 (Lender and Bouchard-Madrelle 1964). Recently, a combination of several methods, including live imaging and functional analyses, along with transgenic animals expressing GFP in most neurons (Dahlberg et al. 2009), revealed that proliferating cells (a potential blastema) accumulated around severed nerve endings. The source of these cells, however, was not identified. The authors speculated that these cells could be progenitor cells already present in the central nervous system since the movement of GFP-positive cells along the axons or migration of undifferentiated cells from other body parts were not detected by confocal time-lapse microscopy. A recent study, however, reports the involvement of proliferating cells, originating in the branchial basket, in brain regeneration (Jeffery 2019). These candidate stem cells supply progenitor cells for regeneration and differentiate into hemocytes, neural, and muscle cells (Jeffery 2019).

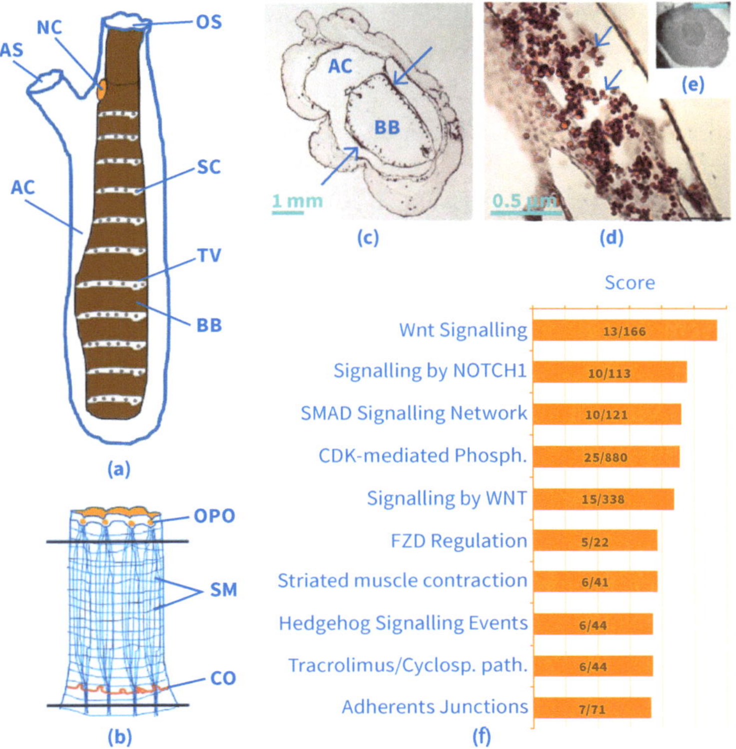

Figure 4. Regeneration and enriched pathways for highly expressed genes associated with stem cell activity in the solitary ascidian *Ciona robusta* endostyle: (**a**) illustration of a young adult individual (sagittal view, dorsal side at left). The branchial basket is perforated by numerous stigmata, delimited by longitudinal and transverse bars where hemocytes flow in vessels; (**b**) illustration of oral siphon. Upper horizontal line: siphon tip; lower horizontal line: siphon base; (**c**) transverse histological section of a juvenile individual (hematoxylin–eosin). Note that in transverse vessels there are hemocyte aggregations (arrows); (**d**) histological section of the same individual shown in (**c**) (hematoxylin–eosin). Detail of hemocytes (arrowheads) in a branchial basket transverse vessel; (**e**) hemoblast, transmission electron microscopy of a juvenile. The hemoblast was recognized at branchial basket level. Scale bar: 2 mm; (**f**) enrichment scores of the top ten pathways of annotated genes in endostyle using GeneAnalytics tool. The gene list used in the analysis is based on all the genes expressed by in situ hybridization (Ogasawara et al. 2002); overall, 185 genes expressed in endostyle were analyzed. In the bars, the number refers to annotated genes out of the list from the total genes in the human indicated pathways. AC: atrial chamber; AS: atrial siphon; CO: coronal organ; G: gut; NC: neural complex; OPO: oral siphon pigmented organ; OS: oral siphon; TV: transverse vessel; SM: siphon muscles; SC: stem cell. Source: Graphic by authors.

5.2. Development, Regeneration, and Stem Cells in Colonial Ascidians

Colonial ascidians possess extreme regenerative capacity known as vascular budding (Sabbadin et al. 1975) or whole-body regeneration (WBR) (Pancer et al. 1995), in which entire colonies regenerate from an aggregation of candidate stem cells in the vasculature (Pancer et al. 1995; Rinkevich et al. 2007a, 2007b; Voskoboynik et al. 2007; Manni et al. 2019; Kassmer et al. 2020). As described above, stem cells have been proven to mediate asexual reproduction in *B. schlosseri* (Laird et al. 2005) and, therefore, most likely also mediate vascular budding in this species.

5.2.1. Embryonic Development

In colonial ascidians, embryos develop inside adult zooids (or outside the parent body, isolated in the tunic) and, depending on the species and the temperature of the water, are released into the water as mature swimming larvae after about a week (Manni et al. 1993; Burighel and Cloney 1997; Winkley et al. 2019). During this development, embryonic stem and progenitor cells divide and generate the primary germ layers (ectoderm, mesoderm, and endoderm). At the morula and blastula stages, a tissue-specific molecular signature can already be detected (e.g., germline, endostyle, nervous system), and these systems subsequently form a swimming larva (Kowarsky et al. 2021). The hatched larva settles and metamorphoses into a sessile oozooid. During *B. schlosseri* embryogenesis, a bud develops within the larva and remains after metamorphosis in the oozooid, initiating asexual reproduction (astogeny) to produce a colony of genetically identical zooids.

The contemporary presence of disparate reproductive strategies (i.e., embryogenesis and blastogenesis) that generate similar individuals (an oozoid from a zygote and zooids from stem cells and progenitor cells), allows colonial ascidians to serve as valuable models to study how stem cells mediate developmental processes (Laird et al. 2005; Manni et al. 2006; Rosner et al. 2014; Voskoboynik and Weissman 2015; Kowarsky et al. 2021). In this context, the origin of hematopoietic stem cells (HSCs) and germline stem cells (GSCs) during embryogenesis and blastogenesis of *B. schlosseri* are of significant research interest (Rinkevich et al. 2013; Rosental et al. 2018; Kowarsky et al. 2021).

HSCs are multipotent stem cells that produce all blood cells in mice and humans (Spangrude et al. 1988). In *B. schlosseri*, candidate HSCs and progenitor cells have been identified, while the endostyle has been identified as their niche (Voskoboynik et al. 2008; Rinkevich et al. 2013; Rosental et al. 2018). During embryogenesis, hemoblasts (undifferentiated cells with a high nucleus–cytoplasm ratio, abundant ribosomes, and cytoplasm with few organelles), and morula cells appear in the early tailbud stage (Kowarsky et al. 2021). By the mid-late tailbud stage, hyaline amoebocytes and pigment cells appear. Macrophage-like cells appear at metamorphosis, and nephrocytes are found in the oozooid. The number of *B. schlosseri* HSC-associated genes present during embryogenesis increases early in development (from the two-cell stage to the morula stage). This gene expression profile includes 239 homologous genes that are known to be expressed in the human hematopoietic bone marrow and 43 with human homologs expressed in HSCs (Kowarsky et al. 2021).

GSCs are the source of the gametes that produce daughter stem and differentiated cells through asymmetric cell division (Spradling et al. 2011). In vertebrates, GSCs segregate early in development producing a small founding population (Ueno et al. 2009). During *B. schlosseri* embryogenesis, cGSCs expressing *vasa* were identified in the early cleavage stage (Brown et al. 2009). Upon isolating the *B. schlosseri* cell populations by FACS, one cell population significantly upregulated 235 genes that were known to be enriched in mammalian germline (Rosental et al. 2018). The genes expressed by this cell population were used in the developmental atlas created by Kowarsky et al. (Kowarsky et al. 2021), to track germline development. The enrichment in GSC-associated genes suggests that, in the embryo, cGSCs develop during the morula stage and proliferate as embryogenesis proceeds. The same study compared the molecular signatures of cHSCs and cGSCs during the embryogenesis and blastogenesis pathways, revealing that both developmental pathways share similar patterns of HSC- and GSC-associated gene enrichment. The same was confirmed for tissue-specific signatures during embryogenesis and blastogenesis. This common trend suggests that tissue-specific stem cells mediate organogenesis with similar molecular dynamics during both sexual and asexual reproduction (Rosner et al. 2019; Kowarsky et al. 2021).

5.2.2. Asexual Reproduction

Colonial ascidian species produce their adult body through asexual reproduction by budding, in a process termed blastogenesis. During embryogenesis, embryonic stem cells differentiate and divide to build the complex adult body of the colony founder, the oozooid. During blastogenesis, asexual reproduction utilizes adult stem cells to clone new bodies and organs. As mentioned above, oozooids derived from metamorphosed larvae carry buds, the precursors for the next generation's zooid. Pharyngeal, stolonal, epicardial, palleal, and vascular budding are various types of blastogenesis described in colonial ascidians (Table 3; Figure 5).

In palleal budding, buds grow out from the body wall, specifically from the epidermis, the epithelium of the peribranchial chamber, and the connective tissue lying between them (Manni et al. 2007, 2014). Morphological studies show that buds form a double vesicle where the outer leaflet will differentiate into the epidermis, while the inner leaflet, originally derived from the peribranchial epithelium, will develop most of the zooidal tissues. This budding mode is used by colonial stolidobranch ascidians and has been mainly studied in *B. schlosseri*.

Figure 5. Regeneration in *B. schlosseri*. (**a–c**) WBR: (**a**) colony before the surgical manipulation; (**b**) colony after the removal of all the zooids of the colony, only the marginal vessel and ampullae are left; (**c**) colony after 5 days from the operation, with an enlargement of the developing vascular bud. (**d–f**) Budectomy induced WBR: (**d**) control colony in which no buds were removed; (**e**) colony after 6 days following the removal of all the buds. When takeover starts, the zooids are only partially resorbed through an attenuated apoptotic process. Tight aggregates of partially absorbed zooids and ampullae are formed. Then, new sporadic transparent elements appear in various sites in the colonial tunic, the new centers of regeneration; (**f**) 20 days after budectomy, functional zooids are differentiated from these regenerating sites. (**g,h**) circulatory system regeneration: (**g**) colony after the removal of a part of the marginal vessel and associated ampullae; (**e**) colony 3 days following partial blood vessel removal. Source: Graphic by authors.

Table 3. Prospective involvement of candidate stem cells in colonial ascidians asexual reproduction modes.

Asexual Reproduction Mode	Species	Candidate Stem Cell Identified and (Methods Used to Identify It)	References
Peribranchial budding	*Botryllus schlosseri*	Multipotent peribranchial epithelia and candidate circulating stem cells (transplantations, labeling, long-term lineage tracing).	(Laird et al. 2005; Voskoboynik et al. 2008; Rinkevich et al. 2013; Rosental et al. 2018)
	Botrylloides violaceus	Candidate circulating vasa expressing cells in buds and vasculature system (ISH)	(Brown and Swalla 2007)
Vascular budding	*Botrylloides leachii/diegensis*	Piwi-positive candidate stem cells lining the vascular epithelium (ISH)	(Rinkevich et al. 2010)
Epicardial budding	*Diplosoma listerianum*	Cells proliferating in the adult and in the bud and high telomerase activity in the buds (BrDU; TRAP)	(Sköld et al. 2011)
Stolonal budding	*Perophora viridis*	Mesenchymal cells form gonads, heart, and cerebral ganglion (morphological studies)	(Lefèvre 1897, 1898)
	Clavelina lepadiformis	Mesenchymal cells form gonads and the nervous system (morphological studies)	(Brien and Brien-Gavage 1928; Brien 1968)

Vascular budding is another budding mode present in colonial stolidobranch ascidians, which occurs under normal conditions or in the aestivation of botryllid colonies. It was first recorded by Savigny (1816) and Giard (1872). Morphological studies suggest that new zooids regenerate from aggregated cells (hemocytes contacting the epidermis lining the hemolymphatic vessels) with the morphological features of undifferentiated cells, such as a small diameter and large round nuclei with packed chromatin (Oka and Watanabe 1957a; Freeman 1964).

Stolonal budding characterizes Clavelinidae and Perophoridae growth (Figure 1k,l). In these taxa, buds develop from the stolon, an outgrowth of the zooid body that connects individual zooids keeping them attached to the substrate. The stolon is bordered by the epidermis and contains 2–3 sinuses (known as vessels), separated by connective tissue (Kott 2001). In *Perophora*, mesenchymal cells accumulate in the growing extremity of the stolon, where they proliferate and develop the bud inner vesicle (Brien and Brien-Gavage 1928; Koguchi et al. 1993). Morphological studies suggest that the outer vesicle originates from the epidermis and will continue to form epidermal layers, while the inner vesicle develops the peribranchial and branchial chambers, as well as the neural gland, gut, and endostyle. Circulating hemocytes participate in the formation of *Perophora's* gonads, heart, and brain (Lefèvre 1897, 1898). Even in Clavelinidae's stolonial budding, mesenchymal cells (also called neurogenital mass) are suggested to be involved in the development of the nervous system and the germline (Brien and Brien-Gavage 1928; Brien 1968). A particular type of stolonal budding, called vasal budding, was recently described in the stolidobranch ascidian *Polyandrocarpa zorritensis* (Figure 1j) (Scelzo et al. 2019). Buds originate from the thickening and invagination of a patch of cells on the epidermis. The invagination leads to the formation of a double vesicle (outer and inner epidermis and the hemolymph between them). Since aggregations of hemoblasts are observed around the forming inner vesicle, it has been suggested that circulating cells also contribute to organogenesis in this budding mode (Scelzo et al. 2019; Alié et al. 2021).

Epicardial budding or strobilation characterizes most colonial Aplousobranchia. In this process, buds derive from epidermal constrictions that enclose part of the epicardium, a tube-like sac originating as an invagination of the pharynx (Sunanaga et al. 2008) and other tissues. Sköld et al. (2011) observed an extensive cell proliferation in growing epicardial buds of *Diplosoma listerianum*.

5.2.3. Whole-Body Regeneration

While some colonial ascidians species continuously develop zooids from their vasculature (Oka and Watanabe 1957a; Freeman 1964; Saito and Watanabe 1985; Okuyama and Saito 2001; Gutierrez and Brown 2017) (Figure 1k,l), other colonial species regenerate the whole body from their vasculature only when injured (e.g., after zooid- and budectomy). This kind of regeneration is known as whole-body regeneration (WBR). *B. schlosseri*, *Botrylloides leachii*, *Botrylloides violaceus*, and *Botrylloides diegensis* are among the species used for WBR studies (Table 4).

In *Botrylloides* species, WBR occurs in isolated fragments of colonial matrix and vasculature (Rinkevich et al. 1995, 2007a, 2007b, 2008; Brown and Swalla 2007; Kassmer et al. 2020). In *B. schlosseri*, WBR can be induced by removing all the individuals from colonies approaching the cyclical generation change or takeover (TO) (Voskoboynik et al. 2007), during which massive apoptosis events occur in adult tissues (Lauzon et al. 1992, 2007; Cima et al. 2010) that are resorbed and succeeded by their primary buds (Figure 5a–c). In this case, WBR requires an intact marginal hemolymphatic vessel of the colony (Milkman 1967; Sabbadin et al. 1975; Voskoboynik et al. 2007; Kürn et al. 2011; Ricci et al. 2016).

Table 4. Regeneration capacity and involvement of candidate stem cells in colonial ascidian regeneration. N/A: not available.

Regenerative Structure	Candidate Stem Cell Description and Identification	Candidate Stem Cell Source	References
Botryllus schlosseri			
Vessels and ampullae	Preexisting vascular tissue-resident cells. Based on vascular cell lineage tracing	Vascular tissue	(Zaniolo and Trentin 1987; Gasparini et al. 2008, 2014; Tiozzo et al. 2008b; Braden et al. 2014)
Whole-body from colonial vasculature	N/A	Hemolymph/colonial vasculature	(Sabbadin et al. 1975; Voskoboynik et al. 2007; Ricci et al. 2016)
Whole-body from body fragments	N/A Circulating cells expressing Pl10	Tissue fragments Hemolymph	(Sabbadin et al. 1975; Majone 1977; Rosner et al. 2019)
Vessels and ampullae	Preexisting vascular tissue-resident cells. Based on vascular cell lineage tracing	Vascular tissue	(Zaniolo and Trentin 1987; Gasparini et al. 2008, 2014; Tiozzo et al. 2008b; Braden et al. 2014)
Botrylloides leachii, Botrylloides diegensis			
Whole-body from colonial vasculature	Candidate stem cells expressing *Piwi*. Based on inhibition of WBR upon injection of siRNA for *Piwi*	Vasculature epithelia	(Rinkevich et al. 1995, 2007b, 2008, 2010; Zondag et al. 2016, 2019)
Botrylloides violaceus			
The whole body from the colonial vasculature	Candidate stem cells in the vasculature expressing Integrin alpha 6. Based on regeneration recovery on colonies treated with mytomycin C and injected with one IA6+ cell and on lineage tracing (EdU)	Circulatory hemocytes	(Brown et al. 2009; Kassmer et al. 2020)
Polyandrocarpa zorritensis			
Whole body	N/A	Hemoblasts, based on morphological data	(Scelzo et al. 2019)

An alternative mode of WBR described in *B. schlosseri* was termed budectomy-induced WBR (Figure 4d–f) (Rosner et al. 2019). Notably, 100% of young colonies (<6 months old) and 50–60% of old colonies (>8 months old) form new zooids within 2–3 weeks following complete budectomy. In this case, adult zooids regularly enter the programmed TO phase. However, the apoptosis process does not culminate in the complete removal of the zooids' debris. Instead, some cells in the degenerating zooids or in the vascular vessels start proliferating to form new zooids. The presence of even a single bud in the colony prevents this mode of regeneration and leads to the full resorption of the zooidal generation and the survival of the single bud, which can reform the colony.

Several members of the *IAP* family of genes, the *PI3K/Akt* pathway, apoptosis signals, as well as signals derived from the buds themselves, are involved in the regulation of this mode of regeneration (Rosner et al. 2019). The tight association between the onset of apoptosis and regeneration may be attributed to a phenomenon called apoptosis-induced compensatory proliferation that has been described in additional animal models (invertebrates and vertebrates) (Bergmann and Steller 2010; Fan and Bergmann 2008). During this process, *caspase 3* activates target genes in a *Xiap* (or its ortholog)-dependent manner.

In stolonal species, such as *P. zorritensis* (Figure 1j), *Clavelina lepadiformis*, and *Perophora viridis* (Figure 1k,l), WBR is induced when part of the stolon is isolated from the remaining colony (Della Valle 1914; Huxley 1921; Brien 1930; Deviney 1934; Ries 1937; Goldin 1948; Scelzo et al. 2019). WBR in *Symplegma reptans*, *P. misakiensis*, and *B. schlosseri* can also occur from isolated bud fragments able to produce new buds before being slowly resorbed (Majone 1977; Sugino and Nakauchi 1987).

In *B. violaceus* (Brown et al. 2009) and *B. leachii* (Rinkevich et al. 2010) WBR, hemocytes adhering to the vasculature epithelium express *piwi*. In *B. violaceus*, the piwi-positive cells show immunopositivity to anti-PCNA antibodies (Brown et al. 2009). Retinoid acid (RA) is required for *B. leachii* WBR: the presence of RA agonist increases the number of buds, whereas RA inhibitors block the process (Rinkevich et al. 2007a). In addition, the development of regenerating buds in *B. leachii* is altered by serine protease inhibitors, suggesting a role to this enzyme during regeneration (Rinkevich et al. 2007b).

Candidate stem cells expressing *integrin-alpha-6* (IA6), *pou3*, and *vasa* have been suggested to mediate WBR in *B. diegensis* (Kassmer et al. 2020). In fragmented *B. diegensis* tissues that were treated by mitomycin C, WBR was triggered by transplantation of a single IA6$^+$ cell. Moreover, when both Notch or canonical Wnt signaling pathways were impaired by treatment with specific drugs, WBR could not be triggered through transplantation of IA6$^+$ cells, suggesting that these pathways play an important role in the process (Kassmer et al. 2020).

This study suggests that a single IA6$^+$ candidate stem cell can mediate WBR in *B. diegensis*. However, long-term tracing of transplanted cells to fully understand their differentiation potential, along with reisolation and transplantation of IA6$^+$ cells from primary recipients to secondary hosts with the same WBR outcomes, will be needed to clarify whether these cells are stem cells, and what is their potency potential (e.g., multipotent, pluripotent cells).

5.2.4. Tunic and Colonial Circulatory System Regeneration

Colonial circulatory system regeneration (CCR) refers to the ability of a colony to regenerate its tunic and the circulatory system following damage. Ascidians possess an open circulatory system containing diverse cells, flowing in hemolymphatic spaces and in sinuses and lacunae of the body wall, delimited by connective tissues (Millar 1953; Kriebel 1968; Monniot et al. 1991). Some colonial ascidians have a system of vessels that cross the tunic and connect between zooids. Tunic vessels originate from the zooid epidermis; therefore, they are not homologous to the mesodermal vertebrate blood vessels. CCR was studied in *B. schlosseri* (Figure 1g; Figure 5g,h), in which full regeneration occurs in a period of time ranging from a few hours to days, depending on the extent of the ablation or the stress that causes vessel degeneration (Zaniolo and Trentin 1987; Gasparini et al. 2008; Qarri et al. 2020; Tiozzo et al. 2008b; Braden et al. 2014). Damage caused by UV exposure to the vasculature was repaired within a few days (Qarri et al. 2020). The incubation with anti-PCNA antibodies revealed that the proliferation of epidermal cells occurs immediately after ablation, as these cells contribute to the synthesis of a new tunic (Gasparini et al. 2008). The regeneration of vasculature is stimulated by the injection into the circulatory system of vertebrate vascular endothelial growth factor (VEGF) and epidermal growth factor (EGF) (Gasparini et al. 2014). Both the knockdown of the VEGF receptor and the inhibition of VEGFR by a chemical agent inhibit vascular regeneration, suggesting the VEGF pathway plays a role in this process (Tiozzo et al. 2008b). Braden et al. (Braden et al. 2014) injected fluorophores to label the cells inside the vasculature in *B. schlosseri* and followed their contribution to vascular regeneration: they identified resident, proliferating cells that expressed homologs of *cd133*, *vegfr*, and *cadherin* and suggested that they contribute to vasculature regeneration.

5.2.5. In and Out of Dormancy

The ability of organisms to become dormant (termed torpor) when rough environmental conditions appear is well documented in marine invertebrates. Some taxa display seasonal torpor where hibernation occurs in the winter, and aestivation occurs in the summer (Storey and Storey 2011). Hibernation and aestivation events are recorded in a wide range of ascidians, including *C. lepadiformis* (De Caralt et al. 2002), several *Perophora* species (Mukai et al. 1983), *Polysyncraton lacazei* (Turon 1992), *Diazona* and *Aplidium* (Nakauchi 1982), *Ecteinascidia turbinata* (Carballo 2000), *Didemnum vexillum* (Valentine 2009), *Pseudodistoma crucisgaster* (Tarjuelo et al. 2004) and botryllid ascidians (Bancroft 1903; Burighel et al. 1976; Rinkevich and Rabinowitz 1993; Rinkevich et al. 1996; Hyams et al. 2017). The torpor states (hibernation/aestivation) were studied in *Botrylloides leachii* on the Levantine coast of Israel (Rinkevich and Rabinowitz 1993; Hyams et al. 2017) and on the Italian coast of the Adriatic Sea (Bancroft 1903; Burighel et al. 1976). The role of stem cells in torpor states of *B. leachii* was first suggested (Rinkevich et al. 1996) as part of the survival budding repertoire of this species, which includes the WBR phenomenon (Pancer et al. 1995). Alongside hibernation, Hyams et al. (2017) revealed high expression levels of genes related to stem cell activity including *piwi, pl10,* and *pcna*, mostly by multinucleated cells, whose numbers were observed to increase during torpor in *B. leachii*. Using in situ hybridization and immunohistochemistry assays, Hyams et al. (2017) documented that *piwi Pl10* and *pcna* expressions during the hibernation processes diverged significantly from normal blastogenesis (asexual growth) related expressions. As the hibernation progressed, the cells that expressed *piwi, Pl10*, and *pcna* significantly increased in numbers, peaking in aroused colonies. In non-hibernating colonies, these markers are highly expressed in the cell islands stem cell niches along the endostyle (Rinkevich et al. 2013).

5.2.6. Stem Cell Aging

As described above, colonial ascidians undergo cyclical formation of new individuals (zooids) by stem-cell-mediated budding (Laird et al. 2005; Voskoboynik et al. 2008). In this cyclical process, zooids die through massive apoptosis as the next generation of buds matures into an entire new replicated zooid body. As the colony ages, both sexual and asexual reproduction methods slow and eventually halt, demonstrating the colonies' reduced regenerative potential (Voskoboynik and Weissman 2015). While the colony can live for years, the zooids live for only a few days, creating unique characteristics that distinguish it from aging in solitary organisms (Rosen 1986; Voskoboynik and Weissman 2015; Rinkevich 2017). The colony ages due to its stem cells that remain and circulate from one generation to the next. As new zooids are formed, the self-renewing stem cells are the cells that are maintained and age throughout the life of the colony (Voskoboynik and Weissman 2015).

Zooid death is part of the botryllids' life cycle and is not indicative of the systemic aging processes that occur within the colony (Borges 2009). A study on the weekly cycle of *B. schlosseri* zooids weekly cycle (Ben-Hamo et al. 2018) has revealed the importance of mortalin (an HSP70 family member that is highly associated with development, cell proliferation, senescence, aging, and apoptosis) for the zooid life cycle. In the planarian *Dugesia japonica, djmot*, the mortalin-like gene is expressed in the neoblasts—the adult stem cells of the animal (Conte et al. 2009).

In *B. schlosseri*, mortalin is highly expressed in the endostyle and putative circulating stem cells, and its expression is reduced in zooids during the takeover stage (Ben-Hamo et al. 2018). It is also expressed in in vitro epithelial monolayers that also express other genes associated with stem cells (Rabinowitz and Rinkevich 2004, 2011).

Lifespans differ between wild *B. schlosseri* colonies grown in the field, compared with colonies reared in the lab. Colonies grown in the field have short, subannual life spans (Grosberg 1988; Chadwick-Furman and Weissman 1995a, 1995b) influenced by seasonal fluctuations of light, nutrients, and temperature. Spring-born colonies have a shorter lifespan of about 3 months, compared with the 8-month lifespan of fall-born colonies (Chadwick-Furman and Weissman 1995a, 1995b). Laboratory-bred colonies exhibit either short (<0.5 years), medium (0.5–2 years), or long (2–20+ years) lifespan (Sabbadin 1969; Boyd et al. 1986; Rinkevich et al. 1992; Lauzon et al. 2000; Voskoboynik and Weissman 2015; Rinkevich 2017; Voskoboynik et al. 2020). However, when an individual *Botryllus* colony is divided into several subclones (clonal replicates), the subclones will often die simultaneously (Rinkevich et al. 1992; Lauzon et al. 2000). This suggests that lifespan in *B. schlosseri* colonies is determined through a heritable factor. Morphological differences are observed in older colonies, such as increased pigmentation, reduced zooid size, and reshaping of the vasculature system (Voskoboynik and Weissman 2015; Voskoboynik et al. 2020; Rodriguez et al. 2021). The diurnal circadian cycle also differs in aged versus young and mid-aged colonies, with younger colonies exhibiting reduced nocturnal heart rate and siphon activity, while aged *B. schlosseri* colonies show no observable circadian changes/changes in heart rate and siphon activity, indicating that diurnal phenotypes diminish with age (Voskoboynik et al. 2020). Using a comprehensive transcriptome sequencing of whole systems, *B schlosseri* colonies were sampled every 3 h over a 24 h period. Samples from three different age groups (36–140 days; 2142–2146 days; 5869–5871 days) (Voskoboynik et al. 2020) revealed that the oscillation patterns of *B. schlosseri* clock and clock-controlled genes declined with age. Age-specific cyclical expressions were found in hundreds of pathways including those associated with known hallmarks of aging (Voskoboynik et al. 2020; López-Otín et al. 2013). Significant age-associated changes were found in the cycling dynamics of genes associated with the *B. schlosseri* enriched HSC and GSC, as well as the endostyle and the central nervous system (Voskoboynik et al. 2020).

A subsequent CNS study (Anselmi et al. 2021) characterized brains from diverse developmental stages and ages, discovering that each week the number of neurons in the zooid brain fluctuates, reaching a maximum of ~1000 cells, and thereafter decreasing while the number of immunocytes increases. Comparing the number of neurons in the brains of young and old colonies, they found that aged brains contain fewer cells. In both weekly degeneration cycles and overall *B. schlosseri* aging, they observed that the decrease in the number of neurons correlates with reduced response to stimuli and with significant changes in the expression of genes for which the mammalian homologous are associated with neural stem cells and neurodegeneration pathways (Anselmi et al. 2021). Among the 411 putative homologous genes that correlate with neurodegenerative diseases (including Alzheimer's, Parkinson's, and dementia), that are expressed in the *B. schlosseri* brain, 71 are differentially expressed between early and late cycle, and 157 are differentially expressed between young and old colonies.

Since stem cells are the only cells that self-renew and are maintained throughout the entire life of the colony, the aging phenotypes described above most likely reflect tissue-specific stem cells exhaustion.

5.2.7. Stem Cell Competition in Development and Aging

As described above, colonial ascidians such as *B. schlosseri* may form natural chimeras with adjacent colonies by vascular fusion if they share one or two alleles in the highly polymorphic histocompatibility gene, BHF (Voskoboynik et al. 2013b). Itinerant GSCs compete in chimeras with heritable winner and loser hierarchies (Stoner et al. 1999; Laird et al. 2005; Rinkevich et al. 2013). These studies revealed fundamental aspects of stem cell biology with relevance to pathological conditions in humans (Weissman 2000, 2015). Studying mammalian stem cells as clones of competing stem cells, the Weissman lab and others discovered that competition between stem cells led to the emergence of myeloid biased HSC clones that dominate aged mice and humans and produced mainly cells from the myeloid lineage when compared to young animals where balanced HSC clones produce cells from both lymphoid and myeloid lineages (Rossi et al. 2005, 2007; Beerman et al. 2010; Pang et al. 2011, 2013). Stem cell competition is also observed in human acute myeloid leukemia where clonal preleukemic progression occurs in the HSC stage and each heritable change increases the competitive competence of the clone vs. normal HSC (Jamieson et al. 2004, 2006; Miyamoto et al. 2000; Jan et al. 2012; Corces-Zimmerman et al. 2014; Jaiswal and Ebert 2014; Sykes et al. 2015), and amongst germline stem cells (Ueno et al. 2009). Understanding the molecular determinants that regulate stem cell competition and the expansion of specific clones throughout an organism's life is now a major area of interest in stem cell aging and cancer and regenerative medicine (Weissman 2015).

6. Adult Stem Cell Niches

The term stem cell niche, originally conceptualized by (Schofield 1978), refers to a discrete anatomical microenvironment where stem cells and their milieu reside, all playing critical roles in maintaining/regulating the stem cell state and self-renewal potential (Fuchs et al. 2004; Saez et al. 2017). Morphologically, all niches hold self-renewal stem cells and their progeny, heterologous cell types, and the surrounding niche-specific extracellular matrix (Chacón-Martínez et al. 2018). Consistent with the strict vertebrate definition in which stem cells are present in their undifferentiated and, in some cases, quiescent states, the vast majority of stem cell niches in ascidians are putative.

6.1. Somatic Stem Cell Niches

In ascidians, prospective stem cell niches have been identified in solitary and colonial species. As previously reported in *Ciona*, putative stem cells residing in pharyngeal sinuses and lymph nodules migrate to the distal regeneration blastema (Figure 4). A short pulse of the DNA synthesis marker 5-ethynyl-2′-deoxyuridine (EdU) labels dividing cells located in the pharyngeal sinuses, while EdU pulse–chase results in the regenerating oral siphon, in an area resembling a blastema. These cells were labeled by anti-piwi antibodies and expressed alkaline phosphatase activity, which is associated with stem cells (Auger et al. 2010; Jeffery 2015b). Furthermore, EdU-labeled cells were observed in *Ciona*-regenerating oral siphons following transplantation of branchial sac fragments, taken from EdU-treated *Ciona* to recipients that were not treated by EdU, but their oral siphon was removed. These results demonstrate the presence of proliferating cells that migrate to regenerating sites and are involved in tissue regeneration (Jeffery 2015b, 2019). Additional putative stem cell niches harboring hemoblast-like cells have been identified in the pharyngeal nodules of *Styela clava* and in the intestinal submucosa of *Styela plicata* (Ermak 1976; Jiménez-Merino et al. 2019).

Using in vivo cell labeling, transplantation experiments, confocal microscopy, and time-lapse imaging, Voskoboynik et al. (2008) found cells with stem cell potentiality in the anterior ventral region of the *B. schlosseri*'s endostyle (subendostylar sinus) (Figure 6a–d). Cells from the endostyle niche divide and migrate to developing organs in buds but do not participate in gonads formation. When a few cells are transplanted from the endostyle niche, they participate in tissue formation and induce long-term chimerism in allogeneic tissues. When a few cells are transplanted from the vasculature, they do not contribute to tissue formation or induce chimerism. Being able to label and monitor cells in vivo by imaging them in their natural niches through the transparent body of this model organism, in combination with the ability to transplant cells between allogeneic colonies, provides a fundamental framework to trace cell differentiation into more mature cell types for studying stem cell development (Voskoboynik et al. 2008). This study was the first to demonstrate the endostyle's role beyond assisting in feeding (secretes mucus) and iodine accumulation (homologous to vertebrate thyroid). Supporting the subendostylar niche as a somatic stem cell niche, a decade later it was shown, through cHSC transplantations and diverse functional essays, that the *B. schlosseri*'s subendostylar niche is harboring cHSC. Furthermore, the endostyle molecular signature further suggested that the vertebrate hematopoietic bone marrow niche evolved from an organ resembling the *B. schlosseri* endostyle (Figure 6e) (Rosental et al. 2018). Specifically, this analysis revealed 337 shared genes with significant upregulation between the *B. schlosseri* endostyle and human hematopoietic bone marrow. These include the genes *foxo3*, needed for hematopoietic stem and progenitor cell maintenance; *notch1*; *smad2*, important for adult murine HSC function; *vwf*, the von Willebrand factor. Analyses of genes expressed in *C. robusta* endostyle based on in situ expression data (Ogasawara et al. 2002) revealed significant similarities with the genes expressed in the endostyle of *B. schlosseri* (Figure 4g) (Rosental et al. 2018). Importantly, other stem cell niches could exist, including ampullae that can regenerate a whole zooid (Voskoboynik et al. 2007) and niches in the branchial sac sinuses (Jeffery 2015a). Considering that many of the candidate stem cell niches found in solitary and colonial ascidian species are associated with sinuses and cells aggregations located in proximity to the branchial sac (e.g., endostyle niche/cell islands/nodule/lymph nodes) a comprehensive study aiming to compare these candidate stem cell niches may reveal conserved elements essential for stem cell maintenance.

6.2. Germ Stem Cells and Their Niches

The oocytes of solitary and colonial ascidians contain a special region called postplasm (Shirae-Kurabayashi et al. 2006; Brown et al. 2009; Rosner et al. 2009), which holds the condensed aggregate of maternal RNA and protein molecules, similar in content and functionality to the germ plasm observed in some organisms (e.g., *C. elegans*, *Drosophila*, Zebrafish). In these organisms, with preformistic modes of germline sequestering, the cells committed to becoming primordial germ cells (PGCs) inherit the germ plasm-like derived maternal components, limiting cell differentiation into the germ lineage to the cell's descendants. In *C. robusta*, postplasm was identified in the posterior-most blastomeres and thereafter in B8.12 cells that were classified as PGCs formed in a postplasm-dependent manner (Wessel et al. 2020). In larvae, those PGCs reside in the ventral side of the tail until metamorphosis, when the PGCs are retracted along with tail tissues into the body trunk and populate the gonads (Shirae-Kurabayashi et al. 2006).

In colonial ascidians, a cGSC-specific transcriptomic signature suggests that PGCs are established at the embryo's morula stage (E1.4) and proliferate as the embryo grows (Kowarsky et al. 2021). At this stage, candidate germ precursor cells expressing vasa were identified (Brown et al. 2009), with candidate PGCs identified in the embryo body trunk but not observed within the tail (Rosner et al. 2009). Examples of postplasm materials that are important for PGC specification include *pem*, *piwi*, and *vasa* gene products. In *C. robusta*, *pem* functions to repress somatic specific gene expressions in early germline at the level of polymerase II (Pol II) activity (Strome and Updike 2015). Pem proteins are only transiently expressed during the early specification of germ cells, while during later stages, their function is replaced by a chromatin repression mechanism (e.g., in *Halocynthia roretzi*) (Zheng et al. 2020).

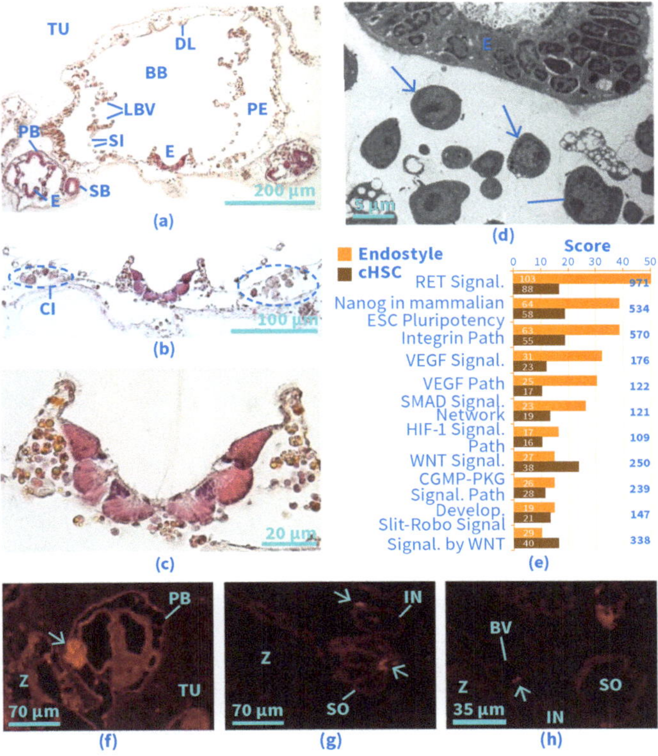

Figure 6. Putative stem cells and stem cell niches in *B. schlosseri*: (**a**–**c**) endostyle and cell island niches. Dotted lines in b: cell islands; (**d**) hemoblasts (arrows) in endostyle niche, transmission electron microscopy; (**e**) enrichment scores of pathways associated with stem cell activity that are expressed in *B. schlosseri* endostyle and enriched HSCs using GeneAnalytics tool. The gene list used in the analysis is based on gene expression data of isolated endostyles and enriched *B. schlosseri* HSCs populations described in Rosental et al. (2018). The number in the bars indicates the number of genes that were significantly upregulated in endostyle/cHSC populations and annotated to human genes in the specific pathway. The numbers on the right indicate the total number of genes in the specific pathway known in humans. Bars indicate the score of the pathway; high and medium scoring pathways associated with stem cells that appear in both gene sets (endostyle and HSCs) were used; (**f**–**h**) immunohistochemical analyses of fixed sections with cy3+coupled BS-Vasa polyclonal antibodies: (**f**) staining of the gonad (arrow) within a primary bud; (**g**) vasa-positive cells aggregate attached to the zooidal stomach and intestine (arrows); (**h**) vasa-positive cells aggregate in the hemolymphatic vessel (arrow) and attach to the zooidal stomach. BB: branchial basket; BV: blood vessel; CI: cell island; DL: dorsal lamina; E: endostyle; LBV: longitudinal branchial vessels; IN: intestine; PB: primary bud; PE: peribranchial chamber; SB: secondary bud; SI: stigmata; SO: stomach; TU: tunic; Z: zooid. Source: Graphic by authors.

Vasa is considered a key marker of PGCs and germ lineages, although its expression was also detected in the somatic cells of various aquatic animals, including ascidians (Rosner et al. 2009). Vasa is an RNA helicase involved in the remodeling of RNA structure and in the regulation of genes translation. Moreover, vasa protein acts on the piwi-interacting RNA (piRNA) metabolic process and, together with piRNA and piwi proteins, governs the transposons methylation needed for their repression to ensure germline integrity (Siomi and Kuramochi-Miyagawa 2009; Kuramochi-Miyagawa et al. 2010). As such, *vasa* and *piwi* expressions were studied in many solitary and colonial ascidians (Fujimura and Takamura 2000; Shirae-Kurabayashi et al. 2006; Brown et al. 2009; Rosner et al. 2009, 2013). In *B. primigenus*, when vasa-expressing cells were depleted from the colony, vasa-expressing germ cells reappeared in the colony to form piwi-expressing candidate germ stem cells (Kawamura and Sunanaga 2009).

Studies suggest that in some marine invertebrate taxa with high regenerative aptitude (e.g., sponges, hydrozoans, and planarians), the adult stem cells can differentiate into both germ and somatic lineages (Buss 1982, 1983; Blackstone and Jasker 2003; Extavour and Akam 2003; Juliano et al. 2010; Alié et al. 2015; Fierro-Constaín et al. 2017; Rosner et al. 2021).

Considering the high regenerative capacity of ascidians, the potential for ascidian adult stem cells to differentiate into germ lineage in an alternative parallel mode of PGC sequestering has been investigated in both solitary and colonial species. In *C. robusta*, PGCs removal by cutting larval tails is compensated by the regeneration of the germ cells from cells that otherwise are assumed to have a somatic fate (Takamura et al. 2002; Yoshida et al. 2017; Wessel et al. 2020). However, since the cells behind this phenomenon were not identified yet, it is not clear whether this mode of sequestering is restricted to a specific time window during development or if PGCs regenerate from soma/germ stem cells or by trans- or de-differentiation of other cells. Experiments performed with TALEN-induced mutations in germ lineage (Yoshida et al. 2017) suggest that this mode of induction might occur without the removal of the original PGCs, and the cells involved in this process might be of epidermal, neural, muscle, or stem cell origin. Opposing conclusions were drawn by Laird et al. (2005) working with *B. schlosseri* and tracing the fate of single cells transplanted into genetically distinct individuals. This research, which was further strengthened by Voskoboynik et al. (2008) and Rinkevich et al. (2013), implies that cells with self-renewing and differentiation abilities can differentiate into somatic or germ cells but not both. These opposing results might reflect differences between species or even solitary versus colonial variations. However, single-cell lineage tracing experiments are needed to solve this discrepancy.

No matter the mode of germ cell sequestering, the germ cell precursors are always formed earlier than the gonads and the PGCs and migrate (passively or actively) to the gonad, which might be relatively far apart from the PGCs. PGC motility is associated with the regulation of the level of their adhesion molecules at the onset and end of the movement and acquisition of amoeboid movement during the migration (Grimaldi and Raz 2020). Molecules that were associated with this movement include G protein-coupled receptors and Dead-end protein (Dnd), an RNA-binding protein involved in cell survival and fate that regulate proteins of the "motility module" (Grimaldi and Raz 2020). Colonial ascidians are characterized by repeated weekly migration of PGCs to the gonads of the newly formed buds. There, apart from the gonads situated in the buds that serve as niches for the germ lineages, it seems that additional "temporary niches" exist in various zooidal tissues including the cell islands (Figure 6b) (Rinkevich et al. 2013; Rosner et al. 2013). In *B. schlosseri*, cell islands were identified as niches for putative germ stem cells (Rinkevich et al. 2013). Expression of genes associated with germ and general stem cells was shown within them. These include *piwi, alkaline-phosphatase, vasa, pl10,* and *pcna.* Transplantation of whole-cell islands induces chimerism in the gonadal tissues. Moreover, labeling of cells in the cell islands leads to the appearance of the stain 10 days later in the gonads, including testis and ovaries of the newly developed zooids (Rinkevich et al. 2013). Isolated by cell sorting, a candidate GSC population, uniquely expressed 80 genes known to be expressed in mammalian germline, migrated to the cell islands following transplantation (Rosental et al. 2018), providing more support to the identification of the cell islands as a germline stem cell niche. Migrating PGCs in *B. schlosseri* were defined as BS-Vasa$^+$-BS-DDX1+BS-cadherin$^+$-γ-H2AX$^+$-phospho-Smad1/5/8$^+$ cell aggregates (Rosner et al. 2013). These PGCs form complexes, mediated by BS-cadherin (Rosner et al. 2013), with follicular cells expressing members of the TGF-b family, which are the migratory unit during PGCs migration to the gonads (Langenbacher and Tomaso 2016). Additionally, changes in the migration of germ cells between old and new gonads in the new generation of buds are due to a chemotactic signal along a sphingosine-1-phosphate gradient (Kassmer et al. 2015) and involves also an ABC transporter-mediated autocrine export of an eicosanoid signaling (Kassmer et al. 2020).

7. Stem Cells as a Unit of Natural Selection

Discussing chimerisms in slime molds, Buss (1982) hypothesized that cells can compete within a chimera and take it over. By studying chimeras of the colonial chordate *Botryllus schlosseri,* the Weissman group discovered that natural selection operates at the level of GSC clones, which compete for niches within the organism's body. Chimeras usually produce only GSC's from one chimeric partner, despite maintaining the soma of both, leading to reproductive pressures toward increasingly competitive GSC's (Stoner and Weissman 1996; Stoner et al. 1999; Weissman 2000, 2015; Laird et al. 2005; Rinkevich et al. 2013). Weissman (2000) further suggested that stem cells are not only units of biological organization, responsible for the development and the regeneration of tissue and organ systems, but are also units in evolution by natural selection. On the other hand, considerations of somatic adult stem cells of animals, including cancer stem cells (another type of adult stem cell not discussed here; Greaves 2013) as units of selection are not trivial, because of the failure to identify the hierarchical level upon which natural selection operates and what exactly is being selected (Rinkevich 2000; Greaves 2013). Indeed, in vertebrates and ecdysozoan invertebrates, adult stem cells are observed as pools of undifferentiated cells capable of self-renewal, proliferation, and production of a number of differentiated but lineage-restricted progenies, all for the general maintenance and various regeneration needs. Yet, the literature on non-ecdysozoan invertebrates (e.g., Gremigni and Puccinelli 1977; Rinkevich et al. 2007b; Ereskovsky et al. 2015; Hyams et al. 2017; Ferrario et al. 2020) suggests that adult stem cells carry a great degree of plasticity in their functions; therefore, the tissue-specific and lineage-restricted adult stem cell view, mainly derived from studies on vertebrates, may need to be expanded. Flexibility in the adult stem cell destiny allows high capabilities for regeneration and changes in cell fates in response to any emerging need; however, experiments that enable long-term lineage tracing of a single cell must be employed before conclusions regarding cell plasticity are made.

In many animal taxa (including sponges, cnidarians, and platyhelminths), the germline is not sequestered from somatic cells early in ontogeny and during the lifespan of the organism germ cells are continuously developing from somatic cells (Buss 1982; Blackstone and Jasker 2003; Müller et al. 2004; Seipel et al. 2004; Rinkevich et al. 2009; Rosner et al. 2009; Gold and Jacobs 2013; Dannenberg and Seaver 2018; DuBuc et al. 2020; Mueller and Rinkevich 2020; Vasquez-Kuntz et al. 2020). In non-chimeric metazoans, somatic and germ cell lineages share a single heritable genotype. In contrast, within a chimera, genotypically different somatic lineages compete for survival, as do germ cell lineages (Buss 1982; Stoner and Weissman 1996; Stoner et al. 1999; Rinkevich 2002a, 2002b, 2004a, 2004b, 2005a, 2011; Rinkevich and Yankelevich 2004; Simon-Blecher et al. 2004; Laird et al. 2005; Voskoboynik et al. 2008; Rinkevich et al. 2013). The genotype that dominates among the somatic cells likely confers some survival advantage and is subject to forces of natural selection. However, heritable germ cell lineages of one genotype may survive within the chimeric entity even though they do not contribute to the somatic tissue (Stoner and Weissman 1996; Stoner et al. 1999; Rinkevich and Yankelevich 2004; Laird et al. 2005; Voskoboynik et al. 2008; Rinkevich et al. 2013). In these cases, the germline is said to hitchhike on or parasitize the soma of a different genotype, transferring heritable traits unseen by natural selection forces to subsequent generations that then express these non-selected "parasitic" traits. The newborn individual carrying a parasitic genotype need not reach sexual maturity to pass on an "unfit" germline genotype to the next generation, as it may quickly fuse with adults or other offspring (e.g., Grosberg 1988) with "fit" somatic cells for continued germline hitchhiking. As a result, superparasitic germ cell genotypes, most capable of dominating foreign soma, may emerge in a population. Thus, germline parasitism may defy the Darwinian paradigm (Rinkevich 2011).

The above notions are further amplified in multichimerism (multipartner associations), where more than two allogeneic adult stem cells form a single botryllid ascidian colony (Rinkevich 1996; Rinkevich and Shapira 1999; Stoner et al. 1999; Paz and Rinkevich 2002). Multipartner chimeras grow faster and produce larger colonies when compared with chimeras made of two partners. They also exhibit other traits associated with more stable entities including fewer cases of morphological resorption or fragmentation events. Following the above, it was proposed that in multichimeras, the different intraspecific conflicts mitigate each other, generating an improved entity (the benefits of the conspecific adult stem cells living in a group exceed the cost of not doing so) where natural selection may act on the level of the whole colony instead of on each conspecific adult stem cell.

As in all stem cells, botryllid ascidian adult stem cells are self-renewing cells capable of differentiation. Five traits highlight these stem cells as genuine units of selection: (a) they efficiently migrate within the organism and between compatible organisms; (b) they compete with the host somatic and/or germline stem cells; (c) they express high and unlimited replication capacity; (d) they may share the soma with conspecific stem cells lineages and commonly determine specific traits for the benefit of the chimeric organism as a whole; (e) they can inhabit several different hosts. As a result, chimerism reflects cases where specific environmental pressures lead to the takeover of the fittest stem cells and their clones (Buss 1982; Rinkevich and Yankelevich 2004).

8. Future Directions on Stem Cells in Ascidians

Although ascidians represent a group of chordates exhibiting astonishing stem cell-mediated processes, most of the significant progress has been made in the last two decades. These advancements were mainly due to the application of unbiased methods translated from mammals to these marine invertebrates and the accessibility of omics methodologies. In the future, the study of ascidians will undoubtedly unravel stem cell potentialities, contributing to the basic knowledge of these cells. In this respect, ascidian simplicity and evolutionary closeness to vertebrates make them unique. Nonetheless, there are still several limitations to exploiting ascidians as a model organism for studies in stem cells biology, mainly due to the lack of methodological tools (such as cell lines, panels of specific monoclonal antibodies) and to the limitations of maintaining colonies in inland facilities, away from seawater supply.

8.1. Cell Cultures and Transgenesis

Immortal cell lines (established cell lines), including cell lines of adult stem cells, may provide an important tool in the research; yet, established cell lines for ascidians, as for all other marine invertebrates, are not yet available (Rinkevich 1999, 2005c). Nevertheless, several attempts in the last three decades have focused on the development of in vitro approaches. The first attempt (Rinkevich and Rabinowitz 1993) concentrated on the development of cell culture from the whole *B. schlosseri* hemocyte populations, followed by the establishment of embryo-derived cell cultures (Rinkevich and Rabinowitz 1994). Then, a series of studies followed the expression of stem cell-associated genes in in vitro cultures of epithelial cells from *B. schlosseri's* buds (Rinkevich and Rabinowitz 1997; Rabinowitz and Rinkevich 2004, 2005, 2011; Rabinowitz et al. 2009). Although well developed in solitary ascidians, transgenic lines are still not available for colonial species. This is mainly due to the difficulties encountered in treating eggs for gene delivery: in colonial ascidians, fertilization is internal, and eggs are enveloped by follicular cells. Yet, the availability of this technique, coupled with the transparency of colonial tissues, facilitates the ability to monitor in vivo the fate of stem cells and to uncover the molecular pathways that control stem cell proliferation and differentiation.

8.2. Monoclonal and Polyclonal Antibodies

Monoclonal antibodies (MAbs) are primary markers in biological sciences. The development of species-specific MAbs is highly valuable in research in general and in stem cells isolation. Initially, several MAbs were developed for experiments performed on *B. schlosseri*. The first sets were target antigens located on *B. schlosseri* hemocyte surface (Schlumpberger et al. 1984a) and embryonic cells (Schlumpberger et al. 1984b). An MAb that recognized all *B. schlosseri* hemocytes and zooids perivisceral epithelium was also developed (Lauzon et al. 1992). Aiming to develop MAbs that recognize epitopes involved in botryllid historecognition, Fagan and Weissman (1998) produced a MAb that labeled an epitope found on the atrial siphon and on the inner surfaces of hemolymphatic vessels. The above sets of experiments further revealed the existence of a MAb that specifically recognized and bound to all somatic cells of one genotype but did not react against somatic cells of another genotype and was used to follow somatic cell movements between partners within chimeras (Rinkevich 2004b). Ballarin et al. produced a monoclonal antibody recognizing a surface epitope on *B. schlosseri's* germ and accessory cells, tunic cells, and hemocytes (Ballarin et al. 2011). However, the majority of previously tested MAbs are no longer available (Rinkevich personal communication), and renewed efforts are needed to establish new panels of MAbs for the research of stem cells in ascidians. In parallel, Lapidot et al. (2003) have established a MAb specific to the *B. schlosseri* pyloric gland cells, and Lapidot and Rinkevich (2005,

2006) developed panels of MAbs specific to cell surface antigens and to intracellular epitopes. Additional polyclonal antibodies were developed against specific stemness proteins of botryllid ascidians and include *B. schlosseri* specific anti-pl10, anti-vasa, anti-cadherin antibodies (Rosner et al. 2006, 2007, 2009), and *B. leachii* specific anti-piwi antibodies (Rinkevich et al. 2010).

8.3. Animal Breeding Methodologies

Established ex situ, inland culturing methods for ascidians species and the development of inbred lines and defined genetic stocks are important prerequisites for research (also alleviating seasonal availability of animals and laboratory acclimatization problems), primarily when dealing with stem cell studies. While research for the cultivation of ascidians under laboratory conditions started decades ago (e.g., Grave 1937), very little has been achieved when considering defined genetic stocks. For colonial species, animal breeding methodologies for the long-term development of inland brood stocks were employed primarily on *B. schlosseri*, representing three various ex situ approaches—one developed in Italy (Brunetti et al. 1984; Sabbadin 1960), another in the USA (Milkman 1967; Boyd et al. 1986) and Israel (Rinkevich and Shapira 1998), and a third, for *Botrylloides simodensis*, in Japan (Kawamura and Nakauchi 1986). Using classical breeding experiments, Yasunori Saito established defined homozygous and heterozygous lines for distinct histocompatibility genotypes (AA, BB, AB, and AX) that were crossed and maintained in the Hopkins Marine Station mariculture for several decades (De Tomaso et al. 1998; Voskoboynik et al. 2013b). These lines added compelling evidence that histocompatibility in *Botryllus* is controlled by a single gene, and they were used to isolate the *Botryllus* histocompatibility factor (BHF).

There were also attempts for inland culturing of other colonial ascidians, such as S. reptans (Sugino and Nakauchi 1987), and Didemnum vexillum (Fletcher and Forrest 2011; Rinkevich and Fidler 2014). Culturing systems have been established for four solitary species: *C. robusta*, *C. intestinalis*, *H. roretzi*, and recently, for *P. mytiligera* (Hendrickson et al. 2004; Joly et al. 2007; Li et al. 2020; Gordon et al. 2020).

8.4. Model Species for Studying Stem Cells and Ascidian Biodiversity

Most of the studies on regeneration and asexual reproduction in ascidians focused on a limited number of species, i.e., the solitary *C. intestinalis* and *C. robusta* and the colonial *B. schlosseri*. Several tools and protocols have been tuned for these animals and different laboratories use them as model species, even in absence of genetically defined lines. However, in recent years, molecular studies suggest the presence of cryptic species with the same nomenclature. In the case of *Ciona* sp., before 2015, the name *C. intestinalis* was used to indicate what is currently known as either *C. intestinalis* or *C. robusta* (Brunetti et al. 2015; Pennati et al. 2015; Gissi et al.

2017). Recently, the species *B. schlosseri* has been redescribed (Brunetti et al. 2017), since five divergent clades have been hypothesized under its name: *B. schlosseri* represents the clade A; *Botryllus gaiae* the clade E (Brunetti et al. 2020); clades B-C have not been determined yet. The uncertainty in species identification for botryllid ascidians has further been discussed (Reem et al. 2018). This equivocal taxonomical determination represents gaps of knowledge at several levels: firstly biological, but also operative, since methods, databases, and tools are developed in a laboratory and cannot be easily applied by other laboratories using different wild-type lines.

It is also worth noting that the ascidians exhibit an extraordinary variety of processes involving stem cells, and many of them are not manifested by *Ciona* sp. or *Botryllus* sp. For example, recently, extraordinary regenerative potentialities, going far beyond what is shown by *Ciona*, have been described in the solitary *P. mitiligera* (Gordon et al. 2021). Some *Botrylloides* species exhibit putative stem cell-based phenomena, such as torpor and constitutive WBR (Hyams et al. 2017; Kassmer et al. 2020) that are not exhibited by *B. schlosseri*. Future studies should consider these attributes.

8.5. Stem Cells and Immunity

The crosstalk between stem cells and immune cells during homeostasis and regeneration is well studied in mammals (Castillo et al. 2007; DelaRosa et al. 2012; Naik et al. 2018) but poorly investigated in aquatic invertebrates (Ballarin et al. 2021b). Studies on allograft rejections in ascidians point to potential relationships between stem and immune cells. For example, in the case of allograft rejections in the solitary ascidian *Styela plicata*, following the initial recruitment of cytotoxic morula cells to the graft area, an increase in the number of hemoblasts in the tunic surrounding the graft is observed (up to 30 days following rejection) (Parrinello 1996; Raftos et al. 1987). Similarly, in *Styela clava*, the injection of allogeneic hemocytes to the tunic induces the proliferation of hemoblasts within 5 days postinjection (Raftos and Cooper 1991).

Several events in the life cycle of colonial ascidians most likely involve interactions between stem and immune cells. These include (i) rejection of an allogeneic colony, (ii) chimerism, and (iii) resorption of zooids when the new generation buds replace them.

As described above, colonial ascidians exhibit natural stem-cell-mediated chimerism (Laird et al. 2005). When two genetically distinct colonies meet, they either anastomose extracorporeal blood vessels to form a chimera with a common vasculature or reject one another (Oka and Watanabe 1957b; Sabbadin 1962; Scofield et al. 1982; Voskoboynik et al. 2013b). In some chimeras, one of the chimeric partners undergoes partial or complete reabsorption (Rinkevich and Weissman 1992; Corey et al. 2016). Circulating germ and/or somatic stem cells of one partner in a chimera can compete with and replace the germline and/or soma of the other partner (Laird et al.

2005; Voskoboynik et al. 2008; Rinkevich et al. 2013). Therefore, stem cell engraftment in colonial ascidians is regulated on four different levels: (1) fusion or rejection; (2) if fusion occurs, the body of the losing partner is resorbed; (3) competition between circulating somatic stem cells to seed buds for asexual whole-body development; (4) stem cell competition among germline stem cells, which determines the genotype of the next generation.

Each level involves immune cell implications: the histocompatibility gene *BHF* controls fusion/rejection and limits stem cell parasitism to kin (Voskoboynik et al. 2013b); rejection is characterized by the extravasation of cytotoxic cells along the contact border and their degranulation and death with the consequent formation of points of rejection (Ballarin et al. 1995; Cima 2004; Rinkevich 2005b; Franchi and Ballarin 2017); resorption is a model for stem cell loss (failure to bud) when the immune system attacks the buds (Corey et al. 2016), and stem cell competitions relate to stem cell transplant engraftability.

The elimination of one partner in a chimera occurs mostly during a developmental period corresponding to a massive wave of programmed cell death and removal (Rinkevich and Weissman 1992; Cima et al. 2010; Corey et al. 2016; Franchi et al. 2016). Each blastogenic cycle in *B. schlosseri* ends in an apoptotic and phagocytic event of parental zooids, concurrent with the rapid development of next-generation primary buds (blastogenic "takeover" stage).

Using differential expression and gene set analysis, Corey et al. (2016) demonstrated that takeover pathways are co-opted by colonies to induce histocompatible partner elimination. These gene profiles show that colonies usurp developmental programs of autophagy, senescence, programmed cell death, and removal to eliminate allogeneic partners. This study also shows that the exposure of asexually propagating tissues to allogeneic cytotoxic and phagocytic populations has clear effects on development, leading to a developmental arrest. These findings suggest that the critical early events of asexual reproduction are dependent on protection from immune damage—a biological theme that emerges in higher vertebrates, where regulatory systems have evolved to create local sites of immune privilege such as for germ cell development or to protect a fetal allograft. The interactions between immune and stem cells in colonial ascidians are also suggested by a marked proliferative response observed following hemocyte xenotransplantation in *Botrylloides* (Simon-Blecher et al. 2004).

The life history of colonial ascidians, in which the interplay between stem and immune cells can be studied in vivo (Voskoboynik et al. 2008; Rinkevich et al. 2013; Corey et al. 2016; Rosental et al. 2018), offers an opportunity to better understand the relationship between immune function and regeneration.

Author Contributions: Conceptual and design: A.V., L.M., and V.V.; writing—review and editing: V.V., C.A., L.B., L.D., F.G., T.G., A.P., B.R., A.R., B.R., A.S., L.M., and A.V. All authors have read and agreed to the published version of the manuscript.

Funding: This study is supported by the European Cooperation in Science and Technology program (EU COST), COST Action 16203: Stem cells of marine/aquatic invertebrates: from basic research to innovative applications (MARISTEM). C.A. is supported by Larry L. Hillblom Foundation and Stanford School of Medicine Dean's Postdoctoral Fellowship and National Institutes of Health Grant R21AG062948. L.M. is supported by the University of Padova, Grant BIRD213252. B.R. is supported by European Research Council (ERC) Grant Number 948476, and by Israel Science Foundation (ISF) Grant Number 1416/19. B.R. and A.V. are supported by a grant from the United States–Israel Binational Science Foundation (BSF no. 2015012). A.V. is supported by the National Institutes of Health Grant R21AG062948 and the Chan Zuckerberg investigator program.

Acknowledgments: We thank Karla Palmeri, Ronnie Voskoboynik, Tom Levy, and Kathi Ishizuka for critical review and edits.

Conflicts of Interest: The authors declare no conflict of interest.

References

Alié, Alexandre, Tetsutaro Hayashi, Itsuro Sugimura, Michaël Manuel, Wakana Sugano, Akira Mano, Nori Satoh, Kiyokazu Agata, and Noriko Funayama. 2015. The Ancestral Gene Repertoire of Animal Stem Cells. *Proceedings of the National Academy of Sciences USA* 112: E7093–100. [CrossRef]

Alié, Alexandre, Laurel S. Hiebert, Marta Scelzo, and Stefano Tiozzo. 2021. The Eventful History of Non embryonic Development in Tunicates. *Journal of Experimental Zoology Part B: Molecular and Developmental Evolution* 336: 250–66. [CrossRef]

Anselmi, Chiara, Mark A. Kowarsky, Fabio Gasparini, Federico Caicci, Katherine J. Ishizuka, Karla J. Palmeri, Rahul Sinhar, Norma F. Neff, Stephen R. Quake, Irving L. Weissman, and et al. 2021. Revealing Conserved Mechanisms of Neurodegeneration in a Colonial Chordate. *BioRxiv*. [CrossRef]

Auger, Hélène, Yasunori Sasakura, Jean-Stéphane Joly, and William R. Jeffery. 2010. Regeneration of Oral Siphon Pigment Organs in the Ascidian *Ciona intestinalis*. *Developmental Biology* 339: 374–89. [CrossRef]

Ballarin, Loriano, Francesca Cima, and Armando Sabbadin. 1995. Morula Cells and Histocompatibility in the Colonial Ascidian *Botryllus schlosseri*. *Zoological Science* 12: 757–64. [CrossRef]

Ballarin, Loriano, Marcello del Favero, and Lucia Manni. 2011. Relationships Among Hemocytes, Tunic Cells, Germ Cells, and Accessory Cells in the Colonial Ascidian *Botryllus schlosseri*. *Journal of Experimental Zoology Part B: Molecular and Developmental Evolution* 316B: 284–95. [CrossRef]

Ballarin, Loriano, Luis du Pasquier, Baruch Rinkevich, and Joachim Kurtz. 2015. Evolutionary Aspects of Allorecognition. *Invertebrate Survival Journal* 12: 233–36.

Ballarin, Loriano, Matteo Cammarata, and Pierangelo Luporini. 2021a. Ancient Immunity. Phylogenetic Emergence of Recognition-Defense Mechanisms. *Biology* 10: 342. [CrossRef]

Ballarin, Loriano, Arzu Karahan, Alessandra Salvetti, Leonardo Rossi, Lucia Manni, Baruch Rinkevich, Amalia Rosner, Ayelet Voskoboynik, Benyamin Rosental, Laura Canesi, and et al. 2021b. Stem Cells and Innate Immunity in Aquatic Invertebrates: Bridging Two Seemingly Disparate Disciplines for New Discoveries in Biology. *Frontiers in Immunology* 12: 688106. [CrossRef]

Bancroft, Frank Watts. 1903. Aestivation of Botrylloides Gascoi Della Valle. *Mark Anniversary* 8: 147–66.

Barker, Nick, Johan H. van Es, Jeroen Kuipers, Pekka Kujala, Maaike van den Born, Miranda Cozijnsen, Andrea Haegebarth, Jeroen Korving, Harry Begthel, Peter J. Peters, and et al. 2007. Identification of stem cells in small intestine and colon by marker gene Lgr5. *Nature* 449: 1003–7. [CrossRef]

Beerman, Isabel, Deepta Bhattacharya, Sasan Zandi, Mikael Sigvardsson, Irving L. Weissman, David Bryder, and Derrick J. Rossi. 2010. Functionally Distinct Hematopoietic Stem Cells Modulate Hematopoietic Lineage Potential During Aging by a Mechanism of Clonal Expansion. *Proceedings of the National Academy of Sciences USA* 107: 5465–70. [CrossRef]

Ben-Hamo, Oshrat, Amalia Rosner, Claudette Rabinowitz, Matan Oren, and Baruch Rinkevich. 2018. Coupling Astogenic Aging in the Colonial Tunicate *Botryllus schlosseri* With the Stress Protein Mortalin. *Developmental Biology* 433: 33–46. [CrossRef]

Bergmann, Andreas, and Hermann Steller. 2010. Apoptosis, Stem Cells, and Tissue Regeneration. *Science Signaling* 3: re8. [CrossRef]

Blackstone, Neil W., and Bryan D. Jasker. 2003. Phylogenetic Considerations of Clonality, Coloniality, and Mode of Germline Development in Animals. *Journal of Experimental Zoology Part B: Molecular and Developmental Evolution* 297: 35–47. [CrossRef]

Blanchoud, Simon, Buki Rinkevich, and Megan J. Wilson. 2018. Whole-body regeneration in the colonial tunicate Botrylloides leachii. In *Marine Organisms as Model Systems in Biology and Medicine*. Edited by Malgorzata Kloc and Jacek Z. Kubiak. Cham: Springer, pp. 337–55. [CrossRef]

Bollner, Tomas, Philip W. Beesley, and Michael C. Thorndyke. 1992. Pattern of Substance P- and Cholecystokinin-Like Immunoreactivity During Regeneration of the Neural Complex in the Ascidian *Ciona intestinalis*. *The Journal of Comparative Neurology* 325: 572–80. [CrossRef]

Borges, Renee M. 2009. Phenotypic Plasticity and Longevity in Plants and Animals: Cause and Effect? *Journal of Biosciences* 34: 605–11. [CrossRef]

Boyd, Heather C., Stephen K. Brown, James A. Harp, and Irving L. Weissman. 1986. Growth and Sexual Maturation of Laboratory-Cultured Monterey *Botryllus schlosseri*. *The Biological Bulletin* 170: 91–109. [CrossRef]

Braden, Brian P., Daryl A. Taketa, James D. Pierce, Susannah Kassmer, Daniel D. Lewis, and Anthony W. De Tomaso. 2014. Vascular Regeneration in a Basal Chordate Is Due to the Presence of Immobile, Bi-Functional Cells. *PLOS ONE* 9: e95460. [CrossRef]

Brien, Paul, and Emilie Brien-Gavage. 1928. Contribution à l'étude De La blastogenèse Des Tuniciers. In *Recherches Sur Le Bourgeonnement De Perophora Listeri Weigm*. Bruxelles: Recueil De l'Institut Zoologique Torley-Rousseau, pp. 123–51.

Brien, Paul. 1930. Contribution À l'étude De La régénération Naturelle Et expérimentale Chez Les Clavelinidae. *Annales de la Société royale zoologique de Belgique* 61: 19–112.

Brien, Paul. 1968. Blastogenesis and Morphogenesis. *Advances in Morphogenesis* 7: 151–203. [CrossRef]

Brown, Federico D., and Billie J. Swalla. 2007. Vasa Expression in a Colonial Ascidian, *Botrylloides violaceus*. *Evolution & Development* 9: 165–77. [CrossRef]

Brown, Federico D., Elena L. Keeling, Anna D. Le, and Billie J. Swalla. 2009. Whole Body Regeneration in a Colonial ascidian, *Botrylloides violaceus*. *Journal of Experimental Zoology Part B: Molecular and Developmental Evolution* 312: 885–900. [CrossRef]

Brunetti, Riccardo, Maria Marin, and Monica Bressan. 1984. Combined Effects of Temperature and Salinity on Sexual Reproduction and Colonial Growth of *Botryllus schlosseri* (Tunicata). *Bollettino Di Zoologia* 51: 405–11. [CrossRef]

Brunetti, Riccardo, Carmela Gissi, Roberta Pennati, Federico Caicci, Fabio Gasparini, and Lucia Manni. 2015. Morphological Evidence That the Molecularly Determined *Ciona intestinalis* Type A and Type B Are Different Species: Ciona Robusta and *Ciona intestinalis*. *Journal of Zoological Systematics and Evolutionary Research* 53: 186–93. [CrossRef]

Brunetti, Riccardo, Lucia Manni, Francesco Mastrototaro, Carmela Gissi, and Fabio Gasparini. 2017. Fixation, Description and DNA Barcode of a Neotype for *Botryllus schlosseri* (Pallas, 1766) (Tunicata, Ascidiacea). *Zootaxa* 4353: 29–50. [CrossRef]

Brunetti, Riccardo, Francesca Griggio, Francesco Mastrototaro, Fabio Gasparini, and Carmela Gissi. 2020. Toward a Resolution of the Cosmopolitan *Botryllus schlosseri* Species Complex (Ascidiacea, Styelidae): Mitogenomics and Morphology of Clade E (*Botryllus* Gaiae). *Zoological Journal of the Linnean Society* 190: 1175–92. [CrossRef]

Brusca, Richard C., Wendy Moore, and Stephen M. Shuster. 2016. *Invertebrates*. Sunderland: Sinauer Associates.

Burighel, Paolo, and Richard A. Cloney. 1997. Urochordata: Ascidiacea. In *Microscopic Anatomy of Invertebrates*. Edited by F. W. Harrison and E. E. Ruppert. New York: Wiley-Liss Inc., pp. 221–347.

Burighel, Paolo, Riccardo Brunetti, and Giovanna Zaniolo. 1976. Hibernation of the Colonial Ascidian *Botrylloides leachi* (Savigny): Histological Observations. *Bollettino Di Zoologia* 43: 293–301. [CrossRef]

Buss, Leo W. 1982. Somatic Cell Parasitism and the Evolution of Somatic Tissue Compatibility. *Proceedings of the National Academy of Sciences USA* 79: 5337–41. [CrossRef]

Buss, Leo W. 1983. Evolution, Development, and the Units of Selection. *Proceedings of the National Academy of Sciences USA* 80: 1387–91. [CrossRef]

Campagna, Davide, Fabio Gasparini, Nicola Franchi, Nicola Vitulo, Francesca Ballin, Lucia Manni, Giorgio Valle, and Loriano Ballarin. 2016. Transcriptome Dynamics in the Asexual Cycle of the Chordate *Botryllus schlosseri*. *BMC Genomics* 17: 1–17. [CrossRef]

Cao, Chen, Laurence A. Lemaire, Wei Wang, Peter H. Yoon, Yoolim A. Choi, Lance R. Parsons, John C. Matese, Michael Levine, and Kai Chen. 2019. Comprehensive Single-Cell Transcriptome Lineages of a Proto-Vertebrate. *Nature* 571: 349–54. [CrossRef]

Carballo, José Luis. 2000. Larval Ecology of an Ascidian Tropical Population in a Mediterranean Enclosed Ecosystem. *Marine Ecology Progress Series* 195: 159–67. [CrossRef]

Castillo, Marianne, Katherine Liu, Larrissa Bonilla, and Pranela Rameshwar. 2007. The Immune Properties of Mesenchymal Stem Cells. *International Journal of Biomedical Science IJBS* 3: 76–80.

Chacón-Martínez, Carlos Andrés, Janis Koester, and Sara A. Wickström. 2018. Signaling in the stem cell niche: Regulating cell fate, function and plasticity. *Development* 145: dev165399. [CrossRef]

Chadwick-Furman, Nanette E., and Irving L. Weissman. 1995a. Life History Plasticity in Chimaeras of the Colonial Ascidian *Botryllus schlosseri*. *Proceedings of the Royal Society of London. Series B: Biological Sciences* 262: 157–62. [CrossRef]

Chadwick-Furman, Nanette E., and Irving L. Weissman. 1995b. Life Histories and Senescence of *Botryllus schlosseri* (Chordata, Ascidiacea) in Monterey Bay. *The Biological Bulletin* 189: 36–41. [CrossRef]

Chan, Charles K., Gunsagar S. Gulati, Rahul Sinha, JustinV. Tompkins, Michael Lopez, Ava C. Carter, Ryan C. Ransom, Andreas Reinisch, Taylor Wearda, Matthew Murphy, and et al. 2018. Identification of the Human Skeletal Stem Cell. *Cell* 175: 43–56. [CrossRef]

Cima, Francesca. 2004. Cellular Aspects of Allorecognition in the Compound Ascidian *Botryllus schlosseri*. *Developmental & Comparative Immunology* 28: 881–89. [CrossRef]

Cima, Francesca, Lucia Manni, Giuseppe Basso, Elena Fortunato, Benedetta Accordi, Filippo Schiavon, and Loriano Ballarin. 2010. Hovering Between Death and Life: Natural Apoptosis and Phagocytes in the Blastogenetic Cycle of the Colonial Ascidian *Botryllus schlosseri*. *Developmental & Comparative Immunology* 34: 272–85. [CrossRef]

Conklin, Edwin G. 1905. Mosaic Development in Ascidian Eggs. *Journal of Experimental Zoology* 2: 145–223. [CrossRef]

Conte, Maria, Paolo Deri, Maria Emilia Isolani, Linda Mannini, and Renata Batistoni. 2009. A Mortalin-Like Gene Is Crucial for Planarian Stem Cell Viability. *Developmental Biology* 334: 109–18. [CrossRef]

Cooper, Edwin L., David A. Raftos, and Karen L. Kelly. 1992. Immunobiology of Tunicates: The Search for Precursors of the Vertebrate Immune System. *Bollettino Di Zoologia* 59: 175–81. [CrossRef]

Corbo, Joseph C., Anna Di Gregorio, and Michael Levine. 2001. The Ascidian as a Model Organism in Developmental and Evolutionary Biology. *Cell* 106: 535–38. [CrossRef]

Corces-Zimmerman, Rayan M., Wan-Jen Hong, Irving L. Weissman, Bruno C. Medeiros, and Ravindra Majeti. 2014. Preleukemic Mutations in Human Acute Myeloid Leukemia Affect Epigenetic Regulators and Persist in Remission. *Proceedings of the National Academy of Sciences USA* 111: 2548–53. [CrossRef]

Corey, Daniel M., Benyamin Rosental, Mark Kowarsky, Rahul Sinha, Katherine J. Ishizuka, Karla J. Palmeri, Stephen R. Quake, Ayelet Voskoboynik, and Irving L. Weissman. 2016. Developmental Cell Death Programs License Cytotoxic Cells to Eliminate Histocompatible Partners. *Proceedings of the National Academy of Sciences USA* 113: 6520–25. [CrossRef]

Dahlberg, Carl, Hélène Auger, Sam Dupont, Yasunori Sasakura, Mike Thorndyke, and Jean-Stéphane Joly. 2009. Refining the *Ciona intestinalis* Model of Central Nervous System Regeneration. *PLoS ONE* 4: e4458. [CrossRef]

Dannenberg, Leah C., and Elaine C. Seaver. 2018. Regeneration of the Germline in the Annelid *Capitella teleta*. *Developmental Biology* 440: 74–87. [CrossRef]

De Caralt, Sonia, Susanna López-Legentil, Isabel Tarjuelo, María Jesús Uriz, and Xavier Turon. 2002. Contrasting Biological Traits of *Clavelina lepadiformis* (Ascidiacea) Populations from Inside and Outside Harbours in the Western Mediterranean. *Marine Ecology Progress Series* 244: 125–37. [CrossRef]

Dehal, Paramvir, Yutaka Satou, Robert K. Campbell, Jarrod Chapman, Bernard Degnan, Anthony De Tomaso, Brad Davidson, Anna Di Gregorio, Maarten Gelpke, David M. Goodstein, and et al. 2002. The Draft Genome of *Ciona intestinalis*: Insights into Chordate and Vertebrate Origins. *Science* 298: 2157–67. [CrossRef]

DelaRosa, Olga, Wilfried Dalemans, and Eleuterio Lombardo. 2012. Toll-Like Receptors as Modulators of Mesenchymal Stem Cells. *Frontiers in Immunology* 3: 182. [CrossRef]

Della Valle, Paolo. 1914. Studi Sui Rapporti Fra Differenziazione E Rigenerazione. Lo Sviluppo Di Segmenti Isolati Di Stolone Di Clavelina Di Lunghezza Diversa E Di Calibro Eguale. *Bollettino della Società dei Naturalisti in Napoli* 27: 195–237.

Delsuc, Frédéric, Henner Brinkmann, Daniel Chourrout, and Hervé Philippe. 2006. Tunicates and Not Cephalochordates Are the Closest Living Relatives of Vertebrates. *Nature Cell Biology* 439: 965–68. [CrossRef]

Delsuc, Frédéric, Hervé Philippe, Georgia Tsagkogeorga, Paul Simion, Marie-Ka Tilak, Xavier Turon, Susanna Lopez-Legentil, Jacques Piette, Patrick Lemaire, and Emmanuel J. P. Douzery. 2018. A Phylogenomic Framework and Timescale for Comparative Studies of Tunicates. *BMC Biology* 16: 1–14. [CrossRef]

De Tomaso, Anthony W., Yasunori Saito, Katharine J. Ishizuka, Karla J. Palmeri, and Irving L. Weissman. 1998. Mapping the Genome of a Model Protochordate. I. A Low Resolution Genetic Map Encompassing the Fusion/Histocompatibility (Fu/HC) Locus of *Botryllus schlosseri*. *Genetics* 149: 277–87. [CrossRef]

Deviney, Eada May. 1934. The Behavior of Isolated Pieces of Ascidian (*Perophora viridis*) Stolon as Compared with Ordinary Budding. *Journal of the Elisha Mitchell Scientific Society* 49: 185–224.

di Maio, Alessandro, Leah Setar, Stefano Tiozzo, and Anthony W. De Tomaso. 2015. Wnt Affects Symmetry and Morphogenesis During Post-Embryonic Development in Colonial Chordates. *EvoDevo* 6: 1–13. [CrossRef]

Dilly, Peter N., and Jerome J. Wolken. 1972. Studies on the Receptors in *Ciona intestinalis*. IV. The Ocellus in the Adult. *Micron (1969)* 4: 11–29. [CrossRef]

DuBuc, Timothy Q., Christine E. Schnitzler, Eleni Chrysostomou, Emma T. McMahon, Febrimarsa, James M. Gahan, Tara Buggie, Sebastian G. Gornik, Shirley Hanley, Sofia N. Barreira, and et al. 2020. Transcription Factor AP2 Controls Cnidarian Germ Cell Induction. *Science* 367: 757–62. [CrossRef]

Ereskovsky, Alexander V., Ilya E. Borisenko, Pascal Lapébie, Eve Gazave, Daria B. Tokina, and Carole Borchiellini. 2015. *Oscarella lobularis* (Homoscleromorpha, Porifera) Regeneration: Epithelial Morphogenesis and Metaplasia. *PLOS ONE* 10: e0134566. [CrossRef]

Ermak, Thomas H. 1975. An Autoradiographic Demonstration of Blood Cell Renewal In *Styela clava* (Urochordata: Ascidiacea). *Experientia* 31: 837–39. [CrossRef]

Ermak, Thomas H. 1976. The Hematogenic Tissues of Tunicates. In *Phylogeny of Thymus and Bone Marrow-Bursa Cells*. Edited by Ronald K. Wright and Emily L. Cooper. Boston: Springer, pp. 45–56.

Extavour, Cassandra G., and Michael Akam. 2003. Mechanisms of Germ Cell Specification across the Metazoans: Epigenesis and Preformation. *Development* 130: 5869–84. [CrossRef]

Fagan, Melinda B., and Irving L. Weissman. 1998. Characterization of a Polymorphic Protein Localized to Vascular Epithelium in *Botryllus schlosseri*: Role in Tunic Synthesis? *Molecular Marine Biology and Biotechnology* 7: 204–13.

Fan, Yun, and Andreas Bergmann. 2008. Apoptosis-Induced Compensatory Proliferation. The Cell Is Dead. Long Live the Cell! *Trends in Cell Biology* 18: 467–73. [CrossRef]

Farley, Emma K.; Katrina M. Olson; Wei Zhang, Alexander J. Brandt, Daniel S. Rokhsar, and Michael S. Levine. 2015. Suboptimization of developmental enhancers. *Science* 350: 325–28. [CrossRef]

Ferrario, Cinzia, Michela Sugni, Ildiko M. L. Somorjai, and Loriano Ballarin. 2020. Beyond Adult Stem Cells: Dedifferentiation as a Unifying Mechanism Underlying Regeneration in Invertebrate Deuterostomes. *Frontiers in Cell and Developmental Biology* 8: 587320. [CrossRef]

Fierro-Constaín, Laura, Quentin Schenkelaars, Eve Gazave, Anne Haguenauer, Caroline Rocher, Alexander Ereskovsky, Carole Borchiellini, and Emmanuelle Renard. 2017. The Conservation of the Germline Multipotency Program, from Sponges to Vertebrates: A Stepping Stone to Understanding the Somatic and Germline Origins. *Genome Biology and Evolution* 9: 474–88. [CrossRef]

Fletcher, Lauren, and Barrie Forrest. 2011. Induced Spawning and Culture Techniques for the Invasive Ascidian *Didemnum vexillum* (Kott, 2002). *Aquatic Invasions* 6: 457–64. [CrossRef]

Forsberg, E. Camilla, Emmanuelle Passegue, Susan S. Prohaska, Amy J. Wagers, Martina Koeva, Joshua Stuart, and Irving L. Weissman. 2010. Molecular Signatures of Quiescent, Mobilized and Leukemia-Initiating Hematopoietic Stem Cells. *PLoS ONE* 5: e8785. [CrossRef]

Franchi, Nicola, and Loriano Ballarin. 2017. Immunity in Protochordates: The Tunicate Perspective. *Frontiers in Immunology* 8: 674. [CrossRef]

Franchi, Nicola, Francesca Ballin, Lucia Manni, Filippo Schiavon, Giuseppe Basso, and Loriano Ballarin. 2016. Recurrent Phagocytosis-Induced Apoptosis in the Cyclical Generation Change of the Compound Ascidian *Botryllus schlosseri*. *Developmental & Comparative Immunology* 62: 8–16. [CrossRef]

Franchi, Nicola, Francesca Ballin, and Loriano Ballarin. 2017. Protection from Oxidative Stress in Immunocytes of the Colonial Ascidian *Botryllus schlosseri*: Transcript Characterization and Expression Studies. *The Biological Bulletin* 232: 45–57. [CrossRef]

Freeman, Gary. 1964. The Role of Blood Cells in the Process of Asexual Reproduction in the Tunicate *Perophora viridis*. *Journal of Experimental Zoology* 156: 157–83. [CrossRef]

Fuchs, Elaine, Tudorita Tumbar, and Geraldine Guasch. 2004. Socializing with the Neighbors: Stem Cells and Their Niche. *Cell* 116: 769–78. [CrossRef]

Fujimura, Miyuki, and Katsumi Takamura. 2000. Characterization of an Ascidian DEAD-Box Gene, Ci-DEAD1: Specific Expression in the Germ Cells and Its MRNA Localization in the Posterior-Most Blastomeres in Early Embryos. *Development Genes and Evolution* 210: 64–72. [CrossRef]

Fujiwara, Shigeki, and Cristian Cañestro. 2018. Reporter Analyses Reveal Redundant Enhancers That Confer Robustness on Cis-Regulatory Mechanisms. In *Results and Problems in Cell Differentiation*. Singapore: Springer Science and Business Media LLC, pp. 69–79. [CrossRef]

Fujiwara, Shigeki, Takaomi Isozaki, Kyoko Mori, and Kazuo Kawamura. 2011. Expression and Function of Myc During Asexual Reproduction of the Budding Ascidian *Polyandrocarpa misakiensis*. *Development, Growth & Differentiation* 53: 1004–14. [CrossRef]

Gasparini, Fabio, Paolo Burighel, Lucia Manni, and Giovanna Zaniolo. 2008. Vascular Regeneration and Angiogenic-Like Sprouting Mechanism in a Compound Ascidian Is Similar to Vertebrates. *Evolution & Development* 10: 591–605. [CrossRef]

Gasparini, Fabio, Sebastian M. Shimeld, Elena Ruffoni, Paolo Burighel, and Lucia Manni. 2011. Expression of a Musashi-Like Gene in Sexual and Asexual Development of the Colonial Chordate *Botryllus schlosseri* and Phylogenetic Analysis of the Protein Group. *Journal of Experimental Zoology Part B: Molecular and Developmental Evolution* 316: 562–73. [CrossRef]

Gasparini, Fabio, Federico Caicci, Francesca Rigon, Giovanna Zaniolo, and Lucia Manni. 2014. Testing an Unusual In Vivo Vessel Network Model: A Method to Study Angiogenesis in the Colonial Tunicate *Botryllus schlosseri*. *Scientific Reports* 4: 6460. [CrossRef]

Giard, Alfred Mathieu. 1872. Recherches Sur Les Ascidies composées Ou Synascidies. *Archives de Zoologie Expérimentale et Générale* 1: 501–704.

Gissi, Carmela, Kenneth E. M. Hastings, Fabio Gasparini, Thomas Stach, Roberta Pennati, and Lucia Manni. 2017. An Unprecedented Taxonomic Revision of a Model Organism: The Paradigmatic Case of *Ciona robusta* and *Ciona intestinalis*. *Zoologica Scripta* 46: 521–22. [CrossRef]

Gold, David A., and David K. Jacobs. 2013. Stem Cell Dynamics in Cnidaria: Are There Unifying Principles? *Development Genes and Evolution* 223: 53–66. [CrossRef]

Goldin, Abraham. 1948. Regeneration in *Perophora viridis*. *The Biological Bulletin* 94: 184–93. [CrossRef]

Gordon, Tal, Lucia Manni, and Noa Shenkar. 2019. Regeneration Ability in Four Stolidobranch Ascidians: Ecological and Evolutionary Implications. *Journal of Experimental Marine Biology and Ecology* 519: 151184. [CrossRef]

Gordon, Tal, Lachan Roth, Federico Caicci, Lucia Manni, and Noa Shenkar. 2020. Spawning Induction, Development and Culturing of the Solitary Ascidian *Polycarpa mytiligera*, an Emerging Model for Regeneration Studies. *Frontiers in Zoology* 17: 1–14. [CrossRef]

Gordon, Tal, Arnav Kumar Upadhyay, Lucia Manni, Dorothée Huchon, and Noa Shenkar. 2021. And Then There Were Three … : Extreme Regeneration Ability of the Solitary Chordate *Polycarpa mytiligera*. *Frontiers in Cell and Developmental Biology* 9: 652466. [CrossRef]

Grave, Caswell. 1937. Notes on the Culture of Eight Species of Ascidians. In *Culture Methods for Invertebrate Animals*. Edited by Paul S. Galtsoff, Frank Lutz, Paul S. Welch and James G. Needham. New York: Dover Publications, pp. 560–64.

Greaves, Mel. 2013. Cancer Stem Cells As 'units of selection'. *Evolutionary Applications* 6: 102–8. [CrossRef]

Gremigni, Vittorio, and Ileana Puccinelli. 1977. A Contribution to the Problem of the Origin of the Blastema Cells in Planarians: A Karyological and Ultrastructural Investigation. *Journal of Experimental Zoology* 199: 57–71. [CrossRef]

Grimaldi, Cecilia, and Erez Raz. 2020. Germ Cell migration—Evolutionary Issues and Current Understanding. *Seminars in Cell & Developmental Biology* 100: 152–59. [CrossRef]

Grosberg, Richard K. 1988. Life-history variation within a population of the colonial ascidian *Botryllus schlosseri*. The genetic and environmental control of seasonal variation. *Evolution* 42: 900–20. [CrossRef]

Gutierrez, Stefania, and Federico D. Brown. 2017. Vascular Budding in *Symplegma brakenhielmi* and the Evolution of Coloniality in Styelid Ascidians. *Developmental Biology* 423: 152–69. [CrossRef]

Hamada, Mayuko, Spela Goricki, Mardi S. Byerly, Noriyuki Satoh, and William R. Jeffery. 2015. Evolution of the Chordate Regeneration Blastema: Differential Gene Expression and Conserved Role of Notch Signaling During Siphon Regeneration in the Ascidian *Ciona*. *Developmental Biology* 405: 304–15. [CrossRef]

Hendrickson, Carolyn, Lionel Christiaen, Karine Deschet, Di Jiang, Jean-Stéphane Joly, Laurent Legendre, Yuki Nakatani, Jason Tresser, and William C. Smith. 2004. Culture of Adult Ascidians and Ascidian Genetics. *Methods in Cell Biology* 74: 143–70. [CrossRef]

Hirschler, Jan. 1914. Über Die Restitutions- Und Involutionsvorgänge Bei Operierten Exemplaren Von *Ciona intestinalis* Flem. (Teil I) Nebst Bemerkungen über Den Wert Des Negativen für Das Potenzproblem. *Archiv für Mikroskopische Anatomie* 85: A205–27. [CrossRef]

Holland, Linda Z. 2016. Tunicates. *Current Biology* 26: R146–52. [CrossRef]

Huxley, Julian S. 1921. Studies in Dedifferentiation. II. Dedifferentiation and Resorption in *Perophora*. *Quarterly Journal of Microscopical Science* 65: 643–97.

Hyams, Yosef, Guy Paz, Claudette Rabinowitz, and Baruch Rinkevich. 2017. Insights into the Unique Torpor of *Botrylloides leachi*, a Colonial Urochordate. *Developmental Biology* 428: 101–17. [CrossRef]

Ilsley, Garth R., Ritsuko Suyama, Takeshi Noda, Nori Satoh, and Nicholas M. Luscombe. 2020. Finding Cell-Specific Expression Patterns in the Early *Ciona* Embryo with Single-Cell RNA-Seq. *Scientific Reports* 10: 4961–10. [CrossRef]

Jaiswal, Siddhartha, and Benjamin L. Ebert. 2014. MDS Is a Stem Cell Disorder After All. *Cancer Cell* 25: 713–14. [CrossRef]

Jamieson, Catriona H. M., Laurie E. Ailles, Scott J. Dylla, Manja Muijtjens, Carol Jones, and James L. Zehnder. 2004. Granulocyte-Macrophage Progenitors as Candidate Leukemic Stem Cells in Blast-Crisis CML. *New England Journal of Medicine* 351: 657–67. [CrossRef]

Jamieson, Catriona H. M., Jason Gotlib, Jeffrey A. Durocher, Mark P. Chao, M. Rajan Mariappan, Marla Lay, Carol Jones, James L. Zehnder, Stan L. Lilleberg, and Irving L. Weissman. 2006. The JAK2 V617F Mutation Occurs in Hematopoietic Stem Cells in *Polycythemia vera* and Predisposes Toward Erythroid Differentiation. *Proceedings of the National Academy of Sciences USA* 103: 6224–29. [CrossRef]

Jan, Max, Thomas M. Snyder, M. Ryan Corces-Zimmerman, Paresh Vyas, Irving L. Weissman, Stephen R. Quake, and Ravindra Majeti. 2012. Clonal Evolution of Preleukemic Hematopoietic Stem Cells Precedes Human Acute Myeloid Leukemia. *Science Translational Medicine* 4: 149ra118. [CrossRef]

Jeffery, William R. 2015a. Regeneration, Stem Cells, and Aging in the Tunicate *Ciona*. *International Review of Cell and Molecular Biology* 319: 255–82. [CrossRef]

Jeffery, William R. 2015b. Closing the Wounds: One Hundred and Twenty Five Years of Regenerative Biology in the Ascidian *Ciona intestinalis*. *Genesis* 53: 48–65. [CrossRef]

Jeffery, William R. 2015c. Distal Regeneration Involves the Age Dependent Activity of Branchial Sac Stem Cells in the Ascidian *Ciona intestinalis*. *Regeneration* 2: 1–18. [CrossRef]

Jeffery, William R. 2015d. The Tunicate *Ciona*: A Model System for Understanding the Relationship Between Regeneration and Aging. *Invertebrate Reproduction & Development* 59: 17–22. [CrossRef]

Jeffery, William R. 2019. Progenitor Targeting by Adult Stem Cells in *Ciona* Homeostasis, Injury, and Regeneration. *Developmental Biology* 448: 279–90. [CrossRef]

Jeffery, William R., and Špela Goricki. 2021. Apoptosis Is a Generator of Wnt-Dependent Regeneration and Homeostatic Cell Renewal in the Ascidian *Ciona*. *Biology Open* 10: 058526. [CrossRef]

Jiménez-Merino, Juan, Isadora Santos De Abreu, Laurel S. Hiebert, Silvana Allodi, Stefano Tiozzo, Cintia M. De Barros, and Federico D. Brown. 2019. Putative Stem Cells in the Hemolymph and in the Intestinal Submucosa of the Solitary Ascidian *Styela plicata*. *EvoDevo* 10: 31–19. [CrossRef]

Madariaga, David Jofré, Marcelo M. Rivadeneira, Fadia Tala, and Martin Thiel. 2014. Environmental Tolerance of the Two Invasive Species *Ciona intestinalis* and *Codium fragile*: Their Invasion Potential Along a Temperate Coast. *Biological Invasions* 16: 2507–27. [CrossRef]

Joly, Jean-Stéphane, Shungo Kano, Terumi Matsuoka, Helene Auger, Kazuko Hirayama, Nori Satoh, Satoko Awazu, Laurent Legendre, and Yasunori Sasakura. 2007. Culture Of *Ciona intestinalis* in Closed Systems. *Developmental Dynamics* 236: 1832–40. [CrossRef]

Juliano, Celina E., S. Zachary Swartz, and Gary M. Wessel. 2010. A Conserved Germline Multipotency Program. *Development* 137: 4113–26. [CrossRef]

Kassmer, Susannah H., Delany Rodríguez, Adam D. Langenbacher, Connor Bui, and Anthony W. De Tomaso. 2015. Migration of Germline Progenitor Cells Is Directed by Sphingosine-1-Phosphate Signalling in a Basal Chordate. *Nature Communications* 6: 8565. [CrossRef]

Kassmer, Susannah H., Shane Nourizadeh, and Anthony W. De Tomaso. 2019. Cellular and Molecular Mechanisms of Regeneration in Colonial and Solitary Ascidians. *Developmental Biology* 448: 271–78. [CrossRef]

Kassmer, Susannah H., Delany Rodriguez, and Anthony W. De Tomaso. 2020. Evidence That ABC-Transporter-Mediated Autocrine Export of an Eicosanoid Signaling Molecule Enhances Germ Cell Chemotaxis in the Colonial Tunicate *Botryllus schlosseri*. *Development* 147: 184663. [CrossRef]

Kawamura, Kazuo, and Mitsuaki Nakauchi. 1986. Establishment of the Inland Culture of the Colonial Ascidian, *Botrylloides simodensis*. *Marine Fouling* 6: 7–14. [CrossRef]

Kawamura, Kaz, and Takeshi Sunanaga. 2009. Hemoblasts in Colonial Tunicates: Are They Stem Cells or Tissue-Restricted Progenitor Cells? *Development, Growth & Differentiation* 52: 69–76. [CrossRef]

Kawamura, Kaz, and Takeshi Sunanaga. 2011. Role of Vasa, Piwi, and Myc-Expressing Coelomic Cells in Gonad Regeneration of the Colonial Tunicate, *Botryllus primigenus*. *Mechanisms of Development* 128: 457–70. [CrossRef]

Kawamura, Kazuhiro, Shigeki Fujiwara, and Yasuo M. Sugino. 1991. Budding-Specific Lectin Induced in Epithelial Cells Is an Extracellular Matrix Component for Stem Cell Aggregation in Tunicates. *Development* 113: 995–1005. [CrossRef]

Kawamura, Kaz, Kohki Takakura, Daigo Mori, Kohki Ikeda, Akio Nakamura, and Tomohiko Suzuki. 2012. Tunicate Cytostatic Factor TC14-3 Induces a Polycomb Group Gene and Histone Modification through Ca^{2+} Binding and Protein Dimerization. *BMC Cell Biology* 13: 3. [CrossRef]

Kawamura, Kaz, Takuto Yoshida, and Satoko Sekida. 2018. Autophagic Dedifferentiation Induced by Cooperation Between TOR Inhibitor and Retinoic Acid Signals in Budding Tunicates. *Developmental Biology* 433: 384–93. [CrossRef]

Kobayashi, Kenji, Lixy Yamada, Yutaka Satou, and Nori Satoh. 2013. Differential Gene Expression in Notochord and Nerve Cord Fate Segregation in The *Ciona intestinalis* embryo. *Genesis* 51: 1–13. [CrossRef]

Koguchi, Saki, Yasuo M. Sugino, and Kazuo Kawamura. 1993. Dynamics of Epithelial Stem Cell in the Process of Stolonial Budding of the Colonial Ascidian, *Perophora japonica*. *Memoirs of the Faculty of Science Kochi University* 14: 7–14.

Kott, Patricia. 2001. The Australian Ascidiacea. Part 4, Aplousobranchia (3), Didemnidae. *Memoirs of the Queensland Museum* 47: 1–407.

Kowarsky, Mark, Chiara Anselmi, Kohji Hotta, Paolo Burighel, Giovanna Zaniolo, Federico Caicci, Benyamin Rosental, Norma F. Neff, Katherine J. Ishizuka, Karla J. Palmeri, and et al. 2021. Sexual and Asexual Development: Two Distinct Programs Producing the Same Tunicate. *Cell Reports* 34: 108681. [CrossRef]

Kriebel, Mahlon E. 1968. Electrical Characteristics of Tunicate Heart Cell Membranes and Nexuses. *Journal of General Physiology* 52: 46–59. [CrossRef]

Kuramochi-Miyagawa, Satomi, Toshiaki Watanabe, Kengo Gotoh, Kana Takamatsu, Shinichiro Chuma, Kanako Kojima-Kita, Yusuke Shiromoto, Noriko Asada, Atsushi Toyoda, Asao Fujiyama, and et al. 2010. MVH in PiRNA Processing and Gene Silencing of Retrotransposons. *Genes & Development* 24: 887–92. [CrossRef]

Kürn, Ulrich, Snjezana Rendulic, Stefano Tiozzo, and Robert J. Lauzon. 2011. Asexual Propagation and Regeneration in Colonial Ascidians. *The Biological Bulletin* 221: 43–61. [CrossRef]

Laird, Diana J., Anthony W. De Tomaso, and Irving L. Weissman. 2005. Stem Cells Are Units of Natural Selection in a Colonial Ascidian. *Cell* 123: 1351–60. [CrossRef]

Lambert, Charles C., and Gretchen Lambert. 1998. Non-Indigenous Ascidians in Southern California Harbors and Marinas. *Marine Biology* 130: 675–88. [CrossRef]

Lambert, Gretchen. 2001. A global overview of ascidian introductions and their possible impact on the endemic fauna. In *The biology of ascidians*. Edited by Hitoshi Sawada, Hideyoshi Yokosawa and Charles C. Lambert. Tokyo: Springer, pp. 249–57. [CrossRef]

Langenbacher, Adam D., and Anthony W. de Tomaso. 2016. Temporally and Spatially Dynamic Germ Cell Niches in *Botryllus schlosseri* Revealed by Expression of a TGF-Beta Family Ligand and Vasa. *EvoDevo* 7: 9. [CrossRef]

Lapidot, Ziva, Guy Paz, and Baruch Rinkevich. 2003. Monoclonal Antibody Specific to Urochordate *Botryllus schlosseri* Pyloric Gland. *Marine Biotechnology* 5: 388–94. [CrossRef]

Lapidot, Ziva, and Baruch Rinkevich. 2005. Development of Panel of Monoclonal Antibodies Specific to Urochordate Cell Surface Antigens. *Marine Biotechnology* 7: 532–39. [CrossRef]

Lapidot, Ziva, and Baruch Rinkevich. 2006. Development of Monoclonal Antibodies Specific to Urochordate Intracellular Epitopes. *Cell Biology International* 30: 190–95. [CrossRef]

Lauzon, Robert J., Katherin J. Ishizuka, and Irving L. Weissman. 1992. A Cyclical, Developmentally-Regulated Death Phenomenon in a Colonial Urochordate. *Developmental Dynamics* 194: 71–83. [CrossRef]

Lauzon, Rob J., Baruch Rinkevich, Chris W. Patton, and Irving L. Weissman. 2000. A Morphological Study of Nonrandom Senescence in a Colonial Urochordate. *The Biological Bulletin* 198: 367–78. [CrossRef]

Lauzon, Robert J., Sarah J. Kidder, and Patricia Long. 2007. Suppression of Programmed Cell Death Regulates the Cyclical Degeneration of Organs in a Colonial Urochordate. *Developmental Biology* 301: 92–105. [CrossRef]

Lefèvre, George. 1897. Budding in Ecteinascidia. *Annals of Anatomy* 13: 474–83.

Lefevre, George. 1898. Budding in *Perophora*. *Journal of Morphology* 14: 367–424. [CrossRef]

Lemaire, Laurence A., Chen Cao, Peter H. Yoon, Juanjuan Long, and Michael Levine. 2021. The Hypothalamus Predates the Origin of Vertebrates. *Science Advances* 7: eabf7452. [CrossRef]

Lemaire, Patrick. 2011. Evolutionary Crossroads in Developmental Biology: The Tunicates. *Development* 138: 2143–52. [CrossRef]

Lemaire, Patrick, William C. Smith, and Hiroki Nishida. 2008. Ascidians and the Plasticity of the Chordate Developmental Program. *Current Biology* 18: R620–R631. [CrossRef] [PubMed]

Lender, Theodore H., and Christiane Bouchard-Madrelle. 1964. Étude expérimentale De La régénération Du Complexe Neural De *Ciona intestinalis* (Prochordé). *Bulletin de la Société zoologique de France* 89: 546–54.

Li, Hanxi, Xuena Huang, and Aibin Zhan. 2020. Stress Memory of Recurrent Environmental Challenges in Marine Invasive Species: *Ciona robusta* as a Case Study. *Frontiers in Physiology* 11: 94. [CrossRef] [PubMed]

López-Otín, Carlos, Maria A. Blasco, Linda Partridge, Manuel Serrano, and Guido Kroemer. 2013. The Hallmarks of Aging. *Cell* 153: 1194–217. [CrossRef]

Majone, Franca. 1977. Regeneration of Isolated Bud Fragments of *Botryllus schlosseri*. *Acta Embryologiae Experimentalis* 1: 11–9.

Manni, Lucia, Giovanna Zaniolo, and Paolo Burighel. 1993. Egg Envelope Cytodifferentiation in the Colonial Ascidian *Botryllus schlosseri* (Tunicata). *Acta Zoologica* 74: 103–13. [CrossRef]

Manni, Lucia, George O. Mackie, Federico Caicci, Giovanna Zaniolo, and Paolo Burighel. 2006. Coronal Organ of Ascidians and the Evolutionary Significance of Secondary Sensory Cells in Chordates. *The Journal of Comparative Neurology* 495: 363–73. [CrossRef]

Manni, Lucia, Giovanna Zaniolo, Francesca Cima, Paolo Burighel, and Loriano Ballarin. 2007. *Botryllus schlosseri*: A Model Ascidian for the Study of Asexual Reproduction. *Developmental Dynamics* 236: 335–52. [CrossRef]

Manni, Lucia, Fabio Gasparini, Kohji Hotta, Katherine J. Ishizuka, Lorenzo Ricci, Stefano Tiozzo, Ayelet Voskoboynik, and Delphine Dauga. 2014. Ontology for the Asexual Development and Anatomy of the Colonial Chordate *Botryllus schlosseri*. *PLoS ONE* 9: e96434. [CrossRef]

Manni, Lucia, Chiara Anselmi, Francesca Cima, Fabio Gasparini, Ayelet Voskoboynik, Margherita Martini, Anna Peronato, Paolo Burighel, Giovanna Zaniolo, and Loriano Ballarin. 2019. Sixty Years of Experimental Studies on the Blastogenesis of the Colonial Tunicate *Botryllus schlosseri*. *Developmental Biology* 448: 293–308. [CrossRef]

Matsumoto, Jun, Chiaki Nakamoto, Shigeki Fujiwara, Toshitsugu Yubisui, and Kazuo Kawamura. 2001. A Novel C-Type Lectin Regulating Cell Growth, Cell Adhesion and Cell Differentiation of the Multipotent Epithelium in Budding Tunicates. *Development* 128: 3339–47. [CrossRef] [PubMed]

Medina, Bianca N. S. P., Isadora Santos de Abreu, Leny A. Cavalcante, Wagner A. B. Silva, Rodrigo N. da Fonseca, Silvana Allodi, and Cintia M. de Barros. 2015. 3-Acetylpyridine-Induced Degeneration in the Adult Ascidian Neural Complex: Reactive and Regenerative Changes in Glia and Blood Cells. *Developmental Neurobiology* 75: 877–93. [CrossRef] [PubMed]

Milkman, Roger. 1967. Genetic and developmental studies on *Botryllus schlosseri*. *The Biological Bulletin* 132: 229–43. [CrossRef] [PubMed]

Millar, Robert H. 1952. The Annual Growth and Reproductive Cycle in Four Ascidians. *Journal of the Marine Biological Association of the United Kingdom* 31: 41–61. [CrossRef]

Millar, Robert H. 1953. Ciona. In *LMBC Memoirs on Typical British Marine Plants and Animals*. Edited by John S. Colman. Liverpool: Liverpool University, pp. 1–123.

Mingazzini, Pio. 1891. Sulla Rigenerazione Nei Tunicati. *Bollettino della Società dei naturalisti in Napoli* 5: 76–79.

Miyamoto, Toshihiro, Irving L. Weissman, and Koichi Akashi. 2000. AML1/ETO-Expressing Nonleukemic Stem Cells in Acute Myelogenous Leukemia With 8;21 Chromosomal Translocation. *Proceedings of the National Academy of Sciences USA* 97: 7521–26. [CrossRef]

Monniot, Claude, Françoise Monniot, and Pierre Laboute. 1991. *Coral Reef Ascidians of New Caledonia*. Paris: Orstom, p. 247.

Morrison, Sean, and Irving L. Weissman. 1994. The Long-Term Repopulating Subset of Hematopoietic Stem Cells Is Deterministic and Isolatable by Phenotype. *Immunity* 1: 661–73. [CrossRef]

Mueller, Werner A., and Baruch Rinkevich. 2020. Cell Communication-Mediated Nonself-Recognition and -Intolerance in Representative Species of the Animal Kingdom. *Journal of Molecular Evolution* 88: 482–500. [CrossRef]

Mukai, Hideo, Hiromichi Koyama, and Hiroshi Watanabe. 1983. Studies on the reproduction of three species of perophora (Ascidiacea). *The Biological Bulletin* 164: 251–66. [CrossRef]

Müller, Werner A., Regina Teo, and Uri Frank. 2004. Totipotent Migratory Stem Cells in a Hydroid. *Developmental Biology* 275: 215–24. [CrossRef]

Naik, Shruti, Samantha Larsen, Christopher J. Cowley, and Elaine Fuchs. 2018. Two to Tango: Dialog Between Immunity and Stem Cells in Health and Disease. *Cell* 175: 908–20. [CrossRef] [PubMed]

Nakauchi, Mitsuaki. 1982. Asexual Development of Ascidians: Its Biological Significance, Diversity, and Morphogenesis. *American Zoologist* 22: 753–63. [CrossRef]

Ogasawara, Michio, Akane Sasaki, Hitoe Metoki, Tadasu Shin-I, Yuji Kohara, Nori Satoh, and Yutaka Satou. 2002. Gene Expression Profiles in Young Adult *Ciona intestinalis*. *Development Genes and Evolution* 212: 173–85. [CrossRef] [PubMed]

Oka, Hidemiti, and Hiroshi Watanabe. 1957a. Vascular budding, a new type of budding in *Botryllus*. *The Biological Bulletin* 112: 225–40. [CrossRef]

Oka, Hidemiti, and Hiroshi Watanabe. 1957b. Colony Specificity in Compound Ascidians as Tested by Fusion Experiments (a Preliminary report). *Proceedings of the Japan Academy* 33: 657–59. [CrossRef]

Oka, Hidemiti, and Hiroshi Watanabe. 1959. Vascular budding in *Botrylloides*. *The Biological Bulletin* 117: 340–46. [CrossRef]

Okuyama, Makiko, and Yasunori Saito. 2001. Studies on Japanese Botryllid Ascidians. I. A New Species of the Genus *Botryllus* from the Izu Islands. *Zoological Science* 18: 261–67. [CrossRef]

Oonuma, Kouhei, Moeko Tanaka, Koki Nishitsuji, Yumiko Kato, Kotaro Shimai, and Takehiro G. Kusakabe. 2016. Revised Lineage of Larval Photoreceptor Cells in Ciona Reveals Archetypal Collaboration Between Neural Tube and Neural Crest in Sensory Organ Formation. *Developmental Biology* 420: 178–85. [CrossRef]

Oren, Matan, Guy Paz, Jacob Douek, Amalia Rosner, Keren Or Amar, and Baruch Rinkevich. 2013. Marine Invertebrates Cross Phyla Comparisons Reveal Highly Conserved Immune Machinery. *Immunobiology* 218: 484–95. [CrossRef]

Pancer, Zeev, Harriet Gershon, and Baruch Rinkevich. 1995. Coexistence and Possible Parasitism of Somatic and Germ Cell Lines in Chimeras of the Colonial Urochordate *Botryllus schlosseri*. *The Biological Bulletin* 189: 106–12. [CrossRef]

Pang, Wendy W., Elizabeth A. Price, Debashis Sahoo, Isabel Beerman, William J. Maloney, Derrick J. Rossi, Stanley L. Schrier, and Irving L. Weissman. 2011. Human Bone Marrow Hematopoietic Stem Cells Are Increased in Frequency and Myeloid-Biased with Age. *Proceedings of the National Academy of Sciences USA* 108: 20012–17. [CrossRef] [PubMed]

Pang, Wendy W., John V. Pluvinage, Elizabeth A. Price, Kunju Sridhar, Daniel A. Arber, Peter L. Greenberg, Stanley L. Schrier, Christopher Y. Park, and Irving L. Weissman. 2013. Hematopoietic Stem Cell and Progenitor Cell Mechanisms in Myelodysplastic Syndromes. *Proceedings of the National Academy of Sciences USA* 110: 3011–16. [CrossRef] [PubMed]

Parrinello, Nicolò. 1996. Cytotoxic Activity of Tunicate Hemocytes. In *Invertebrate immunology*. Edited by Baruch Rinkevich and Werner E. G. Müller. Berlin and Heidelberg: Springer, pp. 190–217. [CrossRef]

Passegué, Emmanuelle, Amy J. Wagers, Sylvie Giuriato, Wade C. Anderson, and Irving L. Weissman. 2005. Global Analysis of Proliferation and Cell Cycle Gene Expression in the Regulation of Hematopoietic Stem and Progenitor Cell Fates. *Journal of Experimental Medicine* 202: 1599–611. [CrossRef] [PubMed]

Paz, Guy, and Baruch Rinkevich. 2002. Morphological Consequences for Multi-Partner Chimerism in *Botrylloides*, a Colonial Urochordate. *Developmental & Comparative Immunology* 26: 615–22. [CrossRef]

Pennati, Roberta, Gentile Francesco Ficetola, Riccardo Brunetti, Federico Caicci, Fabio Gasparini, Francesca Griggio, Atsuko Sato, Thomas Stach, Sabrina Kaul-Strehlow, Carmela Gissi, and et al. 2015. Morphological Differences Between Larvae of the *Ciona intestinalis* Species Complex: Hints for a Valid Taxonomic Definition of Distinct Species. *PLoS ONE* 10: e0122879. [CrossRef]

Qarri, Andy, Amalia Rosner, Claudette Rabinowitz, and Baruch Rinkevich. 2020. UV-B Radiation Bearings on Ephemeral Soma in the Shallow Water Tunicate *Botryllus schlosseri*. *Ecotoxicology and Environmental Safety* 196: 110489. [CrossRef]

Rabinowitz, Claudette, and Baruch Rinkevich. 2004. In Vitro delayed Senescence of Extirpated Buds from Zooids of the Colonial Tunicate *Botryllus schlosseri*. *Journal of Experimental Biology* 207: 1523–32. [CrossRef]

Rabinowitz, Claudette, and Baruch Rinkevich. 2005. Epithelial Cell Cultures from *Botryllus schlosseri* Palleal Buds: Accomplishments and Challenges. *Journal of Tissue Culture Methods* 25: 137–48. [CrossRef]

Rabinowitz, Claudette, and Baruch Rinkevich. 2011. De Novo Emerged Stemness Signatures in Epithelial Monolayers Developed from Extirpated Palleal Buds. *In Vitro Cellular & Developmental Biology - Plant* 47: 26–31. [CrossRef]

Rabinowitz, Claudette, Gilad Alfassi, and Baruch Rinkevich. 2009. Further Portrayal of Epithelial Monolayers Emergent De Novo from Extirpated Ascidians Palleal Buds. *In Vitro Cellular & Developmental Biology - Plant* 45: 334–42. [CrossRef]

Racioppi, Claudia, Ashwani K. Kamal, Florian Razy-Krajka, Gennaro Gambardella, Laura Zanetti, Diego Di Bernardo, Remo Sanges, Lionel Christiaen, and Filomena Ristoratore. 2014. Fibroblast Growth Factor Signalling Controls Nervous System Patterning and Pigment Cell Formation in *Ciona intestinalis*. *Nature Communications* 5: 4830. [CrossRef]

Raftos, David A., Noel N. Tait, and David A. Briscoe. 1987. Cellular Basis of Allograft Rejection in the Solitary Urochordate, *Styela plicata*. *Developmental & Comparative Immunology* 11: 713–25. [CrossRef]

Raftos, David A., and Edwin L. Cooper. 1991. Proliferation of Lymphocyte-Like Cells from the Solitary tunicate, *Styela clava*, in Response to Allogeneic Stimuli. *Journal of Experimental Zoology* 260: 391–400. [CrossRef] [PubMed]

Reem, Eitan, Jacob Douek, and Baruch Rinkevich. 2018. Ambiguities in the Taxonomic Assignment and Species Delineation of Botryllid Ascidians from the Israeli Mediterranean and Other Coastlines. *Mitochondrial DNA Part A* 29: 1073–80. [CrossRef] [PubMed]

Ricci, Lorenzo, Fabien Cabrera, Sonia Lotito, and Stefano Tiozzo. 2016. Redeployment of Germ Layers Related TFs Shows Regionalized Expression During Two Non-Embryonic Developments. *Developmental Biology* 416: 235–48. [CrossRef]

Ries, Erich. 1937. Untersuchungen über den Zelltod II. *W Roux Archiv F Entwicklungsmechanik* 137: 327–62. [CrossRef]

Rinkevich, Baruch, and Claudette Rabinowitz. 1993. In Vitro Culture of Blood Cells from the Colonial Protochordate *Botryllus schlosseri*. *In Vitro Cellular & Developmental Biology - Plant* 29: 79–85. [CrossRef]

Rinkevich, Baruch, and Claudette Rabinowitz. 1994. Acquiring Embryo-Derived Cell Cultures and Aseptic Metamorphosis of Larvae from the Colonial Protochordate *Botryllus schlosseri*. *Invertebrate Reproduction & Development* 25: 59–72. [CrossRef]

Rinkevich, Baruch. 1996. Bi-Versus Multichimerism in Colonial Urochordates: A Hypothesis for Links Between Natural Tissue Transplantation, Allogenetics and Evolutionary Ecology. *Experimental and Clinical Immunogenetics* 13: 61–69.

Rinkevich, Baruch, and Claudette Rabinowitz. 1997. Initiation of Epithelial Cell Cultures from Palleal Buds of *Botryllus schlosseri*, a Colonial Tunicate. *In Vitro Cellular & Developmental Biology - Plant* 33: 422–24. [CrossRef]

Rinkevich, Baruch, and Michal Shapira. 1998. An Improved Diet for Inland Broodstock and the Establishment of an Inbred Line from *Botryllus schlosseri*, a Colonial Sea Squirt (Ascidiacea). *Aquatic Living Resources* 11: 163–71. [CrossRef]

Rinkevich, Baruch. 1999. Cell Cultures from Marine Invertebrates: Obstacles, New Approaches and Recent Improvements. *Journal of Biotechnology* 70: 133–53. [CrossRef]

Rinkevich, Baruch, and Michal Shapira. 1999. Multi-Partner Urochordate Chimeras Outperform Two-Partner Chimerical Entities. *Oikos* 87: 315. [CrossRef]

Rinkevich, Baruch. 2000. A Critical Approach to the Definition of Darwinian Units of Selection. *The Biological Bulletin* 199: 231–40. [CrossRef]

Rinkevich, Baruch. 2002a. Germ Cell Parasitism as an Ecological and Evolutionary Puzzle: Hitchhiking with Positively Selected Genotypes. *Oikos* 96: 25–30. [CrossRef]

Rinkevich, Baruch. 2002b. The Colonial Urochordate *Botryllus schlosseri*: From Stem Cells and Natural Tissue Transplantation to Issues in Evolutionary Ecology. *BioEssays* 24: 730–40. [CrossRef]

Rinkevich, Baruch. 2004a. Will Two Walk Together, Except They Have Agreed? Amos 3:3. *Journal of Evolutionary Biology* 17: 1178–79. [CrossRef] [PubMed]

Rinkevich, Baruch. 2004b. Primitive Immune Systems: Are Your Ways My Ways? *Immunological Reviews* 198: 25–35. [CrossRef]

Rinkevich, Baruch, and Irena Yankelevich. 2004. Environmental Split Between Germ Cell Parasitism and Somatic Cell Synergism in Chimeras of a Colonial Urochordate. *Journal of Experimental Biology* 207: 3531–36. [CrossRef]

Rinkevich, Baruch. 2005a. Natural Chimerism in Colonial Urochordates. *Journal of Experimental Marine Biology and Ecology* 322: 93–109. [CrossRef]

Rinkevich, Baruch. 2005b. Rejection Patterns in Botryllid Ascidian Immunity: The First Tier of Allorecognition. *Canadian Journal of Zoology* 83: 101–21. [CrossRef]

Rinkevich, Baruch. 2005c. Marine Invertebrate Cell Cultures: New Millennium Trends. *Marine Biotechnology* 7: 429–39. [CrossRef] [PubMed]

Rinkevich, Baruch. 2011. Cell Cultures from Marine Invertebrates: New Insights for Capturing Endless Stemness. *Marine Biotechnology* 13: 345–54. [CrossRef] [PubMed]

Rinkevich, Baruch, and Andrew Fidler. 2014. Initiating Laboratory Culturing of the Invasive Ascidian *Didemnum vexillum*. *Management of Biological Invasions* 5: 55–62. [CrossRef]

Rinkevich, Baruch. 2017. Senescence in modular animals: Botryllid ascidians as a unique ageing system. In *The Evolution of Senescence in the Tree of Life*. Edited by Richard P. Shefferson, Owen R. Jones and Roberto Salguero-Gomez. Cambridge: Cambridge University Press, pp. 327–62.

Rinkevich, Baruch, and Irving L. Weissman. 1992. Allogeneic Resorption in Colonial Protochordates: Consequences of Nonself Recognition. *Developmental & Comparative Immunology* 16: 275–86. [CrossRef]

Rinkevich, Baruch, Robert J. Lauzon, Byron W. M. Brown, and Irving L. Weissman. 1992. Evidence for a Programmed Life Span in a Colonial Protochordate. *Proceedings of the National Academy of Sciences USA* 89: 3546–50. [CrossRef] [PubMed]

Rinkevich, Baruch, Zvia Shlemberg, and Lev Fishelson. 1995. Whole-body protochordate regeneration from totipotent blood cells. *Proceedings of the National Academy of Sciences USA* 92: 7695–99. [CrossRef]

Rinkevich, Baruch, Zvia Shlemberg, and Lev Fishelson. 1996. Survival budding processes in the colonial tunicate Botrylloides from the Mediterranean Sea: The role of totipotent blood cells. In *Invertebrate Cell Culture: Looking Towards the 21st Century*. Edited by Karl Maramorosch and Marcia K. J. Loeb. San Francisco: Society for In Vitro Biology, pp. 1–9.

Rinkevich, Yuval, Jacob Douek, Omer Haber, Baruch Rinkevich, and Ram Reshef. 2007a. Urochordate Whole Body Regeneration Inaugurates a Diverse Innate Immune Signaling Profile. *Developmental Biology* 312: 131–46. [CrossRef]

Rinkevich, Yuval, Guy Paz, Baruch Rinkevich, and Ram Reshef. 2007b. Systemic Bud Induction and Retinoic Acid Signaling Underlie Whole Body Regeneration in the Urochordate *Botrylloides leachi*. *PLoS Biology* 5: e71. [CrossRef]

Rinkevich, Yuval, Baruch Rinkevich, and Ram Reshef. 2008. Cell Signaling and Transcription Factor Genes Expressed During Whole Body Regeneration in a Colonial Chordate. *BMC Developmental Biology* 8: 100. [CrossRef]

Rinkevich, Yuval, Valeria Matranga, and Baruch Rinkevich. 2009. Stem Cells in Aquatic Invertebrates: Common Premises and Emerging Unique Themes. In *Stem Cells in Marine Organisms*. Edited by Baruch Rinkevich and Valeria Matranga. Dordrecht: Springer Netherlands, pp. 61–103. [CrossRef]

Rinkevich, Yuval, Amalia Rosner, Claudette Rabinowitz, Ziva Lapidot, Elithabeth Moiseeva, and Buki Rinkevich. 2010. Piwi Positive Cells That Line the Vasculature Epithelium, Underlie Whole Body Regeneration in a Basal Chordate. *Developmental Biology* 345: 94–104. [CrossRef]

Rinkevich, Yuval, Ayelet Voskoboynik, Amalia Rosner, Claudette Rabinowitz, Guy Paz, Matan Oren, Jacob Douek, Gilad Alfassi, Elizabeth Moiseeva, Katherine J. Ishizuka, and et al. 2013. Repeated, Long-Term Cycling of Putative Stem Cells Between Niches in a Basal Chordate. *Developmental Cell* 24: 76–88. [CrossRef]

Rodriguez, Delany, Daryl A. Taketa, Roopa Madhu, Susannah Kassmer, Dinah Loerke, Megan T. Valentine, and Anthony W. De Tomaso. 2021. Vascular Aging in the Invertebrate Chordate, *Botryllus schlosseri*. *Frontiers in Molecular Biosciences* 8: 626827. [CrossRef]

Rosen, Brian R. 1986. Modular Growth and Form of Corals: A Matter of Metamers? *Philosophical Transactions of the Royal Society of London B* 313: 115–42.

Rosental, Benyamin, Mark Kowarsky, Jun Seita, Daniel M. Corey, Katherine J. Ishizuka, Karla J. Palmeri, Shih-Yu Chen, Rahul Sinha, Jennifer Okamoto, Gary Mantalas, and et al. 2018. Complex Mammalian-Like Haematopoietic System Found in a Colonial Chordate. *Nature* 564: 425–29. [CrossRef]

Rosner, Amalia, Guy Paz, and Baruch Rinkevich. 2006. Divergent Roles of the DEAD-Box Protein BS-PL10, the Urochordate Homologue of Human DDX3 and DDX3Y Proteins, in Colony Astogeny and Ontogeny. *Developmental Dynamics* 235: 1508–21. [CrossRef]

Rosner, Amalia, Claudette Rabinowitz, Elizabeth Moiseeva, Ayelet Voskoboynik, and Baruch Rinkevich. 2007. BS-Cadherin in the Colonial Urochordate *Botryllus schlosseri*: One Protein, Many Functions. *Developmental Biology* 304: 687–700. [CrossRef]

Rosner, Amalia, Elizabeth Moiseeva, Yuval Rinkevich, Ziva Lapidot, and Baruch Rinkevich. 2009. Vasa and the Germ Line Lineage in a Colonial Urochordate. *Developmental Biology* 331: 113–28. [CrossRef]

Rosner, Amalia, Elizabeth Moiseeva, Claudette Rabinowitz, and Baruch Rinkevich. 2013. Germ Lineage Properties in the Urochordate *Botryllus schlosseri*—From Markers to Temporal Niches. *Developmental Biology* 384: 356–74. [CrossRef]

Rosner, Amalia, Gilad Alfassi, Elizabeth Moiseeva, Guy Paz, Claudette Rabinowitz, Ziva Lapidot, Jacob Douek, Abraham Haim, and Baruch Rinkevich. 2014. The Involvement of Three Signal Transduction Pathways in Botryllid Ascidian Astogeny, As Revealed by Expression Patterns of Representative Genes. *The International Journal of Developmental Biology* 58: 677–92. [CrossRef]

Rosner, Amalia, Olha Kravchenko, and Baruch Rinkevich. 2019. IAP Genes Partake Weighty Roles in the Astogeny and Whole Body Regeneration in the Colonial Urochordate *Botryllus schlosseri*. *Developmental Biology* 448: 320–41. [CrossRef] [PubMed]

Rosner, Amalia, Jean Armengaud, Loriano Ballarin, Stéphanie Barnay-Verdier, Francesca Cima, Ana Varela Coelho, Isabelle Domart-Coulon, Damjana Drobne, Anne-Marie Genevière, Anita Jemec Kokalj, and et al. 2021. Stem Cells of Aquatic Invertebrates as an Advanced Tool for Assessing Ecotoxicological Impacts. *Science of The Total Environment* 771: 144565. [CrossRef] [PubMed]

Rossi, Derrick J., David Bryder, Jacob M. Zahn, Henrik Ahlenius, Rebecca Sonu, Amy J. Wagers, and Irving L. Weissman. 2005. Cell Intrinsic Alterations Underlie Hematopoietic Stem Cell Aging. *Proceedings of the National Academy of Sciences USA* 102: 9194–99. [CrossRef] [PubMed]

Rossi, Derrick J., Jun Seita, Agnieszka Czechowicz, Deepta Bhattacharya, David Bryder, and Irving L. Weissman. 2007. Hematopoietic Stem Cell Quiescence Attenuates DNA Damage Response and Permits DNA Damage Accumulation During Aging. *Cell Cycle* 6: 2371–76. [CrossRef] [PubMed]

Rowley, Andrew F. 1982. Ultrastructural and Cytochemical Studies on the Blood Cells of the Sea squirt, *Ciona intestinalis*. *Cell and Tissue Research* 223: 403–14. [CrossRef] [PubMed]

Sabbadin, Armando. 1960. Ulteriori Notizie sull'allevamento E Sulla Biologia Dei Botrilli in Condizioni Di Laboratorio. *Arch Oceanology and Limnology* 12: 97–107.

Sabbadin, Armando. 1962. Le Basi Genetiche Della Capacità Di Fusione Fra Colonie in *Botryllus schlössen* (Ascidiacea). *Rend Accad Naz Lincei* 32: 1031–35.

Sabbadin, Armando. 1969. The Compound Ascidian *Botryllus schlosseri* in the Field and in the Laboratory. *Pubbl Staz Zool Napoli* 37: 62–72.

Sabbadin, Armando, and Giovanna Zaniolo. 1979. Sexual Differentiation and Germ Cell Transfer in the Colonial Ascidian *Botryllus schlosseri*. *Journal of Experimental Zoology* 207: 289–304. [CrossRef]

Sabbadin, Armando, Giovanna Zaniolo, and Franca Majone. 1975. Determination of Polarity and Bilateral Asymmetry in Palleal and Vascular Buds of the Ascidian *Botryllus schlosseri*. *Developmental Biology* 46: 79–87. [CrossRef]

Saez, Borja, Rushdia Z. Yusuf, and David T. Scadden. 2017. Harnessing the biology of stem cells' niche. In *Biology and Engineering of Stem Cell Niches*. Edited by Ajaykumar Vishwakarma and Jeffrey M. Karp. Cambridge: Academic Press, pp. 15–31. [CrossRef]

Saito, Yasunori, and Hiroshi Watanabe. 1985. Studies on Japanese Compound Styelid Ascidians -IV. Three New Species of the Genus *Botrylloides* from the Vicinity of Shimoda-. *Publications of the Seto Marine Biological Laboratory* 30: 227–40. [CrossRef]

Satoh, Nori. 2001. Ascidian Embryos as a Model System to Analyze Expression and Function of Developmental Genes. *Differentiation* 68: 1–12. [CrossRef] [PubMed]

Satoh, Noriyuki. 2019. A Deep Dive into the Development of Sea Squirts. *Nature* 571: 333–34. [CrossRef] [PubMed]

Savigny, Jules-César. 1816. *Mémoires Sur Les Animaux Sans vertèbres*. Paris: Doufour, vol. 2. [CrossRef]

Scelzo, Marta, Alexandre Alié, Sophie Pagnotta, Camille Lejeune, Pauline Henry, Laurent Gilletta, Laurel S. Hiebert, Francesco Mastrototaro, and Stefano Tiozzo. 2019. Novel Budding Mode in Polyandrocarpa Zorritensis: A Model for Comparative Studies on Asexual Development and Whole Body Regeneration. *EvoDevo* 10: 7. [CrossRef]

Schlumpberger, Jay M., Irving L. Weissman, and Virginia L. Scofield. 1984a. Separation and Labeling of Specific Subpopulations Of *Botryllus* Blood Cells. *Journal of Experimental Zoology* 229: 401–11. [CrossRef]

Schlumpberger, Jay M., Irving L. Weissman, and Virginia L. Scofield. 1984b. Monoclonal Antibodies Developed Against *Botryllus* Blood Cell Antigens Bind to Cells of Distinct Lineages During Embryonic Development. *Journal of Experimental Zoology* 229: 205–13. [CrossRef]

Schofield, Raymond. 1978. The Relationship Between the Spleen Colony-Forming Cell and the Haemopoietic Stem Cell. *Blood Cells* 4: 7–25.

Scofield, Virginia L., Jay M. Schlumpberger, Lani A. West, and Irving L. Weissman. 1982. Protochordate Allorecognition Is Controlled by a MHC-Like Gene System. *Nature* 295: 499–502. [CrossRef]

Seipel, Katja, Nathalie Yanze, and Volker Schmid. 2004. The Germ Line and Somatic Stem Cell Gene Cniwi in the Jellyfish *Podocoryne carnea*. *The International Journal of Developmental Biology* 48: 1–7. [CrossRef]

Shenkar, Noa, and Tal Gordon. 2015. Gut-Spilling in Chordates: Evisceration in the Tropical Ascidian *Polycarpa mytiligera*. *Scientific Reports* 5: 9614. [CrossRef]

Shirae-Kurabayashi, Maki, Takahito Nishikata, Katsumi Takamura, Kimio J. Tanaka, Chiaki Nakamoto, and Akira Nakamura. 2006. Dynamic Redistribution of Vasa Homolog and Exclusion of Somatic Cell Determinants During Germ Cell Specification in *Ciona intestinalis*. *Development* 133: 2683–93. [CrossRef] [PubMed]

Simon-Blecher, Noa, Yair Achituv, and Baruch Rinkevich. 2004. Protochordate Concordant Xenotransplantation Settings Reveal Outbreaks of Donor Cells and Divergent Life Span Traits. *Developmental & Comparative Immunology* 28: 983–91. [CrossRef]

Siomi, Mikiko C., and Satomi Kuramochi-Miyagawa. 2009. RNA Silencing in germlines—exquisite Collaboration of Argonaute Proteins with Small RNAs for Germline Survival. *Current Opinion in Cell Biology* 21: 426–34. [CrossRef] [PubMed]

Sköld, Helen Nilsson, Thomas Stach, John D. D. Bishop, Eva Herbst, and Michael C. Thorndyke. 2011. Pattern of Cell Proliferation During Budding in the Colonial Ascidian *Diplosoma listerianum*. *The Biological Bulletin* 221: 126–36. [CrossRef] [PubMed]

Sladitschek, Hanna L., Ulla-Maj Fiuza, Dinko Pavlinic, Vladimir Benes, Lars Hufnagel, and Pierre A. Neveu. 2020. MorphoSeq: Full Single-Cell Transcriptome Dynamics Up to Gastrulation in a Chordate. *Cell* 181: 922–35.e21. [CrossRef] [PubMed]

Spangrude, Gerald J., Shelly Heimfeld, and Irving L. Weissman. 1988. Purification and Characterization of Mouse Hematopoietic Stem Cells. *Science* 241: 58–62. [CrossRef]

Spina, Elijah J., Elmer Guzman, Hongjun Zhou, Kenneth S. Kosik, and William C. Smith. 2017. A MicroRNA-MRNA Expression Network During Oral Siphon Regeneration in *Ciona*. *Development* 144: 1787–97. [CrossRef]

Spradling, Allan, Margaret T. Fuller, Robert E. Braun, and Shosei Yoshida. 2011. Germline Stem Cells. *Cold Spring Harbor Perspectives in Biology* 3: a002642. [CrossRef]

Squarzoni, Paola, Fateema Parveen, Laura Zanetti, Filomena Ristoratore, and Antonietta Spagnuolo. 2011. FGF/MAPK/Ets Signaling Renders Pigment Cell Precursors Competent to Respond to Wnt Signal by Directly Controlling Ci-Tcf Transcription. *Development* 138: 1421–32. [CrossRef]

Stolfi, Alberto, and Lionel Christiaen. 2012. Genetic and Genomic Toolbox of the Chordate *Ciona intestinalis*. *Genetics* 192: 55–66. [CrossRef]

Stoner, Douglas S., and Irving L. Weissman. 1996. Somatic and Germ Cell Parasitism in a Colonial Ascidian: Possible Role for a Highly Polymorphic Allorecognition System. *Proceedings of the National Academy of Sciences USA* 93: 15254–59. [CrossRef] [PubMed]

Stoner, Douglas S., Baruch Rinkevich, and Irving L. Weissman. 1999. Heritable Germ and Somatic Cell Lineage Competitions in Chimeric Colonial Protochordates. *Proceedings of the National Academy of Sciences USA* 96: 9148–53. [CrossRef] [PubMed]

Storey, Kenneth B, and Janet M. Storey. 2011. Hibernation: Poikilotherms. In *Encyclopedia of Life Sciences*. Chichester: ohn Wiley & Sons, Ltd. [CrossRef]

Strome, Susan, and Dustin Updike. 2015. Specifying and Protecting Germ Cell Fate. *Nature Reviews Molecular Cell Biology* 16: 406–16. [CrossRef] [PubMed]

Sugino, Yasuo M., and Mitsuaki Nakauchi. 1987. Budding, Life-Span, Regeneration, and Colonial Regulation in the ascidian, *Symplegma reptans*. *Journal of Experimental Zoology* 244: 117–24. [CrossRef]

Sunanaga, Takeshi, Miho Satoh, and Kazuo Kawamura. 2008. The Role of Nanos Homologue in Gametogenesis and Blastogenesis With Special Reference to Male Germ Cell Formation in the Colonial Ascidian, *Botryllus primigenus*. *Developmental Biology* 324: 31–40. [CrossRef]

Sutton, Muriel F. 1953. The Regeneration of the Siphons of *Ciona intestinalis*. *Journal of the Marine Biological Association of the United Kingdom* 32: 249–68. [CrossRef]

Sykes, Stephen M., Konstantinos D. Kokkaliaris, Michael D. Milsom, Ross L. Levine, and Ravindra Majeti. 2015. Clonal Evolution of Preleukemic Hematopoietic Stem Cells in Acute Myeloid Leukemia. *Experimental Hematology* 43: 989–92. [CrossRef]

Takamura, Katsumi, Miyuki Fujimura, and Yasunori Yamaguchi. 2002. Primordial Germ Cells Originate from the Endodermal Strand Cells in the Ascidian *Ciona intestinalis*. *Development Genes and Evolution* 212: 11–18. [CrossRef]

Tarjuelo, I., David Posada, Keith Crandall, Marta Pascual, and X. Turon. 2004. Phylogeography and Speciation of Colour Morphs in the Colonial Ascidian *Pseudodistoma crucigaster*. *Molecular Ecology* 13: 3125–36. [CrossRef]

Tatzuke, Yuki, Takeshi Sunanaga, Shigeki Fujiwara, and Kaz Kawamura. 2012. RACK1 Regulates Mesenchymal Cell Recruitment During Sexual and Asexual Reproduction of Budding Tunicates. *Developmental Biology* 368: 393–403. [CrossRef]

Tiozzo, Stefano, Lionel Christiaen, Carole Deyts, Lucia Manni, and Paolo Burighel. 2005. Embryonic Versus Blastogenetic Development in the Compound Ascidian *Botryllus schlosseri*: Insights From Pitx Expression Patterns. *Developmental Dynamics* 232: 468–78. [CrossRef] [PubMed]

Tiozzo, Stefano, Federico D. Brown, and Anthony W. De Tomaso. 2008a. *Regeneration and Stem Cells in Ascidians, in: Stem Cells*. Dordrecht: Springer Netherlands. [CrossRef]

Tiozzo, Stefano, Ayelet Voskoboynik, Federico D. Brown, and Anthony W. De Tomaso. 2008b. A Conserved Role of the VEGF Pathway in Angiogenesis of an Ectodermally-Derived Vasculature. *Developmental Biology* 315: 243–55. [CrossRef] [PubMed]

Tiozzo, Stefano, Maureen Murray, Bernard M. Degnan, Anthony W. De Tomaso, and Roger P. Croll. 2009. Development of the Neuromuscular System During Asexual Propagation in an Invertebrate Chordate. *Developmental Dynamics* 238: 2081–94. [CrossRef] [PubMed]

Tolkin, Theadora, and Lionel Christiaen. 2016. Rewiring of an Ancestral Tbx1/10-Ebf-Mrf Network for Pharyngeal Muscle Specification in Distinct Embryonic Lineages. *Development* 143: 3852–62. [CrossRef] [PubMed]

Turon, Xavier. 1992. Periods of Non-Feeding in Polysyncraton Lacazei (Ascidiacea: Didemnidae): A Rejuvenative Process? *Marine Biology* 112: 647–55. [CrossRef]

Uchida, Nobuko, and Irving L. Weissman. 1992. Searching for Hematopoietic Stem Cells: Evidence That Thy-1.1lo Lin- Sca-1+ Cells Are the Only Stem Cells in C57BL/Ka-Thy-1.1 Bone Marrow. *Journal of Experimental Medicine* 175: 175–84. [CrossRef] [PubMed]

Uchida, Nobuko, David W. Buck, Dongping He, Michael J. Reitsma, Marilyn Masek, Thinh V. Phan, Ann S. Tsukamoto, Fred H. Gage, and Irving L. Weissman. 2000. Direct isolation of human central nervous system stem cells. *Proceedings of the National Academy of Sciences USA* 97: 14720–25. [CrossRef]

Ueno, Hiroo, Brit B. Turnbull, and Irving L. Weissman. 2009. Two-Step Oligoclonal Development of Male Germ Cells. *Proceedings of the National Academy of Sciences USA* 106: 175–80. [CrossRef]

Valentine, Page. 2009. Larval Recruitment of the Invasive Colonial Ascidian *Didemnum vexillum*, Seasonal Water Temperatures in New England Coastal and Offshore Waters, and Implications for Spread of the Species. *Aquatic Invasions* 4: 153–68. [CrossRef]

Vasquez-Kuntz, Kate L., Sheila A. Kitchen, Trinity L. Conn, Samuel A. Vohsen, Andrea N. Chan, Mark J. A. Vermeij, Christopher Page, Kristen L. Marhaver, and Iliana B. Baums. 2020. Juvenile Corals Inherit Mutations Acquired During the parent's Lifespan. *BioRxiv*. [CrossRef]

Voskoboynik, Ayelet, and Irving L. Weissman. 2015. *Botryllus schlosseri*, an Emerging Model for the Study of Aging, Stem Cells, and Mechanisms of Regeneration. *Invertebrate Reproduction & Development* 59: 33–38. [CrossRef]

Voskoboynik, Ayelet, Noa Simon-Blecher, Yoav Soen, Baruch Rinkevich, Anthony W. De Tomaso, Katherine J. Ishizuka, and Irving L. Weissman. 2007. Striving for Normality: Whole Body Regeneration through a Series of Abnormal Generations. *The FASEB Journal* 21: 1335–44. [CrossRef] [PubMed]

Voskoboynik, Ayelet, Yoav Soen, Yuval Rinkevich, Amalia Rosner, Hiroo Ueno, Ram Reshef, Katherine J. Ishizuka, Karla J. Palmeri, Elizabeth Moiseeva, Baruch Rinkevich, and et al. 2008. Identification of the Endostyle As a Stem Cell Niche in a Colonial Chordate. *Cell Stem Cell* 3: 456–64. [CrossRef] [PubMed]

Voskoboynik, Ayelet, Norma F. Neff, Debashis Sahoo, Aaron Newman, Dmitry Pushkarev, Winston Koh, Benedetto Passarelli, H. Christina Fan, Gary L. Mantalas, Karla J. Palmeri, and et al. 2013a. The Genome Sequence of the Colonial Chordate, *Botryllus schlosseri*. *ELife* 2: e00569. [CrossRef]

Voskoboynik, Ayelet, Aaron M. Newman, Daniel M. Corey, Debashis Sahoo, Dmitry Pushkarev, Norma F. Neff, Benedetto Passarelli, Winston Koh, Katherine J. Ishizuka, Karla J. Palmeri, and et al. 2013b. Identification of a Colonial Chordate Histocompatibility Gene. *Science* 341: 384–87. [CrossRef]

Voskoboynik, Yotam, Aidan Glina, Mark Kowarsky, Chiara Anselmi, Norma F. Neff, Katherine J. Ishizuka, Karla J. Palmeri, Benyamin Rosental, Tal Gordon, Stephen R. Quake, and et al. 2020. Global Age-Specific Patterns of Cyclic Gene Expression Revealed by Tunicate Transcriptome Atlas. *BioRxiv*. [CrossRef]

Weissman, Irving L. 2000. Stem Cells: Units of Development, Units of Regeneration, and Units in Evolution. *Cell* 100: 157–68. [CrossRef]

Weissman, Irving L. 2015. Stem Cells Are Units of Natural Selection for Tissue Formation, for Germline Development, and in Cancer Development. *Proceedings of the National Academy of Sciences USA* 112: 8922–28. [CrossRef]

Wessel, Gary M., Shumpei Morita, and Nathalie Oulhen. 2020. Somatic Cell Conversion to a Germ Cell Lineage: A Violation or a Revelation? *Journal of Experimental Zoology Part B: Molecular and Developmental Evolution* 1: 14. [CrossRef]

Winkley, Konner, Spencer Ward, Wendy Reeves, and Michael Veeman. 2019. Iterative and Complex Asymmetric Divisions Control Cell Volume Differences in *Ciona* Notochord Tapering. *Current Biology* 29: 3466–77. [CrossRef]

Yoshida, Keita, Akiko Hozumi, Nicholas Treen, Tetsushi Sakuma, Takashi Yamamoto, Maki Shirae-Kurabayashi, and Yasunori Sasakura. 2017. Germ Cell Regeneration-Mediated, Enhanced Mutagenesis in the Ascidian *Ciona intestinalis* Reveals Flexible Germ Cell Formation from Different Somatic Cells. *Developmental Biology* 423: 111–25. [CrossRef] [PubMed]

Zaniolo, Giovanna, and Paola Trentin. 1987. Regeneration of the Tunic in the Colonial Ascidian, *Botryllus schlosseri*. *Acta Embryology Morphology Experimental* 8: 173–180.

Zhang, Tengjiao, Yichi Xu, Kaoru Imai, Teng Fei, Guilin Wang, Bo Dong, Tianwei Yu, Yutaka Satou, Weiyang Shi, and Zhirong Bao. 2020. A Single-Cell Analysis of the Molecular Lineage of Chordate Embryogenesis. *Science Advances* 6: eabc4773. [CrossRef] [PubMed]

Zheng, Tao, Ayaki Nakamoto, and Gaku Kumano. 2020. H3K27me3 Suppresses Sister-Lineage Somatic Gene Expression in Late Embryonic Germline Cells of the Ascidian, *Halocynthia roretzi*. *Developmental Biology* 460: 200–14. [CrossRef] [PubMed]

Zondag, Lisa E., Kim Rutherford, Neil J. Gemmell, and Megan J. Wilson. 2016. Uncovering the Pathways Underlying Whole Body Regeneration in a Chordate Model, *Botrylloides leachi* Using De Novo Transcriptome Analysis. *BMC Genomics* 17: 114. [CrossRef]

Zondag, Lisa, Rebecca Clarke, and Megan J. Wilson. 2019. Histone Deacetylase Activity Is Required for *Botrylloides leachii* Whole Body Regeneration. *Journal of Experimental Biology* 222: jeb.203620. [CrossRef]

Improving the Yields of Blood Cell Extractions from *Botryllus schlosseri* Vasculature

Andy Qarri, Yuval Rinkevich and Baruch Rinkevich

Abstract: The tunicate *Botryllus schlosseri* belongs to the Vertebrata's closest living invertebrate group. This colonial species represents an invertebrate model system that maintain high capacity of adult stem cell activity, where various blood cell types, expressing multipotent or totipotent phenotypes, circulate in vasculature throughout life. While isolated *Botryllus* blood cells may serve as indispensable tools for studying stem cells biology, up to date, no single cell line is available. The major bottle-necks for established cultures include the lack of cell division under in vitro conditions as from 24 to 72 h post isolation and enhanced contami-nation rates by bacteria and protists. Moreover, low yields of blood cells are of significant hindrance to the development of long-term cultures since lower numbers of cells eventually lead to poor results. Tackling these two critical technical obstacles, we present here methodologies for improved aseptic conditions and for higher yields of cells extracted from colonial vasculature. This study was performed on two colonial stocks (Israel, laboratory stocks; Helgoland, Germany—field collected stocks) which resulted with a significant difference in the numbers of cell extrac-tions between the two stocks and significantly different blood cell yields between various blastogenic stages (laboratory stocks), further revealing differences between field/laboratory-maintained colonies.

---

1. Introduction

The cosmopolitan tunicate *Botryllus schlosseri* belongs to a taxonomic taxon considered as the closest living invertebrates to the Vertebrata (Delsuc et al. 2006) and is used as an important model species in a wide range of biological disciplines (Ben-Hamo and Rinkevich 2021), such as ecotoxicology (Gregorin et al. 2021; Rosner et al. 2021), immunobiology and allorecognition (Magor et al. 1999; Rinkevich 2004), developmental biology including colony astogeny (Manni et al. 2019; Rosner et al. 2006; Rosner et al. 2019), regeneration (Voskoboynik et al. 2007), senescence (Rabinowitz and Rinkevich 2004a; Rinkevich 2017), evolutionary biology (Rinkevich 2002) and above all—stem cell biology (Ballarin et al. 2021; Voskoboynik et al. 2008). *B. schlosseri* colonies express two modes of reproduction, sexual and asexual (Manni et al. 2019). Sexual reproduction cycles occur weekly, each starting with the fertilization of eggs and progressing through embryonic stages into a tadpole larva featuring chordate characteristics that includes striated musculature, neural tube, notochord and tail (Voskoboynik et al. 2007). The tadpole larva swims for a short period of time and then attaches to a substrate near the mother colony, loses the tail through apoptosis, and then develops into the first zooid (the colonial module), called an oozooid (Berrill 1950). Colonies develop from the oozooids through weekly cycles of growth and death (Manni et al. 2019; Rinkevich 2019) and form several typical star-shaped groups of zooids, each called a system, that are embedded within the tunic, the transparent gelatinous extra cellular matrix (ECM) of the colony which contains cellulose cross-linked with proteins as well as the colonial circulatory system (Figure 1). Colonial systems are connected to each other via common blood vessels, which carry at the periphery of the colony sets of blind vasculature termini, called ampullae (spherical to elongate in structure). Each zooid in the colony possesses an oral siphon (branchial siphon) and an atrial siphon is shared for all zooids in each system (Berrill 1950).

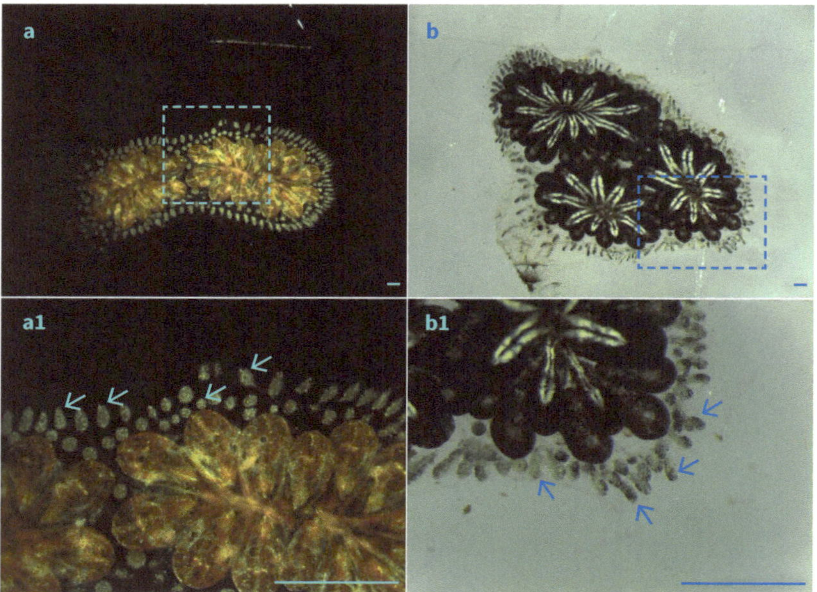

Figure 1. A *B. schlosseri* colony originated from the Israeli stock at the National Institute of Oceanography, Haifa (**a**,**a1**), at blastogenetic stage A, and (**b**,**b1**) a colony from Helgoland Island, Germany in blastogenetic stage C. Cells extracted from the marginal ampullae (arrows in a1, **b1**) of the colonies. Dotted squares represent the enlarged area of a1 and b1, respectively. Bars = 0.1 mm (**a**,**b**) and 1 mm (**a1**,**b1**). Source: Graphic by authors.

The asexual mode of development in *B. schlosseri* is expressed as weekly developmental cycles called blastogenesis, where each blastogenic cycle is composed of four major stages (marked by the letters A to D (Mukai and Watanabe 1976), during which the primary buds mature to adult zooids in concert with the development of the secondary buds from the body wall of each primary bud. A massive apoptotic event concludes each blastogenic cycle with the morphological resorption of all parental zooids, concurrently followed with the development of primary buds to functioning zooids (Lauzon et al. 1993). Thus, the blastogenesis process can be characterized by somatic self-renewal and vasculature regeneration, which demonstrate a model organism that carry out continuous somatic proliferation throughout the organism life span. In other words, the weekly budding process of somatic self-renewal and high vasculature regeneration capacity suggests an invertebrate model organism that maintain high capacity of stem cell activity throughout life (Ben-Hamo and Rinkevich 2021; Qarri et al. 2020; Rinkevich 2019).

Blood cell isolation and culturing are essential tools in the study of stem cells and regeneration in this model organism. Various cell types from *B. schlosseri* possess extensive potentialities such as multipotency and totipotency (Laird et al. 2005; Rinkevich and Rabinowitz 1994; Rinkevich and Rabinowitz 1997; Rosner et al. 2009; Rosner et al. 2021) and may serve as important tools in studying immunology, developmental biology, apoptosis and regeneration (Ballarin et al. 1994; Lauzon et al. 1993; Rosner et al. 2009; Rosner et al. 2021; Voskoboynik et al. 2007). Studies that attempted to develop primary cultures and permanent cell cultures from *B. schlosseri*, commonly used to extract blood cells that are directly collected from the blood vessels (Ballarin et al. 2008; Rinkevich and Rabinowitz 1993). Other studies used cells originated from epithelial layers (Rinkevich and Rabinowitz 1997), which show de novo stemness signatures (Rabinowitz et al. 2009; Rabinowitz and Rinkevich 2011) and cells originating from embryos (Rinkevich and Rabinowitz 1994). However to date, no single *Botryllus* cell line is available and it has repeatedly shown that extracted cells stop dividing in vitro within 24–72 h after their isolation. Moreover, many of the cultures are contaminated with opportunistic organisms including bacteria and protists, such as thraustochytrids (Qarri et al. 2021; Rabinowitz et al. 2006; Rinkevich and Rabinowitz 1993; Rinkevich and Rabinowitz 1994; Rinkevich 1999; Rinkevich 2011.)

The above studies indicate that, in order to establish long-term cell cultures, attempts should approach two critical technical statuses prior cell cultivation, (a) approved methodologies for aseptic conditions and (b) high yields of cell extraction. High yields of cells are of significant importance since lower numbers of cells eventually lead to poor results and fast senescence of extracted cells, primarily when dealing with blood cells that represent a rapid turnover and survival of only several weeks (Raftos et al. 1990; Rinkevich 1999; Rinkevich 2011). These limitations have led to attempts of pooling of blood cells originated from several colonies. For example, Ballarin et al. (2008) extracted 10^6 cells from more than three colonies, and Kamer and Rinkevich (2002) obtained the same cell concentrations (10^6 cells) by cutting the tunic matrix and the zooids without specifying the number of used colonies.

Responding to the above challenges, here we present a general aseptic approach with higher yield for blood cell (including stem cells) extractions employed on *B. schlosseri* colonies originating from two colonial stocks, the long-term established laboratory colonial cultures from Israel and from newly collected colonies originated from Helgoland Island, Germany. The essence of this approach is to improve blood cells yields from a single colony for in vitro applications.

2. Materials and Methods

2.1. Botryllus schlosseri Husbandry

Twenty-two colonies originated from Israeli cultures (long-term cultures maintained at the National Institute of Oceanography, Haifa, Israel) and freshly collected colonies from Helgoland Island, Germany, were selected for cell extraction experiments. Thirteen colonies (blastogenic stages A = 4, B = 3, C = 2 and D = 4) were derived from laboratory stocks reared in the Israeli facility for several years and originated from several USA west coast marinas (Monterey, Half Moon Bay and Moss Landing, California), as from Nelson Marina, New Zealand. The colonies were kept vertically on 5×7.5 cm^2 glass slides in slots of glass staining racks at 20 °C, in a 21-Liter plastic tank under a 12:12 h light:dark regimen, in a standing seawater system, as described (Rinkevich and Shapira 1998). Air stones were continuously used and the seawater was changed twice a week. Colonies were fed daily with freeze-dried rotifers, green unicellular algae and commercial powdered plankton. Nine colonies (blastogenic stages A = 5, B = 2 and C = 2) were collected from the rocky intertidal zone in Helgoland. These colonies were reared at the Biological Institute Helgoland (BAH) of the Alfred Wegener Institute, Helmholtz Centre for Polar and Marine Research—house C and maintained as the Israeli stock colonies under running seawater system and temperatures between 15.7 and 23.4 °C. Colonies were fed daily with dried algae powder. Colonies of both stocks were gently cleaned twice a week using small and soft brushes to remove trapped food particles, fouling organisms and debris. All experimental colonies were in good health and well adapted to their maintenance conditions.

2.2. Aseptic Solution, Instruments and Environment

Washing solution (WS) was used in order to reduce contaminations during the process of cell extraction. Artificial seawater (ASW) was prepared as described in

Rabinowitz and Rinkevich (Rabinowitz and Rinkevich 2004b), autoclaved, sterilized by a 0.2 μm filter membrane (Millipore) and stored at room temperature. For each 50 mL of WS, we used 44 mL of ASW, supplemented with 3 mL of PSA (Biological Industries; Penicillin 10,000 units/mL, Streptomycin sulphate 10 mg/mL and Amphotericin B 25 μm/mL; Cat. 03-033-1B. MP Biomedicals; Cat. 091674049) and 3 mL of Gentamycin Sulfate (Biological Industries; 50 mg/mL; Cat. 03-035-1. Gibco; 50 mg/mL; Cat. 15750037). Only sterilized plasticware was used. In addition to ASW sterilization, glassware was routinely autoclaved. Cell extraction protocols were carefully observed to maintain pathogen-free conditions. Additionally, prior to cell isolation colonies were kept under sterile conditions in a biosafety cabinet within a 20 °C cool room (Israel) and in an incubator of 20 °C (Helgoland).

2.3. Cell Extraction Approach under Aseptic Conditions

Before cell extraction procedure (Figure 2) colonies were taken out from aquaria and photographed under stereomicroscope (SMZ1000, Nikon equipped with DeltaPix digital camera Invenio 3SII, S/N: 3648213012. Leica M125 equipped with a camera Leica IC80 HD). Then colonies were meticulously cleaned by soft brushes, and the glass slides on which they were grown were comprehensively cleaned and wiped with 70% ethanol. Using razor blades under a biosafety cabinet, the colonies were carefully pulled off from the slides and placed in the centers of sterile 60 mm Petri dishes (Greiner bio-one, CELLSTARR 628160, Petri-dish 60 × 15 mm^2) for approximately 20 min in a humidity chamber, containing ASW, to actively attach to the dish substrates (detailed procedure in Rinkevich and Weissman (1987)). Then, 6 mL of WS (Washing solution) was added, and animals were left under sterile conditions for 48 h without food (starvation with antibiotic supplements significantly reduced contaminations of cell cultures; Rinkevich and Rabinowitz 1993). Then, the WS was changed 12 times (every hour for the first 6 h, left for 12 h and then, from the 18th hour, the WS changes protocol was repeated for the next 6 h; total of 12 washes). Following the above, all peripheral ampullae of each colony in a plate were punctured with an insulin syringe needle (28-Gauge) and the WS containing *B. schlosseri* blood cells was dropped into a 15 mL tube, pursued by centrifugation (2000 rpm for 10 min) using Eppendorf (Hamburg, Germany). The plates were then supplemented with WS and left for an addition 1.5 h in a biosafety cabinet, following which the WS from each plate was collected into a tube and cell extraction procedure (described above) was performed again. *B. schlosseri* blood cells in the tubes were centrifuged and the pellets were suspended in 1 mL of WS for further investigations.

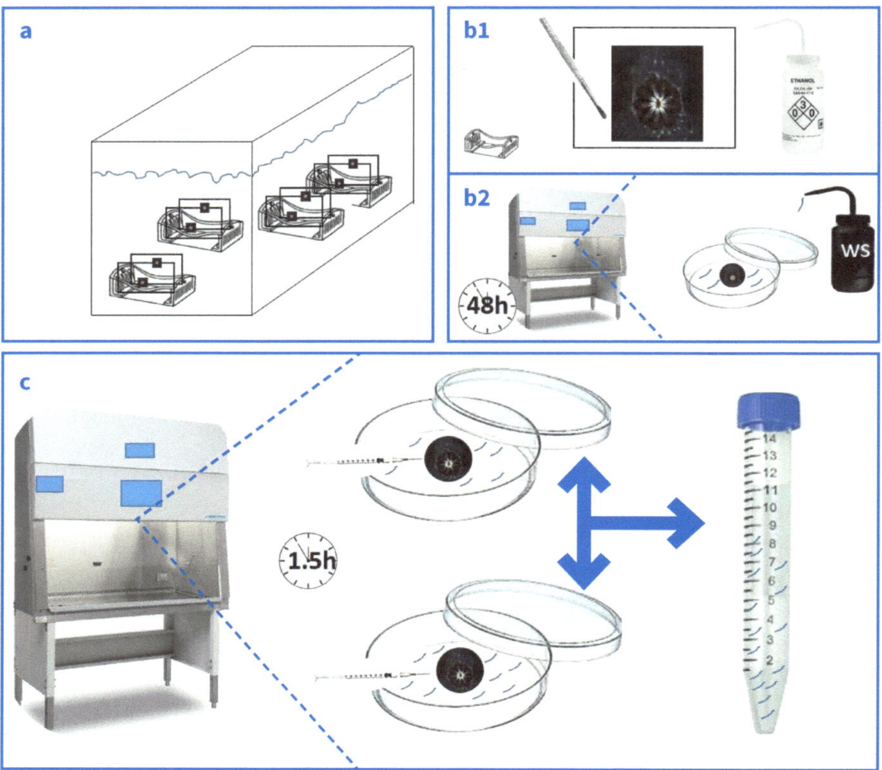

Figure 2. Schematic illustration of the cell extraction approach. (**a**) Colonies maintained under laboratory conditions, in tanks. (**b1**) Prior to cell extraction, colonies with their glass substrates are removed from their growth system, cleaned meticulously by soft brushes and the glass slides are wiped with 70% ethanol. (**b2**) Working in biosafety cabinet, each colony is removed, using a razor blade, from the glass substrate and is transferred to a sterile 60 mm Petri dish until actively attached. Then the Petri dish is being filled with 6 mL of WS. The plates are left in a biosafety cabinet for 48 h following 12 changes of WS. (**c**) Cell extraction is performed within a biosafety cabinet by puncturing the ampullae of each colony with the insulin syringe needle. This procedure is repeated after 1.5 h. Then the WS containing *B. schlosseri* blood cells are collected into a tube for further investigation. Source: Graphic by authors.

2.4. Cell Observation and Counting

All cell extractions were counted, using a hemocytometer, and photographed under the microscope (Olympus inverted system microscope, model I × 70, equipped with DP73 camera. Leica ICC50 HD). Cell viability was determined using Trypan

Blue solution (Biological Industries; Cat. 03-102-1B. Gibco; Cat. 15250061). Obtained values of *B. schlosseri* blood viable cells were between 93.6 and 98.5%.

2.5. Statistical Analyses

Statistical analyses were applied on extracts of two *B. schlosseri* colonial stocks originated from Israeli laboratory cultures and Helgoland Island, Germany using an SPSS V16. An independent-samples T test was performed on two stock cell yields. One-way ANOVA test using post hoc comparison (Bonferroni and Tukey HSD) was applied on blastogenesis of cell extracts of each *B. schlosseri* colonial stocks. Pearson correlation test was performed on zooid numbers of each colony with respect to cell yields.

3. Results

Cell Yields

The cell extraction protocol was performed on the 22 *Botryllus schlosseri* colonies originating from the Israeli stock of colonies (Figure 1a; 13 colonies in blastogenic stages A-D) and from Helgoland, Germany newly established stock (Figure 1b; 9 colonies in blastogenic stages A–C). Zooid numbers of the two stocks varied between 8 and 32 per colony for the Israeli colonies and 8 and 31 for the Helgoland colonies, yet no correlation (Figure 3) was recorded between the number of zooids per colony (of the colonial sizes used in this experiment) and cell yields for each stock ($r_{pearson} = -0.097$, $p > 0.05$; $r_{pearson} = 0.503$, $p > 0.05$; for Israeli/Helgoland stocks, respectively). Cells were extracted (Figure 4) from the marginal ampullae (Figure 1(a1,b1)) and numbers of cells and viability were studied on yields upon cell collections with respect to donors' blastogenic stages. Comparing between the two stocks of colonies, the results revealed a significant difference ($p < 0.0001$; independent-sample T test) in the numbers of cell extractions between the two stocks, where more cells were extracted from the freshly collected Helgoland stock. Within stock analyses revealed significant blastogenic-associated differences in cell yields from the Israeli stock colonies ($p < 0.05$; one-way ANOVA) but not in the freshly collected colonies from Helgoland ($p > 0.05$; one way ANOVA).

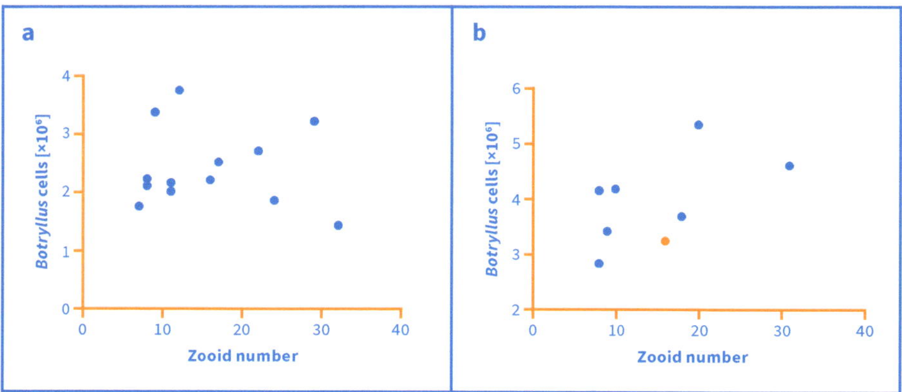

Figure 3. The correlation for cell yields vs. zooid numbers. (**a**) Cell yields vs. zooid numbers for the Israeli stock. (**b**) Cell yields vs. zooid numbers for the Helgoland stock. The red dot in b represents two different colonies with 16 numbers of zooids and cell yields of 3.28 and 3.27 $\times$ 10^6 cells. Source: Graphic by authors.

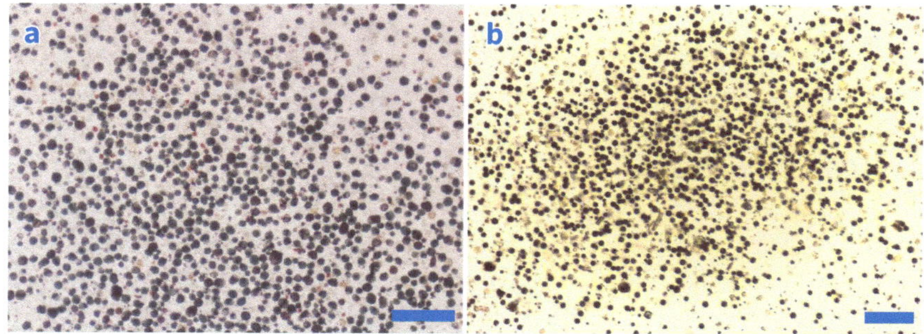

Figure 4. *B. schlosseri* cells under in vitro conditions. (**a**) A primary culture of blood cells from a blastogenesis stage B colony originated from the Israeli stock. (**b**) A primary culture of blood cells from a blastogenesis stage A colony from Helgoland, Germany. Bars = 100 µm. Source: Graphic by authors.

Cell yields from the Israeli blastogenic stages A and C colonies (Figure 5) composed of two significant groups, shared by the cell yields of blastogenic stages B and D colonies (Tukey HSD comparison). Cell yields from blastogenic stage A colonies ($n = 4$) varied between $(2.12 \pm 0.3) \times 10^6$ and $(3.75 \pm 0.2) \times 10^6$ ($p < 0.05$) in the Israeli stocks. For Helgoland colonies ($n = 5$), cell numbers varied between $(2.9 \pm 0.08) \times 10^6$ and $(4.6 \pm 0.69) \times 10^6$ ($p > 0.05$). Cell yields from blastogenic stage B colonies ($n = 3$) varied between $(2.2 \pm 0.13) \times 10^6$ and $(2.7 \pm 0.3) \times 10^6$ ($p > 0.05$) in the Israeli stocks, and for Helgoland colonies ($n = 2$) cell numbers varied between $(2.2 \pm 0.21) \times 10^6$ and $(5.35 \pm 0.3) \times 10^6$ ($p > 0.05$). Cell yields from blastogenic stage

C colonies (n = 2) varied between $(1.76 \pm 0.12) \times 10^6$ and $(1.87 \pm 0.34) \times 10^6$ ($p < 0.05$) in the Israeli stocks, and for Helgoland colonies (n = 2) cell numbers varied between $(3.28 \pm 0.08) \times 10^6$ and $(3.72 \pm 0.83) \times 10^6$ ($p > 0.05$). Cell yields from blastogenic stage D colonies ($n = 4$) varied between $(1.44 \pm 0.11) \times 10^6$ and $(2.17 \pm 0.48) \times 10^6$ ($p < 0.05$) in the Israeli stocks.

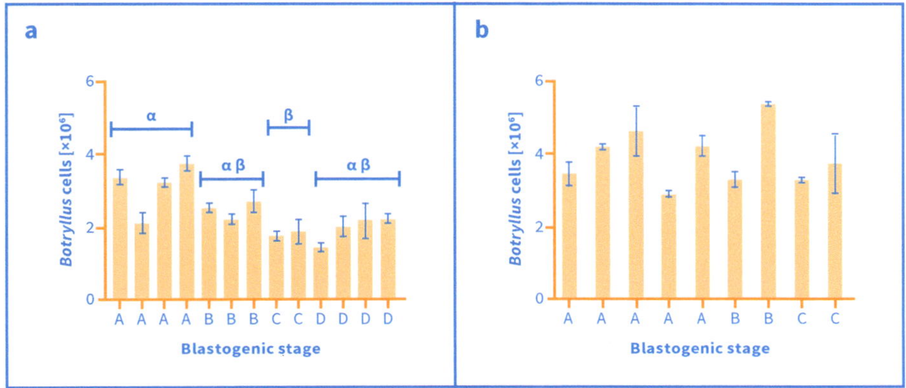

Figure 5. *B. schlosseri* cell yields. (**a**) Cell yields ($\times 10^6$) of colonies at blastogenic stages A–D originated from the Israeli stock. (**b**) Cell yields ($\times 10^6$) of colonies at blastogenic stages A–C originated from Helgoland Island. Each column represents an average of two extractions per colony ($\pm$S.D.). α and β symbolize statistical group differences between the tested blastogenesis stages obtained by Tukey HSD. Source: Graphic by authors.

4. Discussion

The literature reveals that primary cell cultures originated from *B. schlosseri* vasculature stop dividing 24–72 h post isolation (Rinkevich 1999) and further indicates low yields of cells per colony (Table 1) with high contamination rates (Rinkevich 1999; Rinkevich 2011.)

Table 1. Studies on *B. schlosseri* that refer to blood cell yields under laboratory conditions. Abbreviations: AC—Aseptic conditions; C—a colony; LS—laboratory stocks; NS—not specified; ABX—antibiotics; FSW—filtered seawater; FASW—filtered artificial seawater; ?= unknown number of colonies.

No of Colonies	Colony Origin	Cell Extraction under AC	Use of: FSW FASW/ABX	Cell Yields Colony^{-1}	Reference
>3	Italy (Venice)	No	FSW	3×10^5/C	(Ballarin et al. 2008)
NS	Italy (Venice)	No	FSW	5×10^6/?	(Menin and Ballarin 2008)
NS	LS	No	FASW	1×10^6/?	(Kamer and Rinkevich 2002)
NS	Italy (Venice)	No	FSW	$8\text{–}10 \times 10^6$/?	(Ballarin et al. 1994)
NS	LS	No	FSW	$8\text{–}10 \times 10^6$/?	(Ballarin et al. 2011)
NS	Italy (Venice)	No	FSW	1×10^7/?	(Ballarin and Cima 2005)
NS	LS	No	FSW/ABX	1×10^5/?	(Rinkevich and Rabinowitz 1993)
1	Helgoland	Yes	FASW/ABX	$5.3\text{–}5.4 \times 10^6$/C	This study
1	Israel LS	Yes	FASW/ABX	$3.6\text{–}3.9 \times 10^6$/C	This study

Source: By the authors.

Here we present an improved approach for blood cell extractions from *B. schlosseri* vasculature, performed under our aseptic conditions, which showed reduced contamination rates as compared to former outcomes. Yet, this issue was not analyzed in the present study. We used two *B. schlosseri* colonial stocks originated from a long-term laboratory cultures (from Israel) and colonies freshly collected from Helgoland Island, Germany. While at the colonial sizes used in this experiment there was no differences between the blood cell numbers obtained per colony, the results of this study clearly revealed, (1) a significant difference ($p < 0.0001$) in the number of blood cells obtained between the two disparate stocks and (2) changes in the numbers of blood cells obtained from various blastogenic stages (recorded only for the Israeli stock colonies). We obtained two significantly different blood cell yields between colonies at blastogenic stage A vs. stage C colonies. These results point to possible differences in numbers of total blood cells between freshly collected colonies from the field and colonies from established stocks, a result which should be taken into consideration when cell yields are an important component in structuring a research. This is also an interesting result regarding the *B. schlosseri* blood cell (and potentially stem cells) biology that should be studied in further experimentation.

The present study is the first that focuses on cell yields from *B. schlosseri* colonies. The literature (Table 1) reveals that past studies used undefined numbers of colonies in the research, or that the yield was lower than levels detailed in this study. Thus, our approach demonstrates potential for improving the extracting of circulating cells, including stem cells, under aseptic conditions, for any in vitro application, without pooling cells from different genotypes, augmenting the importance of *B. schlosseri* as a model organism in the field of cell biology (Ballarin et al. 2011; Ben-Hamo and Rinkevich 2021; Frizzo et al. 2000; Rosner et al. 2021). As a final point, the recognition of this model organism in the field of cell biology and stem cells biology is associated with its circulating blood cells that hold potentialities such as multipotency and totipotency (Ballarin et al. 2021; Laird et al. 2005; Rosner et al. 2009; Rosner et al. 2021).

Author Contributions: A.Q., Y.R. and B.R. conceived and designed the experiments, B.R. and Y.R. contributed reagents/materials/analysis tools. A.Q. performed the experiments, analyzed the data, prepared figures and authored drafts of the paper. All authors have read and agreed to the published version of the manuscript.

Funding: This research was funded by the United States—Israel Binational Science Foundation (BSF), as part of the joint program with the NSF, the National Science Foundation, USA (NSF/BSF no 2021650; to B.R.), by the BSF no 2015012 (B.R.) and by the European Cooperation in Science & Technology program (EU COST). Grant title: "Stem cells of marine/aquatic invertebrates: from basic research to innovative applications" (MARISTEM).

Acknowledgments: A.Q. thanks to AWI—BAH (Biologische Anstalt Helgoland, Alfred-Wegener-Institut Helmholtz-Zentrum für Polar- und Meeresforschung) for hospitality and to Eva-Maria Brodte and Uwe Nettelmann for technical assistance. A.Q. also thanks the European

STSM CA16203-46154—Stem cells of marine/aquatic invertebrates: from basic research to innovative applications, for the work in Israel.

Conflicts of Interest: The authors declare no conflict of interest.

References

Ballarin, Loriano, and Francesca Cima. 2005. Cytochemical properties of *Botryllus schlosseri* haemocytes: Indications for morpho-functional characterisation. *European Journal of Histochemistry* 49: 255–64. [CrossRef]

Ballarin, Loriano, Francesca Cima, and Armando Sabbadin. 1994. Phagocytosis in the colonial ascidian *Botryllus schlosseri*. *Developmental and Comparative Immunology* 18: 467–81. [CrossRef]

Ballarin, Loriano, Marcello Del Favero, and Lucia Manni. 2011. Relationships among hemocytes, tunic cells, germ cells, and accessory cells in the colonial ascidian *Botryllus schlosseri*. *Journal of Experimental Zoology Part B: Molecular and Developmental Evolution* 316: 284–95. [CrossRef] [PubMed]

Ballarin, Loriano, Arzu Karahan, Alessandra Salvetti, Leonardo Rossi, Lucia Manni, Baruch Rinkevich, Amalia Rosner, Ayelet Voskoboynik, Benyamin Rosental, Laura Canesi, and et al. 2021. Stem cells and innate immunity in aquatic invertebrates: Bridging two seemingly disparate disciplines for new discoveries in biology. *Frontiers in Immunology* 12: 2326. [CrossRef] [PubMed]

Ballarin, Loriano, Adams Menin, Laura Tallandini, Valerio Matozzo, Paolo Burighel, Giuseppe Basso, Elena Fortunato, and Francesca.Cima. 2008. Haemocytes and blastogenetic cycle in the colonial ascidian *Botryllus schlosseri*: A matter of life and death. *Cell and Tissue Research* 331: 555–64. [CrossRef] [PubMed]

Berrill, Norman John. 1950. *The Tunicata with an Account of the British Species*. London: Ray Society.

Ben-Hamo, Oshrat, and Baruch Rinkevich. 2021. Botryllus schlosseri—A model colonial species in basic and applied studies. In *Established and Emerging Marine Organisms in Experimental Biology*. Edited by Agnes Boutet and Bernd Schierwater. Boca Raton: CRC Press, pp. 385–402.

Delsuc, Frederic, Henner Brinkmann, Daniel Chourrout, and Herve Philippe. 2006. Tunicates and not cephalochordates are the closest living relatives of vertebrates. *Nature* 439: 965–68. [CrossRef] [PubMed]

Frizzo, Annalisa, Laura Guidolin, Loriano Ballarin, Barbara Baldan, and Armando Sabbadin. 2000. Immunolocation of phenoloxidase in vacuoles of the compound ascidian *Botryllus schlosseri* morula cells. *Italian Journal of Zoology* 67: 273–76. [CrossRef]

Gregorin, Chiara, Luisa Albarano, Emanuele Somma, Maria Costantini, and Valerio Zupo. 2021. Assessing the Ecotoxicity of Copper and Polycyclic Aromatic Hydrocarbons: Comparison of Effects on *Paracentrotus lividus* and *Botryllus schlosseri*, as Alternative Bioassay Methods. *Water* 13: 711. [CrossRef]

Kamer, Iris, and Baruch Rinkevich. 2002. In vitro application of the comet assay for aquatic genotoxicity: Considering a primary culture versus a cell line. *Toxicology* 16: 177–84. [CrossRef]

Laird, Diana J., Anthony. W. De Tomaso, and Irving L. Weissman. 2005. Stem cells are units of natural selection in a colonial ascidian. *Cell* 123: 1351–60. [CrossRef]

Lauzon, Robert J., Chris W. Patton, and Irving L. Weissman. 1993. A morphological and immunohistochemical study of programmed cell death in *Botryllus schlosseri* (Tunicata, Ascidiacea). *Cell and Tissue Research* 272: 115–27. [CrossRef] [PubMed]

Manni, Lucia, Chiara Anselmi, Francesca Cima, Fabio Gasparini, Ayelet Voskoboynik, Margherita Martini, Anna Peronato, Paolo Burighel, Giovanna Zaniolo, and Loriano Ballarin. 2019. Sixty years of experimental studies on the blastogenesis of the colonial tunicate *Botryllus schlosseri*. *Developmental Biology* 448: 293–308. [CrossRef] [PubMed]

Magor, Brad G., Anthony De Tomoso, Baruch Rinkevich, and Irving L. Weissman. 1999. Allorecognition in colonial tunicates: Protection against predatory cell lineages? *Immunological Reviews* 167: 69–79. [CrossRef] [PubMed]

Menin, Adams, and Loriano Ballarin. 2008. Immunomodulatory molecules in the compound ascidian *Botryllus schlosseri*: Evidence from conditioned media. *Journal of Invertebrate Pathology* 99: 275–80. [CrossRef] [PubMed]

Mukai, Hideo, and Hiroshi Watanabe. 1976. Studies on the formation of germ cells in a compound ascidian *Botryllus primigenus* oka. *Journal of Morphology* 148: 337–61. [CrossRef]

Qarri, Andy, Yuval Rinkevich, and Baruch Rinkevich. 2021. Employing marine invertebrate cell culture media for isolation and cultivation of thraustochytrids. *Botanica Marina* 64: 447–454. [CrossRef]

Qarri, Andy, Amalia Rosner, Claudette Rabinowitz, and Baruch Rinkevich. 2020. UV-B radiation bearings on ephemeral soma in the shallow water tunicate *Botryllus schlosseri*. *Ecotoxicology and Environmental Safety* 196: 110489. [CrossRef]

Rabinowitz, Claudette, Gilad Alfassi, and Baruch Rinkevich. 2009. Further portrayal of epithelial monolayers emergent de novo from extirpated ascidians palleal buds. *Cellular and Developmental Biology-Animal* 45: 334–42. [CrossRef]

Rabinowitz, Claudette, Jacob Douek, R. Weisz, Ariel Shabtay, and Baruch Rinkevich. 2006. Isolation and characterization of four novel thraustochytrid strains from a colonial tunicate. *IJMS* 35: 341–50.

Raftos, David A., Dan L. Stillman, and Edwin L. Cooper. 1990. In vitro culture of tissue from the tunicate *Styela clava*. *In Vitro Cellular and Developmental Biology* 26: 962–70. [CrossRef]

Rabinowitz, Claudette, and Baruch Rinkevich. 2004a. In vitro delayed senescence of extirpated buds from zooids of the colonial tunicate *Botryllus schlosseri*. *Journal of Experimental Biology* 207: 1523–32. [CrossRef] [PubMed]

Rabinowitz, Claudette, and Baruch Rinkevich. 2004b. Epithelial cell cultures from *Botryllus schlosseri* palleal buds: Accomplishments and challenges. *Methods in Cell Science* 25: 137–48. [CrossRef] [PubMed]

Rabinowitz, Claudette, and Baruch Rinkevich. 2011. De novo emerged stemness signatures in epithelial monolayers developed from extirpated palleal buds. *In Vitro Cellular and Developmental Biology-Animal* 47: 26–31. [CrossRef] [PubMed]

Rinkevich, Baruch. 1999. Cell cultures from marine invertebrates: Obstacles, new approaches and recent improvements. *Journal of Biotechnology* 70: 133–53. [CrossRef]

Rinkevich, Baruch. 2002. The colonial urochordate *Botryllus schlosseri*: From stem cells and natural tissue transplantation to issues in evolutionary ecology. *BioEssays* 24: 730–40. [CrossRef]

Rinkevich, Baruch. 2004. Primitive immune systems: Are your ways my ways? *Immunological Reviews* 198: 25–35. [CrossRef]

Rinkevich, Baruch. 2011. Cell cultures from marine invertebrates: New insights for capturing endless stemness. *Marine Biotechnology* 13: 345–54. [CrossRef]

Rinkevich, Baruch. 2017. Senescence in modular animals—botryllid ascidians as a unique aging system. In *The Evolution of Senescence in the Tree of Life*. Edited by Roberto Salguero-Gomez, Richard Shefferson and Owen R. Jones. Cambridge, UK: Cambridge University Press 2017, pp. 220–37.

Rinkevich, Baruch. 2019. The tail of the underwater phoenix. *Developmental Biology* 448: 291–92. [CrossRef]

Rinkevich, Baruch, and Claudette Rabinowitz. 1997. Initiation of epithelial cell cultures from palleal buds of *Botryllus schlosseri*, a colonial tunicate. *Cellular and Developmental Biology-Animal* 33: 422–24. [CrossRef]

Rinkevich, Baruch, and Claudette Rabinowitz. 1993. In vitro culture of blood cells from the colonial protochordate *Botryllus schlosseri*. *Cellular and Developmental Biology-Animal* 29: 79–85. [CrossRef]

Rinkevich, Baruch, and Claudette Rabinowitz. 1994. Acquiring embryo-derived cell cultures and aseptic metamorphosis of larvae from the colonial protochordate *Botryllus schlosseri*. *Invertebrate Reproduction and Development* 25: 59–72. [CrossRef]

Rinkevich, Baruch, and Michal Shapira. 1998. An improved diet for inland broodstock and the establishment of an inbred line from *Botryllus schlosseri*, a colonial sea squirt (Ascidiacea). *Aquatic Living Resources* 11: 163–71. [CrossRef]

Rinkevich, Baruch, and Irving L. Weissman. 1987. A long-term study on fused subclones in the ascidian *Botryllus schlosseri*: The resorption phenomenon (Protochordata: Tunicata). *Journal of Zoology* 213: 717–33. [CrossRef]

Rosner, Amalia, Jean Armengaud, Loriano Ballarin, Stephanie Barnay-Verdier, Francesca Cima, Ana V. Coelho, Isabelle Domart-Coulon, Damjana Drobne, Anne-Marie Genevière, Anita Jemec Kokalj, and et al. 2021. Stem cells of aquatic invertebrates as an advanced tool for assessing ecotoxicological impacts. *Science of the Total Environment* 771: 144565. [CrossRef] [PubMed]

Rosner, Amalia, Olha Kravchenko, and Baruch Rinkevich. 2019. IAP genes partake weighty roles in the astogeny and whole body regeneration in the colonial urochordate *Botryllus schlosseri*. *Developmental Biology* 448: 320–41. [CrossRef] [PubMed]

Rosner, Amalia, Elizabeth Moiseeva, Yuval Rinkevich, Ziva Lapidot, and Baruch Rinkevich. 2009. Vasa and the germ line lineage in a colonial urochordate. *Developmental Biology* 331: 113–28. [CrossRef]

Rosner, Amalia, Guy Paz, and Baruch Rinkevich. 2006. Divergent roles of the DEAD-box protein BS-PL10, the urochordate homologue of human DDX3 and DDX3Y proteins, in colony astogeny and ontogeny. *Developmental Dynamics* 235: 1508–21. [CrossRef] [PubMed]

Voskoboynik, Ayelet, Noa Simon-Blecher, Yoav Soen, Baruch Rinkevich, Anthony W. De Tomaso, Katherine J. Ishizuka, and Irving L. Weissman. 2007. Striving for normality: Whole body regeneration through a series of abnormal generations. *The FASEB Journal* 21: 1335–44. [CrossRef]

Voskoboynik, Ayelet, Yoav Soen, Yuval Rinkevich, Amalia Rosner, Hiroo Ueno, Ram Reshef, Katherine J. Ishizuka, Karla J. Palmeri, Elizabeth Moiseeva, Baruch Rinkevich, and et al. 2008. Identification of the endostyle as a stem cell niche in a colonial chordate. *Cell Stem Cell* 3: 456–64. [CrossRef] [PubMed]

Sweet Tunicate Blood Cells: A Glycan Profiling of Haemocytes in Three Ascidian Species

Fan Zeng, Anna Peronato, Loriano Ballarin and Ute Rothbächer

Abstract: Ascidians are invertebrate chordates and may reveal parallels to vertebrate traits including cellular immunity, tissue rejection, and self-renewal, all functions executed by ascidian blood cells. Understanding their individual properties, functional plasticity, and lineage resemblances among ascidian species is, however, limited by a lack of cytochemical and molecular markers. We performed a lectin-based glycan profiling of haemocytes in three selected ascidian species to compare different blood cell populations and mirror their relatedness. We found differing repertoires of species-specific glycans for blood cells believed to be homologous in their function. Within species, characteristic glycans or glycan combinations mark haemocyte types and support their hematopoietic relatedness or distinguish maturation stages. Strikingly, *Ciona* and *Phallusia* haemoblasts have few carbohydrate decorations and drastically differ from differentiated cells, likewise phagocytes from cytotoxic cells, as compared with *Botryllus*, where a complex role of haemocytes in asexual self-renewal and allorecognition may involve carbohydrates. Cytotoxic cells generally carry most decorations. Within cell types, specific carbohydrates reside on the cell surface including amoeboid extensions, while others are within granules possibly marking molecules important in cytotoxicity and crosslinking. Taken together, these carbohydrate biosensors should further the molecular and functional characterisation of the outstanding properties of the different haemocytes in genetically accessible ascidian species.

---

1. Introduction

Tunicates are the closest relatives to vertebrates and include the ascidians that form swimming larvae with a chordate body plan. Understanding their blood cells is essential for understanding the evolution of the vertebrate immune system and notably the origins of adaptive immunity. Ascidians possess a simple form of allelic self–non-self recognition, particularly important for colonial ascidians to reject non-allelic mates or to fuse with those of allelic resemblance. Genome sequencing of model tunicates and molecular surveys have revealed a clear lack of an orthologous highly polymorphic major histocompatibility complex (MHC) locus used for adaptive immunity in vertebrates but found a less polymorphic fusibility locus (fuhc/BHF) that contains several genes discussed in non-self reactions (reviewed in Taketa and Tomaso 2015). Interestingly, the immune receptors of innate immunity inherent to all metazoans and which normally cooperate with MHC molecules in vertebrates are more numerous in tunicates, and the variable products of the less polymorphic immune loci identified in tunicates may support a primitive resemblance of the immune system to the vertebrate condition (Azumi et al. 2003; Mueller and Rinkevich 2020). It remains a domain of active research to understand how the various immune molecules mediate specific immune functions in tunicates (reviewed in Franchi and Ballarin 2017; Parrinello et al. 2018; Rosental et al. 2020).

As in most metazoans, the immune functions in tunicates are overly executed by haemocytes, and their immune receptors are important for both the evolutionary ancient innate (defensive, non-specific) immunity and the more sophisticated allogeneic immune response (reviewed in Rosental et al. 2020). While tunicate blood cells resemble their vertebrate counterparts in many aspects, their striking features as stem cells attract more attention recently, in parallel to other invertebrates (reviewed in Ballarin et al. 2021a). Of note is their outstanding functional plasticity and regenerative potential, well evident in colonial tunicates where haemocytes can reconstitute an entire animal (reviewed in Manni et al. 2019; Ferrario et al. 2020; Alié et al. 2021). Mature haemocytes exert specialised functions such as immune recognition, phagocytosis, or cytotoxicity, but as a highly dynamic cell population with various differentiation and activation stages, it remains challenging to clearly group them into functional subtypes. Morphological criteria were used to distinguish the different haemocytes, and their resembling characteristics to vertebrates' blood cells were used to categorise their functions (Hartenstein 2006; Arizza and Parrinello 2009; Franchi and Ballarin 2017; Blanchoud et al. 2017; Rosental et al. 2020). When comparing blood cells among tunicates, clear similarities, but also notable differences, are observed between species, both in their morphological diversity and number of prominent subtypes with several questions about their functional homologisation and origin within the hematopoietic lineage remaining open (reviewed in Cima et al. 2016; Parrinello et al. 2018).

Toward understanding the functions of tunicate blood cells at a molecular level, the isolation of distinguishing molecular markers is instrumental to characterise and subgroup them. Such markers have been limiting in the study of haemocytes and for their functional comparisons (Rosental et al. 2020). Classically, proteins are considered major effectors of cellular functions, including immune recognition, and are detected at the level of their coding mRNA or by antibodies for differential expression to be associated with haemocyte functions. An independent but overlapping functional category includes carbohydrate modifications occurring on glycoproteins and glycolipids that strongly influence their maturation, structure, and function.

Furthermore, sugar epitopes often extend far in the extracellular space, giving these structural decorations great relevance to molecular and cellular interactions.

We have previously performed extensive lectin profiling of ascidian (*Ciona intestinalis*) larvae focusing on their sensory adhesive organs (papillae, palps) and have detected interesting similarities in three model organisms (*Ciona*, *Phallusia*, and *Botryllus*), suggesting a possible functional conservation of certain sugar residues, at least related to their papillary function (Zeng et al. 2019a, 2019b). Since we also observed specific lectin binding to migratory cell types, which were suggested to include haemocytes (Cloney and Grimm 1970; Sotgia et al. 1993; Sato et al. 1997; Davidson and Swalla 2002; Jimenez-Merino et al. 2019), we here aimed to profile and compare lectin patterns of the well accessible migratory haemocytes and provide useful markers and tools to access the molecules behind their epitopes.

In an attempt to provide an array of biosensors for tunicate blood cells of *Ciona intestinalis*, *Phallusia mammillata*, and *Botryllus schlosseri*, we fingerprinted their carbohydrate decorations in three model ascidian species using a collection of sixteen biotinylated plant lectins. We obtained glycan patterns of typical combinations in the three species that allow for haemocyte distinction and for mirroring their hematopoietic relatedness. Uniquely binding lectins will further the identification and functional characterisation of the interacting immune receptors within the highly dynamic haemocyte populations. Our glycophenotyping notably identifies the multiple sugar reactive sites for endogenous ascidian lectins on their natural target counterreceptors present on haemocytes. This knowledge is relevant to deciphering the intricate haemocyte functions and crosstalk in simpler chordates.

2. Materials and Methods

2.1. Animal Husbandry

The three ascidian species selected (two solitary and one colonial species) provide well-developed genomics tools and will be amenable for further molecular profiling and functional testing. *Ciona intestinalis* and *Phallusia mammillata* adults were purchased and shipped from the Roscoff Marine Station, France, and kept in aquaria with circulating and oxygenated artificial seawater at 16 °C. *Botryllus schlosseri* colonies were from the Venice Lagoon (provided by A. P. and L. B., Padova, Italy). The colonies were grown on glass slides and maintained in aerated aquaria (temperature 17 °C, salinity 35‰) and fed with Interpet Liquifry Marine (Dorking, UK).

2.2. Haemocyte Preparations from Three Species

Solitary ascidian *Ciona intestinalis* and *Phallusia mammillata* were tissue dried for any excess seawater, then dissected with scissors to expose their hearts. By a small incision, the haemocytes were released and collected into ice-cold Eppendorf tubes with 0.38% Na–citrate in filtered ASWH, pH 7.5 to prevent haemocyte aggregation, then centrifuged at $750\times g$ for 10 min and resuspended in FSW at a final concentration of 5×10^6 cells/ml. 80–100 µL of this haemocyte suspension were placed in the centre of Superfrost glass slides to which they were adhered for 20–30 min, to generate haemocyte monolayers for later lectin staining. For the colonial ascidian, *Botryllus schlosseri*, zooids were torn using a fine tungsten needle causing blood cells leakage, and haemocytes were collected and prepared as above.

2.3. Lectin Staining of Haemocytes from Three Species

Haemocyte monolayers of *Ciona intestinalis*, *Phallusia mammillata*, and *Botryllus schlosseri* were fixed in 4% paraformaldehyde (PFA) for 30 min at room temperature, transferred, and kept in 1× phosphate-buffered saline (PBS) until use. Lectin labelling was performed as described previously (Zeng et al. 2019a). Biotinylated lectins (GSL I, DBA, SBA, PNA, RCA I, SJA, ECL, GSL II, PHA-E, PHA-L, LEL, STL, DSL, PSA, LCA and UEA I, Vector Laboratories, Burlingame) were incubated at a final concentration of 15–25 µg mL^{-1} in 3% BSA–TBS, followed by washes with TBS, then fluorescently labelled with Streptavidin dye light 488 (Vector Laboratories, Burlingame) at 1:300 dilution in 3% BSA–TBS with DAPI (Merck/Sigma, D9542, Darmstadt) added at 0.035 µg mL^{-1} for nuclear staining. Samples were imaged by Leica fluorescent microscopy with filter cube L5 (wavelength 488), DAPI, and DIC; images were analysed with ImageJ (version 1.52 h).

3. Results

3.1. A Common Standard to Compare the Ascidian Haemocytes

The description of the ascidian haemocytes shows considerable variations among species and may exhibit 5–11 morphologically distinguishable cell types (Nette et al. 1999; Hartenstein 2006; Arizza and Parrinello 2009; Blanchoud et al. 2017; Cima et al. 2017; Gutierrez and Brown 2017). Therefore, to gain comparable information on the differing glycans of blood cells, we adopted a common standard for the haemocyte classification in ascidians, with three main groups: haemoblasts, immunocytes, and storage cells (Figure 1).

Figure 1a shows the main groups and subtypes of ascidian blood cells. Haemoblasts (Hbs) are uniquely small haemocytes, with a high nucleus/cytoplasm ratio, considered multipotent cells involved in regeneration and budding phenomena. Among immunocytes described in the three species, the phagocytes (Phcs) may encompass hyaline amoebocytes (HAs), macrophage-like cells (MLCs), and granular amoebocytes (GAs), while cytotoxic cells may include granulocytes with small granules (GSs), with large granules (GLs), morula cells (MCs), and unilocular refractile granulocytes (URGs). Storage cells may include pigment cells (PCs) or bivacuolated cells (BCs) and signet ring cells (SCs).

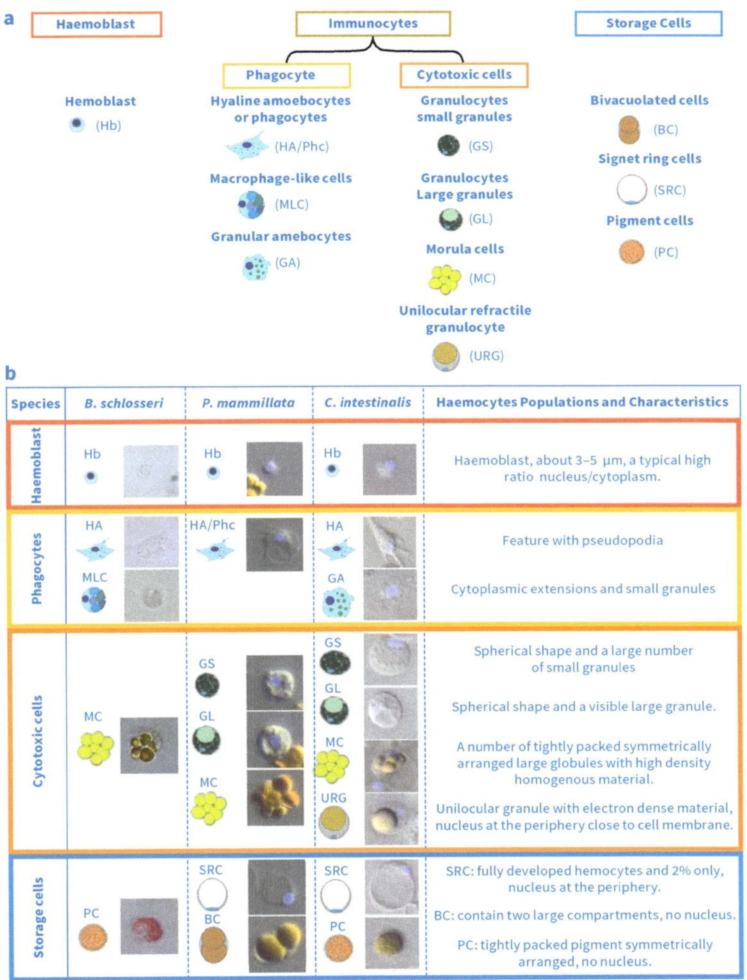

Figure 1. Ascidian blood cell types. (**a**) Scheme of ascidian blood cells: haemoblasts (Hb), immunocytes and storage cells. Immunocytes divide into phagocytes (containing hyaline amoebocytes or phagocytes, HA/Phc, macrophage-like cells, MLC and granular amebocytes, GA) and cytotoxic cells (encompassing granulocytes with small granules, GS, granulocytes with large granules, GL, morula cells, MC, and unilocular refractile granulocyte, URG). Storage cells include bivacuolated cells (BC), signet ring cells (SRC) and pigment cells (PC). (**b**) Haemocyte populations and characteristics in three ascidian species, *Botryllus schlosseri*, *Phallusia mammillata* and *Ciona intestinalis*. Haemocyte colors code: haemoblasts red, phagocytes yellow, cytotoxic cells orange, storage cells blue. Haemocyte schemes are modified from: (Nette et al. 1999, Hartenstein 2006, Arizza and Parrinello 2009, Blanchoud et al. 2017 and Cima et al. 2017). Source: Graphic by authors.

In Figure 1b, according to the three main categories of ascidian blood cells listed above (haemoblast, immunocytes, and storage cells), we grouped the haemocyte subtypes described in each of the three species, *Botryllus schlosseri*, *Phallusia mammillata*, and *Ciona intestinalis*, and summarised their typical characteristics. The blood cells of the colonial *Botryllus* encompass Hbs, phagocytic HAs, and MLCs, while MCs are only cytotoxic cells, and PCs are only storage cells. In contrast, in the solitary *Phallusia* and *Ciona*, several additional cell types are distinguished. They both feature additional cytotoxic granulocytes (GSs, GLs), and *Ciona* has a fourth cytotoxic cell type, the URGs, and the second type of phagocytic GA. As storage cells, they both have SCs and PCs, while *Phallusia* has BCs instead of PCs.

The scheme of ascidian blood cells includes haemoblasts (Hbs), immunocytes, and storage cells. Immunocytes divide into phagocytes, (containing hyaline amoebocytes or phagocytes (HAs/Phcs), macrophage-like cells (MLCs), and granular amoebocytes (GAs)) and cytotoxic cells (encompassing granulocytes with small granules (GSs), granulocytes with large granules (GLs), morula cells (MCs), and unilocular refractile granulocytes (URGs)). Storage cells include bivacuolated cells (BC)s, signet ring cells (SRCs), and pigment cells (PCs). Haemocyte schemes are modified from Blanchoud et al. (2017).

3.2. Carbohydrate Profiling of Haemocytes in Three Model Ascidian Species

The diversity of haemocytes is considered here as a function of their carbohydrate modifications since it was previously shown to play an important role in haemocyte recognition and interactions for immune activation. Plant lectins, in addition to being well-defined biosensors recognising specific carbohydrate moieties, can also be considered as biochemical tools to access the corresponding glycosylated receptors.

To obtain a more complete picture of the carbohydrate moieties in haemocytes of the three ascidian species, we screened 16 lectins featuring various sugar specificities listed in Table 1 (key recognition structures from lectins, Chapters 25 and 45: Yasuda et al. 2014; Kobayashi et al. 2014). We used lectins that recognise derivatives of galactose, glucose, mannose, and fucose. The first group comprises *GSL I*, *SBA*, *DBA*, *PNA*, *SJA*, and *RCA I* recognising galactose (Gal)/N-acetylgalactosamine (GalNAc) and N-acetyllactosamine (LacNAc, Gal-GalNAc)/GalNAc for *ECL*, with *PNA* and several others in this group known to bind O-linked sugars. A second group comprises lectins recognising sugars often found in N-linked protein glycosylation including *N*-acetylglucosamine (GlcNAc), recognised by *GSL II*, Gal/GlcNAc/mannose (Man) by *PHA-L* and *PHA-E* (the latter only for bisections of core Man), chitin (poly-GlcNAc)/GlcNAc/LacNAc by *LEL*, *STL*, *DSL*, and fucosylated glycans containing D-mannose/D-glucose/GlcNAc interacting with *PSA* and *LCA*, or with *UEA I* when in terminal position.

Table 1. Lectin bound glycosylations and key recognition structures.

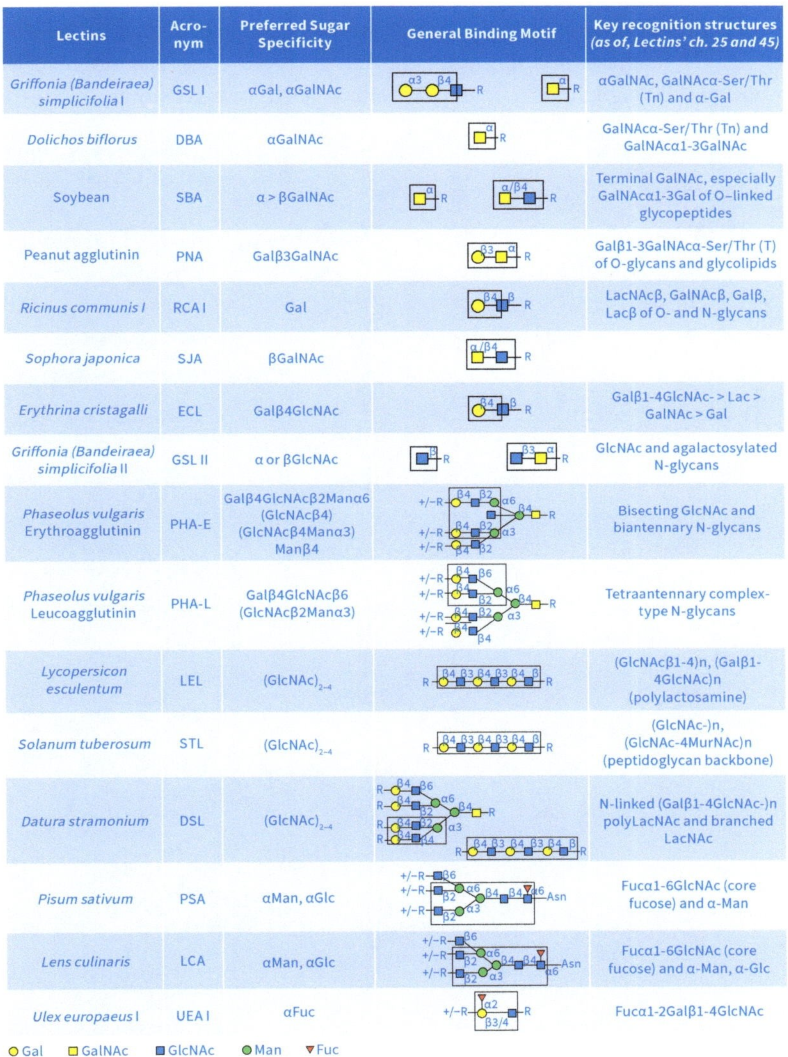

Lectins	Acronym	Preferred Sugar Specificity	General Binding Motif	Key recognition structures (as of, Lectins' ch. 25 and 45)
Griffonia (Bandeiraea) simplicifolia I	GSL I	αGal, αGalNAc		αGalNAc, GalNAcα-Ser/Thr (Tn) and α-Gal
Dolichos biflorus	DBA	αGalNAc		GalNAcα-Ser/Thr (Tn) and GalNAcα1-3GalNAc
Soybean	SBA	α > βGalNAc		Terminal GalNAc, especially GalNAcα1-3Gal of O–linked glycopeptides
Peanut agglutinin	PNA	Galβ3GalNAc		Galβ1-3GalNAcα-Ser/Thr (T) of O-glycans and glycolipids
Ricinus communis I	RCA I	Gal		LacNAcβ, GalNAcβ, Galβ, Lacβ of O- and N-glycans
Sophora japonica	SJA	βGalNAc		
Erythrina cristagalli	ECL	Galβ4GlcNAc		Galβ1-4GlcNAc- > Lac > GalNAc > Gal
Griffonia (Bandeiraea) simplicifolia II	GSL II	α or βGlcNAc		GlcNAc and agalactosylated N-glycans
Phaseolus vulgaris Erythroagglutinin	PHA-E	Galβ4GlcNAcβ2Manα6 (GlcNAcβ4) (GlcNAcβ4Manα3) Manβ4		Bisecting GlcNAc and biantennary N-glycans
Phaseolus vulgaris Leucoagglutinin	PHA-L	Galβ4GlcNAcβ6 (GlcNAcβ2Manα3)		Tetraantennary complex-type N-glycans
Lycopersicon esculentum	LEL	(GlcNAc)$_{2-4}$		(GlcNAcβ1-4)n, (Galβ1-4GlcNAc)n (polylactosamine)
Solanum tuberosum	STL	(GlcNAc)$_{2-4}$		(GlcNAc-)n, (GlcNAc-4MurNAc)n (peptidoglycan backbone)
Datura stramonium	DSL	(GlcNAc)$_{2-4}$		N-linked (Galβ1-4GlcNAc-)n polyLacNAc and branched LacNAc
Pisum sativum	PSA	αMan, αGlc		Fucα1-6GlcNAc (core fucose) and α-Man
Lens culinaris	LCA	αMan, αGlc		Fucα1-6GlcNAc (core fucose) and α-Man, α-Glc
Ulex europaeus I	UEA I	αFuc		Fucα1-2Galβ1-4GlcNAc

Gal   GalNAc   GlcNAc   Man   Fuc

Source: Table created by authors.

Overall, we isolated circulating blood cells of the three ascidian species and performed stainings using conjugates of biotinylated lectins and fluorescent streptavidin. The haemocyte stainings are summarised for the three species in Table 2 and shown for the individual species in Figures 2–4, respectively. The staining pattern of each of the 16 lectins is shown in Appendix A Figures A1–A16 as a comparison in the three species.

Table 2. Summary of lectin profiling for ascidian blood cells: *Botryllus schlosseri*, *Phallusia mammillata*, and *Ciona intestinalis* blood cells lectin fluorescent labelling intensity: (−) no staining, (+) very weak, + weak, ++ intermediate, +++ strong labelling, (?) no corresponding cells identified; colour code: haemoblasts—red; immunocytes—yellow; phagocytes or cytotoxic cells—orange; storage cells—blue.

B. schlosseri

Lectin	Haemoblast	Immunocytes			Storage cells	Sugar Specificity
	Hb	HA	MLC	MC	BC	
GSL I	+	+++	+	++	+	αGal, αGalNAc
DBA	++	+++	+++	+++	+++	αGalNAc
SBA	++	+++	++	+++	+++	α > βGalNAc
PNA	−	−	−	++	−	Galβ3GalNAc
RCA I	?	?	−	−	−	Gal
SJA	−	+	−	+	−	βGalNAc
ECL	?	−	−	++	−	Galβ4GlcNAc
GSL II	?	−	+	−	−	α or βGlcNAc
PHA-E	?	++	+	+	−	Galβ4GlcNAcβ2Manα6 (GlcNAcβ4) (GlcNAcβ4Manα3) Manβ4
PHA-L	−	+	−	++	−	Galβ4GlcNAcβ6 (GlcNAcβ2Manα3) Manα3
LEL	+	?	+++	−	?	(GlcNAc)2–4
STL	?	−	−	−	?	(GlcNAc)2–4
DSL	++	+	+++	+++	+	(GlcNAc)2–4
PSA	−	−	−	++	+	αMan, αGlc
LCA	−	?	−	++	−	αMan, αGlc
UEA I	+/−	?	−	+	+	αFuc

P. mammillata

Lectin	Haemoblast	Immunocytes				Storage cells		Sugar Specificity
	Hb	Phc	GS	GL	MC	SRC	BC	
GSL I	−	−	−	++	++	−	−	αGal, αGalNAc
DBA	−	−	−	−	+++	−	−	αGalNAc
SBA	−	++	++	−	+++	−	−	α > βGalNAc
PNA	−	++	−	−	++	−	−	Galβ3GalNAc
RCA I	−	−	−	−	−	−	−	Gal
SJA	−	+	−	−	+	−	−	βGalNAc
ECL	−	+++	−	−	++	−	−	Galβ4GlcNAc
GSL II	+	+	+	+	−	++	+	α or βGlcNAc
PHA-E	−	−	−	++	+	−	−	Galβ4GlcNAcβ2Manα6 (GlcNAcβ4) (GlcNAcβ4Manα3) Manβ4
PHA-L	−	−	−	+	++	−	−	Galβ4GlcNAcβ6 (GlcNAcβ2Manα3) Manα3
LEL	−	−	−	+	−	−	−	(GlcNAc)2–4
STL	−	−	−	−	−	−	−	(GlcNAc)2–4
DSL	−	−	+	−	+++	−	−	(GlcNAc)2–4
PSA	−	−	+/−	++	++	−	−	αMan, αGlc
LCA	−	−	−	+	++	−	−	αMan, αGlc
UEA I	−	−	−	+	+	−	−	αFuc

C. intestinalis

Lectin	Haemoblast	Immunocytes						Storage cells		Sugar Specificity
	Hb	HA	GA	GS	GL	MC	URG	SRC	PC	
GSL I	−	−	−	+	−	−	+	−	−	αGal, αGalNAc
DBA	−	−	−	−	++	−	−	−	−	αGalNAc
SBA	−	−	−	−	++	−	−	−	−	α > βGalNAc
PNA	−	−	−	−	+++	++	+	−	−	Galβ3GalNAc
RCA I	−	−	−	−	−	++	+	−	−	Gal
SJA	−	−	−	−	−	+	++	−	+	βGalNAc
ECL	−	−	−	+	+	−	++	−	+	Galβ4GlcNAc
GSL II	−	−	−	++	+++	−	++	−	+	α or βGlcNAc
PHA-E	−	−	−	−	−	+	+/−	−	−	Galβ4GlcNAcβ2Manα6 (GlcNAcβ4) (GlcNAcβ4Manα3) Manβ4
PHA-L	−	−	−	+	−	+	−	−	+	Galβ4GlcNAcβ6 (GlcNAcβ2Manα3) Manα3
LEL	−	−	−	−	+++	−	++	−	+	(GlcNAc)2–4
STL	−	−	−	+++	−	−	−	−	−	(GlcNAc)2–4
DSL	−	+	+	−	+++	−	+	−	−	(GlcNAc)2–4
PSA	−	−	+++	+	+	−	−	−	−	αMan, αGlc
LCA	−	−	+/−	++	+++	−	−	−	+/−	αMan, αGlc
UEA I	−	−	++	+	++	−	−	−	−	αFuc

Source: Table created by authors.

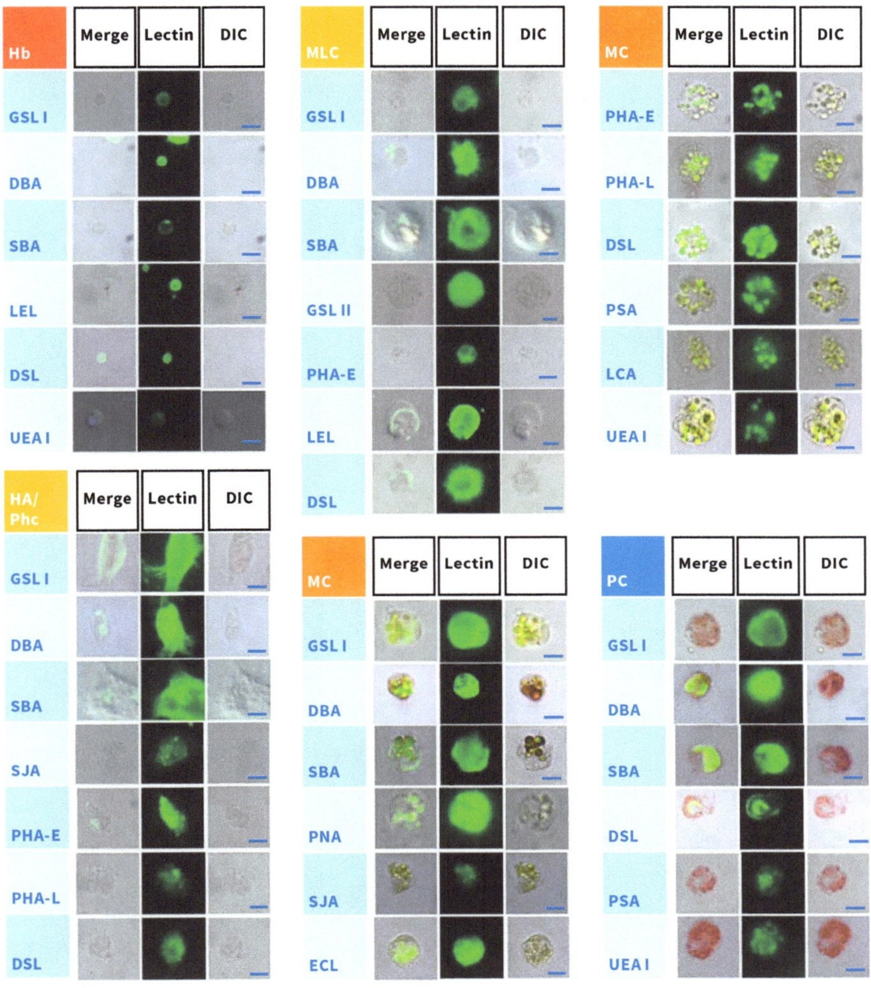

Figure 2. *Botryllus schlosseri* haemocyte types labelled with lectin probes. Lectin fluorescent labelling of *B. schlosseri* haemocytes: haemoblasts (Hbs) with *GSL I, DBA, SBA, LEL, DSL, UEA I*; hyaline amoebocytes or phagocytes (HAs/Phcs) with *GSL I, DBA, SBA, SJA, PHA-E, PHA-L, DSL*; macrophage-like cells (MLCs) with *GSL I, DBA, SBA, GSL II, PHA-E, LEL, DSL*; morula cells (MCs) with *GSL I, DBA, SBA, PNA, SJA, ECL, PHA-E, PHA-L, DSL, PSA, LCA, UEA I*; pigment cells (PCs) with *GSL I, DBA, SBA, DSL, PSA, UEA I*. Each lectin staining compares fluorescence images (green), the merged overlap (lectin with DIC), and DIC (bright field), scale bar: 5 μm. Source: Graphic by authors.

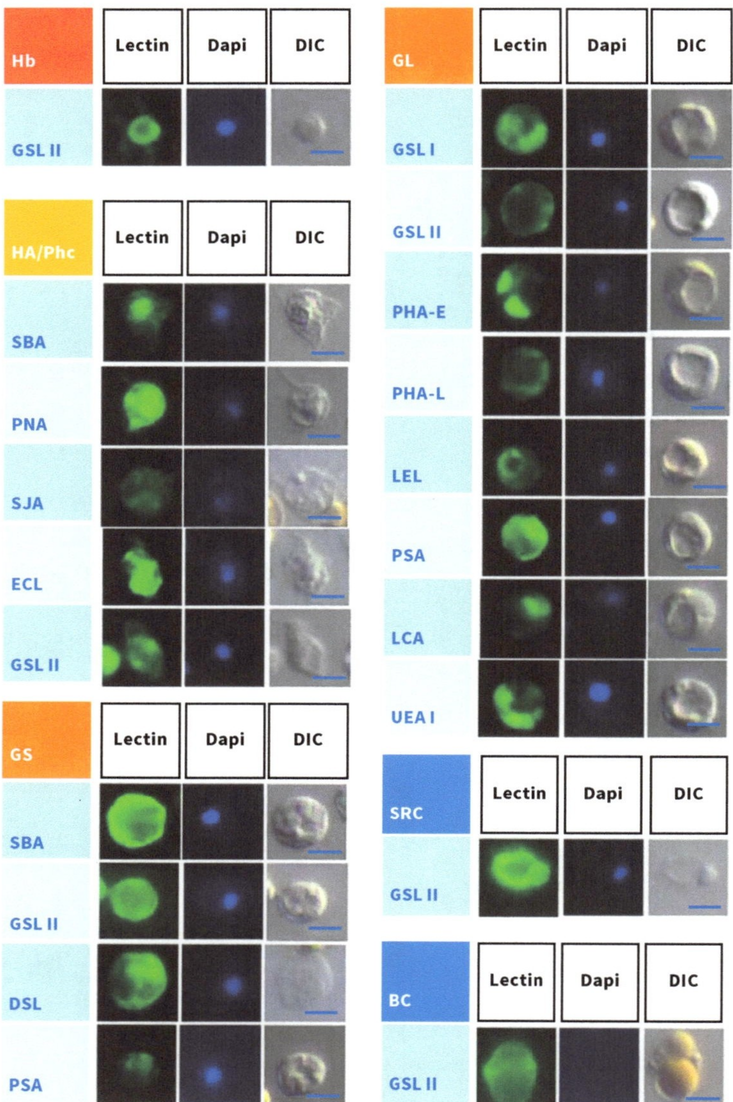

Figure 3. *Phallusia mammillata* haemocyte types labelled with lectin probes. Lectin fluorescent labelling of *P. mammillata* haemocytes: haemoblasts (Hbs) with *GSL II*; hyaline amoebocytes or phagocytes (HAs/Phcs) with *SBA, PNA, SJA, ECL, GSL II*; granulocytes with small granules (GSs) with *SBA, GSL II, DSL, PSA*; granulocytes with large granules (GLs) with *GSL I, GSL II, PHA-E, PHA-L, LEL, PSA, LCA, UEA I*; signet ring cells (SRCs) with *GSL II*; bivacuolated cells (BCs) with *GSL II*. Each lectin staining has fluorescence images (green), Dipa (blue), and DIC (bright filed), scale bar: 5 μm. Source: Graphic by authors.

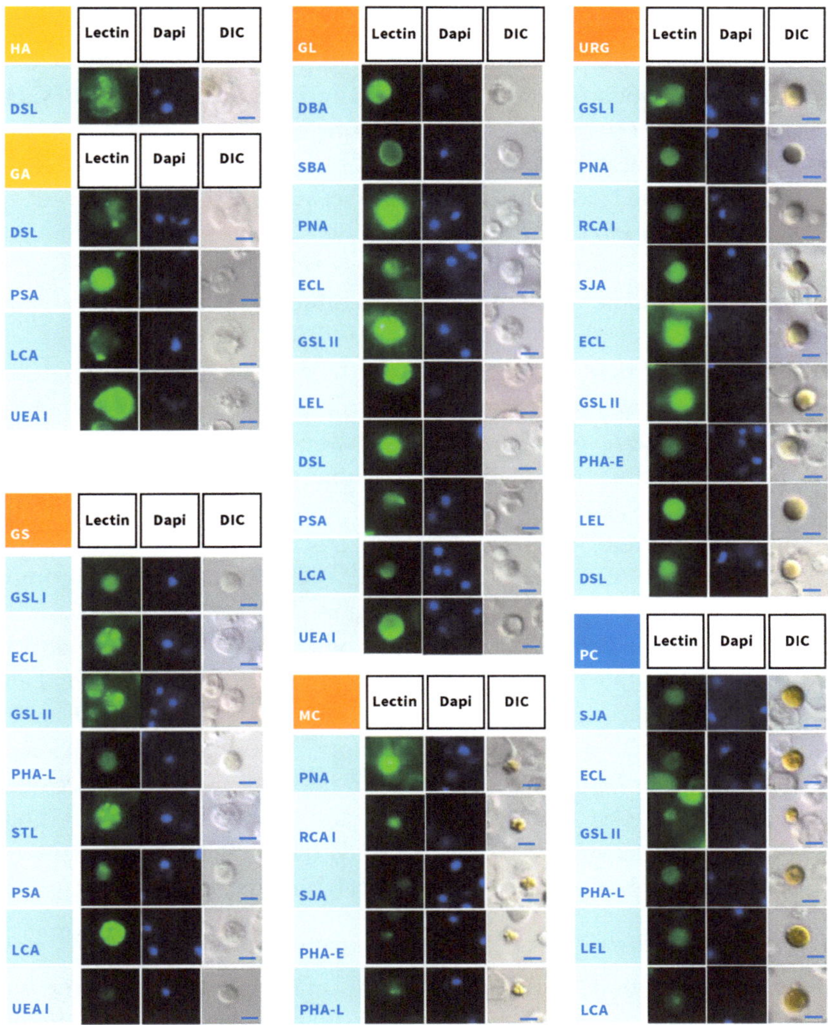

Figure 4. *Ciona intestinalis* haemocyte types labelled with lectin probes. Lectin fluorescent labelling of *C. intestinalis* haemocytes: hyaline amoebocytes (HAs) with *DSL*: granular amoebocytes (GAs) with *DSL, PSA, LCA, UEA I*; granulocytes with small granules (GSs) with *GSL I, ECL, GSL II, PHA-L, STL, PSA, LCA, UEA I*; granulocytes with large granules (GLs) with *DBA, SBA, PNA, ECL, GSL II, LEL, DSL, PSA, LCA, UEA I*; morula cells (MCs) with *PNA, RCA I, SJA, PHA-E, PHA-L*; unilocular refractile granulocytes (URGs) with *GSL I, PNA, RCA I, SJA, ECL, GSL II, PHA-E, LEL, DSL*; pigment cells (PCs) with *SJA, ECL, GSL II, PHA-L, LEL, LCA*. Each lectin staining has Lectin in fluorescence images (green), Dipa (blue), and DIC (bright filed), scale bar: 5 μm. Source: Graphic by authors.

3.2.1. *Botryllus schlosseri* Haemocytes Are Richly Carbohydrated

Circulating *Botryllus schlosseri* blood cells have multiple and diverse carbohydrate decorations on all of their haemocytes (Table 2 and Figure 2). Hbs and PCs bound 6 lectins, phagocytes (HAs and MLCs), 7 lectins each, and cytotoxic cells (MCa) 12 lectins of the 14 positively reacting, of the overall 16 tested lectins.

Four lectins labelled all of the *Botryllus* haemocytes: three were αGalNAc specific (*GSL I*, *DBA*, and *SBA*) and one GlcNAc specific (*DSL*). Interestingly, the first group of sugars (notably *GSL I*) was more abundant in the cellular periphery and well visible on membrane extensions, while the latter was rather enriched in the cytoplasm or in inclusions.

Several shared carbohydrate epitopes were found for haemocyte subgroups: All of the immunocytes (phagocytes and cytotoxic cells) carried the *PHA-E* epitope for complex bisecting N-linked sugars. Among phagocytes, the hyaline amoebocytes (HA) can be distinguished by *SJA* and *PHA-L* staining (complex N-linked sugars) but are negative for *GSL II* and *LEL* which, in turn, label macrophage-like cells (MLCs). Interestingly, the *SJA* and *PHA-L* epitopes of HAs are shared with cytotoxic MCs, while the *LEL* epitope (GlcNAc oligomers) of MLCs is shared with the immature haemoblasts. *UEA I* (fucosyl modifications) occur on Hbs, MCs, and PCs.

Cytotoxic MCs of *Botryllus* were most diversely carbohydrated and bound almost all of the lectins, also recognising other haemocytes, except for *GSL II* and *LEL*. Pigment (storage) cells bound (in addition to the four common lectins) also *PSA* and *UEA I*, both shared with MCs and the latter epitope (fucosylation) with haemoblasts.

Unique sugar specificities associated with a single haemocyte type occurred only on two cell types: only MC-bound *PNA*, *ECL*, and *LCA* on the terminal (O-linked) galactoses and core-fucosylated complex N-linked sugars, while macrophage-like cells (MLCs) uniquely bound *GSL II* for N-acetylglucosamines (GlcNAc).

3.2.2. *Phallusia mammillata* Haemocytes Are Sparsely Carbohydrated

The sugar decorations of *Phallusia mammillata* haemocytes showed the lowest diversity of the three species analysed, particularly for haemoblasts and storage cells (Table 2 and Figure 3). Hbs and storage cells carried a single sugar decoration, common also to all other haemocytes except MCs. Within immunocytes, the phagocytes (Phcs) bound five lectins, while cytotoxic GSs and GLs bound four and eight lectins, respectively. Strikingly, the MCs of *Phallusia* were devoid of any carbohydrate decorations tested.

All blood cells (with the exception of MCs) bound one universal lectin, *GSL II* (for terminal Gal free GlcNAc), and that was also the only one for haemoblasts and storage cells (both BCs and SRCs).

Few overlapping sugars were found on *Phallusia* haemocyte subgroups, while several unique lectins bound to a single haemocyte type within immunocytes:

Phagocytes uniquely featured terminal galactoses on GalNAc or GlcNAc of likely O-linked sugars recognised by *PNA, SJA,* and *ECL.* In contrast, cytotoxic granulocytes presented mostly N-linked sugars on mannose, interacting with *PSA* common to GSs and GLs, which were more complex in GLs, with longer side chains (*LEL*) and terminal decorations including galactose (*PHA-L*), fucose (*UEA I*), or even bisections (*PHA-E*). Finally, GSs uniquely bound *DSL* (GlcNAc chains) and shared *SBA* (terminal GalNAc) with phagocytes only, while GLs uniquely bound *PHA-E, PHA-L, LEL, LCA,* and *UEA I* (bisecting complex and fucosylated N-linked sugars).

3.2.3. *Ciona intestinalis* Haemocytes Are Selectively Carbohydrated

The most diversified haemocyte subtypes of all three ascidian species are reported in *Ciona intestinalis* (Figure 1b, Table 2 and Figure 4). However, only selective subgroups are richly decorated with glycans—namely, all of the cytotoxic cell types (GSs and GLs bound 8 and 10 lectins; MCs and URGs 5 and 9 lectins, respectively) but also the pigment cells (six lectins), while phagocytes bound only one (HA) or four (GA) lectins, respectively). In contrast, Hbs and the storage SRCs did not bind any of the tested lectins.

Sugar residues common to the major haemocyte groups were scarce: Only the storage subtype of PC shared all of their six epitopes with different cytotoxic cells (*SJA, ECL, GSL II, PHA-L, LEL*), representing GalNAc and/or GlcNAc residues on likely core N-linked glycans. Within the group of immunocytes, only a few epitopes (four lectins) were shared between phagocytes and cytotoxic cells, while extremely many (all tested lectins) were variably distributed among the four cytotoxic cell types. More precisely, *Ciona* phagocytes (HAs and GAs) and cytotoxic GSs and GLs commonly featured GlcNAc chains (*DSL* lectin, shared with URGs), but only the GA shared fucosylated N-linked sugars (*PSA, LCA, UEA I* lectins) with the cytotoxic GSs and GLs.

Interestingly, different combinations of all of the 16 tested lectins are found among the four cytotoxic subtypes. The GSs and GLs stained with 8 and 10 lectins, respectively, while MC and URG bound 5 and 9 lectins. Only GS and GL carried fucosylated, N-linked sugars (*PSA, LCA,* and *UEA I*), as mentioned above. Only GLs uniquely bound *DBA* and *SBA* (αGalNAc) but shared Gal-GalNAc (*PNA*) with MCs and URGs. The GSs uniquely bound *STL* but also *PHA-L*-marking Gal-GlcNAc on complex N-linked sugars, which are not shared with GLs. The GLs, instead, stained with *LEL* and *DSL,* for Gal-GlcNAc oligomers but possibly not on complex branches (*PHA-L* negative). Consistently, GSs and GLs, both stained with *ECL* and *GSL II* (GlcNAc lacking Gal). The simpler GL epitopes (*ECL, GSL II*) are shared with URGs, while only the complex N-linked GS epitope (*PHA-L*) is also found on MC. Interestingly, MCs and URGs uniquely share the complex bisecting N-linked sugars (*PHA-E* epitope). In addition, MC and URG were uniquely recognised by *RCA I* and

SJA (galactose and GalNAc), while staining by *PNA* (Gal-GalNAc) was also shared with GLs. URGs, but not MCs, bound *GSL I* (Gal in α-position) shared with GS cells only. Overall, URGs featured no unique markers but significantly overlapped with MCs and GLs, but also PCs, as mentioned above.

Unique sugar decorations on single haemocyte types of *Ciona* were, therefore, only for *STL* on GSs, and *DBA* and *SBA* on GLs, while all the other epitopes were variably shared among the haemocytes.

4. Discussion

Our carbohydrate profiling of circulating haemocytes in three ascidian species revealed that sugar decorations largely differ in composition and complexity between species but also feature some notable commonalities. Interestingly, the circulating haemocytes of the colonial *Bortyllus* are richly sugar modified throughout all subtypes of haemocytes, which contrasts with the solitary *Phallusia* and *Ciona* that, although carrying various decorations on cytotoxic cells, are more scarcely or not glycosylated otherwise. *Botryllus* haemocytes are also the only ones to carry multiple glycan epitopes shared among all of their blood cells. As glycan residues are involved in the molecular interactions of their carriers such differences point to a divergent functional complexity of circulating blood cells in colonial versus solitary ascidians but also to important differences among solitary ascidians.

4.1. Comparing Sugar Decorations on Blood Cells of Different Ascidians

The presented sugar profiling constitutes a more detailed and sensitive carbohydrate fingerprinting, as compared with previous records (Schlumpberger et al. 1984; Cima et al. 2017; Rosental et al. 2018), with unique and overlapping specificities of 16 plant lectins combined to increased signal amplification through biotinylated lectins crosslinking streptavidin of multiple fluorescent residues. The detection of weaker lectin binding and a wider reactivity is expected and observed.

A comparative summary of the lectin profiling for the three species is presented in Table 2, with colour codes for haemocyte subtypes consistently grouped into haemoblasts (red), immunocytes (yellow and orange, for phagocytes and cytotoxic cells, respectively), and storage cells (blue), and further subdivided according to their known morphological and functional relatedness (Figure 1). The plant lectin probes are roughly ordered according to their similar specificities (elaborated in Table 1) with the upper vs. lower groups preferably recognising Gal residues on often O-linked sugars vs. rather N-linked glycans, respectively, separated by the bisecting N-linked modification (*PHA-E*), and at the bottom mostly fucose recognition.

Haemoblasts (Hbs, red in Figures 1–4, Table 2) are small stem cells believed to give rise to all of the other blood cells and are also capable of whole-body regeneration in colonial ascidians (reviewed in Ballarin et al. 2021a, Ballarin et al. 2021a, 2021b).

Interestingly, the Hbs of colonial *Botryllus* carry much richer carbohydrate decorations (6 of 16 tested lectins), as compared with solitary *Phallusia* or *Ciona* (0 or 1 lectin). Such striking difference may reflect the differing degree of importance and variability in functions of Hbs, notably in the circulatory system of colonial versus solitary ascidians. It will be interesting to compare the Hbs of other colonial (vs. solitary) species and consider the bound immune receptors possibly with functions in stemness, regeneration, or plasticity.

Immunocytes (yellow and orange in Figures 1–4, Table 2) constitute the largest and most diverse haemocyte group with a moderate variability in phagocytes and greater diversity defined in cytotoxic cell subtypes. Two types of phagocytes (Figure 1, Table 2, yellow) are presented in *Botryllus* and *Ciona*, but only one in *Phallusia*. The sugar residues of phagocytes differ in the three ascidians with *Phallusia* presenting mostly O-linked sugars, while overly N-linked or bisecting in *Botryllus* and fucosylated in *Ciona*. For *Botryllus* and *Ciona* the phagocyte decorations significantly overlapped with those on cytotoxic cell types.

Cytotoxic cells are the most diverse and richly glycosylated haemocyte subgroup among immunocytes. The considerable differences between species are reflected by a variable presentation of maturation stages in the circulating blood cell populations (Figure 1, Table 2, orange): MCs exist in all three species, the two solitary species also comprise many earlier-stage GSs or GLs and, in *Ciona*, an additional URG. As of plant lectin binding, cytotoxic MCs of *Botryllus* carry abundant O- and N-linked sugars, while those of *Phallusia* are rather N linked (and devoid of sugar for MCs). *Ciona* presents an interesting situation where the various cytotoxic morphotypes carry a combination of all of the 16 lectin epitopes, with typical combinations for subtypes: most strikingly, GSs likely lack O-linked sugars, and GSs and GLs share fucosylation on N-linked sugars that in MCs and URGs lack the fucose but are core bisecting, in turn.

Storage cells (blue in Figures 1–4, Table 2) are much less diverse, and some differences among species exist: PCs are present in *Botryllus* and *Ciona* but absent in *Phallusia*, while they are represented by BCs; only the solitary species feature SRCs. In all three species, their sugar modifications overlap with those of cytotoxic cells. In *Phallusia*, however, their single sugar epitope also occurs on all of the other haemocytes (except the non-glycosylated MCs). In *Botryllus*, the PCs carry fucosylations such as MCs (but also the Hbs), and *Ciona* presents rather only N-linked cores, as shared with many cytotoxic subtypes.

4.2. Functional Implications from Haemocyte Glycophenotyping

The diversity of tunicate blood cells is strikingly amplified by their variable glycan modifications and the idea suggests that they were important drivers of tunicate evolution. Our plant biosensors can 'sugar phenotype' the haemocytes and 'phenocopy' the sugar binding of endogenous lectins to haemocytes.

To understand the enormous combinatorial possibilities offered by carbohydrate residues in biological systems, a defined 'sugar code' is proposed for sugar recognition (Solis et al. 2015). Sugars can be combined in three 'dimensions': linear, branching, and conformational variants. The presentation and recognition of the sugars are rigid on both sides, with conserved residues and little conformational effects upon binding (key-lock principle). Specifically, tailored protein domains, the carbohydrate recognition domains (CRDs), are common denominators of lectins to recognise specific sugars, and 14 different folds are described for animals/humans. Evolutionary diversification of lectins occurred via domain duplications and multiple events are paralleled by a secondary loss of non-functional domains. Lectins may present tandem CRDs and often contain different domains to produce the actual biological effect. The CRDs furthermore read sugar encoded 'postal codes' to reach desired destinations, and haemocytes are thusly targeted, attracted, or activated by endogenous lectins.

Our sugar profiling thus pinpoints to various haemocytic targets for endogenous tunicate lectins. Glycan recognition, indeed, plays a prominent role in both tunicate and vertebrate innate immunity, notably via the group of so-called pattern recognition receptors (PRRs) containing CRD domains to recognise foreign and endogenous carbohydrate residues, elicit interactions with downstream signalling components and trigger a network of crosstalks for a proper inflammatory immune response (Franchi and Ballarin 2017; Parrinello et al. 2018). These include lectins with conserved CRDs such as galectins (binding galactoside residues), RBLs (rhamnose or galactoside binding) or VCBPs (variable chitin-binding proteins, binding poly-GlcNAc) or more variable sugar-binding domains such as C-type lectins, including the collectins MBL/GBL and ficolins (for mannose/glucose and GlcNAc binding, respectively). Such endogenous lectins may have multiple distributions and functions, may be soluble, membrane bound, or intracellular, and may trigger the activation of haemocytes for release of cytokines, phagocytosis, or cytotoxicity, cause their crosslinking, recruitment to specific locations of inflammation, or interact intracellularly in glycan metabolism and proliferation (summary table in Franchi and Ballarin 2017). Not unexpectedly, many endogenous lectins are themselves expressed and also secreted by haemocytes.

Various subcellular locations of sugar epitopes could be detected, pointing to the variable functions of carbohydrates in cell compartments of the haemocytes. Epitopes in the cell periphery were particularly well visible on membrane extensions of phagocytes (Figures 2–4). Extracellular residues are likely involved in cell interactions, migration, and triggering the immune response. Intriguingly, these peripheral epitopes include several sugars shared among all of the *Botryllus* haemocytes (Table 2), and it is tempting to speculate that these are present on membrane-associated molecules that are crucial in the synchronised, weekly generation change in *Botryllus* orchestrated by haemocytes (Cima et al. 2010). Interestingly, these share galactoside residues, suggesting a possible role for galectins or RBLs targeting common haemocyte receptors. Notably, RBLs are known to play important roles in colonial generation change, phagocytic clearance of apoptotic haemocytes, and termination of the blastogenetic cycle, while ascidian galectins are known to play pivotal roles in haemagglutination, as well as the recruitment of migrating haemocytes (reviewed in Ballarin et al. 2013).

In contrast, several sugar epitopes seem enriched in the cytoplasm or in inclusions, well visible within cytotoxic cells (Figures 2–4). These are often N-linked sugars with fucosylations in all three species and bisecting sugars in *Botryllus*. Such inclusions may be degradation products from ingested microbes (pathogen-associated molecular patterns (PAMPs)) often containing high mannose contents. Alternatively, they may represent stored molecules to be released upon inflammatory activation including cytokines, enzymes, or toxic material. It is well known that the vacuolar localisation in haemocytes of inactive phenoloxidase (proPO, related to tyrosinase) causes the oxidation of polyphenol substrates (including the tunichromes, representing L-DOPA and L-TOPA peptides) into microcidal chinones upon release in the seawater/physiologic pH. Intriguingly, the major PO activity and release of microcidal components such as polyphenols is exerted by different cytotoxic subtypes in the three tunicates: MCs in *Botryllus*, URGs in *Ciona*, and GLs in *Phallusia*, rather than MCs (reviewed in Franchi and Ballarin 2017; Parrinello et al. 2018). Strikingly, such distribution roughly coincides with the resembling sugar epitopes in their vacuolar compartments. It will be interesting to identify their molecular carriers. Nevertheless, *Ciona* granulocytes (GSs and GLs) also contain intracellular glycans and could represent glycoproteinaceous maturation stages and targets for intracellular lectins, such as collectins, involved in the complement pathway (Franchi and Ballarin 2017).

Overall, it became evident that the same sugar residues can be followed throughout several related morphotypes, and thus, inversely, haematopoietic lineage relatedness may be concluded by the shared sugar epitopes, thus representing lineage markers. Such lineage identity considerations are of interest when haemocytes are analysed in a tissue context other than the circulating haemolymph, such as microbial-induced inflammation in the pharynx, but also if the host tissue is damaged or stressed (reviewed in Parrinello et al. 2018). Damage-associated molecular patterns (DAMPs) may cause an inflammation-like gene activation repertoire, as was shown for heat-shock proteins or during larval metamorphosis. It will be interesting to determine whether the shared sugar epitopes on DAMPs represent identical molecules or may resemble each other by mere coincidence.

In the future, our sugar profiling and the specificity of the individual biosensors will promote the analysis of glycoproteinaceous interactions of haemocytes and, notably, the various haemocytic targets for the endogenous tunicate lectins. Hybrid synthetic lectin probes prepared from CRDs (Dishaw et al. 2016) of the various lectin families will confirm the binding specificities and the differences detected in the three ascidians. Finally, such probes may also give direct biochemical access to the bound counterreceptors and will enrich our understanding of the intricate life of tunicate haemocytes.

Author Contributions: Conceptualisation, U.R. and F.Z. methodology, F.Z. and A.P.; validation, F.Z., U.R., L.B., and A.P.; formal analysis, F.Z. and U.R.; writing—original draft preparation, U.R. and F.Z.; writing—review and editing, L.B.; funding acquisition, U.R. and F.Z. All authors have read and agreed to the published version of the manuscript.

Funding: This research was funded by the Austrian Academy of Sciences ÖAW, Grant Number DOC 24699 (F.Z.), the University of Innsbruck Vicerectorate for Research Sonderförderung 214947 (U.R.), the Nachwuchsförderung Universität Innsbruck (F.Z.), and the Maristem COST action CA162013 (STSM to A.P. and L.B.).

Acknowledgments: We acknowledge the Roscoff Marine Station, France, for providing *Ciona intestinalis* and *Phallusia mammillata,* and the Chioggia Marine Station, and the Padova University, Italy, for *Botryllus schlosseri.*

Conflicts of Interest: The authors declare no conflict of interest.

Appendix A

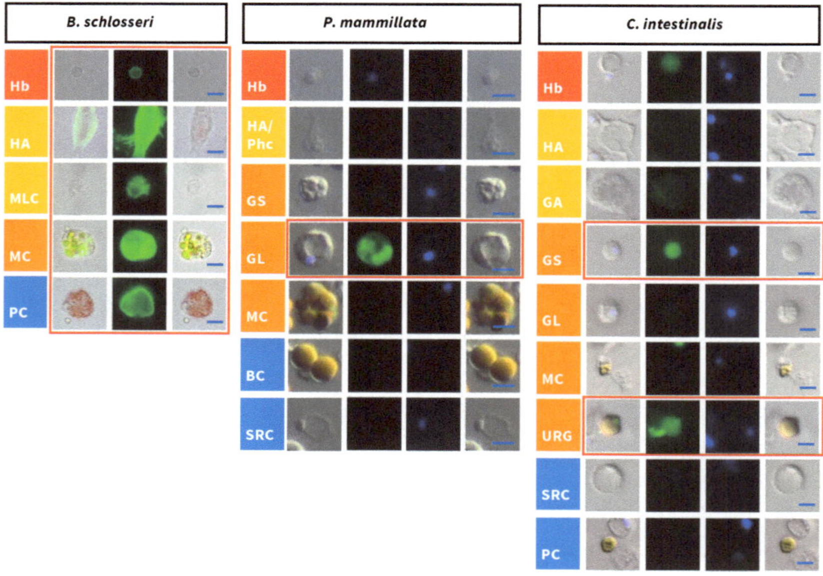

Figure A1. Individual lectin stainings for ascidian blood cells, GSL I. GSL I [αGal, αGalNac]. Brown frames indicate positively stained haemocytes. Scale bar, 5 μm.

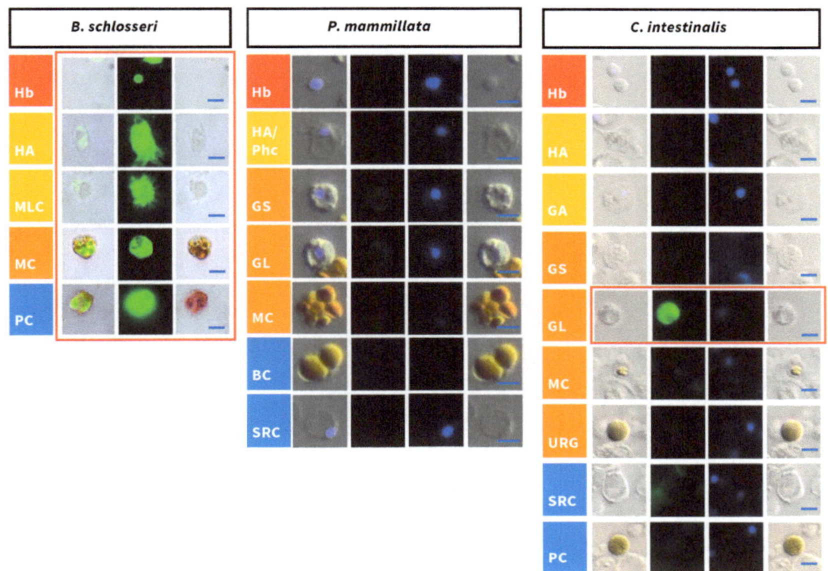

Figure A2. Individual lectin stainings for ascidian blood cells, DBA. DBA [αGalNac]. Brown frames indicate positively stained haemocytes. Scale bar, 5 μm.

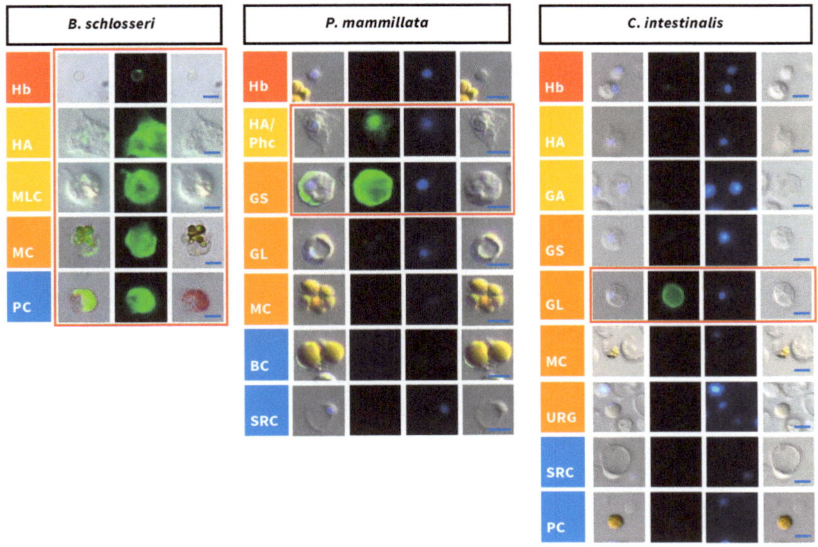

Figure A3. Individual lectin stainings for ascidian blood cells, SBA. SBA [α > βGalNac]. Brown frames indicate positively stained haemocytes. Scale bar, 5 μm.

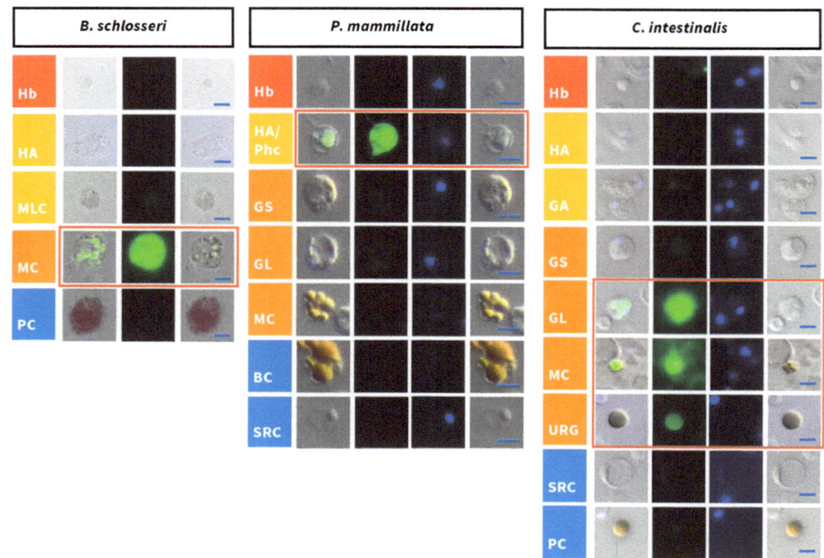

Figure A4.   Individual lectin stainings for ascidian blood cells, PNA. PNA [Galβ3GalNac].   Brown frames indicate positively stained haemocytes.   Scale bar, 5 μm.

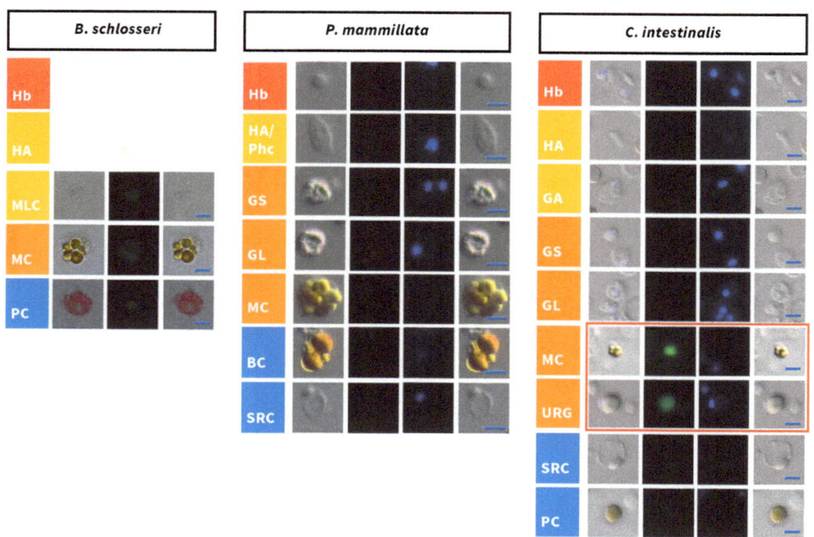

Figure A5. Individual lectin stainings for ascidian blood cells, RCA I. RCA I [Gal]. Brown frames indicate positively stained haemocytes. Scale bar, 5 μm.

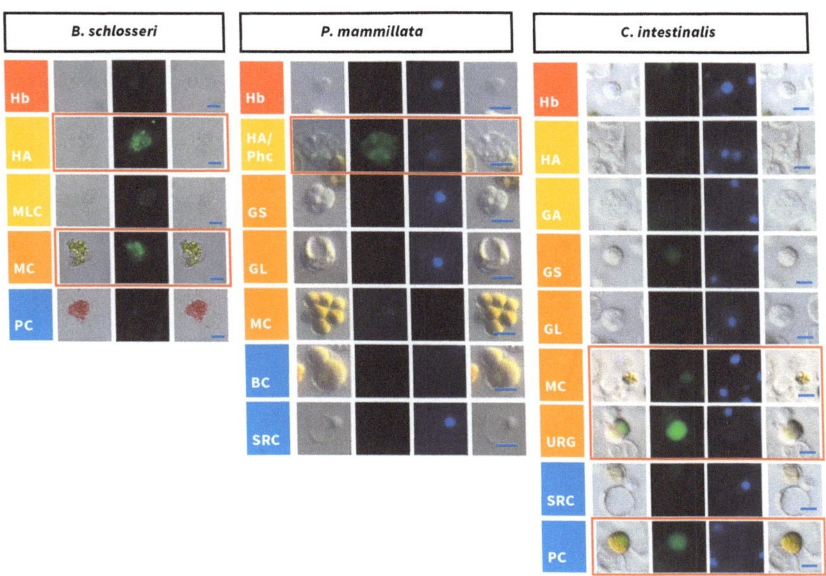

Figure A6. Individual lectin stainings for ascidian blood cells, SJA. SJA [βGalNac]. Brown frames indicate positively stained haemocytes. Scale bar, 5 μm.

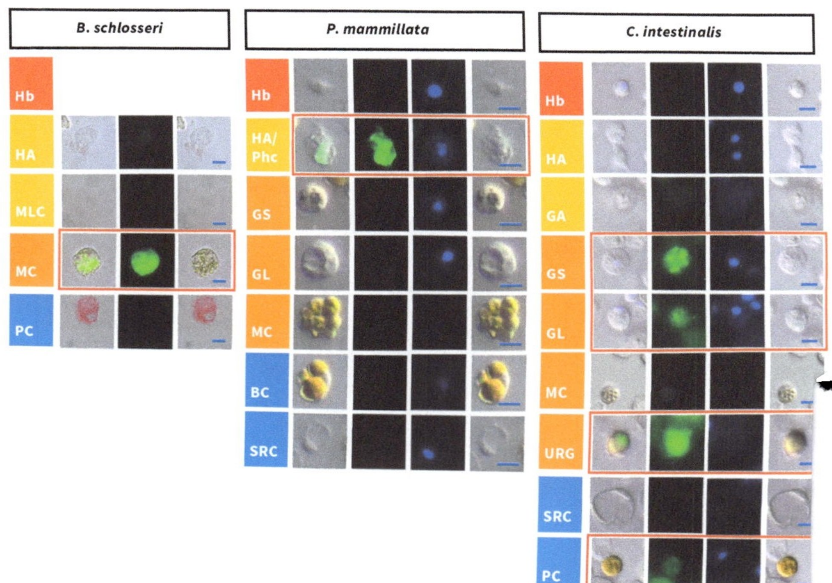

Figure A7. Individual lectin stainings for ascidian blood cells, ECL. EC [Galβ4GlcNAc]. Brown frames indicate positively stained haemocytes. Sca bar, 5 μm.

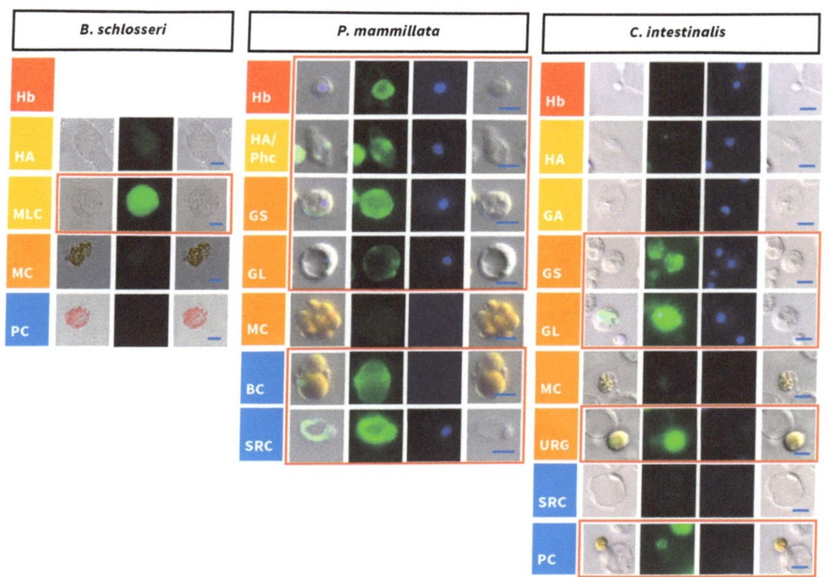

Figure A8. Individual lectin stainings for ascidian blood cells, GSL II. GSL II [α or β GlcNAc]. Brown frames indicate positively stained haemocytes. Scale bar, 5 μm.

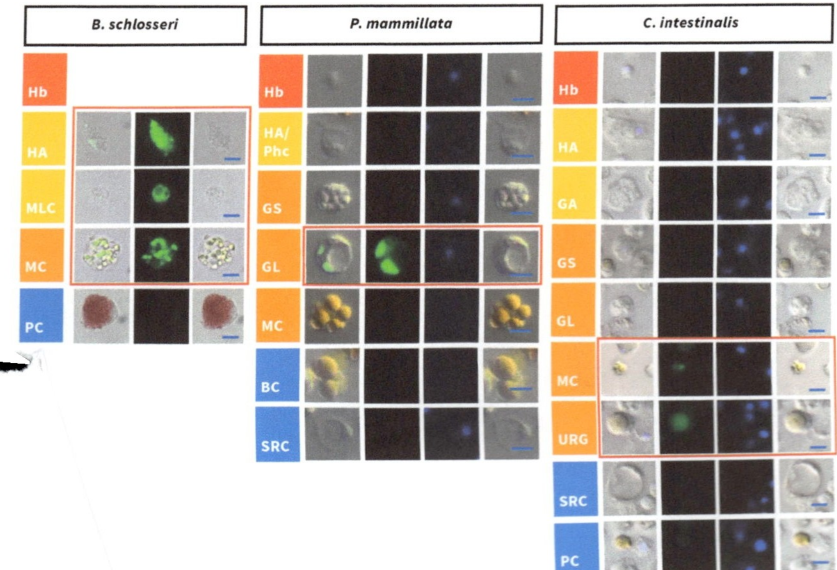

Individual lectin stainings for ascidian blood cells, PHA-E. PHA-E […Acβ2Manα6 (GlcNAcb4) (GlcNAcb4Manα3) Manβ4]. Brown frames …vely stained haemocytes. Scale bar, 5 μm.

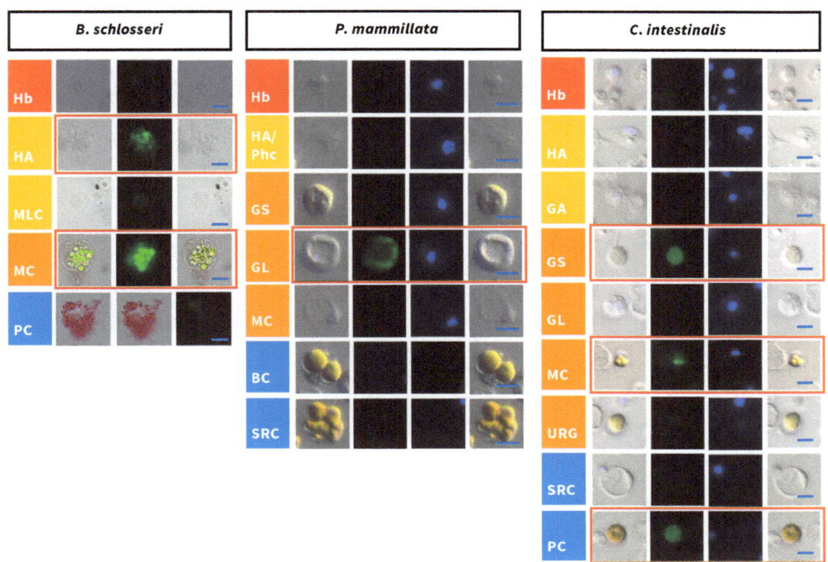

Figure A10. Individual lectin stainings for ascidian blood cells, PHA-L. PHA-L [Galβ4GlcNAcβ6 (GlcNAcβ2Manα3)Manα3]. Brown frames indicate positively stained haemocytes. Scale bar, 5 μm.

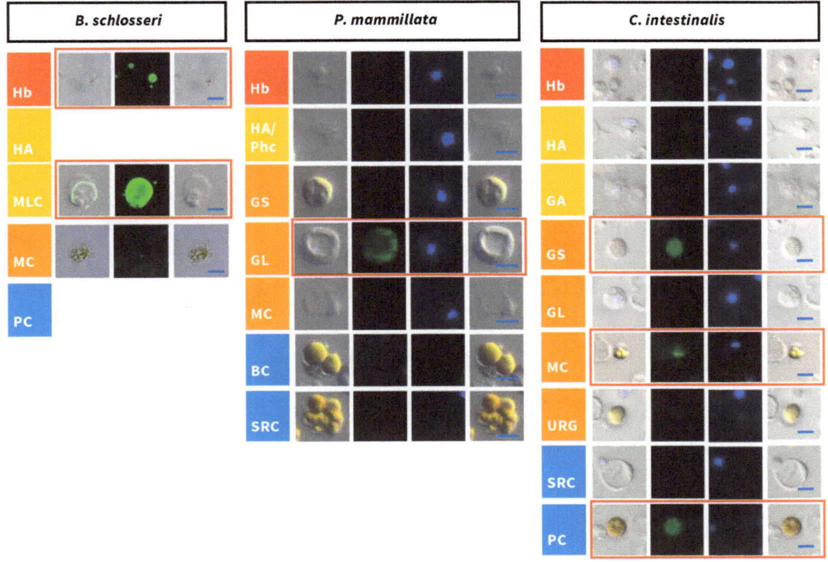

Figure A11. Individual lectin stainings for ascidian blood cells, LEL. LEL [(GlcNAc)₂₋₄]. Brown frames indicate positively stained haemocytes. Scale bar, 5 μm.

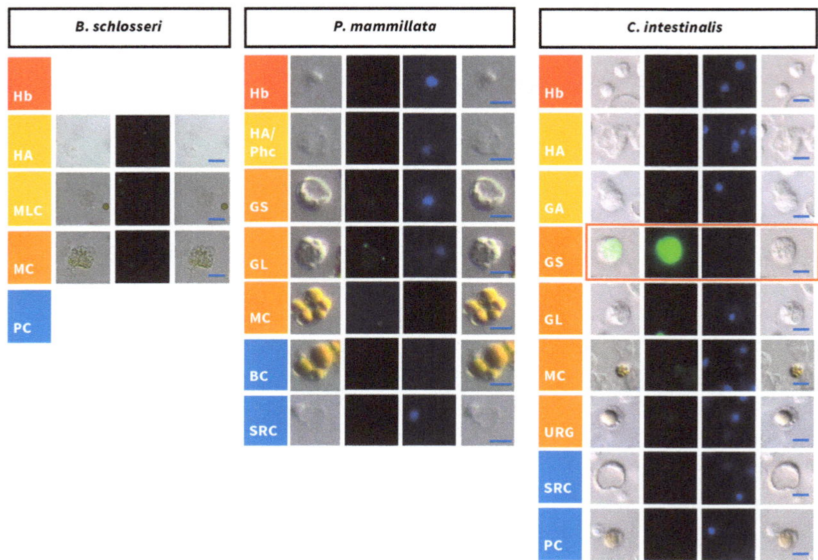

Figure A12. Individual lectin stainings for ascidian blood cells, STL. STL [(GlcNAc)$_{2-4}$]. Brown frames indicate positively stained haemocytes. Scale bar, 5 μm.

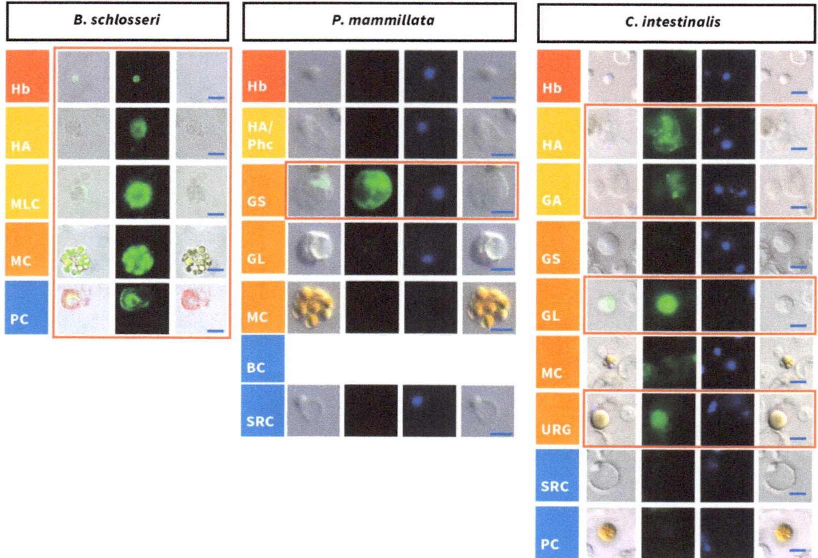

Figure A13. Individual lectin stainings for ascidian blood cells, DSL. DSL [(GlcNAc)$_{2-4}$]. Brown frames indicate positively stained haemocytes. Scale bar, 5 μm.

Ballarin, Loriano, Arzu Karahan, Alessandra Salvetti, Leonardo Rossi, Lucia Manni, Baruch Rinkevich, Amalia Rosner, Ayelet Voskoboynik, Benyamin Rosental, Laura Canesi, and et al. 2021b. Stem Cells and Innate Immunity in Aquatic Invertebrates: Bridging Two Seemingly Disparate Disciplines for New Discoveries in Biology. *Frontiers in Immunology* 12: 688106. [CrossRef] [PubMed]

Blanchoud, Simon, Lisa Zondag, Miles D. Lamare, and Megan J. Wilson. 2017. Hematological Analysis of the Ascidian *Botrylloides leachii* (Savigny, 1816) During Whole-Body Regeneration. *The Biological Bulletin* 232: 143–57. [CrossRef]

Cima, Francesca, Nicola Franchi, and Loriano Ballarin. 2016. Origin and Functions of Tunicate Hemocytes. In *The Evolution of the Immune System Conservation and Diversification*. Amsterdam: Elsevier Inc., pp. 29–49.

Cima, Francesca, Lucia Manni, Giuseppe Basso, Elena Fortunato, Benedetta Accordi, Filippo Schiavon, and Loriano Ballarin. 2010. Hovering between death and life: Natural apoptosis and phagocytes in the blastogenetic cycle of the colonial ascidian *Botryllus schlosseri*. *Developmental & Comparative Immunology* 34: 272–85. [CrossRef]

Cima, Francesca, Anna Peronato, and Loriano Ballarin. 2017. The haemocytes of the colonial aplousobranch ascidian Diplosoma listerianum: Structural, cytochemical and functional analyses. *Micron* 102: 51–64. [CrossRef]

Cloney, Richard A., and Leslie Grimm. 1970. Transcellular emigration of blood cells during ascidian metamorphosis. *Zeitschrift für Zellforschung und Mikroskopische Anatomie* 107: 157–73. [CrossRef]

Davidson, Brad, and Billie J. Swalla. 2002. A molecular analysis of ascidian metamorphosis reveals activation of an innate immune response. *Development* 129: 4739–51. [CrossRef]

Dishaw, Larry J., Brittany Leigh, John P. Cannon, Assunta Liberti, M. Gail Mueller, Diana P. Skapura, Charlotte R. Karrer, Maria R. Pinto, Rosaria De Santis, and Gary W. Litman. 2016. Gut immunity in a protochordate involves a secreted immunoglobulin-type mediator binding host chitin and bacteria. *Nature Communications* 7: 10617. [CrossRef] [PubMed]

Ferrario, Cinzia, Michela Sugni, Ildiko M. L. Somorjai, and Loriano Ballarin. 2020. Beyond Adult Stem Cells: Dedifferentiation as a Unifying Mechanism Underlying Regeneration in Invertebrate Deuterostomes. *Frontiers in Cell and Developmental Biology* 8: 587320. [CrossRef]

Franchi, Nicola, and Loriano Ballarin. 2017. Immunity in Protochordates: The Tunicate Perspective. *Frontiers in Immunology* 8: 674. [CrossRef] [PubMed]

Gutierrez, Stefania, and Federico D. Brown. 2017. Vascular budding in *Symplegma brakenhielmi* and the evolution of coloniality in styelid ascidians. *Developmental Biology* 423: 152–69. [CrossRef] [PubMed]

Hartenstein, Volker. 2006. Blood cells and blood cell development in the animal kingdom. *Annual Review of Cell and Developmental Biology* 22: 677–712. [CrossRef]

Jimenez-Merino, Juan, Isadora Santos de Abreu, Laurel S. Hiebert, Silvana Allodi, Stefano Tiozzo, Cintia M. De Barros, and Federico D. Brown. 2019. Putative stem cells in the hemolymph and in the intestinal submucosa of the solitary ascidian *Styela plicata*. *Evodevo* 10: 31. [CrossRef]

Kobayashi, Yuka, Hiroaki Tateno, Haruko Ogawa, Kazuo Yamamoto, and Jun Hirabayashi. 2014. Comprehensive list of lectins: Origins, natures, and carbohydrate specificities. In *Lectins. Methods in Molecular Biology (Methods and Protocols)*. Edited by Jun Hirabayashi. New York: Humana Press, vol. 1200, pp. 555–77. [CrossRef]

Manni, Lucia, Chiara Anselmi, Francesca Cima, Fabio Gasparini, Ayelet Voskoboynik, Margherita Martini, Anna Peronato, Paolo Burighel, Giovanna Zaniolo, and Loriano Ballarin. 2019. Sixty years of experimental studies on the blastogenesis of the colonial tunicate *Botryllus schlosseri*. *Developmental Biology* 448: 293–308. [CrossRef]

Mueller, Werner A., and Baruch Rinkevich. 2020. Cell Communication-mediated Nonself-Recognition and -Intolerance in Representative Species of the Animal Kingdom. *Journal of Molecular Evolution* 88: 482–500. [CrossRef]

Nette, Geoffrey, Silvia Scippa, Marilena Genovese, and Mario de Vincentiis. 1999. Cytochemical localization of vanadium (III) in blood cells of ascidian *Phallusia mammillata* Cuvier, and its relevance to hematic cell lineage determination. *Comparative Biochemistry and Physiology Part C: Pharmacology, Toxicology and Endocrinology* 122: 231–37. [CrossRef]

Parrinello, Nicolò, Matteo Cammarata, and Daniela Parrinello. 2018. The Inflammatory Response of Urochordata: The Basic Process of the Ascidians' Innate Immunity. In *Advances in Comparative Immunology*. Edited by Edwin L. Cooper. Berlin: Springer International, pp. 521–90.

Rosental, Benyamin, Mark Kowarsky, Jun Seita, Daniel M. Corey, Katherine J. Ishizuka, Karla J. Palmeri, Shih-Yu Chen, Rahul Sinha, Jennifer Okamoto, Gary Mantalas, and et al. 2018. Complex mammalian-like haematopoietic system found in a colonial chordate. *Nature* 564: 425–29. [CrossRef]

Rosental, Benyamin, Tal Raveh, Ayelet Voskoboynik, and Irving L. Weissman. 2020. Evolutionary perspective on the hematopoietic system through a colonial chordate: Allogeneic immunity and hematopoiesis. *Current Opinion in Immunology* 62: 91–98. [CrossRef] [PubMed]

Sato, Yoko, Kiyoshi Terakado, and Masaaki Morisawa. 1997. Test cell migration and tunic formation during post-hatching development of the larva of the ascidian, *Ciona intestinalis*. *Development, Growth & Differentiation* 39: 117–26. [CrossRef]

Schlumpberger, Jay M., Irving L. Weissman, and Virginia L. Scofield. 1984. Separation and labeling of specific subpopulations of *Botryllus* blood cells. *Journal of Experimental Zoology* 229: 401–11. [CrossRef] [PubMed]

Solis, Dolores, Nicolai V. Bovin, Anthony P. Davis, Jesús Jiménez-Barbero, Antonio Romero, Rene Roy, Karel Smetana Jr., and Hans-Joachim Gabius. 2015. A guide into glycosciences: How chemistry, biochemistry and biology cooperate to crack the sugar code. *Biochimica et Biophysica Acta (BBA)-General Subjects* 1850: 186–235. [CrossRef] [PubMed]

Sotgia, C., U. Fascio, G. Ortolani, and F. De Bernardi. 1993. Behavior of endodermal "button cells" during metamorphosis of ascidian larvae. *International Journal of Developmental Biology* 37: 547–53.

Taketa, Daryl A., and Anthony W. De Tomaso. 2015. *Botryllus schlosseri* allorecognition: Tackling the enigma. *Developmental & Comparative Immunology* 48: 254–65. [CrossRef]

Yasuda, Emi, Tomoyuki Sako, Hiroaki Tateno, and Jun Hirabayashi. 2014. Application of lectin microarray to bacteria including Lactobacillus casei/paracasei strains. In *Lectins. Methods in Molecular Biology (Methods and Protocols)*. New York: Humana Press, vol. 1200, pp. 295–311. [CrossRef]

Zeng, Fan, Julia Wunderer, Willi Salvenmoser, Thomas Ederth, and Ute Rothbächer. 2019a. Identifying adhesive components in a model tunicate. *Philosophical Transactions of the Royal Society B* 374: 20190197. [CrossRef]

Zeng, Fan, Julia Wunderer, Willi Salvenmoser, Michael W. Hess, Peter Ladurner, and Ute Rothbächer. 2019b. Papillae revisited and the nature of the adhesive secreting collocytes. *Developmental Biology* 448: 183–98. [CrossRef]

MDPI
St. Alban-Anlage 66
4052 Basel
Switzerland
Tel. +41 61 683 77 34
Fax +41 61 302 89 18
www.mdpi.com
MDPI Books Editorial Office
E-mail: books@mdpi.com
www.mdpi.com/books